APPLIED NONLINEAR ANALYSIS

ACADEMIC PRESS RAPID MANUSCRIPT REPRODUCTION

Proceedings of an International Conference on Applied Nonlinear Analysis, Held at The University of Texas at Arlington, Arlington, Texas, April 20–22, 1978.

APPLIED NONLINEAR ANALYSIS

Edited by

V. Lakshmikantham

Department of Mathematics
The University of Texas at Arlington
Arlington, Texas

ACADEMIC PRESS

NEW YORK SAN FRANCISCO LONDON 1979

A Subsidiary of Harcourt Brace Jovanovich, Publishers

ACADEMIC PRESS, INC.
111 Fifth Avenue, New York, New York 10003

United Kingdom Edition published by
ACADEMIC PRESS, INC. (LONDON) LTD.
24/28 Oval Road, London NW1 7DX

Library of Congress Cataloging in Publication Data

International Conference on Applied Nonlinear Analysis,
3d, University of Texas at Arlington, 1978.
Applied nonlinear analysis.

1. Mathematical analysis—Congresses. 2. Nonlinear theories—Congresses. I. Lakshmikantham, V. II. Texas. University at Arlington. III. Title.
QA300.I48 1978 515 79-10237
ISBN 012-434180-2

PRINTED IN THE UNITED STATES OF AMERICA

79 80 81 82 9 8 7 6 5 4 3 2 1

CONTENTS

**Indicates the author who presented the paper at the conference.*

CONTRIBUTED PAPERS

LIST OF CONTRIBUTORS

Numbers in parentheses indicate the pages on which authors' contributions begin.

ADAMS, E. (3), Institut für Angewandte Mathematik, Universität Karlsruhe, 75 Karlsruhe, Federal Republic of Germany

ADOMIAN, G. (13), Center for Applied Mathematics, The University of Georgia, Athens, Georgia 30602

ALÒ, RICHARD A. (25), Department of Mathematics, Lamar University, Beaumont, Texas 77710

ALVAGER, T. (25), Department of Physics, Indiana State University, Terre Haute, Indiana 47809

AMES, W. F. (3), Center for Applied Mathematics, The University of Georgia, Athens, Georgia 30602

ANDERSON, DAVID H.[1] *(425, 439),* Department of Mathematics, Southern Methodist University, Dallas, Texas 75275

AXELSSON, OWE[2] *(449),* Center for Numerical Analysis, The University of Texas at Austin, Austin, Texas 78712

BALAKRISHNA, M. (451), Rockwool Industries Inc., Southwest Division, P.O. Box 703, Belton, Texas 76513

BANKS, H. T. (47), Division of Applied Mathematics, Brown University, Providence, Rhode Island 02912

BERRYMAN, JAMES G.[3] *(61),* Courant Institute of Mathematical Sciences, 251 Mercer Street, New York University, New York, New York 10012

BUSENBERG, STAVROS N. (67), Department of Mathematics, Harvey Mudd College, Claremont, California 91711

CAREY, G. F. (467), Department of Aerospace Engineering and Engineering Mechanics, The University of Texas at Austin, Austin, Texas 78712

[1]*Also at the Department of Medical Computer Science, The University of Texas Health Science Center, Dallas, Texas 75235.*

[2]*Present address: Department of Computer Science, Chalmers University of Technology, Fack, S-40220 Gothenburg, Sweden.*

[3]*New permanent address: Bell Laboratories, Whippany, New Jersey 07981*

CARISTI, JAMES (479), Department of Mathematics, Texas Lutheran College, Seguin, Texas 78155

CHANDRA, JAGDISH (79), U.S. Army Research Office, Box 12211, Research Triangle Park, North Carolina 27709

CHENG, STEPHEN W. (485), Department of Mathematics, The University of Texas at Arlington, Arlington, Texas 76019

CHIOU, KUO-LIANG (499), Department of Mathematics, Wayne State University, Detroit, Michigan 48202

CHUANG, SUE-LI[4] *(507),* Department of Aerospace Engineering, The University of Texas at Arlington, Arlington, Texas 76019

CLOUTIER, J. R.[5] *(89),* Department of Mathematical Sciences, Rice University, Houston, Texas 77001

COHEN, DONALD S. (95), Department of Applied Mathematics 101-50, California Institute of Technology, Pasadena, California 91125

COOKE, KENNETH L. (67), Department of Mathematics, Pomona College, Claremont, California 91711

CORDUNEANU, C.[6] *(111),* Department of Mathematics, University of Rhode Island, Kingston, Rhode Island 02881

COUGHLIN, PETER (519), Department of Economics, Harvard University, Cambridge, Massachusetts 02138

DANIEL, JAMES W. (123), Department of Mathematics, The University of Texas at Austin, Austin, Texas 78712

DAVIS, PAUL WILLIAM (79), Department of Mathematics, Worcester Polytechnic Institute, Worcester, Massachusetts 01609

DEIMLING, KLAUS (127), Fachbereich 17 der Gesamthochschule, Warburger Straße 100, D-4790 Paderborn, Germany

DE KORVIN, ANDRÉ (25), Department of Mathematics, Indiana State University, Terre Haute, Indiana 47809

DIEKMANN, ODO (529), Mathematisch Centrum, 2e Boerhaavestraat 49, Amsterdam, The Netherlands

DUBAN, M. C. (47), Département de Mathématiques Appliquées, Université de Technologie de Compiègne, 60206 Compiègne, France

EISENFELD, JEROME[7] *(439, 485, 531, 543, 555),* Department of Mathematics, The University of Texas at Arlington, Arlington, Texas 76019

FIFE, PAUL C. (143), Department of Mathematics, The University of Arizona, Tucson, Arizona 85721

FITZGIBBON, W. E. (161), Department of Mathematics, University of Houston, Houston, Texas 77004

[4]Present address: 5603 Surry Avenue, Newport News, Virginia 23605

[5]Present address: Naval Surface Weapons Center, Dahlgren Laboratory, Dahlgren, Virginia 22448

[6]Present address: Department of Mathematics, University of Tennessee, Knoxville, Tennessee 37916

[7]Also affiliated with the Department of Medical Computer Science, University of Texas Health Science Center at Dallas, Dallas, Texas 75235

FORD, COREY C.[8] *(531)*, Department of Physiology, University of Texas Health Science Center, Dallas, Texas 75235

GOH, B. S.[9] *(569, 577)*, Department of Mathematics, The University of British Columbia, Vancouver, B.C., Canada V6T 1W5

GRISELL, R. D. S. (587), Department of Physiology and Biophysics, The University of Texas Medical Branch, Galveston, Texas 77550

GUPTA, CHAITAN P. (175), Department of Mathematical Sciences, Northern Illinois University, DeKalb, Illinois 60115

HALLMARK, JAMES (543), Department of Mathematics, The University of Texas at Arlington, Arlington, Texas 76019

HARRISON, G. W. (605), Department of Mathematics, The University of Georgia, Athens, Georgia 30602

HASTINGS, ALAN (607), Department of Pure and Applied Mathematics, Washington State University, Pullman, Washington 99164

HOLLAND, CHARLES J.[10] *(61)*, Courant Institute of Mathematical Sciences, New York University, 251 Mercer Street, New York, New York 10012

HULLENDER, DAVID A. (451), Mechanical Engineering, The University of Texas at Arlington, Arlington, Texas 76019

JACQUEZ, JOHN A. (185), Department of Physiology, The University of Michigan, Ann Arbor, Michigan 48109

KAGIWADA, HARRIET (207), HFS Associates, 3117 Malcolm Avenue, Los Angeles, California 90034

KALABA, R. E. (619), Biomedical Engineering Department, University of Southern California, Los Angeles, California 90007

KAPER, HANS G. (529), Applied Mathematics Division, Argonne National Laboratory, Argonne, Illinois 60439

KEARFOTT, BAKER (627), Department of Mathematics and Statistics, The University of Southwestern Louisiana, Lafayette, Louisiana 70504

KERNEVEZ, J. P. (47), Département de Mathématiques Appliquées, Université de Technologie de Compiègne, 60206 Compiègne, France

LADDE, G. S. (215), Mathematics Department, The State University of New York at Potsdam, Potsdam, New York 13676

LAKSHMIKANTHAM, V. (219), Department of Mathematics, The University of Texas at Arlington, Arlington, Texas 76019

LEELA, S. (219), Department of Mathematics, State University of New York, Geneseo, New York 14454

LEUNG, K. V. (245), Department of Computer Science, Concordia University, Sir George Williams Campus, 1455 de Maisonneuve Boulevard West, Montreal, Quebec, Canada H3G 1M8

[8]*Present address: 822 G Cabell Avenue, Charlottesville, Virginia 22903*

[9]*Present Address: Mathematics Department, University of Western Australia, Nedlands, W. A. 6009, Australia*

[10]*Present address: Mathematics Department, Purdue University, West Lafayette, Indiana 47907*

LORD, M. E.[11] *(635)* Applied Mathematics Division 2623, Sandia Laboratories, Albuquerque, New Mexico 87185

MAURER, ROBERT N. (657), Department of Mathematics, Worcester Polytechnic Institute, Worcester, Massachusetts 01609

McCANN, ROGER C. (267), Department of Mathematics, P.O. Drawer MA, Mississippi State University, Mississippi State, Mississippi 39762

MIELE, A. (89), Department of Astronautics and Mathematical Sciences, Rice University, Houston, Texas 77001

MURDOCK, J. A. (669) Mathematics Department, Iowa State University, Ames, Iowa 50011

NEGRINI, P. (273), Istituto di Matematica, Università di Camerino, Camerino, Italy

NEUBERGER, J. W. (287), Mathematics Department, North Texas State University, Denton, Texas 76203

NOTESTINE, RONALD D. (657), Department of Pure and Applied Mathematics, Washington State University, Pullman, Washington 99164

OǦUZTÖRELI, M. N. (245), Department of Mathematics, University of Alberta, Edmonton, Alberta, Canada T6G 2G1

PAN, T. T. (467), Department of Aerospace Engineering and Engineering Mechanics, The University of Texas at Austin, Austin, Texas 78712

PAYNE, FRED R. (507, 675), Department of Aerospace Engineering, University of Texas at Arlington, Arlington, Texas 76019

PIANIGIANI, GIULIO[12] *(299),* Institute for Physical Science and Technology, University of Maryland, College Park, Maryland 20742

PLANT, RICHARD E. (309), Department of Mathematics, University of California, Davis, California 95616

POLLARD, HARRY (323), Department of Mathematics, Purdue University, West Lafayette, Indiana 47907

RAO, A. N. V. (325), Department of Mathematics, University of South Florida, Tampa, Florida 33620

REICH, SIMEON (335), Department of Mathematics, University of Southern California, Los Angeles, California 90007

RENKA, R. (467), Department of Aerospace Engineering and Engineering Mechanics, The University of Texas at Austin, Austin, Texas 78712

SALVADORI, L.[13,14] *(273),* Istituto di Matematica, Università di Roma, Roma, Italy

[11]*Present address: Department of Mathematics, The University of Texas at Arlington, Arlington, Texas 76019.*

[12]*Present address: Istituto Matematico "U. Dini", University of Florence, Viale Morgagni 67/A, 50134 Firenze, Italy.*

[13]*Visiting Professor at the Department of Mathematics, The University of Texas at Arlington, Arlington, Texas 76019.*

[14]*Present address: Dipartimento di Matematica, Università di Trento, 38050-POVO (Trento), Italy.*

SANDBERG, S. (439), Department of Medical Computer Science, University of Texas Health Science Center, Dallas, Texas 75235

SCOTT, M. R. (347, 635), Applied Mathematics Division 2623, Sandia Laboratories, Albuquerque, New Mexico 87185

SEIBERT, PETER[15] *(351),* Departamento de Matemáticas y Ciencia de la Computación, Universidad Simón Bolivar, Caracas, Venezuela

SEIFERT, GEORGE (373), Department of Mathematics, Iowa State University, Ames, Iowa 50011

SHOWALTER, R. E. (381), Department of Mathematics, The University of Texas at Austin, Austin, Texas 78712

SINGH, K. L. (689), Department of Mathematics, Texas A & M University, College Station, Texas 77843

SINGH, S. P. (389), Department of Mathematics, Statistics and Computer Science, Memorial University of Newfoundland, St. John's, Newfoundland, Canada A1B 3X7

SONI, B.[16] *(555),* Department of Mathematics, The University of Texas at Arlington, Arlington, Texas 76019

STEIN, R. B. (245), Department of Physiology, University of Alberta, Edmonton, Alberta, Canada, T6G 2H7

TAMBURRO, MICHAEL (705), Department of Mathematics, Georgia Institute of Technology, Atlanta, Georgia 30332

TANABE, KUNIO[17] *(707),* Applied Mathematics Department, Brookhaven National Laboratory, Upton, New York 11973

TAPIA, R. A. (395), Department of Mathematical Sciences, Rice University, Houston, Texas 77001

THOMPSON, RUSSELL C.[18] *(397),* Department of Mathematical Sciences, Northern Illinois University, DeKalb, Illinois 60115

TSOKOS, C. P. (325), Department of Mathematics, University of South Florida, Tampa, Florida 33620

VATSALA, A. S. (219), Department of Mathematics, The University of Texas at Arlington, Arlington, Texas 76019

VISENTIN, F. (721), Istituto di Matematica, "Renato Caccioppoli," Università di Napoli, Via Mezzocannone, 8—Cap. 80134, Italy

WATTS, H. A. (347, 635), Applied Mathematics Division 2623, Sandia Laboratories, Albuquerque, New Mexico 87185

[15]*Present address: Escuela de Ciencias, Universidad Centro Occidental, Apartado 400—Barquisimeto—Estado Lara, Republica de Venezuela.*

[16]*Present address: Bell Helicopter, Department 87, P.O. Box 482, Fort Worth, Texas 76101.*

[17]*Present Address: The Institute of Statistical Mathematics, Minamizatu, Minatoku, Tokyo, Japan.*

[18]*Present address: Department of Mathematics, Utah State University, Logan, Utah 84322.*

WEINBERGER, H. F. (407), School of Mathematics, University of Minnesota, 206 Church Street S.E., Minneapolis, Minnesota 55455

WOLLKIND, DAVID J. (657), Department of Pure and Applied Mathematics, Washington State University, Pullman, Washington 99164

ZAGUSTIN, E. A. (619), Civil Engineering, California State University, Long Beach, California 90840

PREFACE

An international conference on applied nonlinear analysis was held at The University of Texas at Arlington, April 20–22, 1978. This was the third in a series of conferences that were sponsored by The University of Texas at Arlington.

The present volume consists of the proceedings of the conference. It includes papers that were delivered as invited talks and research reports as well as contributed papers.

The aim of the conference was to feature recent advances in applied nonlinear analysis. The emphasis was on the following general areas: (i) reaction–diffusion equations; (ii) optimization theory; (iii) constructive techniques in numerical analysis; and (iv) applications to physical and life sciences.

The contributions to reaction-diffusion equations include basic theory, nonlinear oscillations, rotating spiral waves, stability and asymptotic behavior, comparison results, discrete-time models in population genetics, and predator—prey systems. The group of papers in optimization theory consists of inverse and ill-posed problems with application to geophysics, conjugate gradient, and quasi-Newton methods with applications to large scale optimization, sequential conjugate gradient-restoration algorithm for optimal control problems with nondifferentiable constraints, differential geometric methods in nonlinear programming, and equilibria in policy formation games with random voting. In the area of constructive techniques in numerical analysis, there is a large group of papers consisting of numerical and approximate solutions of boundary value problems for ordinary and partial differential equations, constructive techniques for accretive and monotone operators, computational solutions of nonlinear boundary value problems, and finite element analysis. A group of papers emphasizing linear and nonlinear models of biological systems, neuromuscular systems, compartmental analysis, identification problems, analysis of solidification of a pure metal, turbulent fluid flows, and thought-provocative dynamical systems is also included. There are also papers that deal with stability of general dynamical systems, stability problems for Hopf bifurcation, the current state of the n-body problem, periodic solutions for differential equations, integrodifferential equations, product integral representation of Volterra equations with delay, weak solutions of variational problems, nonlinear integration on measures, fixed point theory, and contracting interval iteration for nonlinear problems.

I wish to express my special thanks to my colleagues Bill Beeman, Steve Bernfeld, Jerome Eisenfeld, R. Kannan, A. R. Mitchell, R. W. Mitchell, and Bennie Williams for assisting me in planning and organizing the conference; to my secretaries Mrs. Gloria Brown, Ms. Debbie Green, and Mrs. Mary Ann Crain for their assistance during the conference; and Mrs. Mary Ann Crain for her excellent typing of the proceedings.

INVITED ADDRESSES
AND RESEARCH REPORTS

ON CONTRACTING INTERVAL ITERATION FOR NONLINEAR PROBLEMS IN $\mathbb{R}^n$: I[1]

E. Adams

Institut für Angewandte Mathematik
Universität Karlsruhe
Federal Republic of Germany

W. F. Ames

Center for Applied Mathematics
University of Georgia
Athens, Georgia

I. INTRODUCTION

The problem

$$\left.\begin{array}{l} f(x) = 0, \quad f: D \subseteq \mathbb{R}^n \to \mathbb{R}^n, \\ D \text{ bounded, } x \in \mathbb{R}^n, \ f \text{ possesses} \\ \text{a Frechet derivative } f', \quad f' \in C(D) \end{array}\right\} \qquad (1)$$

equivalently represented by

[1]*This research was substantially supported by NSF Grant MCS 76-16605. This short version only announces the theorems for Part I.*

ISBN 0-12-434180-2

$$Ax = h(x) \quad \text{on} \quad D$$
$$h(x) \stackrel{\text{def}}{=} - f(x) + Ax, \quad \text{with} \tag{2}$$
$$A \in L(\mathbb{R}^n) \quad \text{or} \quad \hat{\hat{A}} \in L(\mathbb{R}^{2n}), \quad \text{compare (6),}$$

being M-Matrices

is solved by an iterative construction of a sequence $\{H_\nu\}$ of nested intervals. The expression Ax, providing a linear approximation of $f(x)$, must satisfy the following conditions:

(i) $a_{ii} > 0$, $i = 1(1)n$;
(ii) the range of h on D is "small"; (3)
(iii) for every choice of i, $j \in \{1, 2, \ldots, n\}$
sgn $(\partial h_i/\partial x_j)$ on D is fixed independently.

The linear approximation of $f(x)$ need not be optimal in any sense. Property (iii) can be satisfied by a proper choice of A. Because of (ii) A should be chosen such that sgn a_{ij} = sgn $(\partial f_i/\partial x_j)$.

II. ITERATION

Below we need a quasimonotone , i.e. an off-diagonally antitone, matrix.[2] Since A is subject to conditions (3) it will generally not be quasimonotone. To get upper and lower bounds simultaneously, a quasimonotone matrix $\hat{\hat{A}} \in L(\mathbb{R}^{2n})$, constructed from A, is required for the iteration

$$\hat{\hat{A}}\hat{x}^{(\nu+1)} = \hat{h}(\hat{x}^{(\nu)}). \tag{4}$$

With

$$a_{ij}^{+} \stackrel{\text{def}}{=} \begin{cases} a_{ij} & \text{if} \quad a_{ij} \geq 0 \\ 0 & \text{otherwise} \end{cases} \tag{5a}$$

[2] *For other terms see the definitions as given in Ortega and Rheinboldt* [*1*].

$$a_{ij}^{-} \overset{\text{def}}{=} a_{ij} - a_{ij}^{+}, \quad i,j = 1(1)n \tag{5b}$$

the required matrix $\hat{\hat{A}}$ is

$$\hat{\hat{A}} \overset{\text{def}}{=} \begin{pmatrix} a_{11} & -|a_{12}^{-}| \ldots -|a_{1n}^{-}| & 0 & -a_{12}^{+} \ldots -a_{1n}^{+} \\ \vdots & \vdots \qquad \vdots & \vdots & \vdots \qquad \vdots \\ -|a_{n1}^{-}| & -|a_{n2}^{-}| \ldots \; a_{nn} & -a_{n1}^{+} & -a_{n2}^{+} \ldots \; 0 \\ 0 & -a_{12}^{+} \quad -a_{1n}^{+} & a_{11} & -|a_{12}^{-}| \ldots -|a_{1n}^{-}| \\ \vdots & \vdots \qquad \vdots & \vdots & \vdots \qquad \vdots \\ -a_{n1}^{+} & -a_{n2}^{+} \ldots \; 0 & -|a_{n1}^{-}| & -|a_{n2}^{-}| \ldots \; a_{nn} \end{pmatrix} \tag{6}$$

The interval $H_\nu \subseteq D \subset \mathbb{R}^n$, with boundary vectors $\underline{x}^{(\nu)}$, $\overline{x}^{(\nu)} \in D$, is defined to be

$$H_\nu \overset{\text{def}}{=} \{x \mid \underline{x}_i^{(\nu)} \leq x_i \leq \overline{x}_i^{(\nu)}, \quad i = 1(1)n\}, \quad \nu + 1 \in \mathbb{N} \tag{7}$$

whose difference is denoted by

$$w_i^{(\nu)} \overset{\text{def}}{=} \overline{x}_i^{(\nu)} - \underline{x}_i^{(\nu)}.$$

Because of the monotonicity of h the h_i take their maxima and minima at the boundaries of the H_ν. Thus

$$\max_{H_\nu} h_i(x) = h_i(\xi^{(i,\nu)}), \quad i = 1(1)n$$

$$\min_{H_\nu} h_i(x) = h_i(\eta^{(i,\nu)}), \quad i = 1(1)n$$

where

$$\xi_j^{(i,\nu)} = \underline{x}_j^{(\nu)} \text{ or } \overline{x}_j^{(\nu)}, \quad \eta_j^{(i,\nu)} = \underline{x}_j^{(\nu)} \text{ or } \overline{x}_j^{(\nu)}$$

for $i,j = 1(1)n$.

To complete the definition of (4) the vectors

$$\hat{x}^{(\nu)} \overset{\text{def}}{=} (\overline{x}_1^{(\nu)}, \ldots, \overline{x}_n^{(\nu)}, -\underline{x}_1^{(\nu)}, \ldots, -\underline{x}_n^{(\nu)})^T \in \mathbb{R}^{2n} \tag{8}$$

and

$$\hat{h}(\hat{x}^{(\nu)}) = (h_1(\xi^{(1,\nu)}),\ldots,h_n(\xi^{(n,\nu)}),-h_1(\eta^{(1,\nu)}),\ldots$$
$$- h_n(\eta^{(n,\nu)})^T \tag{9}$$

are defined where the superscript T denotes transposition of the vector. The vector $\hat{h}(\hat{x}^{(\nu)})$ is an isotone function of its arguments since $\hat{h}$ increases as $\max_{H_\nu} h_i(x)$ increases and as $\min_{H_\nu} h_i(x)$ decreases.

The existence of starting vectors $\underline{x}^{(0)}$ and $\overline{x}^{(0)} \in D$ is assumed which satisfy the inequalities

$$\left.\begin{aligned}
&\underline{x}_i^{(0)} < \overline{x}_i^{(0)},\\
&a_{ii}\overline{x}_i^{(0)} + \sum_{\substack{j=1\\ j\neq i}}^{n} (a_{ij}^{+}\underline{x}_j^{(0)} + a_{ij}^{-}\overline{x}_j^{(0)}) > \max_{H_0} h_i(x)\\
&\qquad = h_i(\xi^{(i,0)}),\\
&a_{ii}\underline{x}_i^{(0)} + \sum_{\substack{j=1\\ j\neq i}}^{n} (a_{ij}^{+}\overline{x}_j^{(0)} + a_{ij}^{-}\underline{x}_j^{(0)}) < \min_{H_0} h_i(x)\\
&\qquad = h_i(\eta^{(i,0)}).
\end{aligned}\right\} \quad i=1(1)n. \tag{10}$$

The existence of $\underline{x}^{(0)}$ and $\overline{x}^{(0)}$ in (10) follows from Ortega and Rheinboldt [1, p. 460] provided

(i) $\hat{\hat{A}}^{-1}$ exists, $\hat{\hat{A}}^{-1} > \theta \in L(\mathbb{R}^{2n})$ and $-a \leq h(x) \leq a$ for some $a \geq \theta \in \mathbb{R}^n$ and all $x \in \mathbb{R}^n$ <u>or</u>

(ii) h is antitone

The component iteration scheme of (4) is now defined as follows:

$$\left\{\begin{aligned}
&a_{ii}\overline{x}_i^{(\nu+1)} + \sum_{\substack{j=1\\ j\neq i}}^{n} (a_{ij}^{+}\underline{x}_j^{(\nu+1)} + a_{ij}^{-}\overline{x}_j^{(\nu+1)}) = \max_{H_\nu} h_i(x)\\
&\qquad = h_i(\xi^{(i,\nu)}),\\
&a_{ii}\underline{x}_i^{(\nu+1)} + \sum_{\substack{j=1\\ j\neq i}}^{n} (a_{ij}^{+}\overline{x}_j^{(\nu+1)} + a_{ij}^{-}\underline{x}_j^{(\nu+1)}) = \min_{H_\nu} h_i(x)\\
&\qquad = h_i(\eta^{(i,\nu)})
\end{aligned}\right\} \quad \begin{aligned} &i=1(1)n,\\ &\nu+1 \in \mathbb{N}. \end{aligned} \tag{11}$$

If A is quasimonotone, $a_{ij}^{+} = 0$, $i \neq j$, $i,j = 1(1)n$ which implies that the system (11) for the $\overline{x}^{(\nu+1)}$ is decoupled from the system for the $\underline{x}^{(\nu+1)}$. The function $\hat{h}$ in (10) and (11) is isotone.

III. PROPERTIES OF THE ITERATION

A matrix $\hat{A} \in L(\mathbb{R}^n)$ will now be defined and employed to establish a sufficient condition for $\hat{\hat{A}} \in L(\mathbb{R}^{2n})$ to be an M-matrix. $\hat{A} = (\hat{a}_{ij})$ is defined as follows:

$$\hat{a}_{ii} = a_{ii} \text{ for } i = 1(1)n \text{ and } \hat{a}_{ij} = -|a_{ij}| \text{ for } j \neq i.$$

In what follows the term quasimonotonicity will additionally require that $a_{ii} > 0$ for $i = 1(1)n$.

Also needed is the concept of a *positive test vector*. A vector $v \in \mathbb{R}^n$ is called a *positive test vector for* $\hat{A}$, provided

$$\left.\begin{array}{l} v_i > 0 \\ (\hat{A}v)_i = a_{ii}v_i - \sum\limits_{\substack{j=1\\ j\neq i}}^{n} |a_{ij}|v_j > 0 \end{array}\right\} i = 1(1)n \qquad (12)$$

Theorem 1. (Spreuer-Adams [2])

If $\hat{A}$ and $\hat{\hat{A}}$ are quasimonotone and there exists a positive test vector $v \in \mathbb{R}^n$ for $\hat{A}$ then $\hat{A}$ and $\hat{\hat{A}}$ are M-matrices.

The matrix $\hat{A}$ satisfies the strong row sum criterion provided the conditions in (12) can be verified by use of a positive test vector $v \in \mathbb{R}^n$ with $v_i = \alpha > 0$ for $i = 1(1)n$.

Let the following notation

$$\hat{x} = (\overline{x}_1, \ldots, \overline{x}_n, -\underline{x}_1, \ldots, -\underline{x}_n)^T \in \mathbb{R}^{2n}$$

$$\overline{y} = (\overline{y}_1, \ldots, \overline{y}_n, -\underline{y}_1, \ldots, -\underline{y}_n)^T \in \mathbb{R}^{2n}$$

$$H_x = \{x \mid \underline{x}_i \leq x_i \leq \overline{x}_i, \quad i = 1(1)n\}$$

$$H_y = \{y \mid \underline{y}_i \le y_i \le \overline{y}_i, \quad i = 1(1)n\}$$

be employed in Theorem 2.

Theorem 2. If $\hat{\hat{A}}$ is an M-matrix the initial condition (10) (equivalently, $\hat{\hat{A}}\hat{x}^{(0)} > \hat{h}(\hat{x}^{(0)})$) implies

(a) $\hat{\hat{A}}\hat{y} = \hat{h}(\hat{x})$ with $\hat{x} \le \hat{x}^{(0)}$ determines a vector $\hat{y} < \hat{x}^{(0)}$ such that $H_x \subseteq H_0$ is mapped on H_y with $H_y \subset H_0$.

(b) The iteration (10) and (11) determines n strictly monotone decreasing sequences $\{\overline{x}_i^{(\nu)}\}$ and n strictly monotone increasing sequences $\{\underline{x}_i^{(\nu)}\}$, $i = 1(1)n$, $(\nu+1) \in \mathbb{N}$; $\overline{x}_i^{(\nu)}$, $\underline{x}_i^{(\nu)}$ are uniquely defined for each $\nu \in \mathbb{N}$.

(c) $\overline{x}_i^{(\nu)} \to \overline{x}_i^*$, $\underline{x}_i^{(\nu)} \to \underline{x}_i^*$ such that $\underline{x}_i^* \le \overline{x}_i^*$ for $i = 1(1)n$.

The M-matrix property of $\hat{\hat{A}}$ implies the existence of $\hat{\hat{A}}^{-1}$. Hence the iteration (11) is equivalent with $\hat{\hat{I}}\hat{x}^{(\nu+1)} = \hat{\hat{A}}^{-1}\hat{h}(\hat{x}^{(\nu)})$ where $\hat{\hat{I}} \in L(\mathbb{R}^{2n})$ is trivially an M-matrix. This alternative of (11) has disadvantages for proving theorems and will not be employed.

Whereas Brouwer's fixed point theorem implies the existence of solution(s) $x^* \in H_0$, provided (10) is satisfied, the sequences $\{\underline{x}^{(\nu)}\}$ and $\{\overline{x}^{\nu}\}$ restrict these to the subset $\{x \mid \underline{x}_i^* \le x_i^* \le \overline{x}_i^*, \quad i = 1(1)n\}$ of H_0. The construction of $\{\underline{x}^{\nu}\}$ and $\{\overline{x}^{\nu}\}$ therefore is a sweeping process (see Sattinger [3]) to sharpen the existence statement of solution(s) $x^* \in H_0$ guaranteed by Brouwer's theorem.

The sequences $\{\underline{x}^{(\nu)}\}$ and $\{\overline{x}^{(\nu)}\}$ bracket solutions $x^* \in H_0$ by use of intervals, i.e. special subdomains of H_0. However, as the following example shows there may exist more than one solution $x^* \in H_0$ even though (10) is satisfied.

Example 1. Consider $f\colon \mathbb{R} \to \mathbb{R}$ as defined by

$$f(x) = x - g(x) = 0 \text{ with } g(x) = \begin{cases} +\sqrt{x} \text{ if } x \ge 0 \\ -\sqrt{|x|} \text{ if } x < 0. \end{cases} \tag{13}$$

Obviously $x = 0$, $x = 1$, and $x = -1$ are fixed points of g.

H_0 can be defined by the use of any numbers $\underline{x}^0$, $\overline{x}^0$ such that $\underline{x}^{(0)} < -1$ and $\overline{x}^0 > 1$. Because of $\overline{x}^{(0)} > \max_{H_0} g(x) = \sqrt{\overline{x}^{(0)}}$ and $\underline{x}^{(0)} < \min_{H_0} g(x) = -\sqrt{|\underline{x}^{(0)}|}$, condition (10) is satisfied. Theorem 2 then asserts that the monotone sequence $\{\overline{x}^{(\nu)}\}$ converges to $\overline{x}^* = 1$ and the monotone sequence $\{\underline{x}^{(\nu)}\}$ converges to $\underline{x}^* = -1$. Of course $\underline{x}^* = -1$ and $\overline{x}^* = +1$ are fixed points of $x = g(x)$, since $\underline{x}^{(\nu+1)} = g(\underline{x}^{(\nu)})$ and $\overline{x}^{(\nu+1)} = g(\overline{x}^{(\nu)})$.

In the next example $\underline{x}^*$ and $\overline{x}^*$ are *not* fixed points since $\overline{x}^{(\nu+1)} = g(\underline{x}^{(\nu)})$ and $\underline{x}^{(\nu+1)} = g(\overline{x}^{(\nu)})$.

Example 2. Here consider $f\colon \mathbb{R} \to \mathbb{R}$ as defined by

$$f(x) = x - g(x) = 0 \text{ with } g(x) = \begin{cases} -\sqrt{x} \text{ if } x \geq 0 \\ +\sqrt{|x|} \text{ if } x < 0. \end{cases} \tag{14}$$

The only fixed point of $g(x)$ is $x = 0$. Choosing $\underline{x}^{(0)}$ and $\overline{x}^{(0)}$ as in Example 1 it is clear that $\overline{x}^{(0)} > \max_{H_0} g(x) = \sqrt{|\underline{x}^{(0)}|}$ and $\underline{x}^{(0)} < \min_{H_0} g(x) = -\sqrt{\overline{x}^{(0)}}$ so that condition (10) is satisfied. But the sequence $\{\overline{x}^{(\nu)}\}$ converges to $\overline{x}^* = 1$ and $\{\underline{x}^{(\nu)}\}$ converges to $\underline{x}^* = -1$.

A contraction condition to be defined will ensure that $\underline{x}^* = \overline{x}^*$; this condition employs a Lipschitz-condition for each one of the functions h_i, $i = 1(1)n$. With $\nu \in \mathbb{N}$ still arbitrary, this condition is

$$\left.\begin{array}{l} |h_i(x) - h_i(y)| \leq \sum_{j=1}^{n} M_{ij}^{(\nu)} w_j^{(\nu)}, \quad i = 1(1)n \\ \text{every } x,y \in H_\nu, \quad M_{ij}^{(\nu)} \geq 0, \quad \nu+1 \in \mathbb{N}. \end{array}\right\} \tag{15}$$

The $w_j^{(\nu)}$ are non-negative. Because of Theorem 2, the sequence $\{M_{ij}^{(\nu)}\}$ for each $i,j = 1(1)n$ is weakly monotone decreasing and bounded from below by zero. The contraction condition is as follows:

It is assumed that there exists a $\bar{\nu}$ with $\bar{\nu}+1 \in \mathbb{N}$ such that

$$(a_{ii}-M_{ii}^{(\bar{\nu})})\, w_i^{(\bar{\nu})} - \sum_{\substack{j=1\\ j\neq i}}^{n} (|a_{ij}|+M_{ij}^{(\bar{\nu})})w_j^{(\bar{\nu})} > 0, \quad i = 1(1)n. \tag{16}$$

The Lipschitz-constants $M_{ij}^{(\nu)}$ may be calculated by use of $\max|\partial h_i(x)/\partial x_j|$ for $x \in H_\nu$. This maximum generally is not taken on the boundary ∂H_ν of H_ν. Sharp estimates of the $M_{ij}^{(\nu)}$ may be rather difficult to find. Any overestimate of the $M_{ij}^{(\nu)}$ will only make it more difficult to satisfy (16). There are three fixed points of g in Example 2.1, and correspondingly, condition (16) cannot be satisfied.

In case of a linear function h, the constants $M_{ij}^{(\nu)}$ are independent of ν. If there exist vectors $\underline{x}^{(0)}$ and $\bar{x}^{(0)}$ which satisfy (10), the corresponding vector $w^{(0)} = \bar{x}^{(0)} - \underline{x}^{(0)}$ then automatically satisfies the contraction condition (16). The mapping of H_0 into itself due to (10) and Theorem 2, then implies the contraction of the sequences $\{\bar{x}^{(\nu)}\}$ and $\{\underline{x}^{(\nu)}\}$ to the only fixed point $x^* \in H_0$, according to Theorem 3.

The matrix

$$B \stackrel{\text{def}}{=} \hat{A} - M^{(\bar{\nu})} \in L(\mathbb{R}^n) \text{ with } M^{(\bar{\nu})} \stackrel{\text{def}}{=} (M_{ij}^{(\bar{\nu})}) \tag{17}$$

is introduced; B is required to satisfy

$$a_{ii} > M_{ii}^{(\bar{\nu})} \quad \text{for} \quad i = 1(1)n. \tag{18}$$

This may be interpreted as an additional condition for the matrix A.

Theorem 3. (a) If (16) and (18) are satisfied, then the matrix B in (17) is an M-matrix.

(b) If B is an M-matrix and the conditions of Theorem 2 are satisfied, then $\underline{x}_i^* = \bar{x}_i^* \stackrel{\text{def}}{=} x_i^*$ for $i = 1(1)n$ such that $Ax^* = h(x^*)$, and

(c) the error estimate

$$\left.\begin{array}{l}\max\{(\bar{x}_i^{(\nu)}-x_i^*),\ (x_i^*-\underline{x}_i^{(\nu)})\} \le w_i^{(\nu)} \le q^{(\nu-\bar{\nu})}w_i^{(\bar{\nu})} \\ \text{for } i = 1(1)n,\quad \nu \ge \bar{\nu}, \text{ and } x^* \overset{\text{def}}{=} \underline{x}^* = \bar{x}^*\end{array}\right\} \qquad (19)$$

holds true with $\bar{\nu} \in \mathbb{N}$ fixed.

(d) If the inequalities (10) are satisfied by vectors $\underline{y}^{(0)}$, $\bar{y}^{(0)} \in H_0$ with $\underline{y}^{(0)} \le \bar{y}^{(0)}$, the iterates $\underline{y}^{(\nu)}$ and $\bar{y}^{(\nu)}$ as determined by (11) are uniquely defined for all $\nu \in \mathbb{N}$ and the corresponding sequences $\{\underline{y}^{(\nu)}\}$ and $\{\bar{y}^{(\nu)}\}$ converge to the limiting vector y^* with $y^* = x^*$.

The details of these theorems and extensions are available from the authors [4].

REFERENCES

[1] Ortega, J. M., and Rheinboldt, W. C. (1970). "Iterative Solution of Nonlinear Equations in Several Variable," Academic Press, New York.

[2] Spreuer, H. and Adams, E. (1972). "Hinreichende Bedingung für Ausschlus von Eigenwerten in Parameterintervallen bei einer Klasse von linearen homogenen Funktionalgleichungen, *ZAMM 52*, 479-485.

[3] Sattinger, D. H. (1973). "Topics in Stability and Bifurcation Theory," Springer-Verlag, Berlin.

[4] Adams, E., and Ames, W. F. (May, 1978). "On Contracting Interval Iteration for Nonlinear Problems in R^n," Center for Applied Mathematics Report #7, 101 pages.

A CONSTRUCTIVE METHOD FOR LINEAR AND NONLINEAR STOCHASTIC PARTIAL DIFFERENTIAL EQUATIONS

G. Adomian

Center for Applied Mathematics
The University of Georgia
Athens, Georgia

I. LINEAR DETERMINISTIC PARTIAL DIFFERENTIAL EQUATIONS

Let L_t, L_x be linear deterministic partial differential operators (e.g., $\partial/\partial t$, $\partial^2/\partial x^2$ etc.) and consider the equation:

$$L_t u + L_x u = g \tag{1}$$

where g may be a stochastic process. We require that either L_t^{-1} or L_x^{-1} exist and be determinable. Suppose we have L_t^{-1}. Rewrite (1) as $L_{t,x} u = g$. If $L_{t,x}^{-1}$ exists, the solution to (1) is $u = L_{t,x}^{-1} g$. We therefore seek to determine $L_{t,x}^{-1}$. Write (1) as

$$L_t u = g - L_x u \tag{2}$$

Since L_t is invertible

$$u = L_t^{-1} g - L_t^{-1} L_x u \tag{3}$$

or

$$L_{t,x}^{-1} g = L_t^{-1} g - L_t^{-1} L_x L_{t,x}^{-1} g \tag{4}$$

We have therefore the operator equation

$$L_{t,x}^{-1} = L_t^{-1} - L_t^{-1} L_x L_{t,x}^{-1} \tag{5}$$

ISBN 0-12-434180-2

We parametrize (5) with the parameter λ which we will later set equal to unity, thus

$$L_{t,x}^{-1} = \lambda L_t^{-1} - \lambda L_t^{-1} L_x L_{t,x}^{-1} \tag{6}$$

Let

$$L_{t,x}^{-1} = \Sigma\lambda^n H_n \tag{7}$$

thus after we set $\lambda = 1$ we have simply decomposed $L_{t,x}^{-1}$ into the sum $H_0 + H_1 + H_2 + \ldots$

Substituting (7) in (6)

$$\Sigma\lambda^n H_n = \lambda L_t^{-1} - \lambda L_t^{-1} L_x \Sigma\lambda^n H_n \tag{8}$$

Equating equal powers of λ

$$H_0 = 0$$
$$H_1 = L_t^{-1}$$
$$H_2 = -(L_t^{-1} L_x) L_t^{-1}$$
$$H_3 = (L_t^{-1} L_x)(L_t^{-1} L_x) L_t^{-1}$$
$$\vdots$$

Since the desired inverse is given by ΣH_n

$$L_{t,x}^{-1} = \sum_{n=1}^{\infty} (-1)^{n-1} [L_t^{-1} L_x]^{n-1} L_t^{-1} \tag{9}$$

If it is L_x^{-1} which is determinable rather than L_t^{-1} we proceed analogously, thus

$$L_x u = g - L_t u$$
$$u = L_x^{-1} g - L_x^{-1} L_t u$$

from which we derive

$$L_{t,x}^{-1} = \sum_{n=1}^{\infty} (-1)^{n-1} [L_x^{-1} L_t]^{n-1} L_x^{-1} \tag{10}$$

Either (9) or (10), as appropriate, can be used to write the solution $u = L_{t,x}^{-1} g$.

Example: $L_t = \partial/\partial t, \quad L_x = -\partial^2/\partial x^2$

Using (9)

$$u = \Sigma(-1)^{n-1}[L_t^{-1}L_x]^{n-1}L_t^{-1}g$$

or

$$u = (L_t^{-1}g) - (L_t^{-1}L_x)(L_t^{-1}g) + (L_t^{-1}L_x)(L_t^{-1}L_x)(L_t^{-1}g) + \ldots$$

Evaluating the terms,

$$L_t^{-1}g = \int_0^t g(x,\tau)d\tau$$

$$L_xL_t^{-1}g = -(\partial^2/\partial x^2)\int_0^t g(x,\tau)d\tau$$

$$L_t^{-1}L_xL_l^{-1}g = -\int_0^t(\partial^2/\partial x^2)\int_0^\tau g(x,\gamma)d\gamma d\tau$$

$$L_xL_t^{-1}L_xL_t^{-1}g = (\partial^2/\partial x^2)\int_0^t(\partial^2/\partial x^2)\int_0^\tau g(x,\gamma)d\gamma d\tau$$

$$L_t^{-1}L_xL_t^{-1}L_xL_t^{-1}g = \int_0^t(\partial^2/\partial x^2)\int_0^\tau(\partial^2/\partial x^2)\int_0^\gamma g(x,\sigma)d\sigma d\gamma d\tau$$

$$\vdots$$

Consequently

$$u = \int_0^t g(x,\tau)d\tau + \int_0^t(\partial^2/\partial x^2)\int_0^\tau g(x,\gamma)d\gamma d\tau + \int_0^t(\partial^2/\partial x^2)\int_0^\tau(\partial^2/\partial x^2)\int_0^\gamma g(x,\sigma)d\sigma d\gamma d\tau + \ldots$$

II. LINEAR STOCHASTIC PARTIAL DIFFERENTIAL EQUATIONS

$$\mathcal{L}_t u + \mathcal{L}_x u = g \equiv \mathcal{L}_{t,x}u$$

i.e., we replace the deterministic operators L_t, L_x with the stochastic operators $\mathcal{L}_t$, $\mathcal{L}_x$ where $\mathcal{L}_t = L_t + R_t$; $\mathcal{L}_x = L_x + R_x$ with R_t, R_x zero mean random operators. Either L_t^{-1} or L_x^{-1} is assumed to be determinable. Suppose it is L_t^{-1} which exists. Then

$$L_t u = g - R_t u - L_x u - R_x u$$

$$u = L_t^{-1} g - L_t^{-1} R_t u - L_t^{-1} L_x u - L_t^{-1} R_x u \tag{11}$$

If $\mathcal{L}_{t,x}^{-1}$ exists

$$\mathcal{L}_{t,x}^{-1} g = L_t^{-1} g - L_t^{-1} R_t \mathcal{L}_{t,x}^{-1} g - L_t^{-1} L_x \mathcal{L}_{t,x}^{-1} g - L_t^{-1} R_x \mathcal{L}_{t,x}^{-1} g$$

yielding the operator equation

$$\mathcal{L}_{t,x}^{-1} = L_t^{-1} - L_t^{-1} R_t \mathcal{L}_{t,x}^{-1} - L_t^{-1} L_x \mathcal{L}_{t,x}^{-1} - L_t^{-1} R_x \mathcal{L}_{t,x}^{-1} \tag{12}$$

Applying the previous procedure

$$\mathcal{L}_{t,x}^{-1} = \Sigma(-1)^{n-1} [L_t^{-1} (R_t + R_x + L_x)]^{n-1} L_t^{-1} \tag{13}$$

or, if it is L_x^{-1} which exists

$$\mathcal{L}_{t,x}^{-1} = \Sigma(-1)^{n-1} [L_x^{-1} (R_t + R_x + L_t)]^{n-1} L_x^{-1} \tag{14}$$

Then from (13) or (14) we find $u = \mathcal{L}_{t,x}^{-1} g$

III. LINEAR STOCHASTIC DIFFERENTIAL EQUATIONS

$$\mathcal{L} u = g$$

i.e., in I suppose $\mathcal{L}_x = 0$ (i.e., L_x and R_x are both zero). We can now drop subscripts on L_t, R_t to obtain the solution

$$u = \Sigma(-1)^{n-1} [L^{-1} R]^{n-1} L^{-1} g \tag{15}$$

IV. NONLINEAR STOCHASTIC DIFFERENTIAL EQUATIONS

Consider $Fu = g$ where $F = \mathcal{L} + N$, i.e., the previous linear (stochastic) operator plus a nonlinear (stochastic) operator. If L^{-1} exists,

$$u = L^{-1} g - L^{-1} Ru - L^{-1} Nu \tag{16}$$

$$F^{-1} g = L^{-1} g - L^{-1} R F^{-1} g - L^{-1} N F^{-1} g$$

$$\Sigma \lambda^n H_n g = \lambda L^{-1} g - \lambda L^{-1} R \Sigma \lambda^n H_n g - \lambda L^{-1} N \Sigma \lambda^n H_n g$$

Unfortunately because of the nonlinear operator N, the previous procedure is inapplicable in general from this point on i.e., we cannot write the operator equation and resulting series in general. However consider a simple example where $Nu = u^2$. Then

$$\Sigma\lambda^n H_n g = \lambda L^{-1} g - \lambda L^{-1} R \Sigma\lambda^n H_n g - \lambda L^{-1} \Sigma\lambda^n H_n g \Sigma\lambda^n H_n g$$

For $\lambda = 0$, $H_0 g = 0$ consequently $H_0 = 0$ if $g \neq 0$.

For $\lambda = 1$, $H_1 g = L^{-1} g$ consequently $H_1 = L^{-1}$.

For $\lambda = 2$, $h_2 g = -L^{-1} R H_1 g$ consequently $H_2 = -L^{-1} R L^{-1}$.

For $\lambda = 3$, $H_3 g = -L^{-1} R H_2 g - L^{-1} H_1 g H_1 g - L^{-1} H_2 g H_0 g - L^{-1} H_0 g H_2 g$ where the last two terms are zero.

We have then

$$H_3 g = -L^{-1} R H_2 g - L^{-1} H_1 g H_1 g$$

$$H_3 g = (L^{-1} R)(L^{-1} R)(L^{-1} g) - (L^{-1})^2 g (L^{-1} g).$$

For $\lambda = 4$, $H_4 g = -L^{-1} R H_3 g - L^{-1}(H_3 g H_0 g + H_2 g H_1 g + H_1 g H_2 g + H_0 g H_3 g)$ and finally

$$H_4 g = -(L^{-1} R)(L^{-1} R)(L^{-1} R)(L^{-1} g) + (L^{-1} R)(L^{-1})^2 g (L^{-1} g) + (L^{-1})^2 (L^{-1} R)(L^{-1} g)(L^{-1} g) - (L^{-1})^2 g (L^{-1} R)(L^{-1} g)$$

etc.

Thus

$$\begin{aligned} u = F^{-1} g &= H_1 g + H_2 g + H_3 g + \dots \\ &= (L^{-1} g) - (L^{-1} R)(L^{-1} g) + [(L^{-1} R)^2 (L^{-1} g) - L^{-1} (L^{-1} g)^2] \\ &\quad - [(L^{-1} R)^3 (L^{-1} g) - (L^{-1} R) L^{-1} (L^{-1} g)^2 \\ &\quad - (L^{-1})^2 (L^{-1} R)(L^{-1} g)^2 + (L^{-1})(L^{-1} g)(L^{-1} R)(L^{-1} g)] \\ &\quad + \dots \end{aligned} \tag{16}$$

Returning to (16) but supposing $Nu = u^2$

$$u = L^{-1} g - L^{-1} R u - L^{-1} u^2$$

and parametrizing

$$u = L^{-1} g - \lambda L^{-1} R u - \lambda L^{-1} u^2$$

Now let $u = F^{-1}g$ to write

$$F^{-1}g = L^{-1}g - \lambda L^{-1}RF^{-1}g - \lambda L^{-1}F^{-1}gF^{-1}g$$

$$\Sigma H_n g = L^{-1}g - \lambda L^{-1}R\Sigma H_n g - \lambda L^{-1}\Sigma\lambda^n H_n g\Sigma\lambda^m H_m g$$

Writing the series $\Sigma H_n g$ for $F^{-1}g$ or u is equivalent to writing the series $\Sigma(-1)^i\lambda^i y_i$ in the symmetrized method of Adomian and Sibul. Equating powers of λ

$$H_0 g = L^{-1}g$$

$$H_1 g = -L^{-1}[Ry_0 + y_0^2]$$

$$H_2 g = L^{-1}[Ry_1 + (y_0 y_1 + y_1 y_0)]$$

In the Adomian and Sibul form we write $u = y_0 - y_1 + y_2 \cdots$ where

$$y_{n+1} = L^{-1}[Ry_n + (y_0 y_n + y_1 y_{n-1} + \ldots y_n y_0)]$$

Polynomial Nonlinearities: In equation (1) we now consider the case where $N = by^m$. Thus

$$y = L^{-1}x - \lambda L^{-1}Ry - \lambda L^{-1}by^m$$

Let*

$$y = y_0 + \lambda y_1 + \lambda^2 y_2 + \ldots \lambda^n y_n + \ldots$$

and assume

$$y^m = A_0 + \lambda A_1 + \lambda^2 A_2 + \ldots + \lambda^n A_n + \ldots$$

The $A_0, A_1, A_2, \ldots$ were first determined by Hansted[8] in a Danish paper in 1881. The relations are:

$$A_0 = y_0^m$$

$$A_1 = m(y_1/y_0)A_0 = m(y_1/y_0)y_0^m = my_0^{m-1}y_1$$

$$2A_2 = (m-1)(y_1/y_0)A_1 + 2m(y_2/y_0)A_0 = m(m-1)y_0^{m-2}y_1^2 + 2my_2y_0^{m-1}$$

**We have previously assumed* $y = \sum_{i=0}^{\infty}(-1)^i\lambda^i y_i$ *but we get the same series in either case.*

$$3A_3 = (m-2)(y_1/y_0)A_2 + (2m-1)(y_2/y_0)A_1 + 3m(y_3/y_0)A_0$$

etc. to

$$nA_n = (m-(n-1))(y_1/y_0)A_{n-1} + (2m-(n-2))(y_2/y_0)A_{n-2}$$
$$+ (3m-(n-3))(y_3/y_0)A_{n-3} + \dots + nm(y_n/y_0)A_0$$

Thus we have a systematic way of obtaining expansions for larger m. For smaller m we can use the same method or simply multiply out the power series in λ and collect terms of equal powers in λ. Both methods of course yield the same results. As an example let $m = 2$. We obtain

$$A_0 = y_0^2 = y_0y_0$$
$$A_1 = m(y_1/y_0)y_0^2$$
$$A_2 = y_1y_1 + 2y_2y_0$$
$$A_3 = 2y_1y_2 + 2y_3y_0$$
$$A_4 = 2y_3y_1 + y_2y_2 + 2y_4y_0$$
$$A_5 = 2y_5y_0 + 2y_4y_1 + 2y_2y_3$$
$$A_6 = 2y_0y_6 + 2y_1y_5 + 2y_2y_4 + y_3y_3$$

etc.

These can be put into symmetrized form as obtained before where indices of each term add to the index of A_n and all possible sums are taken

$$A_0 = y_0y_0$$
$$A_1 = y_0y_1 + y_1y_0$$
$$A_2 = y_0y_2 + y_1y_1 + y_2y_0$$
$$A_3 = y_0y_3 + y_1y_2 + y_2y_1 + y_3y_0$$
$$A_4 = y_0y_4 + y_1y_3 + y_2y_2 + y_3y_1 + y_4y_0$$
$$A_5 = y_0y_5 + y_1y_4 + y_2y_3 + y_3y_2 + y_4y_1 + y_5y_0$$
$$\vdots$$

$$A_n = y_0y_n + y_1y_{n-1} + \dots + y_{n-1}y_1 + y_ny_0$$

Our series obtained by equating equal powers of λ is

$$y_0 = L^{-1}x$$

$$y_1 = -L^{-1}[Ry_0 + bA_0]$$

$$y_2 = -L^{-1}[Ry_1 + bA_1]$$

$$\vdots$$

$$y_{n+1} = -L^{-1}[Ry_n + bA_n]$$

where the A_n are given above.

In the more general case of y^m,

$$y_0 = L^{-1}x$$

$$y_1 = -L^{-1}[Ry_0 + bA_0] = -L^{-1}[Ry_0 + by_0^m]$$

$$y_2 = -L^{-1}[Ry_1 + bA_1] = -L^{-1}[Ry_1 + bmy_0^{m-1}y_1]$$

$$\vdots$$

$$\begin{aligned} y_{n+1} &= -L^{-1}[Ry_n + bA_n] \\ &= -L^{-1}[Ry_n + b\{(m-(n-1))(y_1/y_0)A_{n-1} \\ &\quad + (2m-(n-2)(y_2/y_0)A_{n-2} \\ &\quad + (3m-(n-3)(y_3/y_0)A_{n-3} \\ &\quad + \dots + nm(y_n/y_0)A_0\} \end{aligned}$$

If $m = 3$

$$y_0 = L^{-1}x$$

$$y_1 = -L^{-1}[Ry_0 + by_0^3]$$

$$y_2 = -L^{-1}[Ry_1 + b(3y_0^2y_1)]$$

$$\vdots$$

$$
\begin{aligned}
y_{n+1} = -L^{-1}[Ry_n + b\{&3-(n-1)(y_1/y_0)A_{n-1} \\
&+ (6-(n-2))(y_2/y_0)A_{n-2} \\
&+ (9-(n-3))(y_3/y_0)A_{n-3} \\
&+ \ldots + 3n(y_n/y_0)A_0\}]
\end{aligned}
$$

It is easily verified that these results for $m = 3$ correspond to the earlier given results in Case 2 and similarly for $m = 4$, we obtain results of Case 3.

We have noted that if we write

$$y_{n+1} = L^{-1}[Ry_n + bA_n]$$

the expression for A_n is simply obtained by writing all terms in which indices add up to n. But now the number of factors in each term is 3, i.e., the same as m.

$$
\begin{aligned}
A_0 &= y_0y_0y_0 \\
A_1 &= y_0y_0y_1 + y_0y_1y_0 + y_1y_0y_0 \\
A_2 &= y_0y_0y_2 + y_0y_2y_0 + y_2y_0y_0 + y_0y_1y_1 + y_1y_0y_1 + y_1y_1y_0 \\
A_3 &= y_1y_1y_1 + y_0y_0y_3 + y_0y_3y_0 + y_3y_0y_0 + y_0y_1y_2 + y_1y_2y_0 \\
&\quad + y_2y_0y_1 + y_1y_0y_2 + y_0y_2y_1 + y_2y_1y_0
\end{aligned}
$$

etc.

Similarly for $m = 4$

$$
\begin{aligned}
A_0 &= y_0y_0y_0y_0 \\
A_1 &= y_0y_0y_0y_1 + y_0y_0y_1y_0 + y_0y_1y_0y_0 + y_1y_0y_0y_0 \\
A_2 &= y_0y_0y_0y_2 + y_0y_0y_2y_0 + y_0y_2y_0y_0 + y_2y_0y_0y_0 \\
&\quad + y_0y_0y_1y_1 + \ldots
\end{aligned}
$$

etc.

Thus if $N = by^m$

$$y_{n+1} = L^{-1}Ry_n + L^{-1}bA_{m,n}$$

where $A_{m,n}$ is given by

$$A_{m,n} = \sum_{i+j+k+\ldots+w} y_i y_j y_k \cdots y_w$$

where $i+j+k+\ldots+w = n$ and the number of multiplicative factors is m. Hence any polynomial nonlinearity is easily handled. Other types of nonlinearity are being studied.

V. NONLINEAR STOCHASTIC PARTIAL DIFFERENTIAL EQUATIONS

Consider the equation for example

$$\mathsf{F}_t u + \mathsf{F}_x u = g \tag{17}$$

where F_t, F_x are nonlinear operators $\mathcal{L}_t + \mathsf{N}_t$, $\mathcal{L}_x + \mathsf{N}_x$. If we have an inverse for either L_t or L_x we proceed as before. Assume we have L_t^{-1}. Then

$$L_t u = g - R_t u - \mathsf{N}_t u - \mathcal{L}_x u - \mathsf{N}_x u$$

$$u = L_t^{-1} g - L_t^{-1} R_t u - L_t^{-1} \mathsf{N}_t u - L_t^{-1} \mathcal{L}_x u - L_t^{-1} \mathsf{N}_x u \tag{18}$$

Rewrite (17) as $\mathsf{F}_{t,x} u = g$ and assume $\mathsf{F}_{t,x}^{-1}$ exists. (18) becomes

$$\mathsf{F}_{t,x}^{-1} g = L_t^{-1} g - L_t^{-1} R_t \mathsf{F}_{t,x}^{-1} g - L_t^{-1} \mathsf{N}_t \mathsf{F}_{t,x}^{-1} g - L_t^{-1} \mathcal{L}_x \mathsf{F}_{t,x}^{-1} g - L_t^{-1} \mathsf{N}_x \mathsf{F}_{t,x}^{-1} g. \tag{19}$$

With the decomposition $\mathsf{F}_{t,x}^{-1} g = \Sigma \lambda^n H_n g$ and parametrizing (19).

$$\Sigma \lambda^n H_n g = \lambda L_t^{-1} g - \lambda L_t^{-1} R_t \Sigma \lambda^n H_n g - \lambda L_t^{-1} \mathsf{N}_t \Sigma \lambda^n H_n g - \lambda L_t^{-1} \mathcal{L}_x \Sigma \lambda^n H_n g - \lambda L_t^{-1} \mathsf{N}_x \Sigma \lambda^n H_n g \tag{20}$$

As before $H_0 = 0$ and $H_1 = L_t^{-1}$. After that we must know the form of the nonlinear operators.

The operator $\mathcal{L}_x$ can of course be further decomposed into $L_x + R_x$; the operator N_t is written $N_t + M_t$ to include both deterministic (N) and stochastic (M) cases; $\mathsf{N}_x = N_x + M_x$. Then one can consider special cases, e.g., R_t or R_x or M_x or M_t equal to zero of L_x or N_x or $N_t = 0$ and various combinations.

Example: $\frac{\partial u}{\partial t} + u\frac{\partial u}{\partial x} = g$

$$\frac{\partial}{\partial t}u + \frac{\partial}{\partial x}(f(u)) = g$$

$$L_t u + L_x f(u) = g$$

$$u = L_t^{-1} g - L_t^{-1} L_x f(u) = L_t^{-1} g - \frac{1}{2} L_t^{-1} L_x u^2$$

Writing u in terms of the inverse operator and parametrizing

$$\Sigma\lambda^n H_n g = \lambda L_t^{-1} g - \frac{1}{2}\lambda L_t^{-1} L_x \Sigma\lambda^n H_n g \Sigma\lambda^m H_m g$$

$$H_1 g = L_t^{-1} g$$

$$H_2 g = -\frac{1}{2} L_t^{-1} L_x [H_1 g H_0 g + H_0 g H_1 g] = 0$$

$$H_3 g = -\frac{1}{2} L_t^{-1} L_x [H_2 g H_0 g + H_1 g H_1 g + H_0 g H_2 g] = -\frac{1}{2} L_t^{-1} L_x (L_t^{-1} g)(L_t^{-1} g)$$

$$\vdots$$

What kind of functions $f(u)$ for which the procedure works needs further study. This and related matters will be reported in several papers by Adomian and Malakian.

ACKNOWLEDGMENTS

Appreciation is expressed to W. F. Ames, L. H. Sibul, and K. Malakian for discussions and the Sloan Foundation for research support.

REFERENCES

[1] Adomian, G. (1978). "Recent Results in Stochastic Equations - The Nonlinear Case", Invited Paper, Nonlinear Equations in Abstract Spaces, edited by V. Lakshmikantham, Academic Press.

[2] Adomian, G. and Sibul, L. H. "Symmetrized Solutions for Nonlinear Stochastic Differential Equations", to appear.

A NON LINEAR INTEGRAL AND A BANG-BANG THEOREM

Richard A. Alo'

Department of Mathematics
Lamar University
Beaumont, Texas

T. Alvager

Department of Physics
Indiana State University
Terre Haute, Indiana

Andre' de Korvin

Department of Mathematics
Indiana State University
Terre Haute, Indiana

INTRODUCTION

Recently non-linear operators on spaces of measurable functions have generated considerable interest. In [4], Batt represented Hammerstein operators from $C(S,E)$ (the space of continuous functions on the compact space S to the Banach space E under Sup. norm) to a Banach space Y. The representation is of the form $T(f) = \int f\, dm$ where the integral does not satisfy linearity. The motivation for looking at Hammerstein operators is pointed out by Krasnosel'skii in [11]. Non-linear operators

ISBN 0-12-434180-2

on L^p spaces have been studied by many authors, we refer the reader to the work of V. Mizel (see [12]) for some interesting results. The representation given in [12] is of the form $T(f) = \int \Psi(f(\delta),\delta)d\mu$ where μ is a positive measure and Ψ is a uniform p-Caratheodory function. In [13] additive operators on Orlicz spaces are studied. Finally we point out the work in [1] where Hammerstein operators on L^p spaces are considered. In contrast to the work of [12], Caratheodory functions are by-passed and the representation is of the form $T(f) = \int f\, dm$ where m is vector-valued.

In systems theory the bang-bang theorem plays a very important role. Many forms of this theorem are available. In [9] a simple version is stated: If y is a column vector-valued function with components $y_1, y_2, \ldots, y_n$ in L^1 then the set $\{\int y\, u\, dt\}$ is compact convex and equal to $\{\int y\, \chi_A\, dt\}$ where $0 \leqq u \leqq 1$ and A ranges over measurable sets. The Liapunov theorem which asserts that the range of an R^n valued vector measure is compact convex is an immediate consequence of the above theorem. In [10] the bang-bang theorem for finite dimensional linear systems is stated as: any point which is reachable by a control taking values in some compact convex set U of R^n is reachable by a con-rol taking values on the extreme points of U. It is well known that in general the theorem is false when the dimension is infinite. For examples of the conditions under which the theorem is true in infinite dimension see the treatment of Liapunov measures in [10] or see [6].

The main purpose of the present article is two-fold: (1) to introduce a new type of non-linear integral that will represent certain non-linear maps on $L^\infty(\mu)$ where μ is finite and (2) to obtain a form of the bang-bang theorem and Liapunov theorem for that integral.

Specifically, if X and Y are Banach spaces, then $U(X,Y)$ will denote all maps from X into Y (not necessarily linear) that are uniformly continuous on bounded sets of X and R will denote the scalar field. μ denotes a positive finite measure and Σ a σ-algebra of subsets of S. m denotes a finitely additive set function from Σ into $U(X,Y)$. Two types of continuity conditions will be considered on m thus generating two sets of additive set functions which we denote by N_1 and N_2. Let T be a Hammerstein operator from $L^{\infty}(\mu)$ into Y, i.e. $T(f_1+f_2) = T(f_1) + T(f_2)$ whenever f_1 and f_2 have disjoint supports. The first theorem shows that if T satisfies certain continuity conditions then there is an algebraic isomorphism between the set of such T and N_1. Moreover the isomorphism is given by $T(f) = \int f\, dm$. Moreover if $\{f_n\} \to f$ in the wk^* topology in the unit ball of $L^{\infty}(\mu)$, then $\{\int f_n\, dm\} \to \int f\, dm$. The second main theorem reaches the same conclusion but T is subjected to different continuity conditions and m is in N_2. There the corresponding convergence of integrals holds if $\{f_n\} \to f$ in the weak topology of $L^2(\mu)$. If $m \in N_2$ and $y \in L^2(\mu)$, then the set $\{\int y\, u\, dm\}$ (where $m\colon \Sigma \to U(R,Y)$ and μ ranges over a specified subset of $L^{\infty}(\mu)$) is compact in Y. Next we shall put some conditions on m (where $m\colon \Sigma \to U(R,R)$) which will be used to show a form of the bang-bang theorem. These conditions will be looked at in a very special case (where m is of the form $m(A)(x) = \mu(A)f(x)$) and properties of f will be deduced. In fact the results will be valid if f satisfies the following equation

$$f(c) = \Sigma x_j(c,q)f(c+\varepsilon d_j(c,q)) = \Sigma x_j(c,q)f(c-\alpha\varepsilon d_j(c,q))$$

where $x_j(c,q) \geqq 0$, $\Sigma x_j(c,q) = 1$ $\qquad c$ bounded away from some set $\{\hat{c}_1,\hat{c}_2,\ldots\}$ and $q > 0$ bounded away from 0. Moreover we require $\Sigma x_j(c,q)d(c,q)$ bounded away from 0. It will be shown that polygonial functions satisfy this condition where $\{\hat{c}_1,\hat{c}_2,\ldots\}$ are corner points.

Recently polygonial functions have been studied in order to represent some operators on the space of functions of bounded variation, as they form a dense subset of that space. For work in this area see [2], [3], [7], [8]. Of course the theorems are shown without the more restrictive assumption that $m(A)(c)$ is of the form $\mu(A)f(c)$. The third theorem will show that with the previous conditions on m, the image of certain subsets of $L^\infty(\mu)$ by T (where $T(\mu) = \int u\, \chi_{A_i}\, dm$, where $1 \leq i \leq n$ and χ_{A_i} is the characteristic function of A_i) may be obtained by restricting u to have very special values. The fourth theorem will generalize this result to the case where χ_{A_i} are replaced by functions $y_i \in L^2(\mu)$. However it will then be assumed that y_i have disjoint supports and that y_i are bounded away from 0.

In case $m\colon \Sigma \to U(R,R^n)$, a corollary similar to the Liapunov corollary will be deduced. In fact if m is subject to the above conditions, then $\{m(P)(1) + m(N)(1-\delta)\}$ will be compact in R^n where P and N range over measurable partitions of S. Finally the case $m\colon \Sigma \to U(R,Y)$, where Y has infinite dimension, will be looked at.

RESULTS

μ represents a positive finite non-atomic measure on Σ, a σ-algebra of subsets of S. We assume moreover that $L^1(\mu)$ is separable. It is well known (see [5] for example) that $(L^1)^* = L^\infty$ and that the unit ball of $L^\infty(\mu)$ is metrizable relative to the wk^* topology. Let d be the corresponding metric. In fact

$$f(f,g) = \sum_{n=1}^{\infty} \frac{1}{2^n} \frac{|\int (f-g)h_n d\mu|}{1+|\int (f-g)h_n d\mu|}$$

where $\{h_n\}$ forms a countable dense subset of $L^1(\mu)$. $U(R,F)$ will denote the set of all functions from the scalars to F, a

Banach space, uniformly continuous on bounded intervals of R and 0 at 0. If $r \in U(R,F)$, r_α will denote the restriction of r to $[-\alpha, +\alpha]$ and $U_\alpha(R,F) = \{r_\alpha: \ r \in U(R,F)\}$. We define

$$D_\delta r_\alpha = \sup\{\|r_\alpha(e) - r_\alpha(e')\|\}$$

where the sup is over $|e| \leqq \alpha$, $|e'| \leqq \alpha$, $|e-e'| \leqq \delta$.

Let $m\colon \ \Sigma \to U(R,F)$ denote a finitely additive set function and let $m_\alpha(B) = m(B)_\alpha$. Define

$$sv[m_\alpha, B] = \sup\{\|\Sigma m(B_i)e_i\|\} \quad \text{with } B_i \subset B$$

where the sup is over disjoint $B_i \in \Sigma$ with $\|\Sigma\chi_{B_i} e_i\|_\infty \leqq \alpha$.

$$sv_\delta[m_\alpha, B] = \sup\{\|\Sigma m(B_i)e_i - \Sigma m(B_j')e_j'\|\}$$

where the sup is over $\left\{B_i\right\}$ and $\left\{B_j'\right\}$ disjoint sets in Σ that are subsets of B and for which

$$\|\Sigma\chi_{B_i} e_i\|_\infty \leqq \alpha, \quad \|\Sigma\chi_{B_j}{}' e_j'\|_\infty \leqq \alpha, \quad \text{and} \quad d(\Sigma\chi_{B_i} e_i, \Sigma\chi_{B_j}{}' e_j') \leqq \delta.$$

We shall call m *wk*-Nemytskii* if in addition $sv[m_\alpha, S] < \infty$ and $\lim_{\delta\to 0} sv_\delta[m_\alpha, S] = 0$. We denote such set functions by $N_1[\Sigma, U(R,F)]$.

We now will define integration with respect to set functions in $N_1[\Sigma, U(r,F)]$. If $f = \Sigma\chi_{B_i} e_i$ where B_i are disjoint, then

$$\int f \, dm = \Sigma m(B_i) \, e_i.$$

It is easy to see that $\int f \, dm$ is well defined. Now let $\{f_n\}$ denote a sequence of simple uniformly bounded functions, say $\|f_n\|_\infty \leqq \alpha$ with f_n converging *wk** to f. Now $\{f_n\}$ is a Cauchy sequence relative to d and it is easy to see that $\lim \int f_n \, dm$ exists. Also if $\left\{f_n\right\} \to f$ and $\left\{g_n\right\} \to f$ in the d metric and the sequences are in an α-ball, then

$$\|\int f_n \, dm - \int g_n \, dm\| \leqq sv_\delta[m_\alpha, S]$$

where δ dominates $d(f_n, g_n)$. It follows that

$$\lim \int f_n \, dm = \lim \int g_n \, dm$$

and again $\int f \, dm$ is well defined by

$$\int f \, dm = \lim \int f_n \, dm.$$

It can easily be shown that

$$\int (f+f_1+f_2) dm = \int (f+f_1) dm + \int (f+f_2) dm - \int f \, dm$$

where f_1 and f_2 are functions with disjoint supports. It follows that

$$\int_{A \cup B} f \, dm = \int_A f \, dm + \int_B f \, dm \quad \text{if} \quad A \cap B = \phi$$

The method is to note that the above formulas work for simple functions and then to approximate $f \in L^\infty(\mu)$, in the norm, by a sequence of simple functions. Let us note that a lifting map ρ exists on $L^\infty(\mu)$. Thus one may work either with equivalence classes in $L^\infty(\mu)$ or with selected functions $\rho(f)$ in $L^\infty(\mu)$.

We now consider a map T: $L^\infty(\mu) \to F$. We call T of type N_1 if T is wk^* continuous on bounded sets (i.e. T is continuous relative to d), $T(0) = 0$, $\|T\|\alpha < \infty$, $\lim_{\delta \to 0} D_\delta T_\alpha = 0$. Here

$$\|T\|_\alpha = \sup\{T(f)/\|f\|_\infty \leq \alpha\}$$

$$D_\delta T_\alpha = \sup\{\|\Sigma T(\chi_{A_i} e_i) - \Sigma T(\chi_{A_j}{}' e_j')\|\}$$

where the sup is over $\|\Sigma \chi_{A_i} e_i\|_\infty \leq \alpha$, $\|\Sigma \chi_{A_j}{}' e_j'\|_\infty \leq \alpha$, $d(\Sigma \chi_{A_i} e_i, \Sigma \chi_{A_j}{}' e_j') \leq \delta$, and $A_i \cap A_j = \phi$ and $A_k' \quad A_f' = \phi$.

Finally we require that

$$T(f+f_1+f_2) = T(f+f_1) + T(f+f_2) - T(f)$$

whenever f_1 and f_2 have disjoint supports.

Theorem 1. There exists an algebraic isomorphism $T \leftrightarrow \hat{m}$ between operators of type N_1 and $N_1[\Sigma, U(R,F)]$ given by

$$T(f) = \int f \, d\hat{m}, \quad \text{for } f \in L^{\infty}(\mu).$$

Moreover

$$\|T\|_{\alpha} = sv[\hat{m}_{\alpha}, S], \quad D_{\delta}T_{\alpha} = sv_{\delta}[\hat{m}_{\alpha}, S].$$

Proof. Let $\hat{m} \in N_1[\Sigma, U(R,F)]$, and $T\colon L^{\infty}(\mu) \to F$ be defined by

$$T(f) = \int f \, d\hat{m}.$$

It is easy to see that $\hat{m} \to T$ is an injection, also $T(0) = 0$ and

$$\|T\|_{\alpha} = \sup \left\| \int f \, d\hat{m} \right\| \leqq sv[\hat{m}_{\alpha}, S] < \infty$$

$$D_{\delta}T_{\alpha} = \sup\{\|\Sigma T(\chi_{A_i} e_i) - \Sigma T(\chi_{A_j}' e_j')\|\}.$$

Thus

$$D_{\delta}T_{\alpha} = sv_{\delta}[\hat{m}_{\alpha}, S]$$

and

$$\lim_{\delta \to 0} D_{\delta}T_{\alpha} = 0$$

Finally the fact that

$$T(f+f_1+f_2) = T(f+f_1) + T(f+f_2) - T(f)$$

whenever f_1 and f_2 have disjoint supports follows from properties of $\hat{m}$. Thus T is of type N_1.

Conversely for T of type N_1 let $\hat{m}(A)(c) = T[\chi_A c]$. It is not difficult to see that $\hat{m}\colon \Sigma \to U(R,F)$. If f is a simple function, then $f = \Sigma \chi_{A_i} e_i$ where $\{A_i\}$ are disjoint and in Σ, and

$$\int f \, d\hat{m} = \Sigma \hat{m}(A_i) e_i = \Sigma T(\chi_{A_i} e_i) = T(f).$$

Now for $f \in L^{\infty}(\mu)$, there exists a sequence $\{f_n\}$ in $L^{\infty}(\mu)$, uniformly bounded by some $\alpha > 0$ such that $\{f_n\} \to f$ in the norm of $L^{\infty}(\mu)$. Thus $\{f_n\}$ converges wk^* to f and $\{\int f_n \, d\hat{m}\} \to \int f \, d\hat{m}$. By wk^* continuity of T, $T(f_n) \to T(f)$. Thus $T(f) = \int f \, d\hat{m}$ for all $f \in L^{\infty}(\mu)$ and $\hat{m} \to T$ is a bijection. It is straightforward to finish the proof.

We now consider the separable space $L^2(\mu)$. The unit ball of $L^2(\mu)$ is metrizable in the weak topology of $L^2(\mu)$ (see [5]). Again denote this metric by d. We now are interested in set functions $m\colon \Sigma \to U(R,F)$ which are finitely additive and satisfy the same formal properties as $\hat{m}$ where $\|(\cdot)\|_{\infty}$ is replaced by $\|(\cdot)\|_2$ (the L^2 norm) and d now is equivalent to the weak topology of $L^2(\mu)$. Such set functions will be denoted by $N_2[\Sigma,U(r,F)]$. Similarly we shall define maps $T\colon L^2(\mu) \to F$ of type N_2. (In particular T is now weakly sequentially continuous in bounded sets of $L^2(\mu)$.) The corresponding semi-variation and δ semi-variation will still be denoted by the symbols $sv[m_\alpha,S]$ and $sv_\delta[m_\alpha,S]$.

Theorem 2. There exists an algebraic isomorphism $T \leftrightarrow m$ between operators of type N_2 and $N_2[\Sigma,U(R,F)]$ given by $T(f) = \int f \, dm$ for all $f \in L^2(\mu)$. Moreover

$$\|T\|_\alpha = sv[m_\alpha,S], \quad D_\delta T_\alpha = sv_\delta[m_\alpha,S].$$

Proof. We use the same arguments as in Theorem 1 with minor changes.

We now are interested in operators on $L^{\infty}(\mu)$ defined by

$$T(u) = \int u \, y \, dm$$

where

$$y \in L^2(\mu) \quad \text{and} \quad m \in N_2[\Sigma,U(R,F)].$$

Proposition 1. T is wk^* continuous on the unit ball of $L^{\infty}(\mu)$ and there exists $m_y \in N_1[\Sigma,U(R,F)]$ such that $T(u) = \int u \, dm_y$.

Proof. Let $\{u_n\}$ be a sequence in the unit ball and let $u_n \to u$ in the wk^* topology. Then by the Hölder inequality $u_n y \to uy$ weakly in $L^2(\mu)$. By properties of $N_1[\Sigma, U(R,F)]$, it follows that $\int un\ y\ dm \to \int u\ y\ dm$. Thus T is wk^* continuous. Clearly $T(0) = 0$, $\|T\|_\alpha < \infty$. Finally

$$D_\delta T_\alpha = \sup\{\|\Sigma T(\chi_{A_i} e_i) - \Sigma T(\chi_{A_j}' e_j')\|\}$$

$$= \sup\{\|\int(\Sigma\chi_{A_i} e_i) y\ dm - \int(\Sigma\chi_{a_j}' e_j') y\ dm\|\}$$

$$\leq sv_{2\delta'}[m_\alpha, S]$$

where

$$\delta' = \sup\|(\Sigma\chi_{A_i} e_i) y - (\Sigma\chi_{A_j}' e_j') y\|_2 \leq \delta\|y\|_2.$$

Thus as $\delta \to 0$, $\delta' \to 0$ and hence

$$D_\delta T_\alpha \to 0.$$

Therefore

$$T \in N_1[\Sigma, U(R,F)].$$

By Theorem 1 there exists m_y such that $T(u) = \int u\ dm_y$. We now define $M_\alpha \subset F^n$ by

$$M_\alpha = \{\int u\ y_i\ dm\}$$

where $1 \leq i \leq n$, $y_i \in L^2(\mu)$, $\|u\|_\infty \leq \alpha$, $m \in N_2[\Sigma, U(R,F)]$.

Proposition 2. M_α is a compact subset of F^n for all $\alpha < 0$.

Proof. This proposition is immediate since T is wk^* continuous and the α-ball of $L^\infty(\mu)$ is wk^* compact.

Form now on we will be looking at $m\colon \Sigma \to U(R,R)$, i.e. $m(A)$ is a real-valued function, uniformly bounded on bounded intervals and 0 at the origin. We are still looking at T

defined over $L^\infty(\mu)$ by $T(u) = \int u\, y_i\, dm$ with $y_i \in L^2(\mu)$ and $m \in N_2[\Sigma, U(R,R)]$.

We now impose a *condition (C)* on m.

(C) Let $\{\hat{c}_1, \hat{c}_2, \ldots\}$ be a finite or infinite set of points in $[-1, +1]$. For every $B \in \Sigma$ such that $\mu(B) > 0$ and for every $\varepsilon > 0$ and every $q > 0$ bounded away from 0 and c bounded away from $\{\hat{c}_1, \hat{c}_2, \ldots\}$, there exists a (finite) partition $\{B_j\}$ of B and a sequence of numbers $\{d_j\}$ such that

(1) $m(B)(c) = \Sigma m(B_j)(c+\varepsilon d_j) = \Sigma m(B_j)(c - \alpha\, \varepsilon d_j)$ and

(2) $\int (\Sigma \chi_{B_j} d_j) d\mu > K_q(\varepsilon)\mu(B)$

where $|d_j| \leq q$, α is a fixed number with $0 < \alpha < 1$, and $K_q(\varepsilon) > 0$ is bounded away from 0 for fixed $\varepsilon < 0$.

Example. We now consider a special case where $m\colon \Sigma \to U(R,R)$. Let $m(A)(c) = \mu(A)f(c)$ where f is some polygonial line passing through the origin with corner points at $c_1, c_2, \ldots, c_n$. Let $f(c) = sc + \alpha$, and let $q > 0$, $\varepsilon > 0$. Let $\mu(B) > 0$. Split B into four sets B_i with $\frac{\mu(B_i)}{\mu(B)} = x_i$. We analyze condition (C) for c bounded from the corner points. If $d_1 = q$, $d_2 = -q^2$, $d_3 = q/2$, $d_4 = -q/4$.

$$s\varepsilon q(x_1 \; qx_2) = 0$$

$$s\varepsilon q/2(x_3 - qx_4) = 0$$

$$x_1 + x_2 + x_3 + x_4 = 1$$

This system is verified when $x_1 = x_3 = \frac{q}{2(q+1)}$ and $x_2 = x_4 = \frac{1}{2(q+1)}$. By the non atomicity of μ, appropriate B_i may be found. Now $\int (\Sigma \chi_{B_j} d_j) d\mu$ in this case is $\mu(B)\frac{q}{8(q+1)}$ and hence is bounded away from 0.

Recently, polygonial functions have received much attention as they have been used to represent the dual space of functions of bounded variations (see for example [2], [3], [7], [8]). In these works the key property of polygonial functions is that they are dense in the set of functions of bounded variation.

It would be of interest to determine the class of all functions satisfying

$$f(c) = \Sigma x_j(c,q)f(c+\varepsilon d_j(c,q)) = \Sigma x_j(c,q)f(c-\alpha\varepsilon d_j(c,q))$$

where $x_j(c,q) \geqq 0$, $\Sigma x_j(c,q) = 1$, and $\Sigma x_j(c,q)d_j(c,q)$ is bounded away from 0 for every c bounded away from $\{\hat{c}_1,\hat{c}_2,\ldots\}$ and for every $q > 0$ bounded away from 0. For such functions $m(A)(c) = \mu(A)f(c)$ would define a measure satisfying condition (C). The partition B_j of B would be defined by

$$\frac{\mu(B_j)}{\mu(B)} = x_j$$

The above example shows polygonial functions satisfy this condition. Thus in the very special case where

$$m(A)(c) = \mu(A)f(c), \text{ where } \mu \text{ is non atomic,}$$

we assume that f satisfies (E) for all c in $(-1, +1)$ bounded away from some set $\{\hat{c}_1,\hat{c}_2,\ldots\}$. The proofs of course are valid without the more restrictive assumption that $m(A)(c)$ is of the form $\mu(A)f(c)$.

From now on $m\colon\ \Sigma \to U(R,R)$ is in $N_2[\Sigma,U(R,R)]$ and is assumed to satisfy (C). For $u \in L^\infty(\mu)$ we define

$$\hat{E}_1(u) = \{x/-1 + \varepsilon \leqq u(x) \leqq 1 - \varepsilon\} \cap \text{Support } (u).$$

Here $\varepsilon > 0$ is fixed.

For the next proposition we will let $y_i = \chi_{A_i}$ and $u = c\chi_A$. We assume c is bounded away from $\{\hat{c}_1,\hat{c}_2,\ldots\}$.

Proposition 3. For every $\varepsilon > 0$ there exists $h^\varepsilon \in L^\infty(\mu)$ such that $h^\varepsilon \neq 0$, $\|h^\varepsilon\|_\infty \leqq 1$ and

$$\int_{\hat{E}_1}(c\chi_A+\varepsilon h^{\varepsilon})\chi_{A_i}dm = \int_{\hat{E}_1}(c\chi_A-\alpha\varepsilon h^{\varepsilon})\chi_{A_i}dm = \int_{\hat{E}_1}c\chi_A\chi_{A_i}dm$$

for $1 \leq i \leq n$. (Without loss of generality we may assume that the support of h^{ε} is a subset of $\hat{E}_1$.) In fact $\int h^{\varepsilon}d\mu > K(\varepsilon)\mu(\hat{E}_1)$ for $K(\varepsilon)$ bounded away from 0 for $-1 < c < +1$, $c \neq 0$ and c bounded away from $\{c_1,c_2,\ldots\}$.

Proof. Let $u = c\chi_A$ for $-1 < c < +1$, $c \neq 0$ and let c be bounded away from $\{\hat{c}_1,\hat{c}_2,\ldots\}$. Then $E_1(u) = A$. We illustrate the procedure when $n = 2$. This procedure will be valid for any n. Let $C_1 = A \cap A_1{}^c \cap A_2{}^c$, $C_2 = A \cap A_1 \cap A_2{}^c$, $C_3 = A \cap A_2 \cap A_1{}^c$, $C_4 = A \cap A_1 \cap A_2$. Clearly C_1, C_2, C_3, C_4 are mutually disjoint. Let $\{D_i\}$, $\{d_i\}$, $\{E_j\}$, $\{e_j\}$, $\{F_k\}$, $\{f_k\}$, $\{G_s\}$, $\{g_s\}$ be the sets and numbers obtained by applying condition (C) to C_1, C_2, C_3, C_4 for $q = 1$.

Let $\{B_\ell\}$, $\{h_\ell\}$ be the resulting sets and numbers obtained by putting together $\{D_i\}$, $\{E_j\}$, $\{F_k\}$, $\{G_s\}$ and $\{d_i\}$, $\{e_j\}$, $\{f_k\}$, $\{g_s\}$. Let $h^{\varepsilon} = \Sigma\chi_{B_\ell}h_\ell$. Clearly $\|h^{\varepsilon}\|_\infty \leq 1$. It is easy to check that

$$\int_{\hat{E}_1}(u+\varepsilon h^{\varepsilon})\chi_{A_i}dm = \int_{\hat{E}_i}(u-\alpha\varepsilon h^{\varepsilon})\chi_{A_i}dm = \int_{\hat{E}_1}u\chi_{A_i}dm$$

for all $1 \leq i \leq n$. Moreover

$$\int_{\hat{E}_1}h^{\varepsilon}d\mu = \int(\Sigma\chi_{D_i}d_i+\Sigma\chi_{E_j}e_j+\Sigma\chi_{F_k}f_k+\Sigma\chi_{G_s}g_s)d\mu \geq K(\varepsilon)\mu(A)$$

where in fact $K(\varepsilon)$ is the least $K(\varepsilon)$ for C_1, C_2, C_3, C_4 $(q = 1)$. In particular $h^{c} \neq 0$.

We now take functions in the unit ball of $L^{\infty}(\mu)$ that are bounded away from $\{\hat{c}_1,\hat{c}_2,\ldots\}$. Assume that for some $\delta > 0$, $\{\hat{c}_1,\hat{c}_2,\ldots\}$ are outside the intervals $[1-2\delta, 1+2\delta]$ and $[-1-2\delta, -1+2\delta]$. Let $U = \{u \in L^{\infty}(\mu)\colon \|u\|_\infty \leq 1$ and $\|u-1\|_\infty \leq \delta$ or $\|u+1\|_\infty \leq \delta\}$.

Actually the proofs that follow work for any set U of the form $U = \{u \in L^{\infty}(\mu)\colon \ \|u-c_i\|_{\infty} \leqq \delta_i$ for some $i\}$ where the sequence c_i is bounded away from $\{\hat{c}_1, \hat{c}_2, \dots\}$ and δ_i are such that U is wk^* compact. To make notations easier we work with $U = \{u \in L^{\infty}(\mu)\colon \ \|u\|_{\infty} \leqq 1$ and $\|u-1\|_{\infty} \leqq \delta$ or $\|u+1\|_{\infty} \leqq \delta\}$. (The results for more general cases go through with obvious modifications).

We now show a bang-bang type theorem for our integral in the case $y_i = \chi_{A_i}$ where $T(f) = \int f\chi_{A_i} dm$, $1 \leqq i \leqq n$. Let $U_0 = \{u \in U\colon u$ takes only the values $1-\delta$, $1+\delta$, -1, $+1\}$.

<u>*Theorem 3.*</u> $T(U) = \{\int u\chi_{A_i} dm\colon \ u \in U_0\}$.

Proof. The first step is to extend Proposition 3 to step functions. Let $u = \Sigma c_j \chi_{\ell_j}$ where ℓ_j are mutually disjoint sets, and $-1 < c_j < +1$, $c_j \neq 0$ and c_j away from $\{\hat{c}_1, \hat{c}_2, \dots\}$. Let $h_1^{\varepsilon}, h_2^{\varepsilon}, \dots$ be functions obtained for $c_1\chi_{\ell_1}, c_2\chi_{\ell_2}, \dots$ by Proposition 3. Let $h^{\varepsilon} = \Sigma h_{\rho}^{\varepsilon}$. Clearly $\hat{E}_1(u) = U\ell_i$ (we can redefine ε to be small enough to have this). Clearly $\|h^{\varepsilon}\|_{\infty} \leqq 1$. We have:

$$\begin{aligned}\int_{\hat{E}_1} (u+\varepsilon h^{\varepsilon})\chi_{A_i} dm &= \Sigma \int_{\hat{E}_1} (c_j\chi_{\ell_j} + \varepsilon h^{\varepsilon})\chi_{A_i} dm \\ &= \Sigma \int_{\hat{E}_1} c_j \chi_{\ell_j} \chi_{A_i} dm \\ &= \int_{\hat{E}_1} u\chi_{A_i} dm.\end{aligned}$$

Similarly

$$\int_{\hat{E}_1} (u-\alpha\varepsilon h^{\varepsilon})\chi_{A_i} dm = \int_{\hat{E}_1} u\chi_{A_i} dm \quad \text{for all } i, \ 1 \leqq i \leqq n.$$

Also,

$$\int h^{\varepsilon} d\mu \geqq \Sigma K(\varepsilon)\mu(\ell_j) = K(\varepsilon)\mu(\hat{E}_1) \quad \text{for an appropriate } K(\varepsilon).$$

Thus if s denotes a simple function we have shown the existence of $h^{\varepsilon} \neq 0$, $\|h^{\varepsilon}\|_{\infty} \leqq 1$ such that

$$\int_{\hat{E}_1} (s+\varepsilon h^{\varepsilon})\chi_{A_i} dm = \int_{\hat{E}_1} (s-\alpha\varepsilon h^{\varepsilon})\chi_{A_i} dm = \int_{\hat{E}_1} s\chi_{A_i} dm$$

where $\hat{E}_1$ denotes the support of s.

Let a denote any point in the range of T restricted to U. Proposition 1 shows that T is wk^* compact. There exists then an extreme point u in the intersection of $T^{-1}(a)$ with U (see [5]). Let $\tilde{E}_1 = \tilde{E}_1(u)$ where $\tilde{E}_1 = \{(1-\delta) + \varepsilon < u < 1 - \varepsilon$ or $-1 + \varepsilon < u < (-1+\delta) - \varepsilon\}$ where of course $\varepsilon < \delta/2$.

Let $\{s_n\}$ be a sequence of simple functions from the unit ball such that $\{s_n\}$ converges in the norm to $\chi_{\tilde{E}_1} u$ where the values of s_n may be taken away from $\{\hat{c}_1, \hat{c}_2, \ldots\}$. By the previous argument we can find a sequence $\{h_n^{\varepsilon}\}$ in the unit ball such that

$$\int_{\tilde{E}_1} s_n \chi_{A_i} dm = \int_{\tilde{E}_1} (s_n+\varepsilon h_n^{\varepsilon})\chi_{A_i} dm = \int s_n \chi_{A_i} dm$$

(Without loss of generality we may assume that all the s_n have $\tilde{E}_1$ as support). By wk^* compactness of the unit ball a subsequence of h_n^{ε} (which we continue to denote by h_n^{ε}) converges wk^* to h^{ε}. Thus $s_n + \varepsilon h_n^{\varepsilon}$ converges wk^* to $\chi_{\tilde{E}_1} u + \varepsilon h^{\varepsilon}$. So

$$\int_{\tilde{E}_1} (s_n+\varepsilon h_n^{\varepsilon})\chi_{A_i} dm \to \int_{\tilde{E}_1} (u+\varepsilon h^{\varepsilon})\chi_{A_i} dm$$

since $m \in N_2[\Sigma, U(R,R)]$ and $(s_n+\varepsilon h_n^{\varepsilon})\chi_{A_i}$ converges weakly to $(u+\varepsilon h^{\varepsilon})\chi_{A_i}$ in $L^2(\mu)$. We have shown that

$$a = T(u) = T(u+\varepsilon h^{\varepsilon}) = T(u-\alpha\varepsilon h^{\varepsilon}).$$

Also since $\int_{\tilde{E}_1} h_n^{\varepsilon} d\mu \geqq K(\varepsilon)\mu(\tilde{E}_1)$, we have $\int_{\tilde{E}_1} h^{\varepsilon} d\mu \geqq K(\varepsilon)\mu\tilde{E}_1$ and $h^{\varepsilon} = 0$ where $K(\varepsilon)$ is bounded away from 0. Thus for ε small enough u is in the interval $(u-\alpha\varepsilon h^{\varepsilon}, u+\varepsilon h^{\varepsilon})$ and this contradicts that u is an extreme point of U. It follows that $\mu(\tilde{E}_1) = 0$. Therefore $u \in U_0$.

Note that h^{ε} has support inside $\tilde{E}_1$ and thus inside $\tilde{E}_1$.

Proposition 4. Let $y_i = q_i \chi_{A_i}$ where $\{A_i\}$ is a finite sequence of disjoint sets, where $1 \geqq q_i > \frac{-1+2\delta}{-1+\delta} + a$ for some $a > 0$ and where $\{\hat{c}_1, \hat{c}_2, \ldots\}$ is outside $[1-2\delta, 1+2\delta]$ and $[-1-2\delta, -1+2\delta]$ with $1 > \delta > \frac{1}{2}$. If m satisfies (C) then for every $\varepsilon > 0$ there exists h_i^{ε} such that

$$\int_{\hat{E}_1} (u+\varepsilon h_i^{\varepsilon}) y_i dm = \int_{\hat{E}_1} (u-\alpha\varepsilon h_i^{\varepsilon}) y_i dm = \int_{\hat{E}_1} u y_i dm$$

with $h \neq 0$ for $1 \leqq 1 \leqq n$, where $u \in U$.

Proof. Define new set functions by

$$\tilde{m}_i(B)(c) = m(B)(q_i c).$$

It is not difficult to check that $\tilde{m}_i \in N_2[\Sigma, U(R,R)]$.

Let $d_j' = \frac{d_j}{q_i}$ where d_j is given by the equation

$$m(B)(cq_i) = \Sigma m(B_j)(cq_i + \varepsilon d_j)$$

for $d_j \in (-q_i, q_i)$ and

$$\int \Sigma \chi_{B_j} d_j \, d\mu \geqq K_{q_i}(\varepsilon)\mu(B).$$

Now d_j exists by condition (C) applied to cq_i, since by hypothesis cq_i is away from $\{\hat{c}_1, \hat{c}_2, \ldots\}$. Now

$$\sum_j \tilde{m}_i(B_j)(c+\varepsilon d_j') = \tilde{m}_i(B)(c)$$

and therefore $\tilde{m}_i$ satisfy (C). Also $|d_j'| \leqq 1$ and

$$\int \Sigma \chi_{B_j} d_j' \, d\mu = \frac{1}{q_i} \int \Sigma \chi_{B_j} d_j \, d\mu \geqq \frac{K_{q_i}(\varepsilon)}{q_i} \mu(B).$$

Because $\frac{K_{q_i}(\varepsilon)}{q_i}$ is bounded away from 0, so is $\int \Sigma \chi_{B_j} d_j' \, d\mu$.

We apply our previous results to $\tilde{m}_i$ for $n = 1$. Thus there exists h_i^ε such that

$$\int h_i^\varepsilon \, d\mu \geqq K_i(\varepsilon)\mu(\hat{E}_1).$$

In fact $h_i^\varepsilon = \Sigma \chi B_j d_j'$ where $\{B_j\}$ is a partition of $\hat{E}_1$ and we may assume that the support of h_i^ε is in A_i. Also

$$K_i(\varepsilon) = \frac{K_{q_i}(\varepsilon)}{q_i},$$

and

$$\int_{\hat{E}_1} (u+\varepsilon h_i^\varepsilon) q_i \chi_{A_i} dm = \int_{\hat{E}_1} (u+\varepsilon h_i^\varepsilon)\chi_{A_i} d\tilde{m}_i = \int u\chi_{A_i} d\tilde{m}_i$$

$$= \int u q_i \chi_{A_i} dm.$$

This completes the proof of the proposition.

The support of h_i is a subset of A_i.

Theorem 4. Assume $\{\hat{c}_1, \hat{c}_2, \ldots\}$ is as in proposition 4, that m satisfies (C) and $y_i \in L^2(\mu)$ $(1 \leqq i \leqq n)$ have disjoint support with $1 \geqq |y_i| > \frac{-1+2\delta}{-1+\delta} + a$, for some $a > 0$, over the support of y_i. Then $T(U) = \{\int u \, y_i \, dm / u \in U_0\}$.

Proof. We first assume $y_i \geqq 0$. Let $y_i = q_i \chi_{A_i}$ with $q_i \geqq \frac{-1+2\delta}{-1+\delta} + a$ where A_i are mutually disjoint. For $\varepsilon > 0$ let h_i^ε be the functions given by proposition 4. Let $h^\varepsilon = h_1^\varepsilon + h_2^\varepsilon + \ldots + h_n^\varepsilon$. Clearly

$$\int (u+\varepsilon h^\varepsilon) q_i \chi_{A_i} dm = \int (u q_i \chi_{A_i} + \varepsilon h_i^\varepsilon q_i \chi_{A_i}) dm$$

$$= \int (u+\varepsilon h_i^\varepsilon)\chi_{A_i} d\tilde{m}_i$$

$$= \int u q_i \chi_{A_i} dm.$$

Similarly

$$\int (u-\alpha\varepsilon h^{\varepsilon})q_i A_i \, dm = \int u q_i \chi_{A_i} dm.$$

Now let y_i be positive simple functions with disjoint supports i.e.

$$y_i = \sum_j \chi_{\ell_{ij}} q_{ij}, \quad q_{ij} \geqq \frac{-1+2\delta}{-1+\delta} + a, \quad \ell_{ij} \text{ mutually disjoint.}$$

By previous arguments there exists h_j^{ε}, $\|h_j^{\varepsilon}\|_{\infty} \leqq 1$ such that for all $1 \leqq i \leqq n$,

$$\int (u+\varepsilon h_j^{\varepsilon})q_{ij}\chi_{\ell_{ij}} dm = \int (u-\alpha\varepsilon h_j^{\varepsilon})q_{ij}\chi_{\ell_{ij}} dm = \int u q_{ij}\chi_{\ell_{ij}} dm.$$

Also

$$\int h_j^{\varepsilon} \, d\mu \geqq K_j(\varepsilon)\mu(\hat{E}_1).$$

Let $h^{\varepsilon} = \sum_j h_j^{\varepsilon}$ with support of h_j^{ε} contained in $U_{i^{\ell_{ij}}}$.

$$\begin{aligned}\int (u+\varepsilon h^{\varepsilon})(\sum_j \chi_{\ell_{ij}} q_{ij})dm &= \int (\sum_j u\chi_{\ell_{ij}} q_{ij} + \varepsilon \sum_j h_j^{\varepsilon}\chi_{\ell_{ij}} q_{ij})dm \\ &= \sum_j \int (u+\varepsilon h_j^{\varepsilon})\chi_{\ell_{ij}} q_{ij} dm \\ &= \sum_j \int u\chi_{\ell_{ij}} q_{ij} dm = \int u \sum_j \chi_{\ell_{ij}} q_{ij} dm\end{aligned}$$

for $1 \leqq i \leqq n$. An identical argument holds when ε is replaced by $-\alpha\varepsilon$. Also

$$\int h^{\varepsilon} \, d\mu \geqq K(\varepsilon)\mu(\hat{E}_1)$$

where

$$K(\varepsilon) = \sum_j K_j(\varepsilon)$$

where

$$K_j(\varepsilon) = \sum_i \frac{K_{q_{ij}}(\varepsilon)}{q_{ij}}$$

(In particular $h^\varepsilon \neq 0$).

Now let y_i denote positive functions of $L^2(\mu)$ with disjoint supports. Let y_i^n converge to y_i in the L^2 norm where y_i^n are positive simple functions with $|y_i^n| \geqq \frac{-1+2\delta}{-1+\delta} + a$. Let h_n^ε be functions in the unit ball of $L^\infty(\mu)$ such that

$$\int_{\hat{E}_1} (u+\varepsilon h_n^\varepsilon) y_i^n \, dm = \int_{\hat{E}_1} (u-\alpha\varepsilon h_n^\varepsilon) y_i^n \, dm = \int_{\hat{E}_1} u y_i^n \, dm. \tag{1}$$

Again by wk^* compactness some subsequence of h_n^ε (which we still denote by h_n^ε) converges to h^ε. Thus $h_n^\varepsilon y_i^n$ converges weakly in $L^2(\mu)$ to $h^\varepsilon y_i$ by the Holder inequality. Since $m \in N_2[\Sigma, U(R,R)]$ by letting $n \to \infty$ in (1) we have

$$\int_{\hat{E}_1} (u+\varepsilon h^\varepsilon) y_i \, dm = \int_{\hat{E}_1} (u-\alpha\varepsilon h) y_i \, dm = \int_{\hat{E}_1} u y_i \, dm.$$

Also

$$\int h_n^\varepsilon \, d\mu \geqq K_n(\varepsilon)\mu(\hat{E}_1)$$

where $\{K_n(\varepsilon)\}$ is bounded away from 0. So $\int h_n^\varepsilon \, d\mu > 0$ and $h^\varepsilon \neq 0$.

We now pick y_i to be negative functions of $L^2(\mu)$ with disjoint supports. Going through previous computations we see that for $y_i = \chi_{\ell_{ij}} q_{ij}$

$$\int h_i^\varepsilon \, d\mu \leqq \frac{K_{q_i}(\varepsilon)}{q_i} \mu(\hat{E}_1) \quad \text{where} \quad |q_i| \geqq \frac{-1+2\delta}{-1+\delta} + a$$

$$= K_i(\varepsilon)\mu(\hat{E}\ ) \quad \text{where} \quad K_i(\varepsilon) < 0.$$

Again for simple functions we set $h^\varepsilon = \Sigma h_i^\varepsilon$ and the proof proceeds as before to show $\int h^\varepsilon \, d\mu < 0$. The theorem then holds if $y_i < -\left(\frac{-1+2\delta}{-1+\delta}\right) - a$ for $1 \leqq 1 \leqq n$ over the support of y_i.

Finally let $y_i \in L^2(\mu)$ have disjoint supports. Let $y_i = y_i^+ + y_i^-$ where $y_i^- \leqq 0$, $y_i^+ \geqq 0$, and $y_i^- y_i^+ = 0$. Let P_i and N_i denote the supports of y_i^+ and y_i^-. Let $P = \bigcup_i P_i$, $N = \bigcup_i N_i$. Then

$$\int uy_i\, dm = \int uy_i^+\, dm + \int uy_i^-\, dm$$

$$= \int u_1 y_i^+\, dm + \int u_2 y_i^-\, dm$$

where u_i takes only the values ± 1 and $1-\delta$, $1+\delta$. Let $u_1 = 0$ on N, $u_1 = 1$ on the complement of $P \cup N$, and $u_2 = 0$ on P. This does not affect the values of the integral and

$$\int uy_i dm = \int (u_1+u_2)(y_i^+ + y_i^-) dm = \int ry_i dm$$

where $r = u_1 + u_2$. Clearly $r \in U_0$ and the theorem is shown.

We now would like to obtain a version of the Liapunov theorem for m. For this we consider $m\colon \Sigma \to U(R,R^n)$. We assume $m \in N_2[\Sigma, U(R,R^n)]$ i.e. in the definitions relative to $N_2[\Sigma, U(R,R^n)]$ we replace the absolute value by the norm in R^n. We assume again that m satisfies condition (C). We call m of orthogonal type if for $x \in R$, $A \in \Sigma$,

$$m(A)(x) = (m_1(A)(x), \ldots m_n(A)(x))$$

where m_i are concentrated on disjoint sets. It is easy to check that if m satisfies condition (C), then $\{m_i\}$ satisfy (C). Since

$$\int c\chi_A\, dm = m(A)(c) = (m_1(A)(c), \ldots m_n(A)(c))$$

$$= \left(\int c\chi_A\, dm_1, \ldots \int c\chi_A\, dm_n\right).$$

It follows by a limit argument that

$$\int f\, dm = \{(\int f\, dm_n)\}$$

for all f in the unit ball of $L^\infty(\mu)$.

Corollary (Liapunov). Let $m \in N_2[\Sigma, U(R,R^n)]$ assume that satisfies (C) and that m is of orthogonal type. Assume $\{\hat{c}_1, \hat{c}_2, \ldots\}$ is as in theorem 4. Then $\{m(P)(1)+m(N)(1-\delta)\}$, as P and N range over measurable partitions of S, is a compact set of R^n.

Proof. Let $\nu = \Sigma m_i$. Then $\nu\colon\ \Sigma \to U(R,R)$. Clearly

$$\int u\chi_A\, dm = \{(\int u\chi_{A_i}\, d\nu)\}$$

for all u in the unit ball of $L^\infty(\mu)$. Let $U' = \{u/\|u\|_\infty \leqq 1$ and $\|u-1\|_\infty \leqq \delta\}$. By the above theorem, $\{(\int u\chi_{A_i}\, d\nu)\}$ ranges over a compact set of R^n as u ranges over U' since ν satisfies (C). Moreover that set is $\{(\int \hat{u}\chi_{A_i}\, d\nu)\}$ where $\hat{u} = 1$ or $1 - \delta$. The last set is also

$$\{(\nu(A_i \cap P)(1) + \nu(A_i \cap N)(1-\delta))\} = \{m(P)(1) + m(N)(1-\delta)\}$$

as P and N range over measurable partitions of S.

For the last result we would like to replace R^n by Y where Y is a general Banach space. We now assume $m\colon\ \Sigma \to U(R,Y)$. Moreover we assume $m \in N_1[\Sigma, U(R,Y)]$. We let $T(f) = \int f\, dm$, let $U^1 = \{u/\|u\|_\infty \leqq 1$ and $\|u-1\|_\infty \leqq \delta\}$.

Theorem 5. Assume that $m \in N_1[\Sigma, U(R,Y)]$, that m satisfies (C) and that $\{\hat{c}_1, \hat{c}_2, \ldots\}$ is as in theorem 4. Then

$$T(U^1) = \{m(P)(1) + m(N)(1-\delta)\}$$

where P and N form a measurable partition of S. Moreover this set is a compact subset of Y provided $\{\Sigma m(A_i)x_i\}$ is precompact in Y where $\|\Sigma\chi_{A_i} x_i\|_\infty \leqq 1$.

Proof. Since $\{\Sigma m(A_i)x_i\}$ is precompact as $\|\Sigma\chi_{A_i} x_i\|_\infty \leqq 1$, it follows that the imate by T of the unit ball of $L^\infty(\mu)$ is precompact. Therefore T is a compact map and hence T^* (the adjoint of T) is compact (see [5]). It can be shown (using the same arguments as in [5]) that T must then send bounded wk^* convergent sequences into norm convergent sequences. Therefore T is wk^* continuous. $T^{-1}(a)$ is therefore compact in U'. Again let u be an extreme point of $T^{-1}(a)$. Repeat the computations done before but with $\int u\,dm$ instead of $\int uy_i\,dm$. We obtain $s_n + \varepsilon h_n^{\varepsilon}$ converging wk^* to $\chi_{\hat{E}_1} u + \varepsilon h^{\varepsilon}$ where $\{s_n\}$ is a sequence of from $\{\hat{c}_1, \hat{c}_2, \ldots\}$. Then $\int(s_n + \varepsilon h_n^{\varepsilon})dm$ converges to $\int(\chi_{\hat{E}_1} u + \varepsilon h^{\varepsilon})dm$ since $m \in N_1[\Sigma, U(R,Y)]$ and again the contradiction to $\mu(E_1) > 0$ is obtained i.e. u is 1 or $1 - \delta$ a.e. The rest of the theorem is shown as earlier.

All of the above theorems may be restated when $U = \{u / \|u\|_\infty \leqq 1$ and $\|u-1\|_\infty \leqq \delta$ or $\|u+1\|_\infty \leqq \delta\}$ or $U' = \{u / \|u\|_\infty \leqq 1$ and $\|u-1\|_\infty \leqq \delta\}$ are replaced by sets of the form $U = \{u / \|u - c_i\| \leqq \delta_i$ for some $i\}$ where the sequence $\{c_i\}$ is bounded away from $\{\hat{c}_1, \hat{c}_2, \ldots\}$ and δ_i are such that U is wk^* compact.

REFERENCES

[1] Alò, R. A., and de Korvin, A. (December 1975). "Representation of Hammerstein operators by Nemytskii measures", *J. of Math. Anal. and Appl.*, *52*, 490-513.

[2] Alò, R. A., and de Korvin, A. (1971). "Functions of bounded variation on idempotent semigroups", *Math. Ann. 194*, 1-11.

[3] Alò, R. A., de Korvin, A., and Easton, R. (October 1972). "Vector valued absolutely continuous functions on idempotent semigroups", *Trans. of the Amer. Math. Soc. 172*, 491-499.

[4] Batt, J. (1973). "Non-linear integral operators on C(S,E)," *Studia Math.*, *48*, 145-177.

[5] Dunford, N., and Schwartz, J. T. (1958). "Linear Operators 1: General Theory", *Pure and Appl. Math VII, Interscience,* New York.

[6] Drobot, V. (1970). "An infinite dimensional version of Liapunov convexity theorem", *Michigan Math. J., 17,* 405-408.

[7] Edwards, J. R., and Wayment, S. (1974). "Extensions of the v-integral", *Trans. Amer. Math. Soc. 191,* 1-20.

[8] Edwards, J. R., and Wayment, S. (1970). "A unifying representation theorem", *Math. Ann.,* 317-328.

[9] Hermes, H., and Lasalle, J. P. (1969). "Functional Analysis and Time Optimal Control", Academic Press, New York.

[10] Kluvánek, I., and Knowles, G. (1976). "Vector Measures and Control Systems", North-Holland.

[11] Krasnosel'skii, M. A. (1964). "Topological Methods in the Theory of Non-linear Integral Equations", (translated by A. H. Armstrong; J. Burlak, Ed.), Macmillan, New York, 20-32.

[12] Mizel, V. J. (1970). "Characterization of non-linear transformations possessing kernels", *Canad. J. of Math., 22,* 449-471.

[13] Woyczynski, W. A. (1968). "Additive functionals on Orlicz spaces", *Colloq. Math.,* 319-326.

OPTIMAL CONTROL OF DIFFUSION-REACTION SYSTEMS*

H. T. Banks

Division of Applied Mathematics
Brown University
Providence, Rhode Island

M. C. Duban
J. P. Kernevez

Département de Mathématiques Appliquées
Université de Technologie de Compiègne
Compiègne, France

We consider control problems governed by the following non-linear diffusion-reaction systems:

$$\begin{aligned} \frac{\partial s}{\partial t} &= \frac{\partial^2 s}{\partial x^2} - \sigma \frac{a}{1+a} \frac{s}{1+s+ks^2} \\ \frac{\partial a}{\partial t} &= \alpha \frac{\partial^2 a}{\partial x^2} \qquad 0 \le x \le 1, \quad 0 \le t \le T, \end{aligned} \tag{1}$$

This research supported in part by the National Science Foundation under grant NSF-GP-28931x3, in part by the U.S. Air Force under contract AF-AFOSR-76-3092, in part by Centre National de la Recherché Scientifique under grant 38097, and in part by the Université de Technologie de Compiègne.

ISBN 0-12-434180-2

$$s(0,t) = s_0(t) \qquad \frac{\partial s}{\partial x}(1,t) = 0$$
$$a(0,t) = a_0(t) \qquad \frac{\partial a}{\partial x}(1,t) = 0 \tag{2}$$

$$s(x,0) = f_0(x) \qquad s(x,T) = f_1(x)$$
$$a(x,0) = g_0(x) \qquad a(x,T) = g_1(x). \tag{3}$$

The control of systems such as (1)-(3) is of importance in the investigation of enzymatically active artificial membranes similar to those employed by D. Thomas and his coworkers in experiments at Université de Technologie de Compiègne (see [2] for more details). In such systems the variables s and a represent respectively normalized variables for substrate and activator concentrations. The nonlinear reaction term in (1) is a Michaelis-Menten-Briggs-Haldane type (see Chapter 1 of [1]) velocity approximation term for a reaction in which one has inhibition by excessive substrate. The boundary conditions are those appropriate for a one dimensional diffusion-reaction medium in contact with a reservoir (at $x = 0$) and an electrode or impermeable wall (at $x = 1$) as depicted in Figure 1.

For the nonlinear system (1)-(2) it can be argued that multiple steady-state solutions exist and the initial and terminal

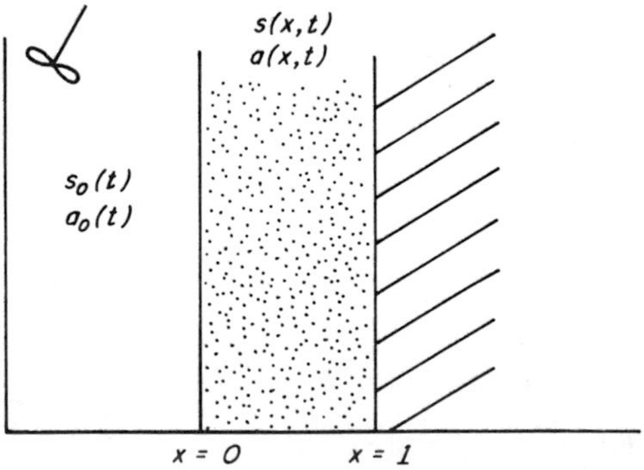

FIGURE 1

functions in (3) are taken to be distinct such steady-states. That is, $f_i, g_i, \ i = 0,1$ are solutions of

$$0 = f_{xx} - \sigma\frac{g}{1+g}\frac{f}{1+f+kf^2}$$

$$0 = g_{xx}$$

$$f(0) = \beta_i \qquad f_x(1) = 0 \tag{4}$$

$$g(0) = \gamma_i \qquad g_x(1) = 0.$$

The basic question we address here is: Given the system in an initial steady-state configuration (f_0,g_0) at time $t = 0$, how does one use boundary controls s_0,a_0 to transfer the system in time $0 \leq t \leq T$ to a second steady-state configuration (f_1,g_1) and do this in an efficient manner. That is, there is some cost associated with adding (or deleting) substrate and/or activator to the system via the boundary controls and one should try to minimize some measure of this cost as the transfer from one steady-state to another is made. We take as cost functional a measure of the total flux (in the L_2 sense) of s and a into the system at the boundary $x = 0$. Thus, we desire to choose control functions s_0,a_0 in some control space $\mathcal{U}$ (e.g. $L_2(0,T)$) so as to minimize

$$J = \int_0^T \{|\frac{\partial s}{\partial x}(0,t)|^2 + |\frac{\partial a}{\partial x}(0,t)|^2\}dt \tag{5}$$

subject to (1)-(3). (In general the system (1)-(3) need not be exactly solvable for a given $f_i,g_i, \ i = 0,1$ (i.e., controllability questions arise) and one must replace the above posed problem by one of transferring f_0,g_0 to a terminal state close to f_1,g_1. One thus actually considers for both theoretical and computational purposes the modified problem of minimizing

$J_\varepsilon \equiv J + \frac{1}{\varepsilon}\int_0^1\{|s(x,T) - f_1(x)|^2 + |a(x,T) - g_1(x)|^2\}dx$ subject to (1), (2) and $s(x,0) = f_0(x)$, $a(x,0) = g_0(x)$.)

The above might appropriately be called a "1-dimensional medium" reaction-diffusion problem. An analogous "0-dimensional medium" problem is of interest in the event that one has
(i) reaction and diffusion separated within the medium or
(ii) very rapid diffusion (i.e., a well-mixed medium for reaction-diffusion). The latter assumption is valid in general models for continuously stirred tank reactors. In the "0-dimensional medium" problem the spatial variable is ignored and one has as control system (for $s = s(t)$, $a = a(t)$)

$$\begin{aligned} \frac{ds}{dt} &= s_0 - s - \sigma\frac{a}{1+a}\,\frac{s}{1+s+ks^2} \\ \frac{da}{dt} &= \alpha\{a_0 - a\} \qquad 0 \le t \le T, \end{aligned} \tag{6}$$

$$\begin{aligned} s(0) &= f_0 \qquad s(T) = f_1 \\ a(0) &= g_0 \qquad a(T) = g_1, \end{aligned} \tag{7}$$

where one still chooses the controls s_0, a_0 from some space U of admissible policies. However, now the initial and terminal states (f_0, g_0), (f_1, g_1) are constants which satisfy

$$\begin{aligned} 0 &= s_0^i - f_i - \sigma\frac{g_i}{1+g_i}F(f_i) \\ 0 &= a_0^i - g_i, \quad i = 0,1, \end{aligned} \tag{8}$$

where $(s_0^0, a_0^0) = (s_0(0), a_0(0))$, $(s_0^1, a_0^1) = (s_0(T), a_0(T))$ and $F(s) \equiv s/(1+s+ks^2)$. The cost functional is taken as

$$J = \int_0^T\{|s_0(t) - s(t)|^2 + |a_0(t) - a(t)|^2\}dt. \tag{9}$$

Just as in the case of the "1-dimensional" problem, one can show that multiple steady-states (i.e., solutions of (8)) are possible for the system (6). Also, one usually must consider a modification of the minimization problem since (6), (7) may not be exactly solvable (i.e., again controllability questions arise).

There are a number of interesting nontrivial theoretical questions (controllability, existence, uniqueness, etc.) associated with the control problems formulated above but we shall not discuss those questions directly here. Our initial interest in these problems arose from an attempt to use computational schemes (i.e., software packages) in connection with experimental efforts. From the descriptions above one might anticipate this to be a rather routine task since the problems would appear tractable using standard ideas from the theory of boundary control of partial differential equations in the case of the "1-dimensional" problems (see [2]) or those from the theory of nonlinear ordinary differential equation control problems in the case of the "0-dimensional" problem (see [3]) along with gradient, conjugate-gradient type numerical techniques. Initial numerical experiments revealed that this is *not* the case and our efforts here will be limited to an explanation of the difficulties along with suggestions as to possible alternative formulations which might lead to problems amenable to solution on the computer.

To facilitate discussions of the above-mentioned difficulties it is helpful to consider the *quasi-steady-state approximation* to the "0-dimensional medium" problem (a similar approximation reveals the inherent difficulties in the "1-dimensional medium" problem). In light of the small transient times found in experimental realizations of these models, one can make a plausible argument that the quasi-steady-state approximations are reasonable approximations to the problems formulated above. We shall not do that here but turn instead to the problem of minimizing J given in (9) subject to the constraint equations (steady-state approximations to (6))

$$s_0(t) - s(t) - \sigma\frac{a(t)}{1+a(t)} F(s(t)) = 0$$
$$a_0(t) - a(t) = 0. \tag{10}$$

Since in this case $a_0 \equiv a$, we define for convenience the variable $\rho \equiv \sigma a/(1+a)$ and consider the problem of minimizing J while transferring a "state" $X^0 = (s_0(0), \rho(0), s(0))$ to a state $X^1 = (s_0(T), \rho(T), s(T))$ subject to the constraint

$$s_0(t) - s(t) - \rho(t)F(s(t)) = 0, \quad 0 \leq t \leq T. \tag{11}$$

A sketch of the surface in (s_0, ρ, s) space described by (11) is given in Figure 2, where one recognizes the well-known "cusp" (catastrophe) surface of Whitney [5] and Thom [4]. In Figure 2 the folds in the cusp surface are projected down into the (s_0, ρ) plane as the (infinite) arcs containing CA and CB. We are thus choosing control strategies (paths in the (s_0, ρ) plane) which yield corresponding "trajectories" that move on this (multi-valued in some regions) surface.

Consider a problem which requires transfer of an initial configuration X^0 to a terminal configuration X^1 as depicted in Figure 2. Two possible distinct control strategies $\{(s_0(t), \rho(t))\}$, $0 \leq t \leq T$, are depicted in Figure 3.

It is clear that two such strategies can be made arbitrarily close (using any reasonable measure of closeness) in the (s_0, ρ) plane while the corresponding "trajectories" $(s_0(t), \rho(t), s(t))$, $0 \leq t \leq T$, lying on the constraint surface will not be close. The trajectory corresponding to strategy 1 (see Fig. 3) "travels" along the lower fold (see Fig. 2) while strategy 2 yields a trajectory which during the corresponding time "travels" along the upper fold of the surface defined by (11). (The heavy lines with arrows in Fig. 2 represent jump discontinuities in s for the quasi-steady-state model. For the original problems, i.e., the non-quasi-steady-state models, these correspond to extremely rapid "motion" from trajectories near the lower surface to trajectories near the upper surface.)

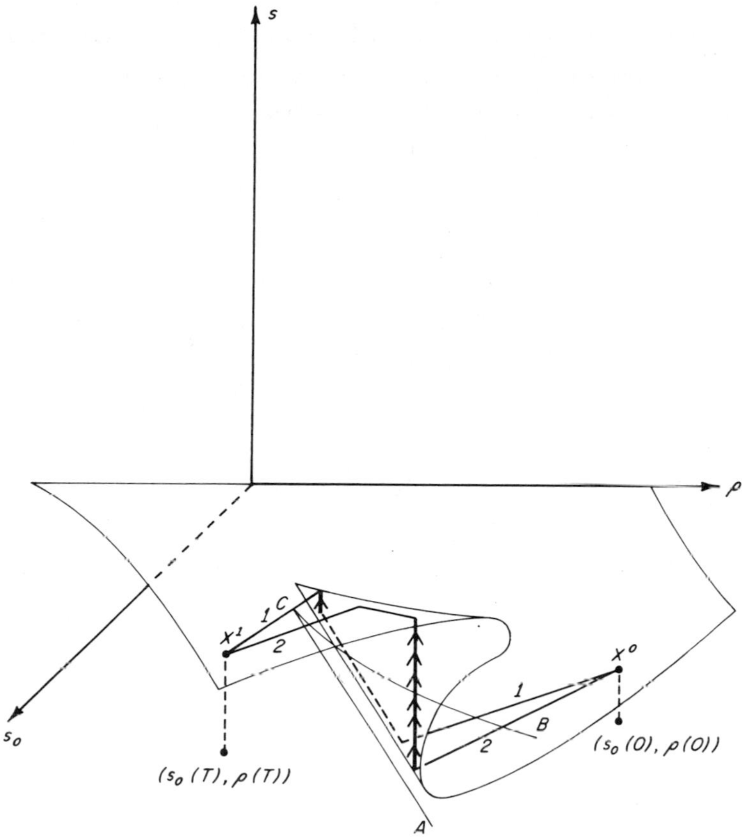

FIGURE 2

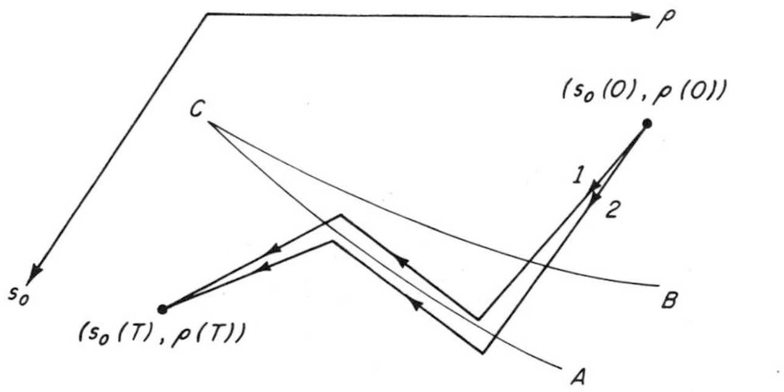

FIGURE 3

From these considerations it is clear that the trajectories for the quasi-steady model are not even continuous *as a function of the control strategies* and hence it is not surprising that methods (e.g. gradient, conjugate-gradient) involving derivatives (with respect to controls) of the cost function are troublesome when applied to the problems governed by (1)-(3) or (6)-(7).

Once one has visualized the problems in this heuristic but informative way, it is apparent that the difficulties are a result of the particular nonlinear reaction velocity approximation found in (1) and subsequent associated versions of this system equation employed above. The models entail a region Γ (for (6) and (11) with transfer from X^0 to X^1 as shown in Figures 2, 3 this region is depicted in Figure 4) in "control" space in which one must choose control strategies with extreme care.

In carrying out laboratory experiments, this region is observed to be one in which the system is highly unstable. Thus from both a theoretical and practical viewpoint, additional constraints on operation of the system in this region are desirable. Careful formulation with additional constraints can lead to

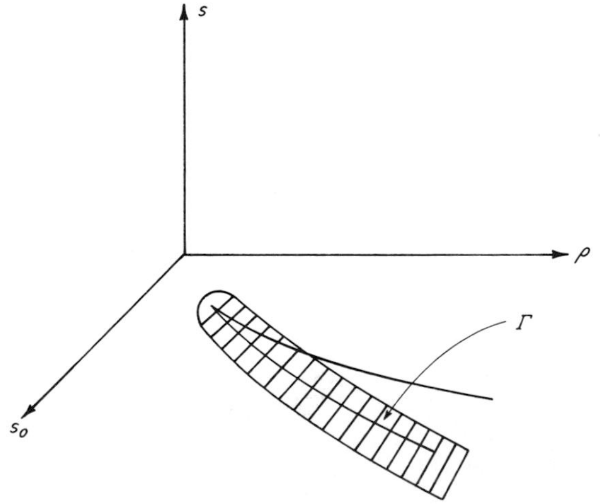

FIGURE 4

tractable problems. We illustrate this first with a sketch of how one might formulate such a control problem for a discretized version of the quasi-steady approximation to the "0-dimensional medium" problem.

Considering the controls s_0, a_0 to be piecewise constant on $[0,T]$, one can reformulate the quasi-steady problem as a multi-stage discrete control problem with "controls" $\{s_0(t_i), a_0(t_i)\}$, $i = 1,\ldots,k$, constrained to lie outside Γ (see Fig. 4) with "states" $\{s(t_i)\}$ given implicitly by

$$s_0(t_i) - s(t_i) - \sigma\frac{a_0(t_i)}{1+a_0(t_i)} F(s(t_i)) = 0.$$

The payoff is then taken as

$$J = \sum_{i=1}^{k} \{s_0(t_i) - s(t_i)\}^2 \Delta t_i.$$

The most natural formulation along these lines leads to immediate difficulties with regard to necessary conditions (multiplier rules or maximum principles are not readily available for discrete control problems with implicit state equations). However, one can reformulate this slightly as a constrained "state" and "control" problem so that necessary conditions are easily obtained. If one identifies s_0, a_0 as "states" and defines a mapping $\Lambda: R^2 \to R^1$ by $x_3 = \Lambda(x_1, x_2)$ where x_3 is a solution (appropriately chosen when multiple solutions exist) to

$$x_1 - x_3 - \sigma\frac{x_2}{1+x_2} F(x_3) = 0$$

and introduces "controls" v_i, w_i (with suitable constraints), the problem becomes one of minimizing

$$J = \sum_{i=1}^{k} \{s_0(t_i) - \Lambda(s_0(t_i), a_0(t_i))\}^2 \Delta t_i$$

subject to state equations

$$s_0(t_i) = s_0(t_{i-1}) + v_i$$

$$a_0(t_i) = a_0(t_{i-1}) + w_i$$

and constraints

$$\phi_j(s_0(t_i), a_0(t_i)) \leq 0$$

defined to prohibit values of s_0, a_0 in the region Γ. In this formulation one can obtain necessary conditions (to use as a basis for computational schemes) via application of the operator theoretic optimization framework with abstract multiplier rule developed by Neustadt (e.g., see Chap. 7 of [3]).

The above formulation essentially involves the assumption that "changes" (or more precisely "rates of changes") in s_0, a_0 are the controls. This can be viewed as a special case of a reformulation for the continuous version problems. Consider the full "0-dimensional medium" problem and adjoin to the state equations (6) additional equations

$$\frac{ds_0}{dt} = v$$

$$\frac{da_0}{dt} = w \tag{12}$$

with control constraints $|v| \leq M_1$, $|w| \leq M_2$. The "states" for the problem are then taken as s_0, a_0, s, a with "controls" given by v, w. In addition to the natural control restraints, one imposes mixed *state-control* (so-called "phase-control" constraints) inequality constraints which restrict the choices of v and w in the event one is in the region Γ in (s_0, ρ) space (see Fig. 4). These constraints are defined so that one rules out control policies that yield paths in the (s_0, ρ) plane that travel along the "singular" arc containing CA (see Figs. 2, 3, 4). That is, one rules out via constraints on v, w policies such as those depicted in Figure 3. Hence, while one does not prohibit crossing of the Γ region, one restricts carefully the types of trajectories one allows while passing through this region. The resulting constraints will thus be joint in the "states" s_0, a_0 and the "controls" v, w.

Finally, we point out that one can also use these reformulation ideas to take a linear programming approach (function minimization problems with linear inequality constraints) to these problems as opposed to the optimal control approach (multiplier rule for constrained control problems) sketched above. For example we illustrate briefly with the quasi-steady approximation to the "0-dimensional medium" problem.

We approximate the arc containing CA above and below by straight lines with slope m. With this fixed slope m, we construct a family of parallel and equidistant lines $\{\rho = ms_0 + b_i\}_{i=0}^{k}$ in the (s_0, ρ) plane as depicted in Figure 5. The admissible

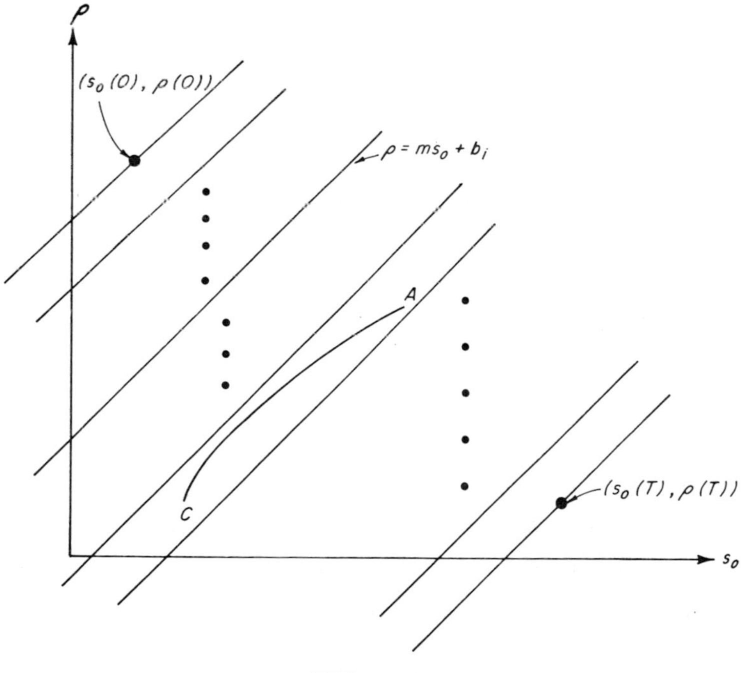

FIGURE 5

"states" are then required to lie on these lines. (The construction is made so that the arc CA lies between two of these lines, $(s_0(0),\rho(0))$ lies on $\rho = ms_0 + b_0$, and $(s_0(T),\rho(T))$ lies on $\rho = ms_0 + b_k$.) A "trajectory" will then consist of a sequence of points $(s_0(t_i),\rho(t_i),s(t_i))$ satisfying (11) with $(s_0(t_i),\rho(t_i))$ belonging to the line $\rho = ms_0 + b_i$. One can take as control policies the collection of sequences

$$U = (\rho_1,t_1,\rho_2,t_2,\ldots,\rho_{k-1},t_{k-1})$$

with $t_0 = 0$, $t_k = T$ and corresponding "state" equations

$$s_0(t_i) = (\rho_i-b_i)/m$$

$$s_0(t_i) - s(t_i) - \rho_i F(s(t_i)) = 0.$$

In addition to making appropriate modifications to the payoff J, one constrains the *analogues* of equations (12), i.e.,

$$\left|\frac{s_0(t_i)-s_0(t_{i-1})}{t_i-t_{i-1}}\right| \le K_1$$
$$\left|\frac{\rho_i-\rho_{i-1}}{t_i-t_{i-1}}\right| \le K_2. \tag{13}$$

One must also add positivity constraints for s_0,ρ given by

$$\rho_i \ge \max\{0,b_i\}. \tag{14}$$

By defining suitable coefficient matrices E and D, one can write the constraints (13), (14) as

$$EU^T \ge D. \tag{15}$$

The problem then becomes one of minimizing a function J subject to the linear inequality constraints (15) and standard computational techniques (e.g., descent methods such as the Davidon-Fletcher-Powell schemes) are applicable.

A more detailed discussion of theoretical aspects of the above different formulations and approximations along with our numerical findings will be presented in a forthcoming manuscript.

REFERENCES

[1] Banks, H. T. (1975). "Modeling and Control in the Biomedical Sciences", *Lec. Notes in Biomath., Vol. 6*, Springer-Verlag.

[2] Kernevez, J. P., and Thomas, D. (1975). "Numerical analysis and control of some biochemical systems", *Appl. Math. Optimization 1*, 222-285.

[3] Neustadt, L. W. (1976). "Optimization: A Theory of Necessary Conditions", Princeton Univ. Press.

[4] Thom, R. (1972). "Stabilité structurelle et morphogènése", W. A. Benjamin, Inc., (English translation 1975).

[5] Whitney, H. (1955). "On singularities of mappings of Euclidean spaces I. Mappings of the plane into the plane", *Ann. Math., 62, Ser. 2*, 374-410.

A NONLINEAR GENERALIZATION OF THE HEAT EQUATION ARISING IN PLASMA PHYSICS

Charles J. Holland[1,2]
James G. Berryman[3,4]

Courant Institute of Mathematical Sciences
New York University
New York, New York

It is well known that if $v^{\tilde{\phi}}(x,t)$ is the solution to the first boundary value problem for the linear diffusion equation

$$v_{xx} = v_t \quad \text{on} \quad (0,1) \times (0,\infty) \tag{1}$$

$$v = 0 \quad \text{for} \quad x = 0,1, \quad 0 \leq t < \infty \tag{2}$$

with positive initial data $\tilde{\phi}$ for $0 < x < 1$, vanishing for $x = 0,1$, then

$$\frac{v^{\tilde{\phi}}(x,t)}{e^{-\pi^2 t}} \to b \sin \pi x \tag{3}$$

[1] *On leave from Purdue University.*

[2] *This research was supported in part by a grant to the Courant Institute from the Sloan Foundation and in part by AFOSR 77-3286.*

[3] *National Science Foundation postdoctoral fellow on leave from Continential Oil Company, Ponca City, Oklahoma 74601.*

[4] *New permanent address: Bell Laboratories, Whippany, New Jersey 07981.*

ISBN 0-12-434180-2

uniformly in x as $t \to \infty$. Here b is the first Fourier coefficient, which depends upon the initial data $\tilde{\phi}$. Note that the function $b \sin \pi x \, e^{-\pi^2 t}$ is a separable solution to (1) with the property that it is positive inside the domain. We wish to establish the analogue of this result in the case of nonlinear (concentration dependent) diffusion where (1) is replaced by

$$((1 + \delta)v^{\delta} v_x)_x = v_t \tag{4}$$

both in the case of "slow" diffusion $\delta > 0$ and in the case of "fast" diffusion $0 > \delta > -1$. The essential difference between the two cases is that for $\delta < 0$ the solution decays to zero in finite time while for $\delta > 0$ the solution decays to zero in infinite time like an inverse power of t.

For this talk we consider only the case $\delta = -\frac{1}{2}$. This corresponds to the scaling predicted for Okuda-Dawson diffusion in plasmas [1]-[5] which motivated the present work. The fast diffusion case will be treated in a more detailed paper [2]. For $\delta = -\frac{1}{2}$ let $u^2 = v$, then u satisfies the equation

$$\begin{aligned} &u_{xx} = 2uu_t, \quad (x,t) \in (0,1) \times (0,T^*) \\ &u = 0 \quad \text{on} \quad \partial\{(0,1)\} \times (0,T^*) \end{aligned} \tag{5}$$

and initial data $u = \phi = \tilde{\phi}^2$. We will consider throughout the latter form (5) of this equation.

Sabinina [6] considered the first boundary value problem for (5), and she showed that there exists T^* depending upon ϕ such that

(i) $u(x,t)$ is a positive classical solution to (5) on $(0,1) \times (0,T^*)$ assuming the boundary values.

(ii) $u(x,T^*) = 0$ for $0 \leq x \leq 1$.

We call T^* the extinction time. Throughout we shall assume that

(A1) The solution u is of class $C^2([0,1] \times (\gamma,T^*))$

for some γ, $0 < \gamma < T^*$.

Let $S(\cdot)$ be the unique solution to $S'' = -2S^2$, $S(0) = S(1) = 0$ with $S(x) > 0$, $0 < x < 1$. Then $S(x)(T^* - t)$ is the unique separable solution to (5) which, for $0 < x < 1$, is positive for $0 < t < T^*$ and vanishes for $t = T^*$. We wish to show that the solution $u^\phi(x,t)$ decays to zero looking like $S(x)(T^* - t)$. To do this we prove

Theorem 1.
$$\lim_{t \to T^{*-}} \left\| \frac{u^\phi(\cdot,t)}{T^*-t} - S(\cdot) \right\| = 0. \tag{6}$$

Above $\|f(\cdot)\| = \sup_{0 \le x \le 1} |f(x)|$. The statement of the theorem is the analogue of the result (3) for the heat equation.

To establish the theorem, we need three lemmas. We shall drop the superscript ϕ; however, recall that T^* always depends upon ϕ.

Lemma 1. $\int_0^1 u^3(x,t)dx \le \left(\int_0^1 u^3(x,0)dx\right)\left(1 - t/T^*\right)^3$ for $0 \le t \le T^*$.

Lemma 2. There exists a positive constant C_1 independent of ϕ such that

$$(T^* - t) \le C_1\left(\int_0^1 u^3(x,t)dx\right)^{1/3} \le C_1\|u(\cdot,t)\|.$$

Proofs of Lemmas 1, 2 will appear in [2]. It should be noted that from Lemmas 1 and 2 we have both upper and lower bounds on the extinction time T^* in terms of $\int_0^1 u^3(x,0)dx$. Better bounds have been found but an exact determination of T^* in terms of the initial data is lacking.

Proof of Theorem 1. Define $v(x,t) = u(x,t)/(T^* - t)$ and $w(x, -\ln(1 - t/T^*)) = v(x,t)$. Then $w = w(x,t)$ satisfies the equation

$$w_{xx} = 2ww_t - 2w^2 \quad \text{on} \quad (0,1) \times (0,\infty) \tag{7}$$

with zero lateral boundary data and initial data $w = \phi/T^*$ when $t = 0$. To prove the theorem we must show that $\|w(\cdot,t) - S(\cdot)\| \to 0$ as $t \to +\infty$.

For functions $h(\cdot) \in H_0^1(0,1)$ let

$$I(h) = \int_0^1 \left(\frac{1}{2}h_x^2 - \frac{2}{3}h^3\right)dx$$

and let $g(t) = I(w(\cdot,t))$. A simple calculation yields

$$g'(t) = -\int_0^1 2ww_t^2 dx.$$

Thus $g' \leq 0$ since $w \geq 0$ and hence $g(t)$ is bounded above. The function I plays the role of a Liapunov function. Now Lemmas 1, 2 give that $\int w^3 dx$ is bounded and hence $g(t)$ is also bounded below.

The boundedness of g and $\int w^3 dx$ has several consequences. First, $\int_0^1 w_x^2 dx < M$ for some M independent of t and therefore $\|w(\cdot,t)\| < N$ for some N independent of t.

Next, since g is monotone decreasing and bounded below, there exists a sequence of times $t_n \to \infty$ such that $g'(t_n) \to 0$. Therefore $\int_0^1 ww_t^2 dx \to 0$ and since $\|w(\cdot,t)\|$ is bounded, then $\int_0^1 w^2 w_t^2 dx \to 0$ as $t_n \to \infty$. For this sequence, $\int_0^1 w_{xx}^2(t_n,x)dx < C$ for an appropriate constant C. Therefore there exists a function $R(x)$ such that for a subsequence of t_n, again labeled t_n, $w(\cdot,t_n) \to R(\cdot)$ in $H_0^1(0,1)$ and hence $\|w(\cdot,t_n) - R(\cdot)\| \to 0$ as $t_n \to \infty$. We now wish to show that R must be S.

Multiply (7) by any $p \in H_0^1(0,1)$ and integrate by parts obtaining

$$\int_0^1 \left(-p_x w_x + 2w^2 p\right)dx = \int_0^1 p\,2ww_t dx.$$

Let $t_n \to \infty$, then we obtain that

$$\int_0^1 \left(-p_x R_x + 2R^2 p\right)dx = 0$$

Since this holds for all $p \in H_0^1(0,1)$ then standard results imply that R satisfies

$$R'' = -2R^2, \quad R(0) = R(1) = 0.$$

Now R must be nonnegative for all x, $0 \leq x \leq 1$ and hence R is either the function S or the function 0. But Lemma 2 rules out the function 0, hence $w(\cdot,t_n) \to S$ in $H_0^1(0,1)$.

We now need to show that $w(\cdot,t) \to S(\cdot)$ in $H_0^1(0,1)$ as $t \to \infty$ and not just for the above sequence t_n. The proof of this depends upon the following Lemma.

Lemma 3. Let $\int_0^1 S^3(x)dx = k$. Consider the problem of minimizing $\int_0^1 h_x^2(x)dx$ subject to $\int_0^1 h^3(x)dx = k$ and $h \in H_0^1(0,1)$. The function S is the unique minimizing function for this problem.

We omit the proof of Lemma 3. Now suppose for the purpose of contradiction that $w(\cdot,t) \not\to S(\cdot)$ in $H_0^1(0,1)$ as $t \to \infty$. Since $w(\cdot,t)$ is bounded in $H_0^1(0,1)$ and $\int w^3 \downarrow \int S^3$ as $t \to \infty$, then there exists a function $R(\cdot)$, $R \neq S$ such that $\int R_x^2 \leq \int S_x^2$ but $\int R^3 = \int S^3 = k$. This contradicts Lemma 3. Hence $w(\cdot,t) \to S(\cdot)$ as $t \to \infty$ in $H_0^1(0,1)$.

For more general results and some outstanding conjectures see [2].

ACKNOWLEDGMENTS

We would like to thank Professor S. R. S. Varadhan for some stiumlating discussions.

REFERENCES

[1] Berryman, J. G. (1977). "Evolution of a stable profile for a class of nonlinear diffusion equations with fixed boundaries", *J. Mathematical Phys.* *18*, 2108-2115.

[2] Berryman, J. G., and Holland, C. J., to be published.

[3] Drake, J. R., and Berryman, J. G. (1977). "Theory of nonlinear diffusion of plasma across the magnetic field of a toroidal multipole", *Phys. Fluids* *20*, 851-857.

[4] Drake, J. R., Greenwood, J. R., Navratil, G. A., and Post, R. S. (1977). "Diffusion coefficient scaling in the Wisconsin levitated octupole", *Phys. Fluids 20*, 148-155.

[5] Okuda, H., and Dawson, J. M. (1973). "Theory and numerical simulation on plasma diffusion across a magnetic field", *Phys. Fluids 16*, 408-426.

[6] Sabinina, E. S. (1962). "A class of nonlinear degenerating parabolic equations", *Sov. Math. Doklady 143*, 495-498.

PERIODIC SOLUTIONS OF DELAY DIFFERENTIAL EQUATIONS ARISING IN SOME MODELS OF EPIDEMICS

Stavros N. Busenberg

Department of Mathematics
Harvey Mudd College
Claremont, California

*Kenneth L. Cooke**

Department of Mathematics
Pomona College
Claremont, California

A number of recent papers have been devoted to the study of the periodic solutions of scalar delay differential equations that arise in certain models of epidemics. One of these models, first studied by Cooke and Kaplan [2], and further elucidated by H. Smith [9,10] and R. Nussbaum [5,6] leads to a delay differential equation of the form $dy(t)/dt = f(t,y(t)) - f(t-T,y(t-T))$ with f periodic in t. We refer to [2], [9], or [10] for a discussion of this model, but note that the basic assumption leading to this special equation is that all individuals who become infectious at time $t-T$ are no longer infectious at time t. A different model places the delay in an incubation period $T > 0$, needed between the infection and the start of

The work of K. L. Cooke was partially supported by N.S.F. Grant #MCS 77-04466.

ISBN 0-12-434180-2

infectiousness of a vector carrier of the disease. This model is described in Busenberg and Cooke [1] and leads to the delay equation $dy(t)/dt = b(t)y(t-T)[1-y(t)] - cy(t)$, where $b(t) > 0$ is periodic in t and $c > 0$. The basic result that these two equations have in common, and with which we are concerned in the present paper, is that there exists a parameter p ($1/T$ in the first case and c in the second) and a positive value p_0, such that, if $p \geq p_0$ the equation has no positive periodic solution, while if $p < p_0$ there exists a positive periodic solution which, under certain restrictions, is unique and has strong stability properties.

In the present paper we consider the functional delay equation

$$dy(t)/dt = f(t,y_t) - g(t,y_t), \quad t > 0, \tag{1}$$

and place conditions on f and g that are suggested by the models discussed above. The form of this equation is obtained by considering $y(t)$ to be the proportion of a population that is infectious, so $0 \leq y \leq 1$. $f(t,y_t)$ represents the rate at which infectious individuals are being added at time t, while $g(t,y_t)$ the rate, at time t, at which infectious individuals are being removed and returned to the pool of susceptibles. As usual, y_t is an element of the space of continuous functions $C[-T,0]$ with y_t: $[-T,0] \to \mathbb{R}$ defined by $y_t(s) = y(t+s)$, $s \in [-T,0]$. There are a number of technical assumptions that we need to impose on f and g, however, we first list those that are suggested by the situation that is being modelled.

(C_0) f and g, as functions of the first variable, are periodic of period $w > 0$.

(C_1) If $0 \leq y_t \leq 1$, then $g(t,y_t) \geq 0$ and $g(t,y_t) > 0$, if $y(t) \geq 1$. Moreover, $g(t,y_t) \leq 0$ if $y(t) < 0$.

(C_2) If $0 \leq y_t \leq 1$, then $f(t,y_t) \geq 0$. Moreover, if $y_t \equiv 0$ or if $y(t) \geq 1$, then $f(t,y_t) \leq 0$. Also, if $y(t) = 0$, $0 \leq y_t \leq 1$ there exists $\varepsilon = \varepsilon(t,y_t) > 0$ such that $f(s,y_s) \geq 0$ for $s \in [t,t+\varepsilon(t,y_t)]$.

The above conditions are quite natural under our present view of f and g provided that it is assumed that the spread of the disease is subject to periodic seasonal fluctuations. We note that it is assumed that the population size is constant and that the disease imparts no immunity.

The technical assumptions we impose on f and g are that $f,g:$ $[-T,\infty) \times C[-T,0] \to \mathbb{R}$ are uniformly continuous on bounded subsets of $[-T,\infty) \times C^+$, $C^+ = \{y \in C[-T,0]: 0 \leq y(t)\}$, and map bounded subsets of $[-T,\infty) \times C^+$ into bounded sets. Moreover, we assume that the initial value problem consisting of (1) and the condition $y_0 = \phi \in C[-T,0]$, $0 \leq \phi \leq 1$, has a unique solution $y(t,\phi)$ which is continuous in its variables and which can be extended to exist for all $t \geq -T$ provided $y(t,\phi)$ remains bounded. In the sequel, these assumptions will be made without their specific mention, and we shall usually write $y(t)$ for $y(t,\phi)$.

Since $y(t)$ represents a proportion, solutions of equation (1) should reflect this fact. This is the content of the following theorem.

Theorem 1. If C_0-C_2 hold, and if $0 \leq y_0 \leq 1$, then the solution $y(t,y_0) = y(t)$ of (1) exists for all $t > 0$ and $0 \leq y(t) < 1$.

This result admits a simple direct proof which we omit. However, we note that Seifert [8] has obtained results of this type of a very general nature.

We now place additional restrictions on f and g in order to compare their relative influence. The conditions take two forms, the first set C_3-C_4 fitting the models of a disease propagated by a vector, and the second set C_3'-C_4' fitting the model of a disease with a fixed period of infectiousness.

(C$_3$) There exists a continuous w-periodic function $c: \mathbb{R} \to \mathbb{R}$ such that if $y: \mathbb{R} \to [0,1]$ is continuous and w-periodic, then $g(t,y_t) \leq c(t)y(t)$.

(C_4) There exist constants $k > 0$ and $r \in (0,1)$, such that if $y\colon \mathbb{R} \to [0,r]$ is w-periodic and continuous, then
$$\int_0^w f(s,y_s)ds \geq k \int_0^w y(s)ds.$$
The alternate conditions C_3' and C_4' are as follows.

(C_3') There exist a constant $r \in (0,1)$, and positive, continuous w-periodic functions $k,k'\colon \mathbb{R} \to \mathbb{R}$ such that $k(t)y(t) \leq f(t,y_t) \leq k'(t)y(t)$, for all $y\colon \mathbb{R} \to [0,r]$ which are continuous and w-periodic.

(C_4') There exists a constant $c > 0$, such that, if $y\colon \mathbb{R} \to [0,1]$, is w-periodic and continuous, then
$$\int_0^w g(s,y_s)ds \leq c \int_0^w y(s)ds.$$
We note that the conditions C_4 and C_4' are quite general. For example, if $f(s,y_s) \geq \sum_{j=1}^{n} b_j(t)y(t-T_j) + \int_{-T}^{0} b_0(t,s)y(t+s)ds$, where $b_j \geq 0$, $j = 0, \ldots, n$, are continuous functions, $0 \leq T_k < \infty$, and if for some i with $0 \leq i \leq n$, $b_i(t) > 0$, for all $t \in \mathbb{R}$, then f satisfies C_4.

In the sequel we shall state our results with the assumption that C_3-C_4 hold, bearing in mind that parallel results can be obtained if C_3'-C_4' were imposed in their stead. We can now give our first theorems on the existence of positive periodic solutions.

Theorem 2. Let C_0-C_4 hold. Then, if $c(t) < k$, (1) possesses a positive, w-periodic solution y with $r \leq y(t) < 1$, for $t \in [-T,\infty)$. If, however, $\int_0^w g(t,y_t)dt > \int_0^w f(t,y_t)dt$ for all continuous, w-periodic $y\colon \mathbb{R} \to [0,1]$, then (1) does not have a non-trivial w-periodic solution with $0 \leq y(t) \leq 1$.

Proof: We construct a compact, continuous operator N on the Banach space $B = \{y\colon \mathbb{R} \to \mathbb{R},\ y$ is w-periodic and continuous, $\|y\| = \max|y(t)|\}$. If $K_1 = \{y \in B\colon\ y(t) \in [0,1]\}$, then the fixed points of N that lie in K_1, are the w-periodic solutions y of (1) that obey $0 \leq y(t) \leq 1$. The operator N is given by

$$(Ny)(t) = \int_0^w K(t,s)G(s,y_s)ds, \quad t \in [0,w] \quad \text{and}$$

$$(Ny)(t+\ell w) = (Ny)(t) \quad \text{for} \quad t \in [0,w], \quad \ell \text{ an integer,}$$

with

$$G(t,y_t) = f(t,y_t) - g(t,y_t) + c(t)y(t), \quad \text{and}$$

$$K(t,s) = \begin{cases} e^{h(w)+h(s)-h(t)}(e^{h(w)}-1)^{-1}, & \text{if } 0 \le s \le t < w, \\ e^{h(s)-h(t)}(e^{h(w)}-1)^{-1}, & \text{if } 0 \le t < s < w, \end{cases}$$

with $h(t) = \int_0^t c(s)ds$. It is not difficult to show that $N\colon K_1 \to B$ is a positive, compact operator possessing the properties that we claim. Moreover, one can show that if $y \in K_1$, if $c(t) < k$, $0 < \lambda < \infty$ and if, $y = Ny + \lambda$, then $\|y\| > r$. For, if $y \in K_1$, $\|y\| \le r$ and $y = Ny + \lambda$, then from the definition of N and condition C_3,

$$y'(t) = f(t,y_t) - g(t,y_t) + \lambda c(t) \ge f(t,y_t) - c(t)y(t) + \lambda c(t).$$

So, $0 = \int_0^w y'(t)dt \ge \int_0^w (k-c(t))y(t)dt + \lambda\int_0^w c(t) > 0$, a contradiction. So, $\|y\| > r$.

Also, if $y \in K_1$, and $y = \lambda Ny$, $0 < \lambda < 1$, then $0 \le y(t) < \alpha$ for some $\alpha \in (0,1)$ and for all $t \in \mathbb{R}$. To see this, we note that $y = \lambda Ny$ implies that

$$y'(t) = \lambda[f(t,y_t) - g(t,y_t)] + (\lambda-1)c(t)y(t).$$

If $y(t^*) = 1$ for some $t^* \ge 0$, then $f(t^*,y_t{}^*) - g(t^*,y_t{}^*) < 0$ by (C_1) and (C_2), and $(\lambda-1)c(t^*)y(t^*) = (\lambda-1)c(t^*) < 0$. So, $y'(t^*) < 0$, which is impossible by the periodicity of y. So, $0 \le y(t) < \alpha$ for some $\alpha \in (0,1)$.

The existence of the claimed positive w-periodic solution now follows by applying the following fixed point theorem (Theorem 4.9 in Schmitt [7]), where K denotes the positive cone $K = \{y \in B\colon \; y(t) \ge 0\}$:

Theorem. Let G_1 and G_2 be open, bounded, neighborhoods of 0 in B, and let $G_1 \subset G_2$. Suppose that $N\colon K \cap (\overline{G}_2 \setminus G_1) \to K$, is completely continuous and satisfies:

(a) There exists $k \in K$, $\|k\| = 1$, and there exists α satisfying $\alpha < \sup\{\|u - Nv\|: u \in K \cap \overline{G}_1, v \in \partial G_1 \cap K\}$, such that all solutions $y \in K \cap (\overline{G}_2 \setminus G_1)$ of $y = Ny + \lambda k$, $0 < \lambda < \alpha$ obey $y \notin \partial G_1$.

(b) All solutions $y \in K \cap (\overline{G}_2 \setminus G_1)$ of $y = \lambda Ny$, $0 < \lambda < 1$, satisfy $y \notin \partial G_2$.

Then N has a fixed point $y \in K \cap (\overline{G}_2 \setminus G_1)$.

Here we take $k \equiv 1$, $G_1 = \{y \in B: \|y\| < r\}$, $G_2 = \{y \in B: \|y\| < 1\}$ and $\alpha = \infty$. Then conditions (a) and (b) of this theorem are implied, if $c(t) < k$, by what we proved above. So, N has a fixed point $y \in K \cap (\overline{G}_2 \setminus G_1)$, hence, (1) has a positive w-periodic solution with $r \leq y(t) \leq 1$, and by Theorem 1, $y(t) < 1$.

Now, if $\int_0^w g(t,y_t)dt > \int_0^w f(t,y_t)dt$, and if $y \in K_1$ satisfies (1), we have $0 = \int_0^w y'(t)dt = \int_0^w [f(t,y_t) - g(t,y_t)]dt < 0$, a contradiction. So, no $y \in K_1$ satisfies (1) and the proof is completed.

We note that, if conditions C_3'-C_4' are substituted for C_3-C_4, then the appropriate operator N leading to a result similar to Theorem 2 is the same as above with $G(t,y_t) = f(t,y_t) - g(t,y_t) - k'(t)y(t)$, and

$$K(t,s) = \begin{cases} e^{h(t)-h(s)+h(w)}(1-e^{h(w)})^{-1}, & \text{if } 0 \leq t < s \leq w, \\ e^{h(t)-h(s)}(1-e^{h(w)})^{-1}, & \text{if } 0 \leq s < t \leq w, \end{cases}$$

$$h(t) = \int_0^t k'(s)ds.$$

Sharper results than Theorem 2 can be obtained if additional restrictions are placed on f and g. These can take the form of conditions on the Frechet derivative of G at $0 \in B$ (recall that G enters in the definition of N). Since these conditions are rather involved, we start by describing the results that we have obtained for the following special case of equation (1):

$$dy(t)/dt = b(t)y(t-T)[1-y(t)] - cy(t), \qquad (2)$$

$b: \mathbb{R} \rightarrow (0,\infty)$, w-periodic; $c > 0$, a constant. We have the following result.

Theorem 3. There exists a positive constant c_T, such that the following hold.

   (a) If $c \geq c_T$, (2) has no non-trivial periodic solutions with initial function $y_0 = \phi$, $0 \leq \phi \leq 1$.

   (b) If $c < c_T$, there exists a unique positive w-periodic solution y_c of (2) satisfying $0 < y_c(t) < 1$.

   (c) The map $c \rightarrow y_c$, with $y_{c_T} \equiv 0$, taking $[0,c_T]$ into B, is continuous and $y_c \rightarrow 1$ as $c \rightarrow 0$. Moreover, $c_1 < c_2$, implies $y_{c_1}(t) > y_{c_2}(t)$.

   (d) $\min\limits_{t \in [0,w]} b(t) \leq c_T < \max\limits_{t \in [0,w]} b(t)$, and if T is an integral multiple of w, $c_T = \frac{1}{T} \int_0^w b(t)dt$. Moreover, c_T is a w-periodic function of T.

Proof: The details of the proof of this theorem will be given elsewhere [1]. The proof hinges on constructing the operator N, as above, with $G(t,y_t) = b(t)y(t) + y(t-T) - y(t)y(t-T)$, and making use of the behavior of the spectrum of $N'(0)$ as a function of the parameter c.

   In the special case of equation (2) the stability properties of the positive periodic solution y_c can also be analyzed. In fact, we have the following result.

Theorem 4. If $c > \max b(t)$, then all solutions y of (2) with $0 \leq y \leq 1$, tend to zero as t tends to positive infinity. Moreover, there exists a constant $c^* > 0$ such that if $c < c^*$, the positive periodic solutions of (2) are locally asymptotically stable. In the special case when T is an integral multiple of ω, all positive periodic solutions are locally asymptotically stable.

Proof: Again we defer the details of the proof to [1]. The method of proof is that Lyapunov functionals.

In order to get a sharp bifurcation result for the general equation (1) we impose the following conditions. Here, if $c^* \in K$, we use the notation $\langle 0,c^*\rangle$ to denote the order interval $\{c \in K: 0 \leq c(t) \leq c^*(t)\}$.

(C_5) There exists a continuous, positive, ω-periodic function $c^* \in K$, such that for all $c \in \langle 0,c^*\rangle$ and each $t \in [-T,\infty)$, there is a map $L_c(t,\cdot): C[-T,0] \to \mathbb{R}$, such that for fixed $c \in \langle 0,c^*\rangle$ $L_c(t,\phi)$ takes bounded subsets of $[-T,\infty) \times C[-T,0]$ into bounded sets, is linear in the variable ϕ, is ω-periodic in t, is uniformly continuous on bounded subsets of $[-T,\infty) \times B$ and satisfies the conditions:

(a) $\|G(t,y_t) - G(t,0) - L_c(t,y_t)\| = o(y_t)$, for $y \in K$ as $\|y_t\| \to 0$. (Here, c is the ω-periodic function entering in the definition of G).

(b) There exist positive constants ℓ_1 and ℓ_2, not depending on $c \in \langle 0,c^*\rangle$, such that $\ell_2 < c^*(t)$, and if $y \in K$, $0 < y(t) \leq 1$, then $L_c(t,y_t) > 0$ and

$$\ell_1\int_0^\omega y(t)dt \leq \int_0^\omega L_c(t,y_t)dt \leq \ell_2\int_0^\omega y(t)dt.$$

(c) The map $c \to L_c(t,\cdot)$ $c \in \langle 0,c^*\rangle$ is continuous in the norm topology, uniformly in $t \in [-T,\infty)$. Also, if $0 \leq c < c'$ $\leq c^*$ (the order being that induced by the cone K), then $L_c(t,y_t) \geq L_{c'}(t,y_t)$ for all $y \in K$ with $0 \leq y(t) \leq 1$ and for all t.

Remark. In the special case of equation (2) described above, if $b(t)$ is fixed, we can take $c^*(t) = c + \max b(t)$, $c(t) = c + b(t)$ (c is the constant entering in (2)), then $G(t,y_t) = b(t)[y(t)+y(t-T)-y(t)y(t-T)]$ and $L_c(t,\phi) = b(t)[\phi(0) + \phi(-T)]$ for

$t \in [-T,\infty)$, $\phi \in C[-T,0]$. Here $L_{c(\cdot)}$ is independent of the constant c, and clearly satisfies condition C_5.

(C_6) If $y \in K$, $0 \le y(t) \le 1$, then there exists $\varepsilon > 0$, dependent on y and $c \in \langle 0,c^* \rangle$, such that $L_c(t,y_t) \ge (1+\varepsilon)G(t,y_t)$.

Theorem 5. Let C_0-C_5 be satisfied. Then there exists a continuous, positive, w-periodic function $c_T \in \langle 0,c^* \rangle$ such that if $c < c_T$ there exists a nontrivial w-periodic solution y^c of (1) with $0 \le y^c(t) < 1$. If in addition, C_6 holds, and if $c \ge c_T$, then (1) does not have a nontrivial w-periodic solution y^c with $0 \le y^c(t) \le 1$. Moreover, $\ell_1 \le c_T(t) \le \ell_2$.

Proof: The proof of this theorem is based on the following fixed point theorem due to Gatica and Smith [3,10]:

Theorem. Let B be a Banach space, let K be a cone in B, and let G be an open, bounded, non-empty neighborhood of zero in B. Let $\overline{G}$ be the closure of G and let $N\colon \overline{G} \cap K \to K$, be a completely continuous operator such that $N(0) = 0$, and N has a Frechet derivative $N'(0)$ with respect to K. Assume the following are satisfied:

(a) All solutions $x \in K$ of $x = \lambda Nx$, $0 < \lambda < 1$, satisfy $x \notin \partial G$.

(b) There exists $y \in K$, $\|y\| = 1$, and $\alpha > 1$ such that $N'(0)y = \alpha y$; and $N'(0)x = x$ for $x \ne 0$ implies $x \notin K$.

Then N has a non-trivial fixed point in $K \cap \overline{G}$.

In applying this theorem we note that condition C_5 implies that N has a Frechet derivative $N'(0)$ at 0 (in the direction of K) given by

$$(N'(0)y)(t) = \int_0^w K(t,s)L(s,y_s)ds, \quad \text{if} \quad t \in [0,w], \tag{3}$$

and $(N'(0)y)(t) = (N'(0)y)(t+kw)$, $t \in [0,w]$ and k an integer, where $K(t,s)$ is the previously defined kernel. It is relatively straightforward to then show that $N'(0)$ is strongly positive, continuous, compact and maps elements $y \in K$ with

$0 \leq y \leq 1$ into K. Moreover, from the definition of $K(t,s)$ and from C_5 it is seen that $N'(0)$ is a strictly decreasing function of $c \in \langle 0,c^* \rangle$ and is continuous in c (using the uniform operator topology). From this and the Krein-Rutman theory [4], it can be shown that the spectral radius of $N'(0)$ is a decreasing function of $c \in \langle 0,c^* \rangle$. The details of these arguments are similar to those used in establishing parallel results for the special case of equation (2) and we defer them to [1]. The verification of the conditions of the fixed point theorem proceed as follows.

Suppose that $0 < \lambda < 1$, $N'(0)y = \lambda y$, and $y \in K - \{0\}$. Then from (3) we get $\lambda y'(t) = L(t,y_t) - \lambda c(t)y(t)$. Since y is w-periodic this implies

$$\lambda\int_0^w y'(t)dt = 0 = \int_0^w L(t,y_t)dt - \lambda\int_0^w c(t)y(t)dt$$
$$\geq \int_0^w (\ell_1 - \lambda c(t))y(t)dt,$$

and this last term is greater than zero if $c(t) \leq \ell_1$. So, if $c(t) \leq \ell_1$, there does not exist $y \in K - \{0\}$ with $N'(0)y = \lambda y$, and $0 < \lambda < 1$. From this and the Krein-Rutman results, there exists a unique $\lambda_c \geq 1$ and some $x \in K - \{0\}$ such that $N'(0)x = \lambda_c x$. This x is in the interior of K since $N'(0)$ is strongly positive. So, $\lambda_c \geq 1$ is the spectral radius of $N'(0)$ if $c(t) \leq \ell_1$.

Next, suppose that $c(t) \geq \ell_2$, $y \in K - \{0\}$, $\lambda > 1$, and $N'(0)y = \lambda y$. Then, as above, $\lambda y'(t) = L(t,y_t) - \lambda c(t)y(t)$, so

$$0 = \int_0^w L(t,y_t)dt - \lambda\int_0^w c(t)y(t)dt \leq \int_0^w (\ell_2 - \lambda c(t))y(t)dt,$$

and this last term is negative since $\ell_2 - \lambda c(t) < 0$, and $\int_0^w y(t)dt > 0$. This is impossible, so again by the Krein-Rutman theorem, if $c(t) \geq \ell_2$ the spectral radius of $N'(0)$ does not exceed one. By the continuity of this spectral radius as a function of $c \in \langle 0,c^* \rangle$, and by the fact that it is a monotonic decreasing function of c, we see that there exists a

$c^T \in \langle 0, c^* \rangle$ such that the spectral radius of $N'(0)$ is equal to one if $c = c^T$, is greater than one if $c < c^T$ and is less than one if $c > c^T$.

Now, if $c < c^T$, the hypothesis (b) of the fixed point theorem is satisfied from what was done above. Hypothesis (a) follows from the fact that if $x \in K$, $x = \lambda N x$, $0 < \lambda < 1$, then x is positive periodic and satisfies

$$x'(t) = \lambda[f(t,x_t) - g(t,x_t)] + (\lambda-1)c(t)y(t).$$

Now, for all $t^* > 0$ with $x(t^*) \geq 1$, it follows from C_1, C_2 and the fact that $\lambda - 1 < 0$, that $x'(t^*) < 0$. Since x is ω-periodic this is impossible, so $x(t) < 1$ for all t. Now, taking $G = \{y \in B: \|y\| < 1\}$, we see that condition (a) is satisfied and there exists a non-trivial ω-periodic solution $y \in K$ of (1) with $0 \leq y(t) \leq 1$.

On the other hand, if $c \geq c^T$, then the spectral radius of $N'(0)$ does not exceed one. So, by C_6 we get $N'(0)y \geq (1+\varepsilon)Ny = (1+\varepsilon)y$, $\varepsilon > 0$, if $y \in K_1 - \{0\}$ with $Ny = y$. The Krein-Rutman theorem now implies that the spectral radius of $N'(0)$ is greater than one. This contradiction shows that no non-trivial, positive ω-periodic solutions exist if $c \geq c^T$, and the proof is complete.

REFERENCES

[1] Busenberg, S., and Cooke, K. "Periodic Solutions of a Periodic Nonlinear Delay Differential Equation", to appear.

[2] Cooke, K., and Kaplan, J. (1976). "A Periodicity Threshold Theorem for Epidemics and Population Growth", *Math. Biosciences 31*, 87-104.

[3] Gatica, J., and Smith, H. "Fixed Point Techniques and Some Applications", preprint.

[4] Krein, M., and Rutman, M. (1950). "Linear Operators Leaving Invariant a Cone in Banach Space", AMS Translations, No. 26.

[5] Nussbaum, R. (1977). "Periodic Solutions of Some Integral Equations from the Theory of Epidemics", in Nonlinear Systems and Applications, (V. Lakshmikantham, editor), Academic Press, New York, 235-255.

[6] Nussbaum, R. "Periodic Solutions of Some Nonlinear Integral Equation", preprint.

[7] Schmitt, K. (1976). "Fixed Points and Coincidence Theorems with Applications to Nonlinear Differential and Integral Equations", *Rapp. #97, Univ. Cath. de Louvain.*

[8] Seifert, G. (1976). "Positively Invariant Closed Sets for Systems of Delay Differential Equations", *J.D.E. 22*, 292-304.

[9] Smith, H. (1977). "On Periodic Solutions of a Delay Integral Equation Modelling Epidemics", *J. Math Biology 4*, 69-80.

[10] Smith, H. (May, 1976). "On Periodic Solutions of Delay Integral Equations Modelling Epidemics and Population Growth", Ph.D. dissertation, University of Iowa.

COMPARISON THEOREMS FOR SYSTEMS OF REACTION-DIFFUSION EQUATIONS

Jagdish Chandra

U.S. Army Research Office
Research Triangle Park, North Carolina

*Paul Wm. Davis**

Department of Mathematics
Worcester Polytechnic Institute
Worcester, Massachusetts

I. INTRODUCTION

Comparison theorems for systems of parabolic differential inequalities are useful tools for the qualitative analysis of the behavior of reacting systems. Unfortunately, many such theorems impose monotonicity requirements on the source terms which are physically unreasonable as well as making other technical restrictions that are not well suited to the problem at hand.

Here we shall briefly illustrate how these restrictions can be systematically circumvented. Using an idea that goes back at least to Müller [6,7], we construct bounding problems whose non-linearities exhibit the monotonicity necessary to the application

**Research supported by the U.S. Army Research Office under grant number DAAG29-76-G-0237.*

ISBN 0-12-434180-2

of a suitable comparison theorem for parabolic systems. We also point out that such theorems for systems depend finally on a *scalar* maximum principle. Hence, the details of the systems comparison theorem (type of boundary conditions, form of differential operator, etc.) are controlled by the scalar maximum principle used.

We shall exhibit two comparison theorems, one based on the usual scalar parabolic maximum principle [8] and the other on two lemmas of classical differential inequalities [5].

As an example of the utility of such comparison results, we shall briefly discuss a prototype combustion model.

Little of what we say here is "new" in the sense of being heretofore unrevealed truth. However, our point of view is somewhat different in that we emphasize how one might construct a useful comparison theorem when confronted with a particular physical system whose structure cannot be hypothesized into submission.

For example, a comparison technique used in the analysis of a prototype combustion problem [9, Theorem 3.1] is a particular application of a general method for obtaining comparison problems with the appropriate monotonicity properties. Bounding theorems used to study certain population models [2, Theorem 1] are, in fact, neither limited to spatially independent bounding problems nor to Neumann boundary data. Indeed, invariance results for such systems may be regarded as consequences of comparison theorems rather than the converse.

II. NOTATION

Let $u(x,t)$, $v(x,t)$ denote the vector-valued functions $(u^1,\ldots,u^n)$, $(v^1,\ldots,v^n)$ for x in some m-dimensional domain D and $t \geq 0$. Let $f^i(x,t,u,u^i_x,u^i_{xx})$ denote a function depending at least continuously upon $x,t,$ the n components $u^1,\ldots,u^n$, the m components of the gradient $u^i_x = (\partial u^i/\partial x_1, \ldots, \partial u^i/\partial x_m)$, and the m^2 components of the Hessian matrix $u^i_{xx} = (\partial^2 u^i/\partial x_k \partial x_\ell)$. Let $f(x,t,u,u_x,u_{xx}) = (f^1,\ldots,f^n)$ with similar notation for g^i and g.

Partial orderings of vectors hold componentwise.

The function $f(x,t,u,p,R)$ is *quasi-monotone nondecreasing* in u if each f^i is nondecreasing in u^j, $j \neq i$, for any fixed $x \in \overline{D}$, $t \geq 0$, m-vector p and $m \times m$ matrix R. The function f is *elliptic* if $f(x,t,u,p,R) \geq f(x,t,u,p,S)$ for any matrices R and S whose difference $R - S$ is nonnegative definite.

Let L^i be a strongly elliptic linear differential operator,

$$L^i \equiv \sum_{k,\ell=1}^{m} a^i_{k\ell}(x)\frac{\partial^2}{\partial x_k \partial x_\ell} + \sum_{k=1}^{m} b^i_k(x)\frac{\partial}{\partial x_k},$$

whose coefficients are bounded in D, and let

$$Lu = (L^1u^1,\ldots,L^nu^n)$$

Solutions are always assumed strong and smooth.

Let B_α denote one of the following boundary operators:

$$B_1u \equiv c(x)u(x,t), \quad c > 0,$$

$$B_2u \equiv d(x)\partial u(x,t)/\partial\nu, \quad d > 0, \quad \text{or}$$

$$B_3u \equiv c(x)u(x,t) + d(x)\partial u(x,t)/\partial\nu,$$

$$c \geq 0, \quad d \geq 0, \quad c^2 + d^2 > 0,$$

for $x \in \partial D$. Here, $\partial/\partial\nu$ denotes any outward directional derivative on ∂D.

III. COMPARISON THEOREMS

Assume for the moment that g depends only upon (x,t,u).

Comparison Theorem 1. Let g be quasi-monotone nondecreasing in u. Furthermore, assume

(i) $u_t \geq Lu + g(x,t,u)$, $x \in D$, $t > 0$; v is the minimal solution of

$$v_t = L_v + g(x,t,v), \quad x \in D, \quad t > 0$$

for prescribed initial and boundary data on v; each L^i is strongly elliptic; each g^i is uniformly Lipschitz continuous in u^i with constant k^i.

(ii) $u(x,0) \geq v(x,0)$, $x \in D$

$$B_\alpha u \geq B_\alpha v, \quad x \in \partial D, \quad t > 0,$$

for one of $\alpha = 1$, 2, or 3; D possesses the interior sphere property (e.g., see [8, Theorem 6, p. 174]).

(iii) v depends continuously upon its initial and boundary data and upon g.

Then $u(x,t) \geq v(x,t)$, $x \in D$, $t \geq 0$.

Proof: For fixed $\varepsilon > 0$, let $w(x,t;\varepsilon)$ denote a solution of the approximate problem.

$$w_t = Lw + g(x,t,w) - \varepsilon, \tag{1}$$

$$B_\alpha w = B_\alpha v - \varepsilon, \quad x \in \partial D, \tag{2}$$

$$w(x,0) = v(x,0) - \varepsilon, \quad x \in D. \tag{3}$$

We claim that $u(x,t) > w(x,t;\varepsilon)$ for $t \geq 0$, $x \in D$, and any $\varepsilon > 0$.

Suppose the claim is false. Then for some index i, there is a finite time $t^* > 0$ at which $u^i(x,t^*) = w^i(x,t^*)$ for some $x \in D$. Since there are but a finite number of components u^j, we may assume i has been chosen so that

$$u^j(x,t) > w^j(x,t), \quad 0 \leq t < t^*, \quad x \in D, \quad j \neq i.$$

By (3) and hypothesis (ii), $t^* > 0$.

Let $\phi = w^i - u^i$. Then hypothesis (i), the preceding inequality, and the quasi-monotonicity of g yield

$$\phi_t \leq (L^i + k^i)\phi, \quad 0 \leq t \leq t^*, \quad x \in D. \tag{4}$$

Furthermore, $\phi(x,t) < 0$, $x \in D$, $0 \leq t < t^*$, while $\phi(x^*,t^*) = 0$ for some $x^* \in D$.

Now the familiar scalar parabolic maximum principle [8, p. 175, Remark (ii)] forces $\phi(x,t^*) = 0$ for all $x \in \overline{D}$ and $\partial\phi(x,t^*)/\partial\nu > 0$, $x \in \partial D$; that is $B_\alpha\phi(x,t^*) \geq 0$, $x \in \partial D$, $\alpha = 1, 2,$ or 3. But this conclusion contradicts (2) and hypothesis (ii). Hence, t^* cannot be finite.

With our claim proven, the theorem follows from hypothesis (iii) and the convergence of solutions of (1-3) to the minimal solution $v(x,t)$. □

Now let g^i depend continuously upon (x,t,u,u^i_x,u^i_{xx}). To construct a second comparison theorem, replace hypotheses (i) and (ii) of Theorem 1 by (i') $u_t \geq g(x,t,u,u_x,u_{xx})$; v is the minimal solution of $v_t = g(x,t,v,v_x,v_{xx})$ for prescribed initial and boundary values; g is elliptic;

(ii') hypothesis (ii) above holds without the requirement that D possess the interior sphere property.

Comparision Theorem 2. Let g be quasi-monotone nondecreasing in u. Assume (i'), (ii'), and (iii) above hold. Then the conclusion of Comparison Theorem 1 remains valid.

Proof: Following the arguments and the notation of the preceding proof, we obtain the analog of (4),

$$\phi_t \leq g(x,t,w,w_x,w_{xx}) - \varepsilon - g(x,t,u,u_x,u_{xx}), \tag{5}$$

$$0 \leq t \leq t^*, \quad x \in D$$

In lieu of the scalar parabolic maximum principle, we employ Lemmas 10.1.1 (for $\alpha = 1$) or 10.1.2 (for $\alpha = 2$ or $\alpha = 3$) of [5, p. 182-183]. Since, by supposition, ϕ assumes a maximum value of zero at (x^*,t^*), $x^* \in D$, $\phi_x(x^*,t^*) = 0$ and $\phi_{xx}(x^*,t^*)$ is nonpositive definite. Then (5) yields $\phi_t < 0$, and [5, Lemma 10.1.1 or 10.1.2] forces $\phi < 0$ throughout $D \times [0,t^*]$, in contradiction of the definition of t^*. □

Note that Theorem 2 admits weakly parabolic operators and domains with corners. These improvements over Theorem 2 are possible because of the use of an alternative to the usual parabolic maximum principle. Roughly speaking, given a scalar maximum principle (e.g., [8,10], or the parabolic analogs of [1]) we may adjust hypothesis (i) to fit the structure of the operator and hypothesis (ii) to fit the form of the boundary conditions and spatial domain in order to obtain a new comparison theorem for systems.

More general boundary conditions and more complicated space-time geometries than the cylinder $D \times [0,t]$ can be easily accommodated within the two theorems we have stated; e.g., see [5, p. 185, p. 149].

Corollary. If $f(x,t,u) \geq g(x,t,u)$, g is quasi-monotone nondecreasing in u, $u_t = Lu + f(x,t,u)$,

$$v_t \leq Lv + g(x,t,v), \quad u(x,0) \geq v(x,0),$$

$$B_\alpha u \geq B_\alpha v, \quad v \in \partial D, \quad \text{for } \alpha = 1, 2, \text{ or } 3, \text{ and}$$

v depends continuously upon its data and its differential equation, then $u(x,t) \geq v(x,t)$ for $x \in D$, $t > 0$.

A parallel corollary follows from Theorem 2 as well.

IV. EXAMPLES

To illustrate the application of these ideas, consider a system of two equations,

$$\Theta_t = \Theta_{xx} + T(\Theta,n) \tag{6}$$

$$n_t = n_{xx} + S(\Theta,n) \tag{7}$$

(The coefficients of the derivative terms could be functions and could vary from equation to equation.)

If we wish to apply the Corollary, we must obtain a quasi-monotone nondecreasing lower bound on (T,S), the nonlinear term in (6-7). Following Müller [6,7], we are led to define $(\underline{T},\underline{S})$ where

$$\underline{T}(\Theta,n) \equiv \inf\{T(\Theta,z):\ n \leq z\}, \tag{8}$$

$$\underline{S}(\Theta,n) \equiv \inf\{S(z,n):\ \Theta \leq z\}, \tag{9}$$

If the infima exist, by construction, $\underline{T}$ and $\underline{S}$ are nondecreasing in their off-diagonal variables; cf. [2].

If $(\underline{\Theta},\underline{n})$ is a solution of (6-7) with $\underline{T},\underline{S}$ replacing (T,S) and $(\underline{\Theta},\underline{n})$ satisfy the same initial and boundary data as (Θ,n), then $(\underline{\Theta},\underline{n}) \leq (\Theta,n)$ for all $x \in D,\ \ t > 0$.

In addition, if

$$T(0,n) \geq 0, \quad S(\Theta,0) \geq 0 \tag{9}$$

and the boundary and initial data is nonnegative, then $(\Theta,n) \geq (0,0)$ for all $x \in D,\ \ t > 0$. We need only note that $T(0,n) > 0$, $\underline{S}(\Theta,0) \geq 0$ to see that

$$(0,0)_t \leq (0,0)_{xx} + (\underline{T}(0,0),\ \underline{S}(0,0)).$$

This inequality and either comparison theorem then yield $(0,0) \leq (\underline{\Theta},\underline{n}) \leq (\Theta,n)$; i.e., the first quadrant is invariant for (6-7) if (9) holds.

Notice that invariance here is a consequence of the comparison theorem and not vice-versa.

By the same arguments, we may also construct bounding solutions which are independent of x.

A parallel analysis using

$$\overline{T}(\Theta,n) = \sup\{T(\Theta,z):\ z \leq n\},$$

$$\overline{S}(\Theta,n) = \sup\{S(z,n):\ z \leq \Theta\},$$

assuming the suprema exist, will yield upper bounds on the solutions of (6-7). If T or S as given does not admit the existence of the necessary supremum or infimum, either may be truncated as required outside the range of the proposed bounds.

As a more concrete illustration, let $T = Hnf(\Theta)$, $S = -\varepsilon nf(\Theta)$, where H, ε are positive constants and $f(\Theta)$ is an increasing function, $1 \leq f(\Theta) \leq 1/\varepsilon$, $f(0) = 1$, $f(\infty) = 1/\varepsilon$. Impose the boundary and initial conditions

$$\Theta = 0, \quad \partial n/\partial v = 0, \quad x \in \partial D \tag{10}$$

$$\Theta(x,0) = \Theta_0 > 0, \quad n(x,0) = n_0 > 0, \quad x \in D. \tag{11}$$

The problem (6-7-10-11) is a nondimensionalized version of a prototype combustion problem proposed by Frank-Kamenetzky [3], in which Θ is temperature, n is species concentration, H is heat of reaction, and ε is derived from activation energy. See [9, §4].

The nonlinear terms in such a model cannot possibly be quasi-monotone because temperature and species concentration must feed back upon one another in contrary ways.

We see immediately that $(\Theta,n) \geq 0$ for all $x \in D$, $t > 0$, and that $\underline{T} = \overline{T} = Hnf(\Theta)$, $\underline{S} = -\varepsilon nf(\infty) = -n$, $\overline{S} = -\varepsilon nf(0) = -\varepsilon n$. Consequently, spatially independent bounds on species concentration are

$$n_0 e^{-t} \leq n(x,t) \leq n_0 e^{-\varepsilon t} \tag{12}$$

while bounds on temperature may be found as solutions of

$$\underline{\Theta}'(t) = Hn_0 e^{-t} f(\Theta), \quad \underline{\Theta}(0) = 0,$$

$$\overline{\Theta}'(t) = Hn_0 e^{-\varepsilon t} f(\Theta), \quad \overline{\Theta}(0) = \theta_0.$$

The bounds of (12) reveal the fastest (e^{-t}) and slowest $(e^{-\varepsilon t})$ rates of combustion one may possibly obtain. Sattinger [9] has analyzed the latter "sub-critical" case in considerable detail.

REFERENCES

[1] Chandra, J., Davis, P. W., and Fleishman, B. A. (1975). "Minimum principles and positive solutions for a class of nonlinear diffusion problems", International Conference on Differential Equations, H. A. Antosiewicz (ed.), Academic Press, New York, N.Y., 149-163.

[2] Conway, E. D., and Smoller, J. A. (1977). "A comparison technique for systems of reaction-diffusion equations", *Comm. in Partial Differential Equations 2*, 679-697.

[3] Frank-Kamenetskii, D. A. (1969). "Diffusion and Heat Exchange in Chemical Kinetics", Plenum.

[4] Ladyženskaja, O. A., Solonnikov, V. A., Ural'ceva, N. N. (1968). "Linear and Quasilinear Equations of Parabolic Type", Translations of Mathematical Monographs, V. 23, American Mathematical Society, Providence, RI.

[5] Lakshmikantham, V., and Leela, S. (1969). "Differential and Integral Inequalities, V. II", Academic Press, New York, NY.

[6] Müller, M. (1927). "Über das Fundamentaltheorem in der Theorie der gewöhnlichen Differentialgleichungen", *Math. Z. 26*, 619-645.

[7] Müller, M. (1927). "Über die Eindeutigkeit der Integrale eines Systems gewöhnlicher Differentialgleichungen und die Konvergenz einer Gattung von Verfahren zur Approximation dieser Integrale", *Sitz.-ber. Heidelberger Akad. Wiss., Math.-Naturw. Kl., 9. Abh.*

[8] Protter, M. H., and Weinberger, H. F. (1967). "Maximum Principles in Differential Equations", Prentice-Hall, Englewood Cliffs, NJ.

[9] Sattinger, D. H. (April, 1975). "A nonlinear parabolic system in the theory of combustion", *Quart. Appl. Math.*, 47-61.

[10] Walter, W. (1970). "Differential and Integral Inequalities", Springer-Verlag, Berlin.

SEQUENTIAL CONJUGATE GRADIENT-RESTORATION ALGORITHM FOR OPTIMAL CONTROL PROBLEMS WITH NONDIFFERENTIAL CONSTRAINTS[1]

A. Miele

Astronautics and Mathematical Sciences
Rice University
Houston, Texas

J. R. Cloutier[2]

Department of Mathematical Sciences
Rice University
Houston, Texas

A sequential conjugate gradient-restoration algorithm is developed in order to solve optimal control problems involving a functional subject to differential constraints, nondifferential constraints, and terminal constraints. The algorithm is composed of a sequence of cycles, each cycle consisting of two phases, a conjugate gradient phase and a restoration phase.

[1]*This research was supported by the Office of Scientific Research, Office of Aerospace Research, United States Air Force, Grant No. AF-AFOSR-76-3075. This paper is based on the investigations presented in Refs. 1-2.*

[2]*Presently, Mathematician, Naval Surface Weapons Center, Dahlgren Laboratory, Dahlgren, Virginia.*

ISBN 0-12-434180-2

The conjugate gradient phase involves a single iteration and is designed to decrease the value of the functional while satisfying the constraints to first order. During this iteration, the first variation of the functional is minimized, subject to the linearized constraints. The minimization is performed over the class of variations of the control and the parameter which are equidistant from some constant multiple of the corresponding variations of the previous conjugate gradient phase. For the special case of a quadratic functional subject to linear constraints, various orthogonality and conjugacy conditions hold.

The restoration phase involves one or more iterations and is designed to restore the constraints to a predetermined accuracy, while the norm of the variations of the control and the parameter is minimized, subject to the linearized constraints. The restoration phase is terminated whenever the norm of the constraint error is less than some predetermined tolerance.

The sequential conjugate gradient-restoration algorithm is characterized by two main properties. First, at the end of each conjugate gradient-restoration cycle, the trajectory satisfies the constraints to a given accuracy; thus, a sequence of feasible suboptimal solutions is produced. Second, the conjugate gradient stepsize and the restoration stepsize can be chosen so that the restoration phase preserves the descent property of the conjugate gradient phase; thus, the value of the functional at the end of any cycle is smaller than the value of the functional at the beginning of that cycle. Of course, restarting the algorithm might be occasionally necessary.

To facilitate numerical integrations, the interval of integration is normalized to unit length. Variable-time terminal conditions are transformed into fixed-time terminal conditions. Then, the actual time at which the terminal boundary is reached becomes a component of a vector parameter being optimized.

Convergence is attained whenever both the norm of the constraint error and the norm of the error in the optimality conditions are less than some predetermined tolerances.

Several numerical examples are presented, some pertaining to a quadratic functional subject to linear constraints and some pertaining to a nonquadratic functional subject to nonlinear constraints. These examples illustrate the feasibility as well as the convergence characteristics of the sequential conjugate gradient-restoration algorithm.

REFERENCES

[1] Cloutier, J. R., Mohanty, B. P., and Miele, A. (1977). "Sequential Conjugate Gradient-Restoration Algorithm for Optimal Control Problems with Nondifferential Constraints, Part 1, Theory", Rice University, Aero-Astronautics Report No. 126.

[2] Cloutier, J. R., Mohanty, B. P., and Miele, A. (1977). "Sequential Conjugate Gradient-Restoration Algorithm for Optimal Control Problems with Nondifferential Constraints, Part 2, Examples", Rice University, Aero-Astronautics Report No. 127.

[3] Miele, A., Cloutier, J. R., Mohanty, B. P., and Wu, A. K. "Sequential Conjugate Gradient-Restoration Algorithm for Optimal Control Problems with Nondifferential Constraints, Part 1", International Journal of Control (to appear).

[4] Miele, A., Cloutier, J. R., Mohanty, B. P., and Wu, A. K. "Sequential Conjugate Gradient-Restoration Algorithm for Optimal Control Problems with Nondifferential Constraints, Part 2", International Journal of Control (to appear).

[5] Lasdon, L. S., Mitter, S. K., and Waren, A. D. (1967). "The Conjugate Gradient Method for Optimal Control Problems", *IEEE* Transactions on Automatic Control, *Vol. AC-12, No. 2.*

[6] Horwitz, L. B., and Sarachik, P. E. (1968). "Davidon's Method in Hilbert Space", *SIAM* Journal on Applied Mathematics, *Vol. 16, No. 4.*

[7] Lasdon, L. S. (1970). "Conjugate Direction Methods for Optimal Control", *IEEE* Transactions on Automatic Control, *Vol. AC-15, No. 2.*

[8] Tripathi, S. S., and Narendra, K. S. (1970). "Optimization Using Conjugate Gradient Methods", *IEEE* Transactions on Automatic Control, *Vol. AC-15, No. 2.*

[9] Sinnott, J. F., Jr., and Luenberger, D. G. (1967). "Solution of Optimal Control Problems by the Method of Conjugate Gradients", Proceedings of the Joint Automatic Control Conference, Philadelphia, Pennsylvania.

[10] Heideman, J. C., and Levy, A. V. (1975). "Sequential Conjugate Gradient-Restoration Algorithm for Optimal Control Problems, Part 1, Theory", Journal of Optimization Theory and Applications, *Vol. 15, No. 2.*

[11] Heideman, J. C., and Levy, A. V. (1975). "Sequential Conjugate Gradient-Restoration Algorithm for Optimal Control Problems, Part 2, Examples", Journal of Optimization Theory and Applications, *Vol. 15, No. 2.*

[12] Pagurek, B., and Woodside, C. M. (1968). "The Conjugate Gradient Method for Optimal Control Problems with Bounded Control Variables", Automatica, *Vol. 4, Nos. 5-6.*

[13] Miele, A., Damoulakis, J. N., Cloutier, J. R., and Tietze, J. L. (1974). "Sequential Gradient-Restoration Algorithm for Optimal Control Problems with Nondifferential Constraints", Journal of Optimization Theory and Applications, *Vol. 13, No. 2.*

[14] Miele, A., Pritchard, R. E., and Damoulakis, J. N. (1970). "Sequential Gradient-Restoration Algorithm for Optimal Control Problems", Journal of Optimization Theory and Applications, *Vol. 5, No. 4.*

[15] Miele, A. (1968). "Method of Particular Solutions for Linear, Two-Point Boundary-Value Problems", Journal of Optimization Theory and Applications, *Vol. 2, No. 4.*

[16] Heideman, J. C. (1968). "Use of the Method of Particular Solutions in Nonlinear, Two-Point Boundary-Value Problems", Journal of Optimization Theory and Applications", *Vol. 2, No. 6.*

[17] Miele, A., and Iyer, R. R. (1970). "General Technique for Solving Nonlinear, Two-Point Boundary-Value Problems via the Method of Particular Solutions", Journal of Optimization Theory and Applications, *Vol. 5, No. 5.*

[18] Miele, A., and Iyer, R. R. (1971). "Modified Quasilinearization Method for Solving Nonlinear, Two-Point Boundary-Value Problems", Journal of Mathematical Analysis and Applications, *Vol. 36, No. 3.*

[19] Miele, A., and Cantrell, J. W. (1969). "Study on a Memory Gradient Method for the Minimization of Functions", Journal of Optimization Theory and Applications", *Vol. 3, No. 6.*

[20] Miele, A. (1975). "Recent Advances in Gradient Algorithms for Optimal Control Problems", Journal of Optimization Theory and Applications, *Vol. 17, Nos. 5/6.*

[21] Ralston, A. (1960). "Numerical Integration Methods for the Solution of Ordinary Differential Equations", Mathematical Methods for Digital Computers, Vol. 1, Edited by A. Ralston and H. S. Wilf, John Wiley and Sons, New York, New York.

ROTATING SPIRAL WAVES AND OSCILLATIONS IN REACTION-DIFFUSION EQUATIONS*

Donald S. Cohen

Department of Applied Mathematics
California Institute of Technology
Pasadena, California

I. INTRODUCTION

During the past few years there has been a great deal of work on a series of recently occurring problems involving chemical reactor theory and the theory of chemically or biochemically reacting mixtures. As a result of many studies of various specific problems, general properties and results valid for large classes of problems have started to emerge. Reviews of some of the results together with a list of references can be found in the papers of D. S. Cohen [1], [2] and N. Kopell and L. N. Howard [3], [4]. Many outstanding problems remain. Some of the more interesting and important involve waves and oscillatory type of phenomena, and this will be the subject of this paper.

In order to describe both the observed phenomena and the analytical techniques developed for studying the problems we shall discuss several different problems involving the dynamic behavior

Supported in part by the U.S. Army Office under Contract DAHC-04-68-0006 and the National Science Foundation under Grant GP32157X2.

ISBN 0-12-434180-2

of certain classes of chemical reactors and various properties of certain chemical and biochemical reactions. Our results are based on the papers of D. S. Cohen, J. C. Neu, and R. R. Rosales [5], J. C. Neu [6], [7], and D. S. Cohen and S. Rosenblat [8]. We shall simply formulate the problems and state the main results; all proofs and details together with several studies we do not pursue here can be found in these references and several others given later in this paper.

II. ROTATING SPIRAL WAVES

Rotating spiral waves arise naturally and as models of spatially organized activity in various chemical and biochemical processes. The Belousov-Zhabotinsky reaction [9], [10] provides a classic example. Experiments with this reaction in a two-dimensional medium (i.e., a thin layer in a Petri dish) produce spiral concentration patterns which rotate with constant frequency about a fixed center [11], [12]. A. T. Winfree [10], [13] proposes that these waves result from an interplay between the chemical process of reaction and the physical process of molecular diffusion. D. S. Cohen, J. C. Neu, and R. R. Rosales [5] have proved the existence of these waves, derived various analytical and asymptotic properties of these waves, and presented concrete computations for specific parameter values.

The demonstration of the existence of such rotating spiral waves which are smooth from the origin (the fixed center of the spiral) to infinity resolves the following important issue: Previous authors [14], [15] have found asymptotic solutions which represent spiral waves far from a fixed origin, but no analysis is given to show that these asymptotic spirals correspond to solutions that are smooth at the origin. In view of this failure, arguments have been advanced that a mechanism in addition to reaction and diffusion must be present to produce and possible maintain spiral waves in the core of the spiral. The results of

Cohen, Neu, and Rosales [5] show that a rotating spiral wave can be maintained by reaction and diffusion alone. However, whether an additional mechanism (e.g., local precipitation) occurs in the actual chemistry is of course still an open question.

N. Kopell and L. N. Howard [3] have introduced a simple mathematical model of a reaction-diffusion process called a λ-ω system. The equations are

$$\begin{aligned} U_t &= \nabla^2 U + \lambda(R)U - \omega(R)V, \\ V_t &= \nabla^2 V + \omega(R)U + \lambda(R)V, \end{aligned} \tag{2.1}$$

where λ and ω are given functions of $R = \sqrt{U^2+V^2}$. $\lambda(R)$ is assumed to be a decreasing function that passes through zero when $R = 1$, so that the spatially independent solutions of (1.1) asymptotically approach a limit cycle with amplitude $R = 1$ and frequency $\omega = \omega(1)$. Cohen, Neu, and Rosales [5] rigorously prove the existence of smooth spiral wave solutions for a certain class of λ-ω systems. Although it is commonly claimed that the λ-ω systems do not actually correspond to any particular physical situation, they show that in fact, a λ-ω system arises naturally as the dominant system in the asymptotic analysis of more general reaction-diffusion systems actually describing specific physical processes.

It is convenient to introduce polar variables (R,Θ) via the change of variables $U = R\cos\Theta$, $V = R\sin\Theta$. Then, system (2.1) becomes

$$\begin{aligned} R_t &= \Delta^2 R - R|\nabla\Theta|^2 + R\lambda(R), \\ \Theta_t &= \nabla^2\Theta + \frac{2\nabla R.\nabla\Theta}{R} + \omega(R). \end{aligned} \tag{2.2}$$

We seek solutions of the form

$$R = \rho(r), \quad \Theta = \Omega t + \theta + \psi(r), \tag{2.3}$$

where (r, θ) are polar coordinates of the plane. Such solutions correspond to rotating waves in the concentrations. The corresponding values of U and V given by

$$U(r,\theta,t) = \rho(r)\cos(\Omega t + \theta + \psi(r)),$$
$$V(r,\theta,t) = \rho(r)\sin(\Omega t + \theta + \psi(r)), \tag{2.4}$$

represent a spiral wave that rotates with frequency Ω about $r = 0$. Upon substituting (2.3) into (2.2), we find that $\rho(r)$ and $\psi(r)$ must satisfy

$$\rho'' + \frac{\rho'}{r} - \rho(\psi'^2 + \frac{1}{r^2}) + \rho\lambda(\rho) = 0$$
$$\psi'' + (\frac{1}{r} + \frac{2\rho'}{\rho})\,\psi' = \Omega - \omega(\rho). \tag{2.5}$$

Physical considerations dictate the proper boundary conditions. We week solutions with ρ and ψ' bounded, so that the concentrations U, V given in (2.4) will have bounded values and gradients. Solutions of (2.5) that are regular at $r = 0$ have $\rho(0)$, $\psi'(0) = 0$. These are boundary conditions at $r = 0$. As $r \to \infty$, we demand that $\rho(r)$ asymptotes to a non-zero constant value $\rho(\infty)$. Furthermore, we assume

H.1. $\lambda = \lambda(\rho)$ is defined and continuously differentiable on $0 \le \rho \le a$ for some $a > 0$, $\lambda(\rho) > 0$ for $0 \le \rho < a$, $\lambda(a) = 0$, and $\lambda'(a) < 0$.

H.2. $\omega = \omega(\rho)$ is defined and continuous for $0 \le \rho \le a$, and furthermore, there exist $\varepsilon \ge 0$ and $\mu > 0$ such that

$$|\omega(a) - \omega(\rho)| \le \varepsilon(a - \rho)^{1+\mu}, \quad 0 \le \rho \le a. \tag{2.6}$$

Under these conditions Cohen, Neu, and Rosales [5] have proved the following result:

Theorem 2.1. *For ε sufficiently small there exist a number $\Omega = \omega(\rho(\infty)) = \omega(a)$ and functions $\rho = \rho(r)$ and $\psi = \psi(r)$, twice continuously differentiable on $0 \le r \le \infty$, satisfying* (2.5), *and*

$$0 < \rho(r) < a \quad \text{for} \quad 0 < r < \infty, \tag{2.7}$$

$$\rho(r) = \begin{cases} O(r) & \text{as} \quad r \to 0, \\ a + O(r^{-2}) & \text{as} \quad r \to \infty, \end{cases} \tag{2.8}$$

$$\psi'(r) = \begin{cases} O(r) & \text{as} \quad r \to 0, \\ cr^{-1} + O(r^{-1-2\mu}) & \text{as} \quad r \quad , \end{cases} \tag{2.9}$$

where

$$c = \int_0^\infty s\rho^2(s)[\omega(a) - \omega(\rho(s))]ds.$$

Remark I. In the proof it is shown that from (2.7) - (2.9) it follows that

$$\rho'(0) > 0, \quad \rho''(0) = 0, \quad \rho(r) = 1 + \frac{1+c^2}{\lambda'(a)r^2} + o(r^{-2})$$
$$\rho'(r) = o(r^{-2}) \quad \text{and} \quad \rho''(r) = o(r^{-2}) \quad \text{as} \quad r \to \infty, \tag{2.10}$$

$$\psi(r) = c\ell nr + \text{constant} + O(r^{-2\mu}) \quad \text{as} \quad r \to \infty. \tag{2.11}$$

Upon substituting (2.8) and (2.11) into (2.4), we obtain

$$U \sim \rho(\infty)\cos(\Omega t + \theta + c \log r),$$
$$V \sim \rho(\infty)\sin(\Omega t + \theta + c \log r). \tag{2.12}$$

Equations (2.12) represent a rotating spiral wave with U and V constant along the logarithmic spirals $\Omega t + \theta + c \log r =$ constant.

Remark II. In addition to the rotating logarithmic spiral waves, Cohen, Neu, and Rosales [5] present numerical compuations of both rotating logarithmic spiral waves and rotating Archimedean spiral waves for certain specific function $\lambda(\rho)$, $\omega(\rho)$. Graphs of the concentrations, amplitude, and phase are given in that paper.

III. MULTI-SPECIES INTERACTIONS

We now present the results of D. S. Cohen and S. Rosenblat [8]. We study the effect of spatial diffusion on oscillatory states in arbitrary multi-species growth models having hereditary terms. Specifically we shall be concerned with models which in the absence of diffusion admit equilibrium states in the form of (orbitally asymptotically) stable periodic oscillations. For a variety of reasons this type of equilibrium occurs commonly and quite naturally in ecological communities. The monograph [16] by J. M. Cushing constitutes an excellent and useful survey of results and problems for diffusionless growth models with hereditary terms. However, there is essentially no theoretical work on such models when the various species, besides evolving in time, are allowed to diffuse spatially. We shall show below that in the presence of a small amount of diffusion the spatial and temproal evolution of a periodic equilibrium state is virtually independent of the details of the model and is in accordance with a generalized Burgers' equation.

Our results apply to very general nonlinear systems [17]. However, to understand the effects in multi-species growth models we study the specific system

$$\frac{\partial N_i}{\partial t} = \varepsilon\nabla^2 N_i + N_i[f_i(N_1,\ldots,N_n) + \sum_{j=1}^{n}\int_{-\infty}^{t} K_{ij}(t-s)F_{ij}(N_1(s),\ldots,N_n(s))ds], \quad (3.1)$$

where $N_i = N_i(\underline{x},t)$ with $\underline{x} = (x_1,\ldots,x_m)$ and where ∇^2 represents the m-dimensional Laplacean. Hence, N_i, $i = 1,\ldots,n$, represents some measure of the size or density of the ith species (for example, biomass, population density, etc.) at position x at time t, and ε, with $0 < \varepsilon << 1$, represents the effect of a small amount of diffusion acting equally on all species. The explicit forms of the delay kernels $K_{ij}(s)$, $i,j = 1,\ldots,n$, and of the nonlinear functions f_i and F_{ij} are not essential for our results, so we leave them arbitrary. In fact, any of the forms commonly used to describe predator-prey or competing species situations are included in the model (1.1). All we require is that the K_{ij}, f_i and F_{ij} are such that the system (1.1), *in the absence of diffusion* $(\varepsilon = 0)$, possesses an orbitally asymptotically stable T-periodic solution $N_i = U_i(t)$.

The main result of Cohen and Rosenblat [8] is that for small ε the system (3.1) possesses, to a first approximation, a periodic wave-like solution given by

$$N_i = U_i(t + \psi(x,\tau)), \tag{3.2}$$

where $\tau = \varepsilon t$, and where the phase $\psi(\underline{x},\tau)$ evolves according to the nonlinear equation

$$\frac{\partial\psi}{\partial\tau} = \nabla^2\psi + \frac{\alpha}{2}|\nabla\psi|^2. \tag{3.3}$$

Where α is a constant. Hence, the gradient of the phase $\vec{V} \equiv \nabla\psi$ satisfies

$$\frac{\partial\vec{v}}{\partial\tau} = \nabla \vec{v} + \alpha(\vec{v}.\nabla)\vec{v}, \tag{3.4}$$

which is the m-dimensional analog of Burgers' equation. In one dimension equation (3.4) reduces to the well-known Burgers' equation. In the special case of one dimension and of two species and with $F_{ij} \equiv 0$ the system (3.1) reduces to a coupled pair of reaction-diffusion equations. For this case the results implied by (3.3) and (3.4) were first discovered independently and from entirely different viewpoints by J. C. Neu [6] and L. N. Howard and N. Kopell [4]. Neu's work motivated the studies of Cohen and Rosenblat discribed in this paper and in [17].

The equation (3.4), or (3.3), is the key to understanding the effect of spatial diffusion on oscillatory equilibria in an ecological community. Near any given position $\underline{x}$ the solution looks like a travelling wave with frequency $\approx \frac{2\pi}{T}$ and wave-number $\frac{2\pi}{T}\frac{\partial\psi}{\partial x}$. Thus (3.4) shows that the wave-number evolves slowly in time according to a generalized Burgers' equation. Cohen and Rosenblat [8] show that *it is a general principle that the addition of spatial diffusion to a stable oscillatory ecological community induces a periodic diffusion wave in which the original wave-number (phase) evolves according to a nonlinear equation of generalized Burgers' type.*

For a concrete example of the implications of our result consider a one-dimensional problem and the well-known simple shock solution of Burgers' equation [18]. By setting $X = x + c\tau$ it is well-known that the Burgers' equation (i.e., the one-dimensional form of (3.4)) has a solution of the form

$$v = v_1 + \frac{v_2 - v_1}{1+\exp[\frac{\alpha}{2}(v_2-v_1)X]}, \tag{3.5}$$

with the properties

$$v \to \begin{cases} v_1 & \text{as } X \to +\infty, \\ v_2 & \text{as } X \to -\infty, \end{cases} \tag{3.6}$$

provided $c = \frac{1}{2}(v_1+v_2)$. Hence

$$\psi = \int v dx \to \begin{cases} v_1 x & \text{as } X = x + c\tau \to +\infty, \\ v_2 x & \text{as } X = x - c\tau \to -\infty. \end{cases} \tag{3.7}$$

Therefore,

$$N(x,t) \approx U(t+\psi(x,\tau)) \to \begin{cases} U(t+v_1 x) & \text{as } x+\varepsilon ct \to +\infty, \\ U(t+v_2 x) & \text{as } x+\varepsilon ct \to -\infty, \end{cases} \tag{3.8}$$

Hence, as illustrated in Figure 1, we see a wave train propagating with speed $\frac{1}{v_1}$ and wavelength $\frac{1}{v_1}$ as $x + \varepsilon ct \to +\infty$, and a wave train propagating with speed $\frac{1}{v_2}$ and wavelength $\frac{T}{v_2}$ as $x + \varepsilon ct \to -\infty$. The interface between the two regimes propagates with speed $\varepsilon c = \frac{\varepsilon}{2}(v_1+v_2)$.

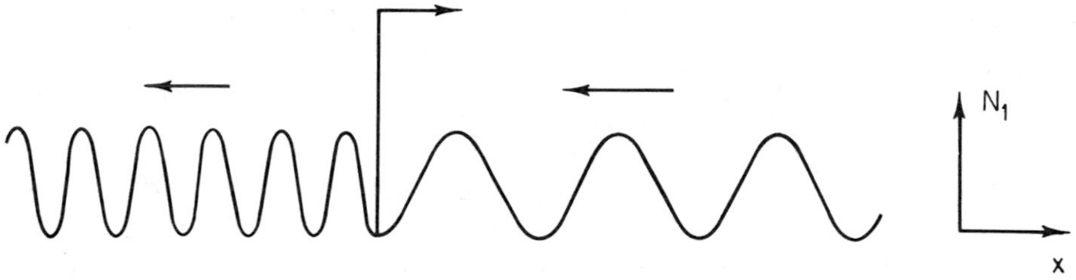

FIGURE 1

IV. COUPLED CHEMICAL OSCILLATORS

M. Marek and I. Stuchl [19] have performed a series of experiments designed to observe the interaction between two continuous stirred tank reactors, in each of which a Belusov-Zhabotinskii reaction with different parameters is taking place. The coupling of the reactors occurs via an exchange of materials through a performated wall that separates the reactors.

In order to account for the observations and further investigate the situation, J. C. Neu [7] has studied the system

$$\begin{aligned}
\mathring{x}_1 &= F(x_1,y_1) + \varepsilon\{k(x_2-x_1) + \lambda f(x_1,y_1)\} \\
\mathring{y}_1 &= G(x_1,y_1) + \varepsilon\{k(y_2-y_1) + \lambda g(x_1,y_1)\} \\
\mathring{x}_2 &= F(x_2,y_2) + \varepsilon k(x_1-x_2) \\
\mathring{y}_2 &= G(x_2,y_2) + \varepsilon k(y_1-y_2),
\end{aligned} \qquad (4.1)$$

where $0 < \varepsilon << 1$. The parameter k is a positive coupling constant. When $\varepsilon = 0$, we have two identical uncoupled oscillators described by

$$\begin{aligned}
\mathring{x}_i &= F(x_i,y_i) \\
\mathring{y}_i &= G(x_i,y_i), \quad i = 1,2.
\end{aligned} \qquad (4.2)$$

We assume that (4.2) has a stable, T-periodic limit cycle given by

$$x_i = X(t+\psi_i), \quad y_i = Y(t+\psi_i), \quad i = 1,2, \tag{4.3}$$

where the ψ_i are arbitrary constants. When $\lambda = 0$, $\varepsilon \neq 0$, two identical oscillators are coupled. When $\lambda \neq 0$, $\varepsilon \neq 0$, two different oscillators are coupled. Notice that the forcing of the oscillators upon each other is proportional to the differences $x_2 - x_1$ and $y_2 - y_1$. This derives from the physical assumption that the coupling is due to mass transfer, which is the case in the experiments of Marek and Stuchl [19].

Using multiscale asymptotics, Neu [7] has showed that to lowest order in ε, the solution of (4.1) is

$$x_i = X(t+\psi_i(\tau)), \quad y_i = Y(t+\psi_i(\tau)), \quad i = 1,2, \tag{4.4}$$

where $\tau = \varepsilon t$ and the phase shift $\Psi(\tau) = \psi_2(\tau) - \psi_1(\tau)$ satisfies

$$\frac{d\Psi}{d\tau} = kP(\Psi) + \lambda\beta. \tag{4.5}$$

Here, β is a constant determined from certain integrals involving the functions X, Y, f and g, and $P(\Psi)$ is a T-periodic function determined from X and Y which has value 0 and slope -2 when Ψ is an integer multiple of T.

If the parameters for both reactors are nearly identical, so that their autonomous frequencies are nearly the same, then the phase difference between the oscillations of each reactor tends to a constant value as time passes. This phenomenon is called phase locking. If the parameters of the reactors are altered so that the difference of their autonomous frequencies is sufficiently large, the phase locking cannot be maintained. Long time intervals of slow variation in the phase difference are punctuated by brief intervals of rapid fluctuations. This behavior is called rhythm splitting.

Following Neu [7] we account for these observations by studying the time evolution of the phase shift $\Psi(\tau)$ which is governed by (4.5). Phase locking occurs at $\Psi = \Psi_0$ for which the right hand side of (4.5) is zero. The stability of the zeros is easily determined. Values of Ψ_0 at which the derivative of the right hand side is negative are stable, and values of Ψ_0 at which the derivative is positive are unstable. Thus, for $\lambda = 0$ (coupled identical oscillators) we see from Figure 2 that $\Psi \equiv 0$ is a stable solution. *The coupling sychronizes identical oscillators.* If $\lambda \neq 0$, $\varepsilon \neq 0$, we see that $kP(\Psi) - \lambda\beta$ has at least two roots in $0 \leq \psi \leq T$ if $\min P(\Psi) < \frac{\lambda\beta}{k} < \max P(\psi)$. The root Ψ_0 with negative slope is stable, and the system will evolve to stable oscillations with constant phase shift Ψ_0.

For $\min P(\Psi) < \frac{\lambda\beta}{K} < \max P(\Psi)$, the zeros of $kP(\Psi) - \lambda\beta$ depend continuously on $\frac{\lambda\beta}{k}$. At $\frac{\lambda\beta}{k} = \underset{(\max)}{\min} P(\Psi) = P(\Psi_m)$, an interesting bifurcation takes place; namely, the change from phase locking to rhythm splitting. To show this, Neu [7] writes (4.5) as

$$\frac{d\Psi}{d\tau} = kP(\Psi) - kP(\Psi_m) + kP(\Psi_m) - \lambda\beta$$

$$= k\{P(\Psi) - P(\Psi_m)\} + \delta^2, \tag{4.6}$$

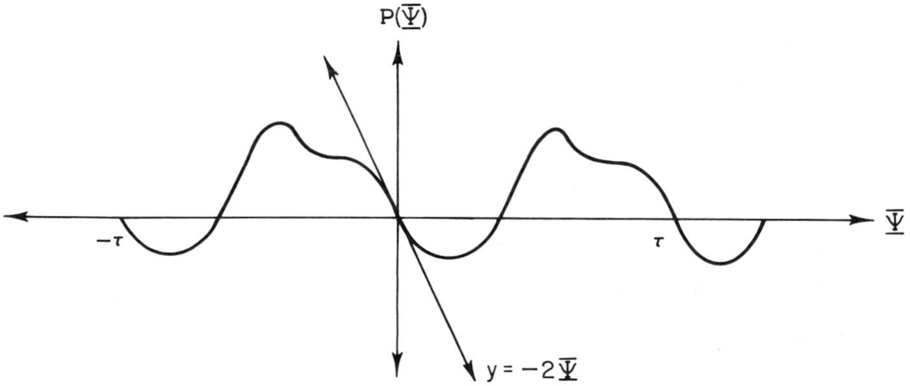

FIGURE 2

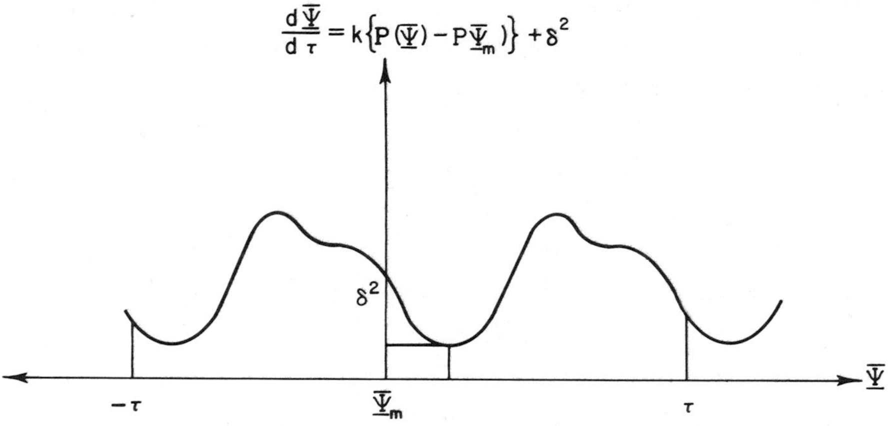

FIGURE 3

where $\delta^2 = kP(\Psi\ ) - \lambda\beta$, and $\frac{\lambda\beta}{k}$ is slightly less than min $P(\Psi) = P(\Psi_m)$, so that $0 < \delta^2 << 1$. Figure 3 illustrates the situation with a graph of $k\{P(\Psi) - P(\Psi_0)\} + \delta^2$ vs. Ψ. Equation (4.6) constitutes a singularly perturbed problem in the small parameter δ. The analysis is straightforward; we omit the details and present the results. The solution is nearly constant at the values $\Psi = \Psi_m + nT$, $n =$ integer, where the right hand side is $O(1)$. Away from $\Psi = \Psi_m + nT$, the right hand side is $O(1)$. These regions of rapid variation in $\Psi(\tau)$ correspond to boundary layers that join the constant values $\Psi = \Psi_m + nT$, as shown in Figure 4.

This is the behavior of the phase shift $\Psi(\tau)$ that corresponds to rhythm splitting. Once Ψ is known, we find ψ_1 and ψ_2 defined in (4.4) from formulas derived by Neu [7]. The solution for x_i, y_i, $i = 1,2$ are then given by

$$x_i = X(t+\psi_i), \quad y_i = Y(t+\psi_i), \quad i = 1,2. \tag{4.7}$$

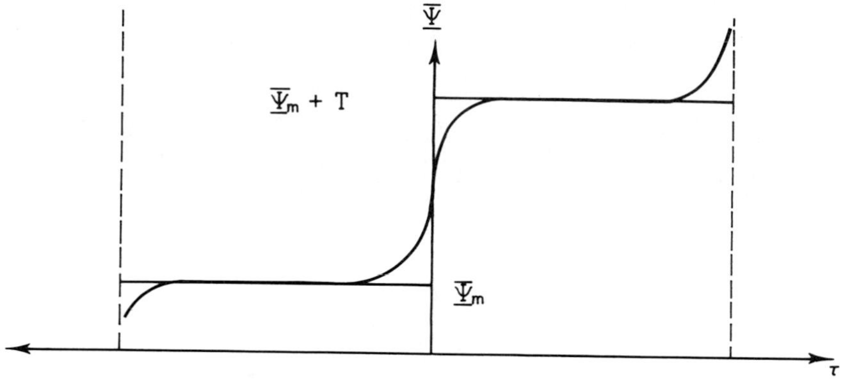

FIGURE 4

The frequency of the oscillations in x_2 is

$$\upsilon = \frac{d}{dt}(t + \psi_2) = 1 + \varepsilon kH(-\Psi). \tag{4.8}$$

Figure 5 shows a graph of υ vs. t and Figure 6 shows the behavior of the concentration $x_2(t)$. The peaks in Figure 5 correspond to the regions of rapid change in $\Psi(\tau)$.

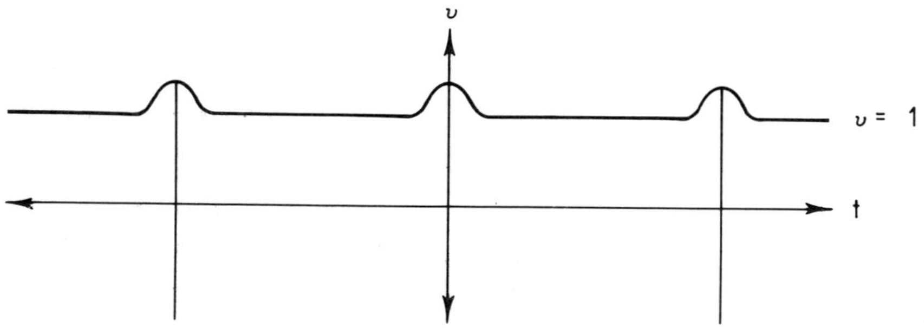

FIGURE 5

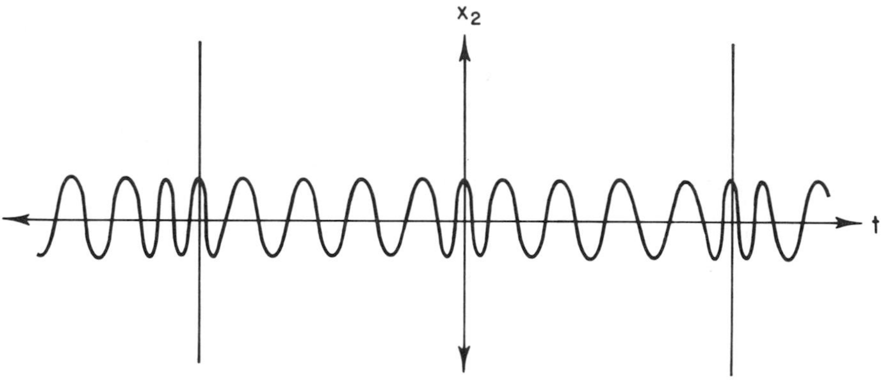

FIGURE 6

REFERENCES

[1] Cohen, D. S. (1974). "Some chemical and biochemical problems involving nonlinear diffusion processes", Proc. Seventh U.S. National Congress of Appl. Mech., Boulder, Colorado, 93-98.

[2] Cohen, D. S. (1976). "Oscillations in nonlinear diffusion processes", Proc. Fourteenth International Congress of Theoretical and Applied Mechanics, Delft, The Netherlands, North-Holland Publishing Co.

[3] Kopell, N., and Howard, L. N. (1973). "Plane wave solutions to reaction-diffusion equations", *Studies in Appl. Math. 52,* 291-328.

[4] Howard, L. N., and Lopell, N. (1977). "Slowly varying waves and shock structures in reaction-diffusion equations", *Studies in Appl. Math. 56,* 95-145.

[5] Cohen, D. S., Neu, J. C., and Rosales, R. R. (1978). "Rotating spiral wave solutions of reaction-diffusion equations", *SIAM J. Appl. Math.*

[6] Neu, J. C. "Chemical waves and the diffusive coupling of limit cycle oscillators", *SIAM J. Appl. Math.*

[7] Neu, J. C. "Coupled chemical oscillators", *SIAM J. Appl. Math.*, to appear.

[8] Cohen, D. S., and Rosenblat, S. "Multi-species interactions with hereditary effects and spatial diffusion", to appear.

[9] Winfree, A. T. (1972). "Spiral waves of chemical activity", Science *175*, 634-636.

[10] Winfree, A. T. (June, 1974). "Rotating chemical reactions", Scientific American, 82-95.

[11] Winfree, A. T. (1973). "Scroll-shaped waves in three dimensions", Science *181*, 937-939.

[12] Zhabotinsky, A. M., and Zaikin, A. N. (1973). "Autowave processes in a distributed chemical system", *J. Theor. Bio.* *40*, 45-61.

[13] Winfree, A. R. (1974). "Rotating solutions to reaction-diffusion equations in simply-connected media", Proceedings of Conference on Mathematical Aspects of Chemical and Biochemical Problems and Quantum Chemistry, *SIAM-AMS Proceedings, Vol. 8*, D. S. Cohen, editor.

[14] Greenberg, J. M., Hassard, B. D., and Hastings, S. P. "Pattern formation and periodic structures in systems modeled by reaction-diffusion equations", to appear.

[15] Greenberg, J. M. (1976). "Periodic solutions to reaction-diffusion equations", *SIAM J. Appl. Math.* *30*, 199-205.

[16] Cushing, J. M. (1977). "Integrodifferential Equations and Delay Models in Population Dynamics", Lecture Notes in Biomathematics, No. 20, Springer-Verlag.

[17] Rosenblat, S., and Cohen, D. S. "Diffusive effects in dynamical oscillatory systems", to appear.

[18] Whitham, G. B. (1974). "Linear and Nonlinear Waves", John Wiley and Sons.

[19] Marek, M., and Stuchl, I. (1975). "Synchronization in two interacting oscillatory systems", *Biophysical Chem.* *3*.

Applied Nonlinear Analysis

SOME APPLICATIONS OF ROTHE'S METHOD TO PARABOLIC AND RELATED EQUATIONS

C. Corduneanu

Department of Mathematics
University of Rhode Island
Kingston, Rhode Island

In this paper we shall consider parabolic equations of the form

$$u_t = u_{xx} + f(t,x,u), \tag{1}$$

and the system of two equations

$$v_t = v_{xx} + f(v) - u, \quad u_t = \sigma v - \gamma u, \tag{2}$$

known as "FitzHugh-Nagumo equations" (see, for instance, [6]).

Using the lines approximation method of E. Rothe [7], and certain qualitative properties of the solutions of approximating equations, we shall establish some properties for the solutions of (1) or (2), valid in an unbounded domain.

We are not going to consider in this paper the existence problem of solutions for (1) and (2), though the lines method is one of the most frequent tool used in this case (see, for instance, [9]). The main purpose of this paper is to show how qualitative properties of the ordinary differential systems, occurring in the approximating scheme, can be extended to the corresponding partial differential equations.

ISBN 0-12-434180-2

Let us remark that L. I. Kamynin [4] has investigated the approximation of solutions of the heat equation, by the method of lines, in the case of an unbounded domain. He discretized the variable x, while the author ([1], [2]) has given some results in the case of an unbounded domain, when the discretization regards the variable t.

A comprehensive bibliography concerning the method of lines, for the period up to 1965, can be found in [5]. Further references, concerning the parabolic case, are given in [8].

W. Walter [9] considered the problem of existence of periodic solutions in t for equation (1), and extensively used the lines method associated with differential inequalities.

Let us consider first the equation (1), with $f(t,x,u)$ continuous on $R_+ \times [0,1] \times R$, and values in R. The following boundary value problem will be associated to equation (1):

$$u(0,x) = \phi(x), \quad x \in [0,1], \quad u(t,0) = u(t,1) = 0, \quad t \in R. \tag{3}$$

Of course, the solution of the problem is sought in $R_+ \times [0,1]$.

The discretized system (with respect to x) associated with (1), (3), for a given natural n, will be

$$\frac{du_k}{dt} = (n+1)^2(u_{k+1} - 2u_k + u_{k-1}) + f(t,x_k,u_k), \tag{4}$$
$$k = 1,\ 2,\ \ldots,\ n; \quad u_0(t) = u_{n+1}(t) = 0.$$

In (4), $u_k(t)$ is the approximating value for $u(t,x_k)$, where $x_k = k/(n+1)$, $k = 1,\ 2,\ \ldots,\ n$. The initial conditions for (4) are

$$u_k(0) = \phi(x_k), \quad k = 1,\ 2,\ \ldots,\ n. \tag{5}$$

With obvious notations, we can rewrite the system (4) in the vector form

$$\frac{du}{dt} = (n+1)^2 A_n u + f(t,u). \tag{6}$$

The matrix A_n is a square matrix of order n and has been considered in connection with approximating schemes by

V. N. Faddeeva [3]. The following basic properties of A_n can be found in [3]:

(1) The eigenvalues of A_n are

$$\lambda_s = -4\sin^2\frac{\pi s}{2(n+1)}, \quad s = 1, 2, \ldots, n. \tag{7}$$

(2) The unitary matrix $B_n = B_n^* = B_n^{-1}$, such that $B_n A_n B_n^{-1} = \mathrm{diag}(\lambda_1, \lambda_2, \ldots, \lambda_n)$, is $B_n = (b_{ks})$, with

$$b_{ks} = (-1)^{k+s}\sqrt{\frac{2}{n+1}}\sin\frac{\pi ks}{n+1}, \quad k,s = 1, 2, \ldots, n. \tag{8}$$

Consequently, the eigenvalues of the matrix $(n+1)^2 A_n$ are

$$\lambda'_s = -4(n+1)^2\sin^2\frac{\pi s}{2(n+1)} \leq -4(n+1)^2\sin^2\frac{\pi}{2(n+1)}, \tag{9}$$

$s = 1, 2, \ldots, n.$

It is of basic importance the fact that

$$4(n+1)^2\sin^2\frac{\pi}{2(n+1)} \to \pi^2, \quad \text{as } n \to \infty. \tag{10}$$

This shows that the characteristic roots of the matrix $(n+1)^2 A_n$ are located on the negative half-axis, left to $-\pi^2 + \varepsilon$, $\varepsilon > 0$, as soon as n is sufficiently large. In other words, the approximating systems (4) or (6) posses strong stability properties.

Indeed, the fundamental matrix of the homogeneous system associated to (6) is

$$U_n(t) = B_n\, \mathrm{diag}(e^{\lambda'_1 t}, e^{\lambda'_2 t}, \ldots, e^{\lambda'_n t})B_n^{-1}, \tag{11}$$

with B_n and λ'_s given by (8) and (9) respectively. Since B_n is a unitary matrix, from (11) we find the following estimate for the operator-norm of the matrix $U_n(t)$:

$$\|U_n(t)\| \leq e^{-\lambda t}, \quad t \geq 0, \tag{12}$$

where $\lambda < \pi^2$ and n is sufficiently large.

From (6) and (5) we find the vector integral equation

$$u(t) = U_n(t)\phi_n + \int_0^t U_n(t-s)f(s,u(s))ds, \tag{13}$$

with $u = \mathrm{col}(u_1, u_2, \ldots, u_n)$, $f(t,u) = \mathrm{col}(f(t,x_1,u_1), f(t,x_2,u_2), \ldots, f(t,x_n,u_n))$ and $\phi_n = \mathrm{col}(\phi(x_1), \phi(x_2), \ldots, \phi(x_n))$.

The integral equation (13) leads easily to some boundedness results, if adequate hypotheses are placed on $f(t,u)$. Let us remark first that it can be rewritten as

$$u(t) = U_n(t)\phi_n + \int_0^t U_n(t-s)f(s,0)ds + \int_0^t U_n(t-s)\bar{f}(s,u(s))ds, \quad (14)$$

where $\bar{f}(t,u) = f(t,u) - f(t,0)$. Taking into account (12), and assuming

$$f(t,0) \in M(R_+,C([0,1],R)), \quad (15)$$

$$|f(t,x,u) - f(t,x,v)| \leq L|u-v|, \quad L < \pi^2, \quad (16)$$

we shall easily find an upper estimate for $u(t)$, valid on the half-axis R_+.

Let us remind first that the space $M = M(R_+,E)$, where E is a Banach space, consists of those locally integrable maps from R_+ into E, such that

$$\sup_{t\geq 0} \int_t^{t+1} \|x(s)\| ds = |x|_M < \infty. \quad (17)$$

Finally, $x \in M$ if and only if

$$\sup_{t\geq 0} \int_0^t e^{-\alpha(t-s)} \|x(s)\| ds < +\infty, \quad (18)$$

for some $\alpha > 0$.

From (14), (15), (16) and (12) we find the integral inequality

$$\|u(t)\| \leq \sqrt{n}\, K + L\int_0^t e^{-\lambda(t-s)} \|u(s)\| ds, \quad (19)$$

where K is such that

$$\sup_{x\in[0,1]} |\phi(x)| + \sup_{t\geq 0} \int_0^t e^{-\lambda(t-s)} |f(s,0)| ds \leq K, \quad (20)$$

and $L < \lambda < \pi^2$. Accordingly, n has to be sufficiently large, i.e.

$$4(n+1)^2 \sin^2 \frac{\pi}{2(n+1)} > \lambda. \quad (21)$$

Multiplying both sides in (19) by $e^{\lambda t}$, we obtain

$$e^{\lambda t}\|u(t)\| \le \sqrt{n}Ke^{\lambda t} + L\int_0^t e^{\lambda s}\|u(s)\|ds, \tag{22}$$

from which one derives using Gronwall's lemma

$$\|u(t))\| \le \frac{\sqrt{n}K\lambda}{\lambda - L}, \quad t \in R_+. \tag{23}$$

The estimate (23) shows that the solution of the approximating system (4) is bounded on the half-axis R_+, for each n. This remark is interesting enough, if we take into account the fact that (15) does not imply the boundedness of $f(t,x,0)$ in $R_+ \times [0,1]$.

Of course, the approximating scheme we used above is valid only in the case

$$\varepsilon_k(t) = u(t,x_k) - u_k(t), \quad k = 1, 2, \ldots, n, \tag{24}$$

are arbitrarily small, for large enough n. A similar approach to the one used above will give the needed estimates for $\varepsilon_k(t)$.

First, let us remark that $\varepsilon(t) = \mathrm{col}(\varepsilon_1(t), \varepsilon_2(t), \ldots, \varepsilon_n(t))$ satisfies the following system

$$\frac{d\varepsilon}{dt} = (n+1)^2 A_n\varepsilon + f(t,u+\varepsilon) - f(t,u) + r(t), \tag{25}$$

with $\varepsilon_0(t) = \varepsilon_{n+1}(t) = 0$ and $r(t) = \mathrm{col}(r_1(t), r_2(t), \ldots, r_n(t))$, where

$$r_k(t) = u_{xx}(t,x_k) - (n+1)^2[u(t,x_{k+1}) - 2u(t,x_k) + u(t,x_{k-1})], \; k = 1, 2, \ldots, n. \tag{26}$$

The initial condition associated to (25) is, according to (3) and (5),

$$\varepsilon(0) = \theta = \text{the null vector in } R^n. \tag{27}$$

The system (25), under condition (27), leads to the following integral equation:

$$\varepsilon(t) = \int_0^t U_n(t-s)r(s)ds + \int_0^t U_n(t-s)[f(s,u+\varepsilon) - f(s,u)]ds. \tag{28}$$

As above, with the Euclidean norm for vectors and the operator norm for the matrices, we obtain from (28):

$$\|\varepsilon(t)\| \leq \int_0^t e^{-\lambda(t-s)} \|r(s)\| ds + L\int_0^t e^{-\lambda(t-s)} \|\varepsilon(s)\| ds. \quad (29)$$

From (26) there results

$$|r_k(t)| \leq \omega\left(\frac{1}{n+1} \right), \quad k = 1, 2, \ldots, n, \quad (30)$$

where $\omega(\delta)$ *is the modulus of continuity* for $u_{xx}(t,x)$ in $R_+ \times [0,1]$. To be more specific, $\omega(\delta)$ is such that

$$|u_{xx}(t,x) - u_{xx}(t,y)| \leq \omega(\delta), \quad t \in R_+, \quad (31)$$

with $0 \leq x,y \leq 1$ and $|x-y| < \delta$. Therefore, the Euclidean norm of $r(s)$ in (29) satisfies

$$\|r(s)\| \leq \sqrt{n}\, \omega\left(\frac{1}{n+1} \right), \quad s \in R_+. \quad (32)$$

The inequality (29) becomes now

$$\|\varepsilon(t)\| \leq \lambda^{-1}\sqrt{n}\, \omega\left(\frac{1}{n+1} \right) + L\int_0^t e^{-\lambda(t-s)} \|\varepsilon(s)\| ds, \quad (33)$$

and it can be handled in the same manner as (19). There follows,

$$\|\varepsilon(t)\| \leq (\lambda-L)^{-1}\sqrt{n}\, \omega\left(\frac{1}{n+1} \right), \quad t \in R_+. \quad (34)$$

Consequently, the error $\varepsilon(t)$ can be made arbitrarily small for large n, provided

$$\sqrt{n}\, \omega\left(\frac{1}{n+1} \right) \to 0, \quad \text{as} \quad n \to \infty. \quad (35)$$

Let us summarize the above discussion in the following Theorem.

Theorem 1. *Consider the scalar parabolic equation (1) under conditions (3), and assume that:*

(1) $f(t,x,u)$ *is a continuous mapping from* $R_+ \times [0,1] \times R$ *into* R, *such that (15) and (16) are verified;*

(2) $\phi(x)$ *is continuous from* $[0,1]$ *into* R *and* $\phi(0) = \phi(1) = 0$;

(3) *there exists a solution* $u(t,x)$ *of (1), (3), such that* $u_{xx}(t,x)$ *has a modulus of continuity* ω *in* $R_+ \times [0,1]$, *satisfying (35).*

Then the solution of the system (4), with initial conditions (5), approximates $u(t,x)$ *in* $R_+ \times [0,1]$, *the error being estimated by (34).*

Corollary. *If the solution* $u(t,x)$ *is uniformly continuous on* $R_+ \times [0,1]$, *then it is necessarily bounded there.*

This easily follows from the boundedness of solutions of the approximating system of ordinary differential equations and the possibility to approximate $u(t,x)$ as shown above. The following formula is the key of the proof:

$$u(t,x) = [u(t,x) - u(t,x_k)] + \varepsilon_k(t) + u_k(t), \tag{36}$$

where $x_k = k/(n+1)$ is the closest number to x in the considered division of $[0,1]$. Of course, the existence of the continuity modulus gives the boundedness of the first term in the right hand side of (36).

Remark 1. If condition (15) is replaced by a stronger condition, say $f(t,0) \in L^p(R_+,C([0,1],R))$, $1 \leq p < \infty$, then instead of boundedness for $u_k(t)$ and $u(t,x)$, we get $u_k(t) \to 0$ and $u(t,x) \to 0$ in $C([0,1],R)$, as $t \to \infty$.

Remark 2. Condition (35) is satisfied when

$$|u_{xx}(t,x) - u_{xx}(t,y)| \leq K|x-y|^\alpha, \quad \alpha > 1/2, \tag{37}$$

in $R_+ \times [0,1]$. In other words, $u_{xx}(t,x)$ needs not to be Lipschitz continuous in $R_+ \times [0,1]$, a Holder continuity of order $\alpha > 1/2$ being sufficient for the validity of (35).

Let us remark that (1) gives $u_{xx} = u_t - f(t,x,u)$, from which (37) can be easily derived if similar conditions are imposed on u and u_t.

Finally, it is worth to point out that (37) would suffice to reach the conclusion of Theorem 1 when it takes place in $R_a \times [0,1]$, with $R_a = [a,\infty)$, $a > 0$.

We shall consider now the system (2), under initial conditions

$$v(0,x) = \phi(x), \quad u(0,x) = \psi(x), \quad x \in R, \tag{38}$$

where $\phi(x)$ and $\psi(x)$ are scalar continuous periodic functions, of period ℓ.

The problem we shall deal with is the following: let us associate to the system (2), under initial conditions (38), the equations

$$\left.\begin{aligned} &\frac{d^2v_k}{dx^2} = h^{-1}(v_k - v_{k-1}) - f(v_{k-1}) + u_k, \\ &u_k = \sigma h v_{k-1} + (1-\gamma h)u_{k-1}, \\ &k = 1, 2, \ldots, n, \quad v_0(x) = \phi(x), \quad u_0(x) = \psi(x), \end{aligned}\right\} \tag{39}$$

where $nh = T > 0$. The system (30) is obtained from (2) and (38), substituting to $v_t(t_k,x)$ the approximate value $h^{-1}[v(t_k,x) - v(t_{k-1},x)]$, and similarly for u_t. It is aimed to provide approximate values for $v(t_k,x)$ and $u(t_k,x)$, $k = 1, 2, \ldots, n$. In other words, we look for an approximation of the solution $(v(t,x), u(t,x))$ in the strip $0 \le t \le T$, $x \in R$. Moreover, we want $v_k(x)$ and $u_k(x)$, $k = 1, 2, \ldots, n$, to be periodic, with period ℓ.

The system (39) is defining uniquely $v_k(x)$ and $u_k(x)$, $k = 1, 2, \ldots, n$, as periodic functions of period ℓ. Indeed, the first equation of (39) is of the form $y'' - \alpha y = g(x)$, $\alpha > 0$, where $g(x)$ is periodic, of period ℓ. It is worth to remark that $u_k(x)$ is found from the second equation of (39), as soon as $v_{k-1}(x)$ and $u_{k-1}(x)$ are known. For details, see [1].

Let us denote now

$$\varepsilon_k(x) = v(t_k,x) - v_k(x), \quad \eta_k(x) = u(t_k,x) - u_k(x), \quad k = 0, 1, 2, \ldots, n. \tag{40}$$

From (39) and the equations (2) taken for $t = t_k$, $k = 1, 2, \ldots, n$, one easily finds

$$\left.\begin{aligned}\frac{d^2\varepsilon_k}{dx^2} &= h^{-1}(\varepsilon_k-\varepsilon_{k-1}) - f(v(t_k,x)) + f(v_{k-1}(x)) \\ &\qquad + u(t_k,x) - u_k(x) + r_k(x), \\ h^{-1}[\eta_k-\eta_{k-1}] &= \sigma\varepsilon_{k-1} - \gamma\eta_{k-1} + \sigma[v(t_k,x) - v(t_{k-1},x)] \\ &\qquad - \gamma[u(t_k,x) - u(t_{k-1},x)] + s_k(x), \\ k &= 1,\ 2,\ \ldots,\ n,\end{aligned}\right\} \tag{41}$$

where $\varepsilon_0(x) = \eta_0(x) = 0$ and

$$\left.\begin{aligned}r_k(x) &= v_t(t_k,x) - h^{-1}[v(t_k,x) - v(t_{k-1},x)], \\ s_k(x) &= h^{-1}[u(t_k,x) - u(t_{k-1},x)] - u_t(t_k,x).\end{aligned}\right\} \tag{42}$$

Taking into account that the unique periodic solution of $y'' - \alpha y = g(x)$, $\alpha > 0$, satisfies $\|y(x)\| \le \|g(x)\|/\alpha$, with $\|y(x)\| = \sup|y(x)|$, from (41) we get the system of inequalities

$$\left.\begin{aligned}\|\varepsilon_k\| &\le (1 + hM)\|\varepsilon_{k-1}\| + h\|\eta_k\| + h\tilde{\omega}(h), \\ \|\eta_k\| &\le h\sigma\|\varepsilon_{k-1}\| + (1 + h\gamma)\|\eta_{k-1}\| + h\tilde{\omega}(h), \\ k &= 1,\ 2,\ \ldots,\ N,\end{aligned}\right\} \tag{43}$$

where M is a Lipschitz constant for f and $\tilde{\omega}(h) \to 0$ as $h \to 0$. More precisely, $\tilde{\omega}(h)$ is given by $\tilde{\omega}(h) = \omega_1(h) + K\omega(h)$, where $\omega_1(h)$ represents a common continuity modulus for $v_t(t,x)$ and $u_t(t,x)$, $\omega(h)$ is a common continuity modulus for $v(t,x)$ and $u(t,x)$, in the rectangle $[0,T] \times [0,\ell]$. The constant K can be chosen as $K = \max\{M,\sigma+\gamma\}$.

The system (43) can be manipulated in the standard manner and an estimate of the form

$$\|\varepsilon_k\| + \|\eta_k\| \le N\omega^*(h), \quad k = 1,\ 2,\ \ldots,\ n, \tag{44}$$

can be derived. In (44), N is a positive constant depending only on M, σ, γ, T and ℓ, while $\omega^*(h)$ is a function that tends to zero with h.

We will summarize now the result of the discussion of the system (2) in the following theorem.

Theorem 2. *Consider the system (2), with initial conditions (38), and assume that:*

(1) *$f(v)$ is Lipschitz continuous from R into R, with constant M;*

(2) *σ and γ are positive constants;*

(3) *the functions $\phi(x)$ and $\psi(x)$ are continuous periodic functions, with period ℓ;*

(4) *there exists a solution $(v(t,x), u(t,x))$ of (2), satisfying (38), such that it is defined in $[0,T] \times R$, $T > 0$, and is periodic in x, with period ℓ.*

Then, the approximating scheme described by equations (39) is convergent.

Remark 1. If in the first equation (39) one takes $f(v_k)$ instead of $f(v_{k-1})$, the existence and uniqueness of the periodic solution is still guaranteed if $f'(v) \geq 0$ for all $v \in R$.

Remark 2. It would be interesting to apply for (2) the same approximation procedure as in the case of equation (1). The characteristic equation for the linear system of ordinary differential equations is now more intricate. The periodicity conditions should be $v_0(t) = v_{n+1}(t)$, $u_0(t) = u_{n+1}(t)$.

REFERENCES

[1] Corduneanu, C. (1961). "Approximation des solutions d'une équation parabolique dans un domaine non borné", *Mathematica 3*, 217-224.

[2] Corduneanu, C. (1967). "Solutions presque-périodiques de certaines équations paraboliques", *Mathematica 9*, 241-244.

[3] Faddeeva, V. N. (1949). "The method of lines applied to certain boundary value problems (Russian)", *Trudy Mat. Inst. im. V. A. Steklova, 28*, 73-103.

[4] Kamynin, L. I. (1953). "On the applicability of a finite difference method to the solution of the heat equation",

Izvestija AN SSSR, Ser. Math., *17*, 163-180, 249-268.

[5] Liskovets, O. A. (1965). "The method of lines (review)", *Differential Equations*, *1*, 1308-1323.

[6] Rauch, J., and Smoller, J. (1978). "Qualitative theory of the FitzHugh-Nagumo equations", *Advances in Mathematics*, *27*, 12-44.

[7] Rothe, E. (1930). "Zweidimensionale parabolische Randwertaufgaben als Grenzfall eindimensionaler Randwertaufgaben", *Math. Annalen*, *102*, 650-670.

[8] Walter, W. (1970). "Differential and Integral Inequalities", Springer-Verlag, Berlin.

[9] Walter, W. (1974). "The line method for parabolic differential equations. Problems in boundary layer theory and existence of periodic solutions", Lecture Notes in Mathematics, No. 430, Springer-Verlag.

A COARSE-RESOLUTION ROAD MAP TO METHODS FOR APPROXIMATING SOLUTIONS OF TWO-POINT BOUNDARY-VALUE PROBLEMS

James W. Daniel

Departments of Mathematics and of Computer Science, and Center for Numerical Analysis
The University of Texas
Austin, Texas

I intentionally avoided calling this paper a "survey" because, having once worked as a surveyor, I know that a survey of a city gives an extremely detailed description of the precise layout of the property in that city and is not very helpful to someone trying to find his or her way around town. Analogously, presenting all the details of various implemented methods for solving boundary-value problems can obscure the concepts. It is also true, however, that a coarse aerial photograph of a city is a poor guide for the lost traveler, and, analogously, a very abstract model representing all methods for boundary-value problems is too general to impart much information. What both the traveler and the student of numerical methods need is a useful roadmap with not only enough detail to show the various points of interest but also enough perspective to show where these sites lie in relation to one another. At the Working Conference on Codes for Boundary-value Problems in ODEs in Houston in May of 1978 I will present my own such roadmap of what some numerical methods for boundary-value problems *are*, of how they relate to one

ISBN 0-12-434180-2

another, and of what areas need development in order to improve methods. In this present lecture I present a roadmap with a more coarse resolution, concentrating mainly on a classification of what the methods are; the lecture is based on the paper for the Houston conference.

Now, what kinds of boundary-value problems are we to consider? I want to present neither a single abstract problem including all cases nor a vast list of specific special problems. I will discuss instead a couple of model problems for which the solution methods will share many features with methods for the panorama of distinct problem types: eigenvalues, non-linear boundary conditions, m-th order equations, systems of equations for vector-valued functions, mixed-order systems, infinite-intervals for the independent variable, singular problems, singular-perturbation problems, multi-point boundary conditions, et cetera. I will consider both the *first-order system* for the $n \times 1$ vector valued function $\underline{\underline{y}}$:

$$\underline{\underline{y}}'(t) = \underline{\underline{f}}(t, \underline{\underline{y}}(t)) \quad \text{for} \quad 0 < t < 1 \tag{1.1}$$

and the *second-order scalar equation:*

$$y''(t) = f(t, y(t), y'(t)) \quad \text{for} \quad 0 < t < 1 \tag{1.2}$$

since numerical methods on the first-order system equivalent to (1.2) usually are dramatically less efficient than methods directly intended for second-order problems; note that I restrict myself to a finite range for the independent variable and I use $0 < t < 1$ as a canonical interval. *Boundary conditions* for (1.1) are given by n nonlinear equations for $\underline{\underline{y}}(0)$ and $\underline{\underline{y}}(1)$:

$$\underline{\underline{b}}(\underline{\underline{y}}(0), \underline{\underline{y}}(1)) = \underline{\underline{0}} \tag{1.3}$$

in vector notation. For (1.2) we give two nonlinear equations relating $y(0)$, $y'(0)$, $y(1)$, and $y'(1)$, which we can express in vector notation:

$$\underline{\underline{c}}(y(0), y'(0), y(1), y'(1)) = \underline{\underline{0}} \tag{1.4}$$

In many cases the boundary conditions will in fact be *linear*, in which case we replace (1.3) with

$$B_0 y(0) + B_1 y(1) = e \tag{1.5}$$

where B_0 and B_1 are $n \times n$ and e is $n \times 1$, while we replace (1.4) with

$$c_0 y(0) + d_0 y'(0) + c_1 y(1) + d_1 y'(1) = a \tag{1.6}$$

where c_0, d_0, c_1, d_1 and a are all 2×1; these special forms can be useful computationally. Another common and computationally advantageous situation is that in which the boundary conditions are *separated*, so that conditions at $t = 0$ and at $t = 1$ do not interact. In this case we can write (1.3) and (1.5) as

$$\tilde{B}_0 y(0) = e_0, \quad \tilde{B}_1 y(1) = e_1 \tag{1.7}$$

where $\tilde{B}_0$ is $q \times n$, $\tilde{B}_1$ is $(n - q) \times n$, e_0 is $q \times 1$ and e_1 is $(n - q) \times 1$, for some integer q with $1 < q < n$. Similarly (1.4) and (1.6) become in the separated case

$$c_0 y(0) + d_0 y'(0) = a_0, \quad c_1 y(1) + d_1 y'(1) = a_1. \tag{1.8}$$

Thus we will be considering either (1.1) with one of the boundary conditions (1.3), (1.5), (1.7) or (1.2) with one of the boundary conditions (1.4), (1.6), (1.8). In the interest of time and space we often will discuss a method as applied to *either* the first-order *or* the second-order problem when the analogous use of the idea of the method for the other standard problem is fairly straightforward.

The next task and the main task of this paper is to describe how to classify various methods. The "aerial photograph" approach would be to note that the problem is simply to solve $F(x) = 0$ for x in some appropriate abstract space and F some nonlinear operator, while numerical methods eventually solve some discretization $F_h(x_h) = 0_h$ for x_h is some discretized (finite-dimensional) space; this doesn't really tell us much about the structure of various specific methods. The "survey" approach would be to describe computer codes implementing various specific methods; this gives us more detail than we can absorb. Instead I will give a "road map" approach which defines a complete method as having three aspects:

(1) a *Transformed Problem*,

(2) a *Discrete Model* of the Transformed Problem, and

(3) a *Solution Technique* for the Discrete Model.

First I describe various Transformed Problems equivalent to (1.1) or (1.2) and their boundary conditions: (i) Original problem, (ii) Variational problem, (iii) Shooting and its variants, (iv) Quasi-linearization, (v) Continuation and embedding, and (vi) Integral equations. Then I present a few ways of creating Discrete Models for the Transformed Problem: (i) finite differences, and (ii) projections. Finally I sketch some Solution Techniques: (i) Optimization, (ii) Gauss elimination, et cetera.

I thank the many colleagues, especially Victor Pereyra and Andy White, who over the years have influenced my view of numerical methods for boundary-value problems. Although I would gladly blame mistakes on my colleagues and take credit for any insights, unfortunately I must accept responsibility for all the views expressed in this paper.

CONE-VALUED PERIODIC SOLUTIONS OF ORDINARY DIFFERENTIAL EQUATIONS

Klaus Deimling

Fachbereich 17 der Gesamthochschule
Paderborn, GERMANY

Let X be a real Banach space, $K \subset X$ a cone, f and $g: \mathbb{R}^+ \times K \to X$ ω-periodic in time. We look for conditions on X, K, f and g such that the differential equation $u' = f(t,u) + g(t,u)$ has an ω-periodic solution. We shall assume that f satisfies an estimate of the type

$$\alpha(f(t,B)) \leq \phi_1(t,\alpha(B)) \text{ for } t \in [0,\omega],\ B \subset K \text{ bounded,} \tag{1}$$

where $\alpha(B)$ denotes Kuratowski's measure of noncompactness, i.e.

$$\alpha(B) = \inf\{d > 0 : B \text{ admits a finite cover by sets of diameter} \leq d\}.$$

Concerning g, we shall assume that

$$(g(t,x) - g(t,y),x-y)_- \leq \phi_2(t,|x-y|)|x-y| \text{ for } t \geq 0 \text{ and } x,y \in K, \tag{2}$$

the semi-inner products $(\cdot,\cdot)_{\mp}: X \times X \to \mathbb{R}$ being defined by

$$(x,y)_{\mp} = \underset{(\max)}{\min}\ \{x^*(x) : x^* \in Fy\},$$

$$Fy = \{x^* \in X^* : x^*(y) = |y|^2 = |x^*|^2\},$$

where X^* denotes the dual of X. The functions ϕ_i will be such that $\phi_1 + \phi_2$ is uniqueness function, i.e. such that the

ISBN 0-12-434180-2

IVP $\rho' = \phi_1(t,\rho) + \phi_2(t,\rho)$, $\rho(0) = 0$ admits the trivial solution only. The results will be illustrated by certain Markov processes and an integro-differential equation.

I. AUXILIARY RESULTS

We shall need an existence theorem for the IVP

$$u' = f(t,u) + g(t,u), \quad u(0) = x_0. \tag{3}$$

Proposition 1. Let X be a Banach space, $D \subset X$ closed bounded and convex, $J = [0,\omega]$. Suppose that

(i) $f: J \times D \to X$ satisfies (1) for $B \subset D$, $g: J \times D \to X$ satisfies (2) in D, $f + g$ is uniformly continuous and
$|f(t,x) + g(t,x)| \leq c$ in $J \times D$.

(ii) $\phi_1: J \times \mathbb{R}^+ \to \mathbb{R}^+$ and $\phi_2: J \times \mathbb{R}^+ \to \mathbb{R}$ are continuous and $\phi_1 + \phi_2$ is a uniqueness function.

(iii) $\liminf_{\lambda \to 0+} \lambda^{-1}\rho(x+\lambda f_0(t,x),D) = 0$ for $t \in J$ and $x \in \partial D$, where $f_0 = f + g$ and $\rho(x,D)$ denotes the distance from x to D.

Then (3) has a solution on J, for every $x_0 \in D$.

This result is Theorem 2 of [4]. In this paper D will be a subset of a *cone* $K \subset X$, i.e. of a closed convex subset K such that $\lambda K \subset K$ for every $\lambda \geq 0$ and $K \cap (-K) = \{0\}$. In case D is closed convex, the boundary condition (iii) is equivalent to

$$t \in J, \quad x \in \partial D, \quad x^* \in X^* \text{ and}$$

$$x^*(x) = \sup_D x^*(y) \Rightarrow x^*(f_0(t,x)) \leq 0. \tag{4}$$

For $D = K$ condition (4) becomes

$$t \in J, \quad x \in \partial K, \quad x^* \in K^* \text{ and}$$

$$x^*(x) = 0 \Rightarrow x^*(f_0(t,x)) \geq 0, \tag{5}$$

where $K^* = \{x^* \in X^* : x^*(x) \geq 0 \text{ for every } x \in K\}$; see e.g. Lemma 4.1 and Example 4.1 in [2].

Let us recall that $f_0: J \times D \to X$ is said to be *quasimonotone* with respect to the cone K if

$$t \in J, \quad y - x \in K, \quad x^* \in K^* \quad \text{and}$$

$$x^*(y-x) = 0 \Rightarrow x^*(f_0(t,x) - f_0(t,y)) \leq 0 \tag{6}$$

holds; see e.g. § 5.3 of [2]. In case $D = K$ we notice that (5) implies $f_0(t,0) \in K$ on J, since $x^*(f_0(t,0)) \geq 0$ for all $x^* \in K^*$. If, on the other hand, f_0 is quasimonotone and $f_0(t,0) \in K$ on J then f_0 satisfies (5).

For our examples we shall need the following representations of $(\cdot,\cdot)_\mp$ in case $X = l^1$ or $X = C(G) = \{x : G \to \mathbb{R} \text{ continuous}\}$ with the max-norm, where $G \subset \mathbb{R}^n$ is compact.

<u>*Proposition 2.*</u> (i) Let $X = l^1$. Then

$$(x,y)_\pm = |y| \left[\pm \sum_{i \in A} |x_i| + \sum_{i \notin A} x_i \operatorname{sgn} y_i \right],$$

where $A = \{i : y_i \neq 0\}$.

(ii) Let $X = C(G)$ with $G \subset R^n$ compact. Then

$$Fx = \overline{\operatorname{conv}}^{w^*}\{\delta_\xi \operatorname{sgn} x(\xi) : \xi \in M_x\}|x|$$

with $M_x = \{\eta \in G : |x(\eta)| = |x|\}$, and

$$(x,y)_\pm = \max_{\min}\{x(\xi)\operatorname{sgn} y(\xi) : \xi \in M_y\}|y|,$$

where δ_ξ denotes the Dirac measure defined by $\delta_\xi(x) = x(\xi)$.

<u>*Proof.*</u> Part (i) has been shown in Example 3.1 of [2]. To understand (ii), we note that Fx is convex and closed in the w^*-topology; see e.g. Lemma 3.1 in [2]. Therefore Fx is w^*-compact and the Theorem of Krein/Milman implies that Fx is the w^*-closed convex hull of its extreme points. Since the extreme points of the unit ball in X^* are signed Dirac measures, see e.g. [6], it is easy to see that the extreme points of Fy are given by $\delta_\xi \operatorname{sgn} y(\xi)$ with $\xi \in M_y$. Now, the formulas for $(\cdot,\cdot)_\pm$ follow immediately.

q.e.d.

II. PERIODIC SOLUTIONS VIA THE POINCARÉ OPERATOR

The following result is Theorem 4 in [3].

Theorem 1. Let X be a Banach space; $D \subset X$ closed bounded and convex; $f: \mathbb{R}^+ \times D \to X$ and $g: \mathbb{R}^+ \times D \to X$ be ω-periodic and satisfy (i)-(iii) of Proposition 1. Suppose also that

(iv) For every $x \in D$, the IVP $u' = f_0(t,u)$, $u(0) = x$ has at most one solution on $[0,\omega]$.

(v) The maximal solution $\rho^*(\cdot,\lambda)$ of the IVP $\rho' = \phi_1(t,\rho) + \phi_2(t,\rho)$, $\rho(0) = \lambda$ exists on $[0,\omega]$ and $\rho^*(\omega,\lambda) < \lambda$ for every $\lambda \in (0,\alpha(D)]$.

Then $u' = f(t,u) + g(t,u)$ has an ω-periodic solution.

By Proposition 1 and (iv), the Poincaré operator U_ω: $x \to u(\omega,x)$ for $u' = f_0(t,u)$ maps D into D. It is easy to see that U_ω is continuous and it can be shown that $\alpha(U_\omega B) < \alpha(B)$ for $B \subset D$ such that $\alpha(B) > 0$. Therefore, U_ω has a fixed point and $u' = f_0(t,u)$ has an ω-periodic solution. It is, however, hard to see whether Theorem 1 remains true without the hypothesis (iv). Let us consider an example to Theorem 1, where D is the intersection of the unit sphere and the standard cone of l^1.

Example 1. Consider the denumerable system of ordinary differential equations

$$x_i' = -a_{ii}(t)x_i + \sum_{j \neq i} a_{ij}(t)x_j \quad \text{for } i \geq 1. \tag{7}$$

The model is as follows. Given a system S which at every time $t \geq 0$ is in one of the countable states $i = 1, 2, \ldots$, let $x_i(t) = \text{prob}\{S(t) = i\}$, the probability of the event "S is in i at time t". Assume that

$$\text{prob}\big(S(t+h) = i \,|\, S(t) = j\big) = \begin{cases} a_{ij}(t)h + o(h) & \text{for } i \neq j \\ 1 - a_{jj}(t)h + o(h) & \text{for } i = j \end{cases} \quad \text{as } h \to 0.$$

Then, the total probability formula yields (7), see e.g. Example 3 in the introduction to [2], where the $a_{ij}(t)$ are nonnegative and $\sum_{i\neq j} a_{ij}(t) = a_{jj}(t)$. We shall also assume that the a_{ij} are ω-periodic and we look for ω-periodic solutions x of (7) satisfying $\sum_{i\geq 1} x_i(t) \equiv 1$.

Let us note that the right hand side of (7) defines a bounded linear operator $A(t)$ from l^1 into l^1 iff $\sup_i a_{ii}(t) < \infty$. By Theorem 1 we have

Corollary 1. Let $\sup_i a_{ii}(t) \leq c$ in $[0,\omega]$, a_{ij} be ω-periodic and such that $A: [0,\omega] \to L(l^1)$ is continuous, where $A(t)$ is defined by the right hand side of (7). Let $\phi(t) = \inf_i a_{ii}(t)$,

$$\psi(t) = \lim_{m\to\infty} \sup_j \sum_{\substack{i>m \\ i\neq j}} a_{ij}(t) \quad \text{and} \quad \int_0^\omega [\psi(t) - \phi(t)]dt < 0.$$

Then (7) has an ω-periodic solution satisfying $\sum_{i\geq 1} x_i(t) \equiv 1$.

Proof. Let $X = l^1$, $K = \{x \in l^1 : x_i \geq 0$ for every $i \geq 1\}$ and $D = \{x \in K : \sum_{i\geq 1} x_i = 1\}$. The function $f_0(t,x) = A(t)x$ is uniformly continuous on $\mathbb{R}^+ \times X$, satisfies a Lipschitz condition and the boundary condition (5) for K; see Example (i) in §4.4 of [2]. Therefore, $x \in K$ implies that $u' = f_0(t,u)$ has a unique solution $u(t;s,x) \in K$ on $[s,\infty)$ such that $u(s;s,x) = x$. This implies $u(t;s,x) \in D$ if $x \in D$, since

$$\frac{d}{dt}|u(t;s,x)| = \sum_{i\geq 1} u_i'(t) = -\sum_{j\geq 1} a_{jj}(t)u_j + \sum_{j\geq 1}\Big(\sum_{i\neq j} a_{ij}\Big)u_j = 0.$$

Therefore, f_0 satisfies the boundary condition for D, since $\rho(x+\lambda f_0(s,x),D) \leq |u(s+\lambda;s,x) - x - \lambda f_0(s,x)| = o(\lambda)$ as $\lambda \to 0+$. We define $g: \mathbb{R}^+ \times l^1 \to l^1$ by $g_i(t,x) = -a_{ii}(t)x_i$ and $f: \mathbb{R}^+ \times l^1 \to l^1$ by $f_i(t,x) = \sum_{j\neq i} a_{ij}x_j$ for $i \geq 1$. By Proposition 2(i) we have

$$(g(t,x) - g(t,y),x-y)_- = -|x-y|\sum_{i\geq 1} a_{ii}(t)|x_i-y_i| \leq -\phi(t)|x-y|^2.$$

Let $R_n: l^1 \to l^1$ be defined by $R_n x = (0,\ldots,0,x_{n+1},x_{n+2},\ldots)$. Then $\alpha(B) = \alpha(R_n B)$ for every bounded $B \subset l^1$ and every $n \geq 1$, in particular $\alpha(R_n f(t,B)) = \alpha(f(t,B))$. Now, we have

$$|R_n f(t,x) - R_n f(t,y)| \le \sum_{i\ge n+1} \sum_{j\ne i} a_{ij}(t)|x_j - y_j|$$
$$\le \sup_j \sum_{n+1\le i\ne j} a_{ij}(t)|x-y|.$$

Therefore, $\alpha(f(t,B)) \le \psi(t)\alpha(B)$. Hence, we have $\phi_1(t,\rho) = \psi(t)\rho$, $\phi_2(t,\rho) = -\phi(t)\rho$ and the solution ρ^* of $\rho' = (\phi_1+\phi_2)(t,\rho)$, $\rho(0) = \lambda$ satisfies $\rho^*(\omega,\lambda) = \lambda\exp(\int_0^\omega(\psi(t) - \phi(t))dt) < \lambda$ for $\lambda > 0$.

q.e.d.

It would be interesting to prove a similar result in case $\sup_i a_{ii}(t) = \infty$ for some t; concerning existence of solutions to the initial value problem satisfying $\sum_{i\ge 1} x_i(t) \equiv 1$ in the unbounded constant case we refer to §§ 7,8 of [2] and the references given there.

III. EXISTENCE WITHOUT UNIQUENESS

We shall start with an existence theorem without the uniqueness condition (iv) of Theorem 1, provided that the interior $\mathring{D} \ne \emptyset$. In this case the Poincaré "operator" $U_\omega\colon D \to 2^D$ is multivalued and we shall apply the following fixed point theorem for such maps; see Théorème 7 in [5].

Proposition 3. Let $C \subset X$ be compact convex, $T\colon C \to 2^C$ pseudo-acyclic. Then T has a fixed point, i.e. there exists an $x \in C$ such that $x \in Tx$.

A multivalued map $T\colon C \to 2^C$ is pseudo-acyclic if there exist a metric space Y, an upper semi-continuous multivalued map $T_0\colon C \to 2^Y$ such that T_0x is compact and acyclic (w.r. to the cohomology of Čech) for every $x \in C$, and a continuous map $R\colon Y \to C$ such that $T = R\circ T_0$; see Définition 5 in [5]. For the compact set T_0x to be acyclic it is sufficient to show that T_0x is the limit in the Hausdorff-metric of sets $A_n \supset T_0x$ such that A_n is homeomorphic to a compact convex set B_n.

Theorem 2. Theorem 1 remains true if condition (iv) is replaced by condition "$\mathring{D} \neq \emptyset$".

Proof. We may assume $0 \in \mathring{D}$, without loss of generality. We fix $\varepsilon > 0$ and let $f_\varepsilon = f_0 - \varepsilon I$. Then f_ε satisfies the boundary condition

$$t \in J, \quad x \in \partial D, \quad |x^*| = 1 \text{ and}$$

$$x^*(x) = \sup_D x^*(y) \Rightarrow x^*(f_\varepsilon(t,x)) \leq -\varepsilon\delta \tag{8}$$

where $\delta > 0$ is such that $K_\delta(0) \subset D$. For $x \in D$, let Sx be the set of all solutions on $J = [0,\omega]$ of $u' = f_\varepsilon(t,u)$, $u(0) = x$ and $U_\omega x = \{u(\omega) : u \in Sx\}$. We know that Sx is a compact subset of $C_D(J)$ and that $\alpha(U_\omega B) < \alpha(B)$ if $\alpha(B) > 0$. Let $D = \overline{\text{conv}}\, U_\omega D$ and $D_n = \overline{\text{conv}}(U_\omega D_{n-1})$ for $n \geq 2$. Then $C = \bigcap_{n \geq 1} D_n$ is compact and convex and $U_\omega C \subset C$. The multivalued map $S\colon C \to C_D(J)$, defined by $x \to Sx$ is easily seen to be upper semi-continuous, and $R\colon u \to u(\omega)$ is continuous. Therefore, $U_\omega = R \circ S\colon C \to 2^C$ has a fixed point if Sx is acyclic, by Proposition 3. To prove that Sx is acyclic we extend f_0 continuously to $\mathbb{R}^+ \times X$, choose $\eta_n > 0$ such that $2\eta_n \leq \varepsilon\delta$ and $g_n\colon \mathbb{R}^+ \times X \to X$ locally Lipschitz and such that $|g_n(t,x) - f_0(t,x)| \leq \eta_n$ in $\mathbb{R}^+ \times X$; see Lemma 1.1 in [2]. Let $\tilde{f}_n = g_n - \varepsilon I$ and $y \in C_X(J)$ such that $|y| \leq \eta_n$. Then the IVP

$$v' = \tilde{f}_n(t,v) + y(t), \quad v(0) = x \tag{9}$$

has a unique solution on $\mathbb{R}^+$ with range in D, since $\tilde{f}_n + y$ satisfies (8) with $-\varepsilon\delta$ replaced by 0. Now, we follow the proof to Théorème C in [1]. Let $B_n = \overline{\text{conv}}\{\tilde{f}_n(\cdot,u) - f_\varepsilon(\cdot,u) : u \in Sx\}$ and let Hy be the solution on J of (9), for $y \in B_n$. Since B_n is compact, H is a homeomorphism from B_n onto $A_n = H(B_n)$. Evidently, we also have $Sx \subset A_n$ and $d(A_n, Sx) = \sup\{\rho(v,Sx) : v \in A_n\} \to 0$ as $n \to \infty$. Therefore, Sx is acyclic.

Thus, we have found an ω-periodic solution of $u' = f_0(t,u) - \varepsilon u$, and we may let $\varepsilon \to 0$ to find a solution of $u' = f_0(t,u)$.

q.e.d.

Example 2. Consider the integro-differential equation

$$\frac{\partial u(t,\xi)}{\partial t} = h(t,\xi,u) + \int_G k(t,\xi,\zeta,u(t,\zeta))d\zeta \tag{10}$$

for $t \geq 0$, $x \in G$,

where $G \subset \mathbb{R}^n$ is compact and all functions are real-valued. By means of Theorem 2 we have

Corollary 2. Let $h\colon \mathbb{R}^+ \times G \times \mathbb{R}^+ \to \mathbb{R}$ and $k\colon \mathbb{R}^+ \times G \times G \times \mathbb{R}^+ \to \mathbb{R}^+$ be continuous and ω-periodic in t. Suppose also that

(i) $(h(t,\xi,x) - h(t,\xi,y))\operatorname{sgn}(x-y) \leq \phi_2(t,|x-y|)$, where ϕ_2 is such that $\phi_1 + \phi_2$, with $\phi_1 = 0$, satisfies the hypotheses of Theorem 1; $h(t,\xi,0) \geq 0$ in $[0,\omega] \times G$.

(ii) There exists an $r > 0$ such that $h(t,\xi,r) + \max\limits_{[0,r]} \int_G k(t,\xi,\zeta,\rho)d\zeta \leq 0$ in $[0,\omega] \times G$.

Then (10) has an ω-periodic solution u such that $0 \leq u(t,\xi) \leq r$ on $\mathbb{R}^+ \times G$.

Proof. We regard (10) as the differential equation $u' = g(t,u) + f(t,u)$ in $X = C(G)$ with the max-norm, where $g(t,x)(\xi) = h(t,\xi,x(\xi))$ and $f(t,x)(\xi) = \int_G k(t,\xi,\zeta,x(\zeta))d\zeta$.

Evidently f and g are uniformly continuous on bounded sets and they satisfy the relevant estimates with $\phi_1(t,\rho) \equiv 0$, by Proposition 2 (ii). The standard cone $K = \{x \in C(G) : x(\xi) \geq 0 \text{ in } G\}$ has nonempty interior. $f + g$ satisfies the boundary condition for K since $x \in \partial K$ and $\mu(x) = \int_G x(\xi)d\mu = 0$ for a positive Radon measure (i.e. $\mu \in K^*$) imply

$$\int_G \{h(t,\xi,x(\xi)) + \int_G k(t,\xi,\zeta,x(\zeta))d\zeta\}d\mu(\xi) \geq \int_G h(t,\xi,0)d\mu(\xi) \geq 0,$$

and this, together with (ii), implies that $f + g$ satisfies the boundary condition for $D = \{x \in K : |x| \leq r\}$.

q.e.d.

The proof to Theorem 2 shows that we also have

Theorem 3. Theorem 1 remains true if condition (iv) is replaced by the conditions "X^* is uniformly convex" and

"The metric projection $P\colon X \to D$ exists and is continuous."

Proof. We may assume $0 \in D$. Consider $\delta_n > 0$ such that $\delta_n \to 0$ as $n \to \infty$ and $D_n = \{x\colon \rho(x,D) \leq \delta_n\}$. Then $f_0(t,Px) - \varepsilon x$ satisfies the boundary condition

$$t \in J,\quad x \in \partial D_n,\quad |x^*| = 1 \quad \text{and}$$

$$x^*(x) = \sup_{D_n} x^*(y) \Rightarrow x^*(f_0(t,Px) - \varepsilon x) \leq -\varepsilon\delta_n.$$

Now, we follow the proof to Theorem 2, choose $\eta_n > 0$ such that $2\eta_n \leq \varepsilon\delta_n$, g_n locally Lipschitz such that $|g_n(t,x) - f_0(t,Px)| \leq \eta_n$, and so on.

q.e.d.

The extra condition in Theorem 3 is evidently satisfied if X and X^* are uniformly convex. For general X we do not see how we could prove that Sx is acyclic and therefore we need stronger conditions on f and g in the following theorems.

Theorem 4. Let X be a real Banach space, $K \subset X$ a cone, $f\colon \mathbb{R}^+ \times K \to K$ and $g\colon \mathbb{R}^+ \times K \to X$ uniformly continuous on bounded sets and ω-periodic. Suppose also that

(i) f satisfies (1) with $\phi_1(t,\rho) = L_1(t)\rho$, g satisfies (2) with $\phi_2(t,\rho) = L_2(t)\rho$, where L_1 and L_2 are continuous in $[0,\omega]$, $L_1(t) + L_2(t) \leq 0$ in $[0,\omega]$ and $\int_0^\omega (L_1(t) + L_2(t))dt < 0$.

(ii) g satisfies the boundary condition (5) for K, and g maps bounded sets into bounded sets.

Then $u' = f(t,u) + g(t,u)$ has an ω-periodic solution provided that one of the following conditions holds.

(iii) There exists an $r > 0$ such that $(g(t,x) + f(t,y),x)_- \leq 0$ for $t \in [0,\omega]$, $|x| = r$ and $|y| \leq r$.

(iv) $(f(t,x),x)_+ \leq L_3(t)|x|^2 + L_4|x|$, where L_3 is continuous and $\int_0^\omega (L_2(t) + L_3(t))dt < 0$.

Proof. Let $v\colon \mathbb{R}^+ \to K$ be continuous and ω-periodic and consider the IVP

$$u' = g(t,u) - \varepsilon u + f(t,v),\ u(0) = x \in K\ (\varepsilon > 0 \text{ fixed}). \tag{11}$$

Since $f\colon \mathbb{R}^+ \times K \to K$ and g satisfies (ii), the right hand side of (11) satisfies the boundary condition for K. Therefore, (11) has a unique solution $U(t)x$ on $\mathbb{R}^+$. By (i), $U_\omega\colon K \to K$ is a strict contraction and therefore $u' = g(t,u) - \varepsilon u + f(t,v)$ has a unique ω-periodic solution Tv.

Suppose first that (iii) holds, let $D = \{x \in K : |x| \leq r\}$ and v as above but with range in D. If we consider (11) with $x \in D$, we see that this IVP has a unique solution in D. Therefore, Tv has its range in D.

Since f and g are bounded on $\mathbb{R}^+ \times D$, there exists $c > 0$ such that $|(Tv)(t) - (Tv)(s)| \leq c|t-s|$ for all v as above. Let $\Omega = \{v\colon \mathbb{R}^+ \to D$ continuous, ω-periodic, $|v(t)-v(s)| \leq c|t-s|$ for all $t,s \in [0,\omega]\}$. Then $T\colon \Omega \to \Omega$ is continuous, and since f and g are uniformly continuous on bounded sets, we have, with the notation $\psi(t) = \alpha\{(Tv)(t) : v \in B\}$ and $\rho(t) = \alpha\{v(t) : v \in B\}$ for $B \subset \Omega$

$$\psi'(t) \leq (L_2(t) - \varepsilon)\psi(t) + L_1(t)\rho(t) \text{ a.e. and } \psi(0) = \psi(\omega),$$

and therefore, with $M(t) = \exp(\int_0^t L_2(\tau)d\tau - \varepsilon t)$

$$\psi(t) \leq M(t)(1-M(\omega))^{-1}\int_0^\omega M(\omega)(M(s))^{-1}L_1(s)\rho(s)ds$$
$$+ \int_0^t M(t)(M(s))^{-1}L_1(s)\rho(s)ds. \tag{12}$$

Let $\mu(t)$ denote the right hand side of (12). Then

$$\mu' = (L_2(t) - \varepsilon)\mu + L_1(t)\rho(t) \text{ and } \mu(0) = \mu(\omega).$$

Suppose that μ is maximal for $t = t_0$. Then

$$(L_2(t_0) - \varepsilon)\mu(t_0) + L_1(t_0)\rho(t_0) = 0.$$

Since $L_1(t) + L_2(t) - \varepsilon < 0$ in $[0,\omega]$, this implies $\mu(t_0) < \max_{[0,\omega]} \rho(t)$ if this max is > 0. Therefore, $\alpha_0(TB) < \alpha_0(B)$ if $\alpha_0(B) > 0$, where α_0 denotes the Kuratowski measure of noncompactness for $C_X([0,\omega])$. Hence, T has a fixed point, i.e. $u' = g(t,u) - \varepsilon u + f(t,u)$ has an ω-periodic solution in D. Now, we may let $\varepsilon \to 0$ to get an ω-periodic solution of $u' = g(t,u) + f(t,u)$.

Suppose that (iv) holds and that u is an ω-periodic solution of $u' = g(t,u) + f(t,u)$. Then we have

$$\begin{aligned}(|u(t)|^2)' &\leq 2L_2(t)|u(t)|^2 + 2|g(t,0)||u(t)| \\ &\quad + 2L_3(t)|u(t)|^2 + 2L_4|u(t)| \quad \text{a.e..}\end{aligned} \tag{13}$$

Let $\phi(t) = |u(t)|^2$ and $\varepsilon > 0$. Then (13) implies

$$\begin{aligned}\phi(t) &\leq \phi(0)\exp(2\int_0^t\{L_2(s)+L_3(s)+\varepsilon^2\}ds) \\ &\quad + \frac{1}{\varepsilon^2}\int_0^t \exp(2\int_s^t\{L_2(\tau)+L_3(\tau)+\varepsilon^2\}d\tau)\left[|g(s,0)|^2+L_4{}^2\right]ds.\end{aligned} \tag{14}$$

Since $\phi(0) = \phi(\omega)$ and $\int_0^\omega (L_2(s) + L_3(s))ds < 0$, this estimate implies that there exists an $r > 0$, independent of u, such that $|u(t)| \leq r$ in $\mathbb{R}^+$, i.e. we have an a-priori bound for all ω-periodic solutions. Let $D_r = \{x \in K : |x| \leq r\}$ and $P: K \to D_r$ be defined by $Px = x$ if $|x| \leq r$ and $Px = \frac{r}{|x|}x$ if $|x| > r$. The retraction P satisfies $|Px-Py| \leq 2|x-y|$ and $\alpha(PB) \leq \alpha(B)$ for every bounded B. As before, $u' = g(t,u) + f(t,Pv(t))$ has a unique ω-periodic solution Tv if $v: \mathbb{R}^+ \to K$ is continuous and ω-periodic. For $\phi(t) = |Tv(t)|^2$ we obtain with $c_1 = \sup\{|f(t,x)|: \ t \in [0,\omega], \ x \in D_r\}$

$$\phi'(t) \leq 2L_2(t)\phi(t) + 2\varepsilon^2\phi(t) + \frac{1}{\varepsilon^2}(|g(t,0)|^2 + c_1{}^2)$$

a.e. and $\phi(0) = \phi(\omega)$.

Hence, there exists an $R \geq r$ such that $|Tv(t)| \leq R$ on $[0,\omega]$ for every v as above. Now we may proceed as in the first part, with D in the definition of Ω replaced by D_R, to find an ω-periodic solution u of $u' = g(t,u) + f(t,Pu)$ in D_R. Suppose that $|u(t)| > r$ for some $t \in [0,\omega]$. Since the duality map F is positively homogeneous, we then have

$$\begin{aligned}(f(t,Pu(t)),u(t))_+ &= r^{-1}|u(t)|(f(t,Pu(t)),Pu(t))_+ \\ &\leq r^{-1}|u(t)|(L_3(t)r^2 + L_4 r) \\ &\leq L_3(t)|u(t)|^2 + L_4|u(t)|.\end{aligned}$$

Evidently this estimate is also true if $|u(t)| \leq r$. Therefore, we have (14) and this implies $|u(t)| \leq r$ i.e. $Pu = u$, and therefore u is an ω-periodic solution of $u' = g(t,u) + f(t,u)$.

q.e.d.

IV GALERKIN APPROXIMATIONS

Let X be a real Banach space with a projectional scheme $\{X_n,P_n\}$, where the X_n are finite dimensional subspaces, $P_n\colon X \to X_n$ is a linear projection with $|P_n| = 1$ for every $n \geq 1$ and such that $P_n x \to x$ as $n \to \infty$, for every $x \in X$. Let $K \subset X$ be a cone and suppose that $P_n(K) \subset K$ for every $n \geq 1$. It is immediately verified that many of the common separable spaces and their standard cones of nonnegative elements meet these hypotheses.

Consider the differential equation $u' = f(t,u) + g(t,u)$, where $f\colon \mathbb{R}^+ \times K \to X$ and $g\colon \mathbb{R}^+ \times K \to X$ are ω-periodic and such that the boundary condition (5) is satisfied for $f_0 = f + g$. We look for ω-periodic solutions via ω-periodic solutions of the finite systems

$$v' = P_n f_0(t,v) \quad \text{for} \quad t \geq 0, \quad v \in K_n = K \cap X_n. \tag{15}$$

Since $P_n K \subset K$ we have $P_n K = K_n$ and

$$\rho(x+\lambda P_n f_0(t,x),K_n) \leq \rho(x+\lambda f_0(t,x),K) \quad \text{for} \quad x \in \partial K_n.$$

Therefore $P_n f_0$ satisfies the boundary condition for K_n. Let f_0 be continuous and suppose that there exists an $r > 0$ such that $(f_0(t,x),x)_+ \leq 0$ on $\{x \in K : |x| = r\}$. Since $|P_n| = 1$, we have $(P_n x,y)_+ \leq (x,y)_+$ for $x \in X$ and $y \in X_n$; see e.g. Proposition 7.1 in [2]. Therefore, $(P_n f_0(t,x),x)_+ \leq 0$ for $x \in K_n$ such that $|x| = r$. This condition together with the boundary condition for K implies that $v' = P_n f_0(t,v), v(0) = x \in D_n = \{x \in K_n: |x| \leq r\}$ has a solution on $\mathbb{R}^+$ with range in D_n. Therefore $P_n f_0$ satisfies the boundary condition (4) for D_n. Since D_n is compact convex, Theorem 1 in [3] implies that $v' = P_n f_0(t,v)$ has an ω-periodic solution v_n with range in D_n.

Now, we have to find properties of f_0 sufficient for uniform convergence of a subsequence of (v_n) to a solution of $u' = f_0(t,u)$. We are not able to find such a subsequence under the hypotheses (1), (2) and (v) of Theorem 1, but we have

Theorem 5. Let X be a real Banach space and $K \subset X$ a cone such that X has a projectional scheme $\{X_n, P_n\}$ satisfying $|P_n| = 1$ and $P_n K \subset K$ for every $n \geq 1$. Suppose also that

(i) $f: \mathbb{R}^+ \times K \to X$ is uniformly continuous on bounded sets, ω-periodic in t, and f maps bounded sets into relatively compact sets.

(ii) $g: \mathbb{R}^+ \times X \to X$ is continuous, ω-periodic in t and such that $(g(t,x)-g(t,y),x-y)_+ \leq L(t)|x-y|^2$ for $t \geq 0$ and $x,y \in X$, where $L: [0,\omega] \to \mathbb{R}$ is continuous and such that $\int_0^\omega L(s)ds < 0$, and g maps bounded sets into bounded sets.

(iii) $f_0 = f + g$ satisfies the boundary condition (5) for K and $(f_0(t,x),x)_+ \leq 0$ on $\{x \in K: |x| = r\}$ for some $r > 0$.

Then $u' = f_0(t,u)$ has an ω-periodic solution in $\{x \in K : |x| \leq r\}$.

Proof. We have ω-periodic solutions v_n in D of

$$v_n' = P_n f(t,v_n) + P_n g(t,v_n).$$

By (i) we may assume, without loss of generality, that $P_n f(t,v_n(t)) \to w(t)$ as $n \to \infty$, uniformly on $\mathbb{R}^+$, for some continuous ω-periodic w. Let v be the unique ω-periodic solution of $v' = g(t,v) + w(t)$, $z_n = v_n - P_n v$ and $\rho_n(t) = |z_n(t)|$. Then we have

$$\begin{aligned} D^- \rho_n(t) &\leq L(t)\rho_n(t) + \{|g(t,P_n v) - g(t,v)| + |P_n f(t,v_n) \\ &\qquad - w(t)| + |w(t) - P_n w(t)|\} \\ &= L(t)\rho_n(t) + \delta_n(t) \quad \text{with} \quad \delta_n(t) = \{\ \} \to 0, \\ &\qquad \text{uniformly on } R^+, \end{aligned}$$

and $\rho_n(0) = \rho_n(\omega)$ for every $n \geq 1$. Therefore,

$$\rho_n(t) \leq e^{H(t)}\rho_n(0) + \int_0^t e^{H(t)-H(s)}\delta_n(s)ds, \quad H(t) = \int_0^t L(\tau)d\tau.$$

These inequalities imply $\overline{\lim_{n\to\infty}}\, \rho_n(\omega) = \overline{\lim_{n\to\infty}}\, \rho_n(0) = 0$ and therefore $\rho_n(t) \to 0$ uniformly in $[0,\omega]$. Since $P_n v(t) \to v(t)$ and $v_n(t) \in D$, we have $v(t) \in D$ in $[0,\omega]$ and $v' = f_0(t,v)$.

q.e.d.

Theorem 6. Let X, K, $\{X_n,P_n\}$ be as in Theorem 5, $f\colon \mathbb{R}^+ \times K \to X$ and $g\colon \mathbb{R}^+ \times K \to X$ be uniformly continuous on bounded sets and ω-periodic. Suppose also that

(i) $f_0 = f + g$ satisfies the boundary conditions in (iii) of Theorem 5 and f is bounded on $[0,\omega] \times D$, where $D = \{x \in K : |x| \leq r\}$.

(ii) f satisfies (1) for $B \subset D$, g satisfies (2) for $x,y \in D$, where ϕ_1 and ϕ_2 satisfy conditions (ii) of Proposition 1 and (v) of Theorem 1.

(iii) $\alpha\{R_n f_0(t,x_n) : n \geq 1\} = 0$ in $[0,\omega]$, for $R_n = P_n - I$ and every $(x_n) \subset D$ such that $x_n \in D \cap X_n$ for every $n \geq 1$.

Then $u' = f_0(t,u)$ has an ω-periodic solution in D.

Proof. We have $v_n' = P_n f_0(t,v_n) = f(t,v_n) + R_n f_0(t,v_n)$, $v_n(0) = v_n(\omega)$ and $v_n(t) \in D \cap X_n$. Condition (ii) and (iii) imply that $\alpha\{v_n(t) : n > 1\} \leq \rho^*(t,\alpha\{v_n(0) : n \geq 1\})$ with ρ^* from (v) in Theorem 1. Therefore $\alpha\{v_n(0) : n \geq 1\} = 0$, and this implies that $u' = f_0(t,u)$ has an ω-periodic solution in D.

q.e.d.

Let us note that Theorem 6 is interesting for certain countable systems

$$u_i' = \tilde{f}_i(t,u) + \tilde{g}_i(t,u), \quad i \geq 1. \tag{16}$$

Suppose that X is a sequence space, P_n and X_n are defined by means of a base for X and K is the standard cone $\{x \in X: x_i \geq 0 \text{ for every } i\}$. Suppose also that

$$\tilde{f}_i(t,u) = \tilde{f}_i(t,u_1,\ldots,u_{i-1}), \quad \tilde{g}_i(t,u) = \tilde{g}_i(t,u_i,u_{i+1},\ldots) .$$

Then $R_n(\tilde{f}(t,x^n) + \tilde{g}(t,x^n)) = R_n\tilde{f}(t,x^n) + R_n\tilde{g}(t,0)$, and therefore (iii) is satisfied if $\tilde{f}$ is compact, in particular if (16) is upper diagonal (i.e. $\tilde{f} = 0$). Applications to fixed point theorems for operators on cones will be considered elsewhere.

REFERENCES

[1] Aronszajn, N. (1942). "Le correspondant topologiques de l'unicité dans la théorie des equations différentielles", *Ann. of Math. 43*, 730-738.

[2] Deimling, K. (1977). "Ordinary Differential Equations in Banach Spaces", Notes in *Math. Vol. 596*, Springer-Verlag.

[3] Deimling, K. (1978). "Periodic solutions of differential equations in Banach spaces", *Manuscripta Math. 24*, 31-44.

[4] Deimling, K. (1978). "Open problems for ordinary differential equations in Banach spaces", *Proc. Equa. Diff. 78 Florence, Centro 2P Firenze*, 127-137.

[5] Lasry, M., and Robert, R. (1976). "Degré topologique pour certains couples de fonctions et applications aux équations différentielles multivoques", *C. R. Acad. Sci. Paris 283, Serie A,* 163-166.

[6] Phelps, R. R. (1966). "Lect. on Choquet Theory", *Stud. in Math. Vol. 7,* Van Nostrand.

THE BISTABLE NONLINEAR DIFFUSION EQUATION: BASIC THEORY AND SOME APPLICATIONS

*Paul C. Fife**

Department of Mathematics
The University of Arizona
Tucson, Arizona

I. INTRODUCTION

The equation

$$u_t = u_{xx} + f(u), \tag{1}$$

in the case when f has two or more zeros, is of considerable mathematical interest. This largely stems from the fact that although the equation is parabolic, it admits bounded traveling front solutions, $u(x,t) = U(x - ct)$. These solutions "join" two zeros of f in the sense that one of the zeros is approached by $U(z)$ as $z \to -\infty$, and the other as $z \to \infty$. Our concern here is with the case when the two zeros (which we label 0 and 1, for convenience) are both stable as rest states of the simple equation $du/dt = f(u)$. More specifically, we suppose that f is continuously differentiable, $f'(0) < 0$, $f'(1) < 0$ (Figure 1), and that f has only one (it will have at least one in any case) intermediate zero (which we call α) in the interval $(0,1)$. In this case, there is a unique traveling front joining 0 on the

**Supported by N.S.F. under Grant MPS-74-06835-A01.*

ISBN 0-12-434180-2

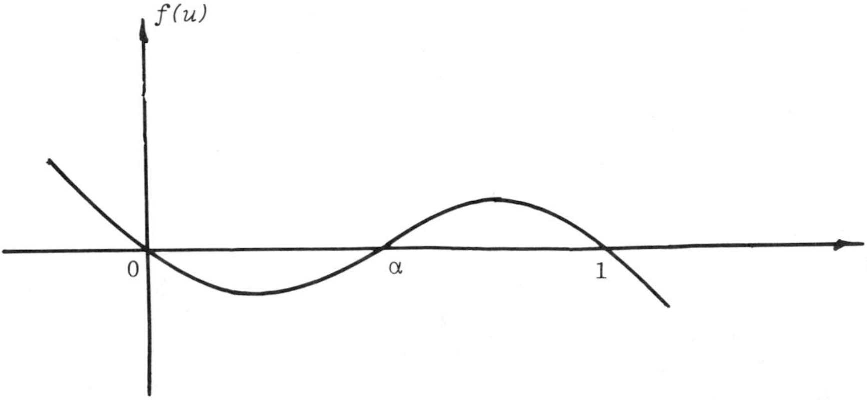

FIGURE 1.

left (say) with 1 on the right, with a unique velocity c. These facts were proved by Kanel' [16] and Aronson and Weinberger [1, 2].

A second reason for the mathematical interest in (1), at least in the bistable case described above, is that the long-time asymptotic theory of bounded solutions is both nontrivial and tractable. In fact, under reasonable restrictions on the initial data, we can categorize all possible long-time behaviors of solutions. There turn out to be (under those restrictions) only a few possible asymptotic solution "forms." The details will be given in the next section.

Besides its strictly mathematical interest, (1) arises also in interesting applications, a few of which will be considered in this paper.

On the most primitive level of model-building, (1) can be characterized as the simplest continuous space-time model of systems with spatial diffusion and two stable equilibria. Systems with these two properties are of interest in population and chemical dynamics, and in fact (1) has been used as a simplistic model, at least in the former area.

Of course, any real system one might consider modeling would apparently be too hopelessly complex to be adequately modeled by so simple an equation as (1). If such a system is chemical, for example, there will be many reacting species, never just one; and the proper model would typically be a system of reaction-diffusion equations. If the system is ecological, again many equations presumably would be needed to represent the many species and extraneous effects influencing the phenomenon studied. Complex models, however, are often formally simplified by the use of standard mathematical techniques such as asymptotics (the pseudo-steady-state hypothesis is an example). Crudely speaking, what this amounts to is discarding those quantities in the equations which appear, and which one hopes, to be small under conditions of interest. The resulting model, being removed from the original one by one or more steps of mathematical processing, may be less intuitively clear, subject to more restrictions in its range of applicability, and less accurate. But it may now be mathematically tractable. In this paper I shall illustrate this process by three examples in which (1) arises as, or at least is relevant to, a final model obtained through simplification of higher order reaction-diffusion systems.

But first, I shall outline the main points in the asymptotic theory of bounded solutions of (1).

II. ASYMPTOTIC THEORY OF THE BISTABLE NONLINEAR DIFFUSION EQUATION

The results described here were proved in [1], [2], [13], and [12]. See [12] for a statement and a discussion of them in their most complete form.

As mentioned before, there exist only a few possible eventual solution forms, when reasonable restrictions are placed on the initial data $\phi(x) \equiv u(x,0)$. To describe them, some notation is needed. It is known ([2], [16]) that there exists a wave front solution

$$u(x,t) = U(x - ct)$$

with $U(-\infty) = 1$, $U(\infty) = 0$. It has its unique velocity c, and the wave profile $U(z)$ is unique except for translation of the independent variable z. Its reflection, $U(-x - ct)$, is a wave front with opposite boundary conditions; it travels in the opposite direction. The velocity c has the following sign:

$$c \gtreqless 0 \quad \text{according as} \quad \int_0^1 f(u)du \gtreqless 0. \tag{2}$$

If $c \neq 0$, solutions representing a diverging pair of wave fronts exist. This phenomenon is conveniently represented by the function

$$V(x,t) = \begin{cases} U(x - ct), & x > 0 \\ U(-x - ct), & x < 0, \end{cases}$$

when $c > 0$, with an analogous definition when $c < 0$.

Finally, again when $c \neq 0$, there exists a special nonconstant stationary solution $W(x)$ of (1) which is even in x and approaches a limit as $x \to \pm\infty$. If $c > 0$, $W(\pm\infty) = 0$ and if $c < 0$, $W(\pm\infty) = 1$.

Theorem. Let $u(x,t)$ be a bounded solution of (1) for all x and all $t \geq 0$. Let $\alpha \in (0,1)$ be the intermediate zero of f. Assume $\phi(x) \equiv u(x,0)$ satisfies

$$\liminf_{|x|\to\infty} |\phi(x) - \alpha| > 0.$$

Then one of the following five asymptotic relations holds:

(1) $\lim_{t\to\infty} u(x,t) = 0$ uniformly in x,

(2) $\lim_{t\to\infty} u(x,t) = 1$ uniformly in x,

(3) for some z_0, $\lim_{t\to\infty} |u(x,t) - U(\pm x - ct - z_0)| = 0$ uniformly in x,

(4) for some x_0 and some t_0,
$\lim_{t\to\infty} |u(x,t) - V(x - x_0, t - t_0)| = 0$ uniformly in x,

or

(5) for some n and some functions $\xi_i(t)$, $i = 1,\dots,n$ with $\xi_i'(t) \to 0$ and $|\xi_i(t) - \xi_j(t)| \to \infty$, $i \neq j$, the following relation holds if $c > 0$, a similar one holding if $c < 0$: $\lim_{t\to\infty} |u(x,t) - \sum_{i=1}^{n} W(x-\xi_i(t))| = 0$, uniformly in x.

Various sufficient conditions on ϕ may be given which will ensure one or another of these alternatives to hold. Alternative (5) is not uniformly stable, so would be the least important in applications.

The proof of this result is involved. It is somewhat interesting, partly because it uses a Lyapunov functional whose domain does not necessarily include the solution at any time.

III. AN EXAMPLE FROM POPULATION GENETICS

This model is historically significant, because it was in this context that an equation of the form (1) (but in which f has only two, rather than three, zeros) was first introduced to the applied mathematical world. This was done in 1937 by R. A. Fisher [14].

We envisage a population of individuals differing genetically at only one gene locus. That gene can assume one of two possible forms (alleles). The gene at this locus, as well as the densities of the different genetic groups in the population, determine an organism's fitness to survive and reproduce. If spatial migration is also somehow accounted for, we have a simple selection-migration model. Such models can and have been made mathematical in various ways. In particular, arguments have been given to reduce the model to the single equation (1). For a careful derivation of this equation via a "stepping stone" model, in which the population occupies discrete homogeneous colonies, see Sawyer (in preparation).

In [1], [5], continuous space-time single locus migration-selection models are written down for a diploid population (the gene locus in question occurs on two chromosomes, so that the population is split into three genotypes). The models take the form of a system of three reaction-diffusion equations. Assuming the selection mechanism is weak, in the sense that the three genotypes are approximately equally fit, a small parameter is introduced. In [5], this small parameter may affect the carrying capacities, fecundities, mating preferences, etc. After appropriate time and space rescalings, the small parameter appears in new positions in the three equations. At this point a Fisher-type equation (1) appears to be a plausible first approximation. The meaning of the dependent variable u in this equation is the frequency of one of the two alleles in the population. The function f vanishes at $u = 0$, $u = 1$, and possibly at other values of u as well; in particular, the bistable case is relevant.

By "plausible first approximation," what I mean is that given any bounded solution of the reduced equation (1) which has uniformly bounded derivatives of the orders appearing in the equation, there corresponds an associated approximate solution of the original system of three equations. This approximation is in the sense that it is a solution of a system which differs from the original system only by uniformly small terms. Proceeding further, it appears plausible that if the original solution of (1) is stable in the uniform norm (in the bistable case this means it belongs to one of the first four categories listed in the theorem of Section 2), then there would exist a corresponding stable solution of the reaction-diffusion system which is uniformly close to the former, for all time.

For example, in [5] it is shown that in the bistable case, when a stable wave front exists, there exists a corresponding wave front also for the reaction-diffusion system. Other general perturbation results of this sort are obtained as well.

IV. STATIONARY PATTERN FORMATION FOR SYSTEMS

It sometimes happens (see, for example, [24], [22], [15], [3], [23], [9], [11], [17], and many other papers) that reaction-diffusion systems support stable patterned solutions without there being any explicit pattern structure in the equation itself (clearly there won't be if the system is homogeneous). This has often been proposed as one possible mechanism to explain the diversity and patchiness of ecological communities (in which individuals "diffuse" by migration and "react" through reproductive and death processes, [18], [19]) and patterns in the physiological structure of organisms (diffusion and reaction of organic molecules). Again, we have a crude model at best.

Let us explain the mathematics involved, and then show how the scalar equation (1) is relevant to the question of which types of small-amplitude structures appear.

We consider a reaction-diffusion system in one space dimension which depends on a numerical parameter λ:

$$u_t = Du_{xx} + f(u,\lambda); \quad u \in R^n, \quad D \text{ a matrix.} \tag{3}$$

Suppose $u \equiv 0$ is a solution; we call it the homogeneous or uniform solution. Then $f(0,\lambda) \equiv 0$.

We linearize (3) about this uniform solution in order to get some idea of the nature of the small structured states. More or less, bifurcation theory tells us that this linearization procedure is justified. Suppose the linear part of $f(u,\lambda)$ is $(A + \lambda B)u$. Then there exist solutions

$$u = \phi e^{\pm ikx}$$

of the linearized equation whenever the vector ϕ is a nullvector of the matrix

$$H(k^2,\lambda) \equiv -k^2D + A + \lambda B.$$

Let $\mu(k^2,\lambda)$ be an algebraically simple eigenvalue of H for λ small and for k in some neighborhood of some number $k_0 \neq 0$. Then μ depends analytically on λ and k in such a domain. Our first basic assumption on D, A, and B is that there exists such an eigenvalue μ which is real and which is the unique eigenvalue of H with maximal real part. We further assume that all eigenvalues of H are negative for all k and all $\lambda < 0$, that $\mu(k^2,0)$ has a unique maximum of 0 at $k^2 = k_0^2$, that $\frac{\partial\mu}{\partial\lambda}(k_0^2,0) > 0$, and that $\frac{\partial^2\mu}{\partial(k^2)^2}(k_0^2,0) < 0$. (See Figure 2.)

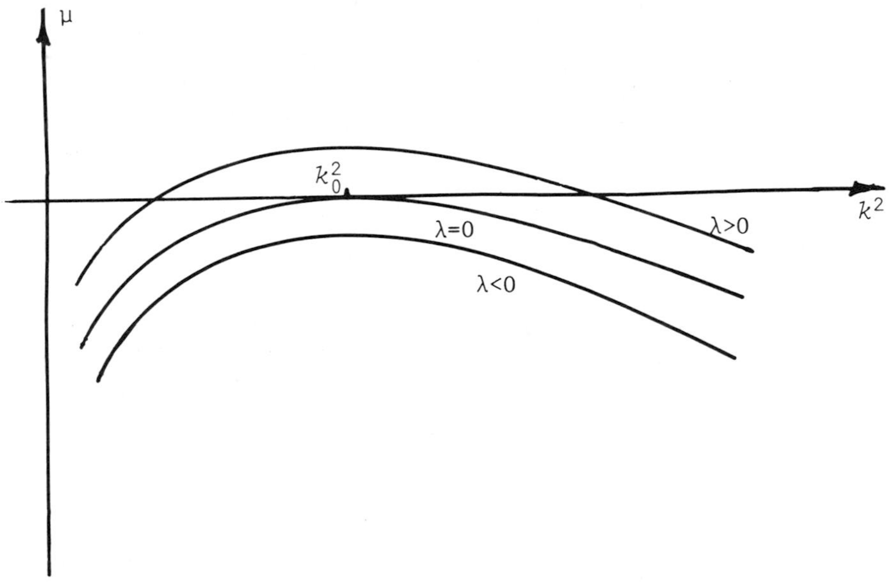

FIGURE 2. Level curves of $\mu(k^2,\lambda)$ *for fixed* λ.

Then stationary solutions, periodic in x, with wave number near k_0, exist for $\lambda > 0$ small [11]. If the solutions are characterized by some specific measure of amplitude, as well as by k, then a typical (but not universal) bifurcation picture is as follows in Figure 3. Points in the shaded region correspond to periodic solutions with wave number near k_0.

Once the existence of these small amplitude periodic solutions is established, two natural questions arise:

1. Which are stable, and in what sense?
2. Are other stable nonuniform solutions possible?

To shed some light on the second of these questions, we describe an "amplitude function" approach which has been used in connection with other nonlinear wave problems ([6], [21], etc.). The idea of using this approach for a stability analysis goes back, at least, to the investigations of Eckhaus [7]. Somewhat related analyses of other reaction-diffusion problems occur in several other papers; see [4] for example.

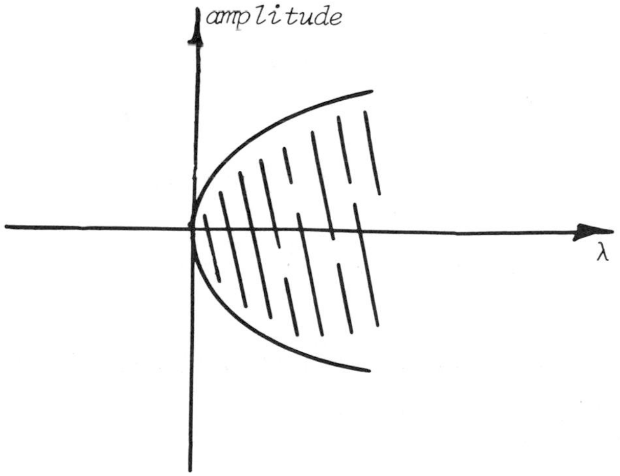

FIGURE 3.

Let us assume that the nonlinear part of f is strictly quadratic:

$$u_t = Du_{xx} + (A + \lambda B)u + g(u,u) \tag{4}$$

where g is a bilinear function from $R^n \times R^n$ to R^n.

We retain the same assumptions described above; in particular, λ is small. We set $\lambda = \varepsilon^2$ with $\varepsilon << 1$, and look for "small" formal solutions in the form of series

$$u = \varepsilon u^1 + \varepsilon^2 u^2 + \ldots . \tag{5}$$

The type of solutions we shall look for are slowly varying periodic solutions, i.e., periodic solutions of the type described already, enveloped by a function with larger characteristic space and time scales. The approach is simply a straightforward asymptotic analysis using two space and time scales. Accordingly, we introduce the new long-space variable $\xi = \varepsilon x$ as well as x. The form of the equations (4) suggests that time should also be rescaled, in the fashion $\tau = \varepsilon^2 t$. We now rewrite (4) in terms of the variables t, τ, x, and ξ:

$$L^0 u - g(u,u) + \varepsilon L^1 u + \varepsilon^2 L^2 u = 0, \tag{6}$$

where

$$L^0 \equiv \partial_t - D\partial_x^2 - A, \quad L^1 \equiv -2D\partial_x\partial_\xi, \quad L^2 \equiv \partial_\tau - D\partial_\xi^2 - B.$$

From (5) and (6) we have (formally)

$$L^0 u^1 = 0 \tag{7}$$

$$L^0 u^2 = g(u^1,u^1) - L^1 u^1, \tag{8}$$

$$L^0 u^3 = g(u^1,u^2) + g(u^2,u^1) - L^1 u^2 - L^2 u^1, \tag{9}$$

etc.

As we saw before, the only bounded solutions of (7) are

$$u^1 = a\phi e^{ik_0 x} + b\phi e^{-ik_0 x},$$

except for possible additional terms which decay to zero as $t \to \infty$. We disregard these transient terms since we are interested in possible long-time asymptotic structures. At this point, we restrict consideration to real solutions. Since ϕ is real, then necessarily $b = \bar{a}$. Here a may perfectly well depend on the other two variables ξ and τ, so

$$u^1 = a(\xi,\tau)\phi e^{ik_0x} \sim,$$

the final symbol meaning "plus the complex conjugate of the preceding term." Of course, a has yet to be determined.

Substituting this expression for u^1 into the right side of (8), we obtain a polynomial in e^{ik_0x} and e^{-ik_0x}. In accordance with the assumed asymptotic validity of the expansion (5), we want each of the u^i to be a bounded function of x. For u^2 to be bounded, we must avoid resonance by ensuring that any term in $e^{\pm ik_0x}$ appearing in the above-mentioned polynomial satisfies an appropriate orthogonality condition. To dwell on this point a bit further, note that (again disregarding transients) all bounded solutions of

$$L^0u = Ve^{ink_0x} \quad (V \in C^n) \tag{10}$$

are

$$H_n^{-1}Ve^{ink_0x} + C_1e^{ik_0x} + C_2e^{-ik_0x},$$

with C_i arbitrary. Here we denote $H_n \equiv H(n^2k_0^2,0)$. For $n \neq \pm 1$, this matrix is nonsingular by our original hypothesis. For $n = \pm 1$, H_n^{-1} exists only on the orthocomplement of ψ, the nullvector of H_1^* such that $(\phi,\psi) = 1$. And if V is not in this orthocomplement, there are no bounded solutions. Therefore when $n = \pm 1$, (10) has bounded solutions only if

$$(V,\psi) = 0. \tag{11}$$

We express the right side of (8) as a polynomial in e^{ik_0x} and e^{-ik_0x}, and check this orthogonality condition for the terms in the polynomial involving $e^{\pm ik_0x}$. The only such term comes from L^1u^1, and the condition is seen to hold automatically by virtue of the fact that $(D\phi,\psi) = 0$; this latter equation follows from the assumed condition $\frac{\partial\mu}{\partial(k^2)}(k_0^2,0) = 0$ (μ has a maximum of 0 at $k^2 = k_0^2$).

At this stage we still know nothing about a, but may solve (8) for u^2, in the form

$$u^2 = a^{(2)}(\xi,\tau)\phi e^{ik_0x} \sim + Q(ae^{ik_0x}) \sim,$$

Q being a quadratic vector-valued polynomial in ae^{ik_0x}, and $a^{(2)}$ being a second coefficient, as yet undetermined.

We now substitute this expression into the right side of (9), and again check the orthogonality condition. This time, it turns out to be satisfied only if a satisfies the following equation:

$$a_\tau - Ka_{\xi\xi} + \alpha a - \beta a|a|^2 = 0, \qquad (12)$$

where $K = 4k_0^2(DH_1^{-1}D\phi,\psi)$, and α and β depend on g, D, A, and B. Our assumption that $\frac{\partial^2}{\partial(k^2)^2}\mu(k_0^2,0) < 0$ turns out to guarantee that $K > 0$.

The function $a(\xi,\tau)$ can be pictured as an envelop of the basic small amplitude periodic function, for to lowest order,

$$u \approx \varepsilon a(\varepsilon x,\varepsilon^2 t)\phi e^{ik_0x} \sim.$$

If a is real at some instant of time (such as $t = 0$), then by (12) it remains real, and then clearly

$$u \approx \varepsilon a(\varepsilon x,\varepsilon^2 t)\phi \cos k_0 x.$$

The possible stable asymptotic states for the amplitude function a are then found from the theorem in Section 2, applied to (12). In this case we have $c = 0$, and stable uniform rest states $a = \pm\sqrt{\alpha/\beta}$, provided that $\alpha < 0$ and $\beta < 0$.

So if these inequalities are fulfilled, the only possible ultimate configurations for the amplitude function are

$$a \equiv \pm\sqrt{\alpha/\beta}$$

and

a = a stationary front approaching $\sqrt{\alpha/\beta}$ as $x \to -\infty$ and $-\sqrt{\alpha/\beta}$ as $x \to +\infty$, or vice versa.

In the latter case, the effect of the slowly varying amplitude is to change the phase of the periodic solution by the amount π as x proceeds from $-\infty$ to $+\infty$. An estimate of the gradient of the amplitude function can be obtained from the parameters K, α and β.

If we carried through the expansion initiated above to higher order, we would find that in fact the front is not necessarily stationary, but may move slowly.

V. SHARP FRONTS AND SINGULAR PERTURBATIONS

Here I shall illustrate still a third way in which the study of reaction-diffusion systems may be reduced to the study of the scalar equation (1). It should be emphasized that the example below is only one of many types of systems to which this technique is applicable. In this vein, see also [9], [8], [10].

We consider a system of two reaction-diffusion equations, the first reaction being "rapid" compared to the second:

$$u_t = u_{xx} + k^2 f(u,v),$$

$$v_t = v_{xx} + g(u,v),$$

where k is a large parameter, expressing the rapidity of the first reaction.

We assume the regions of positivity and negativity of the two functions f and g are as follows:

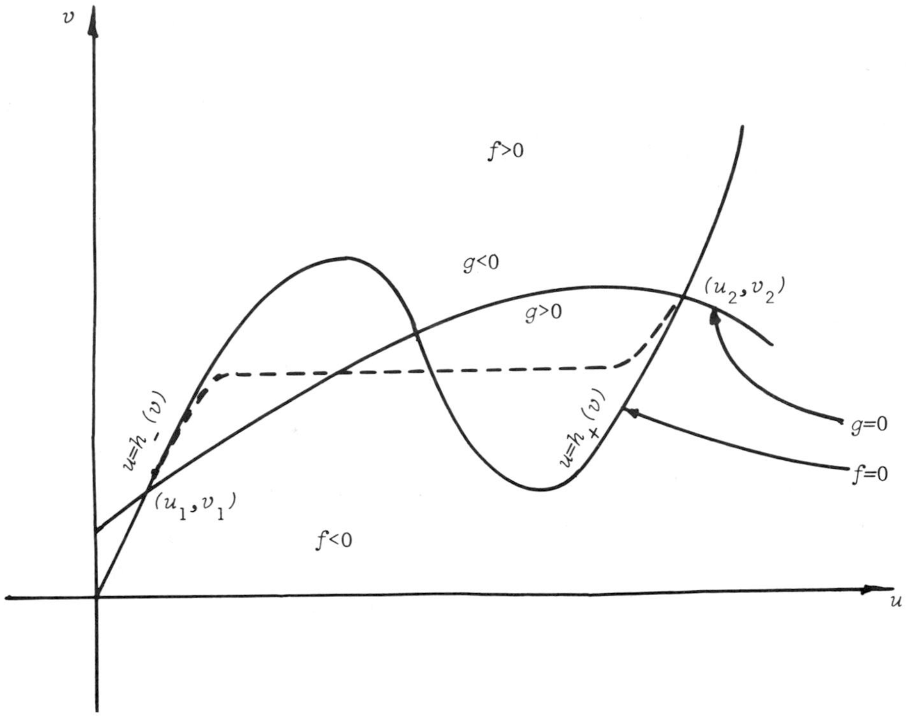

FIGURE 4.

There are two stable uniform rest states for the system: (u_1,v_1) and (u_2,v_2). We now ask whether a front exists which joins these two rest states. If so, then it is represented by $u = U(x - ct)$, $v = V(x - ct)$, where U and V satisfy the following ordinary differential equations in $z = x - ct$:

$$U'' + cU' + k^2 f(U,V) = 0, \quad U(-\infty) = u_1, \quad U(+\infty) = u_2, \tag{13}$$

$$V'' + cV' + g(U,V) = 0, \quad V(-\infty) = v_1, \quad V(+\infty) = v_2. \tag{14}$$

The fact that k is large, together with (13), suggests that approximately $f(U,V) = 0$ except on portions of the trajectory where a different length scale is needed. We look for monotone solutions (U,V). Because of the S-shaped nature of the isocline $f = 0$, there must be a (rapid) transition from one ascending branch of this isocline to the other, at some value

v^*. This means that the wave front's phase portrait must be as shown by the dotted line in Figure 4.

Let $u = h_-(v)$ and $u = h_+(v)$ be these two ascending branches. Postulating that V does not undergo an abrupt change at that point, we may approximate $g(U,V)$ in (14) by

$$g_{v^*}(V) \equiv \begin{cases} g(h_-(V),V), & V < v^*, \\ g(h_+(V),V), & V > v^*, \end{cases}$$

Thus

$$V'' + cV' + g_{v^*}(V) = 0. \tag{15}$$

For each $v^* \in (v_1,v_2)$, the function $g_{v^*}(v)$ has the properties of the function f in (1): the points $V = v_1$ and $V = v_2$ are stable zeros of the associated kinetic equation

$$\frac{dv}{dt} = g_{v^*}(v).$$

Even though g_{v^*} is discontinuous, the theory of (1) still holds, and we obtain the existence of a stable front solution $V_{v^*}(z)$ of (15) satisfying $V_{v^*}(-\infty) = v_1$, $V_{v^*}(\infty) = v_2$. Of course the speed c will depend on v^*, and we shall have to use other information to determine that constant.

For this, we look at (13), and stretch the variable z in a neighborhood of the point where the sharp transition of u from $h_-(v^*)$ to $h_+(v^*)$ occurs. Since the location of fronts may always be adjusted, by translation, to any value of z, we shall assume the position of the sharp part of the front to be at $z = 0$, and therefore define the stretched variable $\zeta = kz$. Then (13) becomes

$$k^2U_{\zeta\zeta} + kcU_\zeta + k^2f(U,V) = 0.$$

To lowest order in k^{-1},

$$U_{\zeta\zeta} + f(U,v^*) = 0, \quad U(-\infty) = h_-(v^*), \quad U(\infty) = h_+(v^*).$$

This is also the equation of a wave front, but with velocity 0. For its existence, we see from (2) that necessarily

$$\int_{h_-(v^*)}^{h_+(v^*)} f(U,v^*)dU = 0. \tag{16}$$

Considering the meaning of this integral in relation to Figure 4, we see that if the isoclines are as shown, typically there will exist a value (possibly unique) of v^* for which (16) holds. This, then, determines v^*, and hence $c = c(v^*)$.

We have constructed the desired wave front to lowest order, using properties of (1) at two crucial places in the argument. Approximations to any higher order may also be constructed. For details of these and analogous problems, see Fife [10].

REFERENCES

[1] Aronson, D. G., and Weinberger, H. F. (1975). "Nonlinear diffusion in population genetics, combustion and nerve propagation", in Proceedings of the Tulane Program in Partial Differential Equations and Related Topics, Lecture Notes in Mathematics *446*, Springer, Berlin, 5-49.

[2] Aronson, D. G., and Weinberger, H. F. (1978). "Multidimensional nonlinear diffusion arising in population genetics", Advances in Math., to appear.

[3] Boa, J. A., and Cohen, D. S. (1976). "Bifurcation of localized disturbances in a model biochemical reaction", *SIAM J. Appl. Math. 30*, 123-135.

[4] Cohen, D. S., Hoppensteadt, F. C., and Miura, R. M. (1977). "Slowly modulated oscillations in nonlinear diffusion processes", *SIAM J. Appl. Math. 33*, 217-229.

[5] Conley, C., and Fife, P., in preparation.

[6] Diprima, R. C., Eckhaus, W., and Segel, L. A. (1971). "Nonlinear wave-number interaction in near-critical two-dimensional flows", *J. Fluid Mech. 49*, 705-744.

[7] Eckhaus, W. (1965). "Studies in Nonlinear Stability Theory", Springer, New York.

[8] Feinn, D., and Ortoleva, P. (1977). "Catastrophe and propagation in chemical reactions", *J. Chem. Phys. 67*, 2119.

[9] Fife, P. C. (1976). "Pattern formation in reacting and diffusing systems", *J. Chem. Phys. 64*, 854-864.

[10] Fife, P. C. (1977). "Asymptotic analysis of reaction-diffusion wave fronts", *Rocky Mountain J. Math. 7*, 389-415.

[11] Fife, P. C. (1977). "Stationary patterns for reaction-diffusion equations", in Nonlinear Diffusion, *Res. Notes in Math. 14*, Pitman, London.

[12] Fife, P. C. (1979). "Long time behavior of solutions of bistable nonlinear diffusion equations", *Arch. Rational Mech. Anal.*, to appear.

[13] Fife, P. C., and McLeod, J. B. (1977). "The approach of solutions of nonlinear diffusion equations to travelling front solutions", *Arch. Rational Mech. Anal. 65*, 335-361. Also: *Bull. Amer. Math. Soc. 81*, (1975), 1075-1078.

[14] Fisher, R. A. (1937). "The advance of advantageous genes", *Ann. of Eugenics 7*, 355-369.

[15] Gmitro, J. I., and Scriven, L. E. (1966). "A physicochemical basis for pattern and rhythm", in Intracellular Transport, K. B. Warren, ed., Academic Press, New York and London.

[16] Kanel', Ya. I. (1962). "On the stabilization of solutions of the Cauchy problem for the equations arising in the theory of combustion", *Mat. Sbornik 59*, 245-288.

[17] Lefever, R., Herschkowitz-Kaufman, M., and Turner, J. W. (1977). "Dissipative structures in a soluble nonlinear reaction-diffusion system", Physics Letters *60A*, 389-391.

[18] Levin, S. A. (1977). "Pattern formation in ecological communities", NATO School on Spatial Pattern in Plankton Communities, Sicily.

[19] Levin, S. A. (1976). "Spatial patterning and the structure of ecological communities", in Some Mathematical Questions in Biology, ed., S. A. Levin, *7*, 1-36. Lectures on

Mathematics in the Life Sciences, Vol. 8, *Am. Math. Soc.*, Providence.

[20] Newell, A. C. (1974). "Envelop equations", Lectures in Applied Mathematics, Vol. 15, Nonlinear Wave Motion, A. C. Newell, ed., *Amer. Math. Soc.*, Providence.

[21] Newell, A. C., and Whitehead, J. A. (1969). "Finite bandwidth, finite amplitude convection", *J. Fluid Mech. 38*, 279-303.

[22] Nicolis, G., and Prigogine, I. (1977). "Self-organization in Nonequilibrium Systems", Wiley-Interscience, New York.

[23] Segel, L. A., and Jackson, J. L. (1972). "Dissipative structure: an explanation and an ecological example", *J. Theor. Biol. 37*, 545-559.

[24] Turing, A. M. (1953). "The chemical basis of morphogenesis", *Phil. Trans. Roy. Soc. Lon. B237*, 37-72.

PRODUCT INTEGRAL REPRESENTATION OF SOLUTIONS TO SEMILINEAR VOLTERRA EQUATIONS WITH DELAY

W. E. Fitzgibbon

Department of Mathematics
University of Houston
Houston, Texas

In what follows we shall utilize the theory of nonlinear evolution operators to represent and approximate a class of semilinear Volterra equations which involve delay and have a linear evolution operator as kernel. More specifically we consider equations of the form

$$x(\phi)(t) = W(t,\tau)\phi(0) + \int_{\tau}^{t} W(t,s)F(s,x_s(\phi))ds \tag{1.1}$$
$$x_\tau(\phi) = \phi \in C.$$

Here X denotes a Banach space and $\{W(t,s)|0 \leq s \leq t < T\}$ is a family of linear evolution operators defined on X. C is the space of bounded uniformly continuous functions mapping an interval of the form $I = [-r,0]$ or $(-\infty,0]$ to X. We endow the space $C = C(I,X)$ with the supremum norm, $\|\ \|_C$. If z is a continuous function mapping $I \cup [0,T]$, z_t is that element of C having pointwise definition $z_t(\theta) = z(t + \theta)$ for $\theta \in I$. If the linear evolution operator is generated by a family of linear operators $\{A(t)|t \in [0,T]\}$ then (1.1) is seen to provide a variation of parameters representation of solutions to functional differential equation:

ISBN 0-12-434180-2

$$\dot{x}(\phi)(t) = A(t)x(\phi)(t) + F(t,x_t(\phi)). \tag{1.2}$$

One might say that (1.1) provides mild solutions to (1.2). We use the existence of solutions to (1.1) to define a nonlinear solution operator. The Crandall-Pazy product integral representation theorem is then used to represent the evolution operator. The representation theorem is then used to provide criteria for the convergence of solutions to approximating equations.

Much recent work is concerned with abstract functional differential equations. Travis and Webb [19], [20], [21] apply the theory of nonlinear semigroups; a treatment of stability appears in [23]. Dyson and Villella Bressan [7] connect the work on product integration of Crandall and Pazy [5] with functional differential equations. The work presented here is closely related to the work of Dyson and Villella Bressan.

In [3] Crandall defines a generalized domain for a nonlinear quasi-dissipative operator A. If $\{A(t)|t \in [0,T]\}$ is a family of nonlinear quasi-dissipative operators defined on a Banach space $X.$, Crandall and Pazy [5] show that one can define a family of nonlinear evolution operators, $\{V(t,\tau)|0 \leq \tau \leq t \leq T\}$, on X via the product integral:

$$V(t,\tau)x = \lim_{n\to\infty} \Pi^n (I - (t-\tau)/nA(\tau + it/n))^{-1}x \tag{1.3}$$

Basically they require that the closure of the domain of $A(t)$ is constant, that the range of the resolvent contains the closure of the domain of $A(t)$ and that the resolvent $(I - \lambda A(t))^{-1}x$ satisfies a Lipschitz-type condition on t. One denotes the generalized domain of $A(t)$ by $\overline{D(A(t))}$ and associates with each $x \in \overline{D(A(t))}$ a possibly infinite positive real number $|A(t)x|$ the Lipschitz constant for $(I - \lambda A(t))^{-1}x$ depends both on $\|x\|$ and $|A(t)x|$. Crandall and Pazy also develop an approximation result. They give conditions sufficient to guarantee the uniform convergence

$$\lim_{n\to\infty} V_n(t,\tau)x = V(t,\tau)x$$

where for each $n \in Z^+$ $\{V_n(t,\tau) | 0 \leq \tau \leq t \leq T\}$ is the family of nonlinear evolution operators constructed from $\{A(n,t) | t \in [0,T]\}$ via formula (1.3). We remark that the Crandall-Pazy [5] results have much more generality than we have discussed. We have limited our discussion to versions which are tailored to our needs.

We now turn our attention to families of linear operators. From this point on the notation $A(t)$ shall denote a linear operator acting on the general Banach space X. The nonlinear operators which appear subsequently will be denoted by $\hat{A}(t)$. These operators will work on the space of initial functions $C = C(I,X)$. We introduce the following restrictions for a time dependent family of linear operators which map X to itself.

(L.1) For each $t \in [0,T]$; $A(t)$ is closed, $D(A(t))$ is independent of t and $\overline{D(A(t))} = X$.

(L.2) There exists a real ω so that for all $\tau \in [0,T]$, $A(t)$ is the infinitesimal generator of a strongly continuous semigroup of linear operators $\{e^{tA(\tau)} | t \geq 0\}$ such that

$$\|e^{tA(\tau)}\| \leq e^{\omega t}$$

(L.3) There exists an $L > 0$ so that for all $x \in D(A(t))$ and $t,\tau \in [0,T]$

$$\|A(t)x - A(\tau)x\| \leq |t - \tau| L(\|x\|)(1 + \|A(\tau)x\|).$$

It is well known (cf. [25]) that conditions (L.1) through (L.3) imply that $\{A(t) | t \in [0,T]\}$ generates a family of linear evolution operators $\{W(t,\tau) | t,\tau \in [0,T]\}$ which have product integral representation (1.3). Moreover we can set $v(t) = W(t,\tau)x$ to obtain the unique solution to the intitial value problem

$$v'(t) = A(t)v(t)$$

$$v(\tau) = x.$$

We shall place the following hypothesis on the functional portion of (1.2). We require that for $F\colon [0,T] \times C \to X$ there exists a continuous $\beta(\ )[0,T] \to R^+$ and $M > 0$ so that

(F.1) $\|F(t,\phi) - F(t,\psi)\| \le \beta(t)\|\phi - \psi\|_C \quad t \in [0,T] \quad \phi,\psi \in C$

(F.2) $\|F(t,\phi) - F(t,\psi)\| \le |t - \tau| M \|\phi\|_C$.

It is not difficult to guarantee unique solutions to (1.1), We have:

Proposition 1.4. Let $\{A(t) | t \in [0,T]\}$ satisfy (L.1) through (L.3) and let $F\colon [0,T] \times C \to X$ satisfy (F.1) and (F.2). Then, for each $\phi \in C$ and $\tau \in [0,T]$ there exists a unique $x(\phi)\colon [\tau,T] \to X$ which satisfies

$$x(\phi)(t) = W(t,T)\phi(0) + \int_\tau^t W(t,s)F(s,x_s(\phi))ds; \tag{1.5}$$

$$x_\tau(\phi) = \phi \in C.$$

This proposition is established via a classical Picard iteration; the reader is referred to [12] for a proof.

Unique solutions to (1.5) give rise to nonlinear evolution operators. If $x(\phi)$ is the solution to (1.5) we define $U(t,\tau)\colon C \to C$

$$U(t,\tau)\colon \quad C \to C$$

pointwise by the equation

$$U(t,\tau)\phi = x_t(\phi) \qquad t \ge \tau. \tag{1.6}$$

It is not difficult to check that $U(t,\tau)$ so defined is continuous in t and τ and the property that $U(t,s)U(s,\tau) = U(t,\tau)$ follows from the uniqueness of solutions. We observe that $(U(t,\tau)\phi)(0) = x(\phi)(t)$ and we may think of $U(t,\tau)\phi$ as providing segments of the solutions to (1.3).

We now introduce a time dependent family of nonlinear operators which map a subset of the Banach space C to C. For each $t \in [0,T]$ we define $\hat{A}(t)\colon C \to C$ by the equations:

$$A(t)\phi(\theta) = \dot{\phi}(\theta)$$

$$D(\hat{A}(t)) = \{\phi | \dot{\phi} \in C,\ \phi(0) \in D(A(t)) \tag{1.7}$$

$$\text{and } \dot{\phi}(0) = A(t)\phi(0) + F(t,\phi)\}$$

$\hat{A}(t)$ is nonlinear by virtue of its nonlinear domain and it is quite possible that its domain varies in t. We have the following theorem:

Theorem 1. Assume the conditions of Proposition (1.4) are satisfied. If $\{\hat{A}(t) | t \in [0,T]\}$ and $\{U(t,\tau) | 0 \leq t \leq \tau \leq T\}$ are defined via (1.7) and (1.6), then $\{\hat{A}(t) | t \in [0,T]\}$ generates $\{U(t,\tau) | 0 \leq \tau \leq t \leq T\}$, in other words, $\hat{A}(t)$ is the operator defined as the limit

$$\hat{A}(t)\phi = \lim_{h \to 0^+} (U(t + h,t)\phi - \phi)/h$$

This theorem is established by noting that each $\hat{A}(t)$ is the infinitesimal generator of a nonlinear semigroup, (cf. [20], [16]). A formula appearing in [26] allows comparison of the infinitesimal generator of a nonlinear semigroup and the genreator of a nonlinear evolution operator.

$\hat{A}(t)$ is examined in detail in [20] for the case of finite delay; infinite delays are treated in [10]. It is shown that $D(\hat{A}(t))$ is dense in X; it is further shown that if $\lambda > 0$ is sufficiently small then $R(I - \lambda\hat{A}(t)) = C$. We now wish to introduce an approximation scheme for $(I - \lambda\hat{A}(t))^{-1}\phi$. For each $n \in z^+$ let $\{A(n,\tau) | \tau \in [0,T]\}$ and F_n: $[0,T] \times C \to X$ satisfy (L.1) through (L.3) and (F.1) and (F.2) with constants independent of n. Further assume that:

(A.1) There exists a $\lambda_0 > 0$ so that for

$0 < \lambda < \lambda_0$, $\tau \in [0,T]$, $x \in X$,

$$\lim_{n\to\infty} (I - \lambda A(n,\tau))^{-1}x = (I - \lambda A(\tau))^{-1}x$$

(A.2) For all $\phi \in C$ and $\tau \in [0,T]$ $\lim_{n\to\infty} F_n(\tau,\phi) = F(\tau,\phi)$.

We use $\{A(n,\tau) | t \in [0,T]\}$ and F_n: $[0,T] \times C \to X$ to define nonlinear operators $\{\hat{A}(n,\tau) | \tau + [0,T]\}$ on C via formula (1.7). The following simple observation will become the lynchpin of our subsequent discussion.

Lemma. If λ is sufficiently small, $\phi \in C$ and $\tau \in [0,T]$ then $\lim_{n\to\infty} (I - \lambda\hat{A}(n,\tau))^{-1}\phi = (I - \lambda\hat{A}(\tau))^{-1}\phi$.

The proof of this lemma consists of a straightforward but complicated computation involving the following representation of the resolvent.

$$\begin{aligned}(I - \lambda\hat{A}(t))^{-1}\phi(\theta) \\ &= e^{\theta/\lambda}(I - \lambda A(t))^{-1}(\phi(0) + \lambda F(t,(I - \lambda\hat{A}(t))^{-1}\phi)) \\ &+ e^{\theta/\lambda}/\lambda\int_{\theta}^{0} e^{-s\lambda}\phi(s)ds\end{aligned}$$

The preceding formula is established in [20].

We can now obtain a product integral representation for the nonlinear evolution operator associated with solutions to (1.5). This turns out to be an application of the approximation theory of nonlinear evolution equations appearing in [5]. We have the following theorem:

Theorem 2. Let $\{A(t) | t \in [0,T]\}$ be a family of linear operators which satisfy (L.1) through (L.3) and suppose that $F: [0,T] \times C \to X$ satisfies (F.1) and (F.2). The limit

$$U(t,\tau)\phi = \lim_{n\to\infty} \prod_{i=1}^{n} (I - (t - \tau)/n\ \hat{A}(\tau + i(t - \tau)/n))^{-1}\phi \qquad (1.8)$$

exists for all $\phi \in C$ and $(U(t,\tau)\phi)(0)$ provides the unique solution to the integral equation:

$$x(\phi)(t) = W(t,\tau)\phi(0) + \int_{\tau}^{t} W(t,s)F(s,x_s(\phi))ds \qquad (1.9)$$

$$x_\tau(\phi) = \phi \in C$$

Indication of Proof. It is shown that $\{\hat{A}(t) | t \in [0,T]\}$ satisfies the conditions of the Crandall-Pazy product integral theorem. Thus one can define a family of nonlinear evolution operators on C which are represented as

$$V(t,\tau)\phi = \lim_{n\to\infty} \prod_{i=1}^{n} (I - (t-\tau)/n\ \hat{A}(\tau + i(t-\tau)/n))^{-1}\phi.$$

It is necessary to show that $V(t,\tau)\phi = U(t,\tau)\phi$ where $U(t,\tau)$ is the nonlinear evolution operator associated with solutions to

(1.5). We let $A_n(t)$ denote the Yosida approximations $A(t)(I - t/nA(t))^{-1}$ and form $\{\hat{A}(n,t) | t \in [0,T]\}$ on C via (1.7). The operators $\{\hat{A}(n,t) | t \in [0,T]\}$ also satisfy the Crandall-Pazy conditions and thus we can define:

$$U_n(t,\tau)\phi = \lim_{n\to\infty} \Pi_{i=1}^{n} (I - (t - \tau)/n \, \hat{A}(n,t + i(t - \tau)/n))^{-1}\phi$$

Furthermore the continuity of $A_n(t)$ and Theorem 2 of [7] imply that $(U_n(t,\tau)\phi)(0)$ is the unique solution of

$$\dot{x}(n,\phi)(t) = A_n(t)x(n,\phi)(t) + F(t,x_t(n,\phi))$$

$$x_\tau(n,\phi) = \phi \in C.$$

Solutions to the above abstract functional differential equation have variation of parameters representation:

$$x(n,\phi)(t) = W_n(t,\tau)\phi(0) + \int_\tau^t W_n(t,s)F(s,x_s(n,\phi))ds \tag{1.10}$$

The approximating nonlinear evolution operators $U_n(t,\tau)$ satisfy the hypotheses of the Crandall-Pazy approximation theorem [5]. Thus we have $\lim_{n\to\infty} U_n(t,\tau)\phi = V(t,\tau)\phi$ and can apply a straightforward convergence argument to the approximating equations (1.10) to deduce that $V(t,\tau)\phi = U(t,\tau)\phi$. A detailed version of the proof will appear in [11].

We now introduce an approximation scheme: for each $n \in Z^+$ let $\{A(n,\tau) | \tau \in [0,T]\}$ and $F_n\colon [0,T] \times C \to X$ satisfy (L.1) through (L.3) and (F.1) and (F.2) with constants independent of n. Further assume that

(A.1) There exists a $\lambda_0 > 0$ so that for $0 < \lambda < \lambda_0$, $\tau \in [0,T]$, $x \in X$, $\lim_{n\to\infty} (I - \lambda A(n,\tau))^{-1}x = (I - \lambda A(\tau))^{-1}x$.

(A.2) For all $\phi \in C$ and $\tau \in [0,T]$, $\lim_{n\to\infty} F_n(\tau,\phi) = F(\tau,\phi)$.

We now obtain the approximation result:

<u>*Theorem 3.*</u> For $n \in Z^+$, let $\{A(n,t) | t \in [0,T]\}$ and $F_n\colon [0,T] \times C \to X$ approximate $\{A(t) | t \in [0,T]\}$ and $F\colon [0,T] \times C \to X$ in the fashion described by (A.1) and (A.2). For all $\phi \in C$

$$\lim_{n\to\infty} U_n(t,\tau)\phi = U(t,\tau)\phi \text{ uniformly for } t \in [0,T].$$

Indication of Proof. The proof consists of a lemma establishing that $\lim_{n\to\infty} (I - \lambda\hat{A}(n,t))^{-1}\phi = (I - \lambda\hat{A}(t))^{-1}\phi$ and observing that the conditions of the Crandall-Pazy approximation theorem are satisfied.

We remark that the convergence criterion (A.1) can be obtained by requiring $A(n,t)x$ to converge to $A(t)x$ on a dense subset of X.

The product integral representation facilitates the examination of the behavior of solutions. Questions of stability and asymptotic behavior are considered in [7], [20], and [12]. A result appearing in [7] is immediately modified to yield.

Proposition 1.11. Assume that the conditions of Theorem 2 are satisfied and

$$x(\phi)(t) = W(t,\tau)\phi(0) + \int_\tau^t W(t,s)F(s,x_s(\phi))ds, \qquad t \in [\tau,T].$$

If X is a reflexive Banach space and $\phi(0) \in D(A(t))$ then for a.c. $t \in [\tau,T]$, $\dot{x}(\phi)(t)$ exists and satisfies

$$\dot{x}(\phi)(t) = A(t)x(\phi)(t) + F(t,x_t(\phi)).$$

We shall conclude by applying our theory to partial functional differential equations.

We first consider a semilinear hyperbolic equation of the form:

$$\partial u(x,t)/\partial t = -a(t,x)\partial u(x,t)/\partial x + f(t,x,u(x,\omega(t))) \tag{1.12}$$

$$t \in [0,T], \quad -\infty < x < \infty$$

$$u(x,0) = u(x)$$

$$u(x,s) = \psi(x,s) \quad -\infty < x < \infty \quad S \in [-r,0]$$

$$t - r \leq \omega(t) \leq t.$$

Here we require that $a(t,x) > a_0 > 0$, $a(t,x)$ is bounded and uniformly continuous in x and Lipschitz in t; f is bounded in x and Lipschitz in its first and third places; and that $\omega(t)$ is Lipschitz. If X denotes the space of bounded uniformly continuous functions $(-\infty,\infty)$ we realize (4.1) as the Banach space differential equation:

$$\dot{x}(\phi)(t) = A(t)x(\phi)(t) + F(t,x_t(\phi)) \tag{1.13}$$

where $A(t)u$ is defined to be the operator on X satisfying

$$A(t)u = -a(t,x)u'$$

$$D(A(t)) = \{u|u' \in X\}.$$

We take C to be the space $C([-r,0],X)$ and F is the nonlinear function defined pointwise by the equation,

$$F(t,\psi(s,x)) = f(t,x,\psi((\omega(t) - t,x)).$$

It is not difficult to see that F so defined satisfies the Lipschitz properties (F.1) and (F.2). The operator $A(t)$ can easily be shown to be the infinitesimal generator of a group of translations on X explicitly given by

$$(e^{hA(\tau)})u(x) = u(x - ha(\tau,x)).$$

Since $(e^{hA(\tau)})$ is an isometry on A it clearly satisfies (A.2). The continuity requirement for $A(t)u$ is obviously satisfied and Proposition (1.4) provides the existence of mild solutions to (1.12). We approximate these solutions in a straightforward manner. Let $\varepsilon_n \downarrow 0$ then

$$A(\varepsilon_n,t)u(x) = (e^{\varepsilon_n A(t)} - I)/\varepsilon_n = (u(x-\varepsilon_n a(t,x))-u(x))/\varepsilon_n. \tag{1.15}$$

If $\phi \in C$ we extend the domain to $[-\varepsilon_n-r,0]$ by

$$\phi(s,x) = \begin{cases} \phi(s,x) & \text{if } s \in [-r,0] \\ \phi(-r,x) & \text{if } s \in [r-\varepsilon_n,r]. \end{cases}$$

We now define

$$\phi(\varepsilon_n)(s,x) = \phi(s - \varepsilon_n,x) \quad \text{for } s \in [-r,0]$$

and the operators

$$F_{\varepsilon_n}(t,\phi) = F(t,\phi(\varepsilon_n))$$

It is immediate that F_{ε_n} satisfies our convergence requirements.

Theorem (3.14) guarantees the convergence of the approximate solutions:

$$x(\phi)(\varepsilon_n,t) = W_{\varepsilon_n}(t,\tau)\phi(0) + \int_\tau^t W_{\varepsilon_n}(t,s)F_{\varepsilon_n}(s,x_s(\varepsilon_n))ds \quad (1.16)$$

to solutions of (4.2). Furthermore the delay of equation (4.4) is positive definite and consequently the finite difference approximations can be solved via the method of steps. We remark that the hyperbolic equation arises in problems of stream modeling. The quantity $u(x,t)$ denotes the concentration of dissolved oxygen in a stream at position x and time t, $a(t,x)$ denotes the velocity of the stream at position x and t and the non-linear delay term will denote a source distribution of dissolved oxygen. (cf. [6]).

As a second example we apply our theory to demonstrate the continuous dependence of solutions of a parabolic delay equation on its coefficients. We consider a problem of the form:

$$\begin{aligned} \partial u(x,t)/\partial t = & \sum_{i,j=1}^{n} a_{ij}(t)\partial^2 u(x,t)/\partial x_i \partial x_j \quad (1.17)\\ & + \sum_{i=1}^{n} b_i(t)\partial u(x,t)/\partial x_i + c(t)u(x,t) \\ & + f(t,u(x,t-r)) \quad r > 0, \quad t \in [0,T] \end{aligned}$$

$$u(x,\theta) = (x,\theta) \quad \text{for} \quad -r, \le \theta \le 0$$

where $x = (x_1,\ldots,x_n) \in R^n$. We require that the coefficient matrix $[a_{ij}(t)]$ be positive definite; each of $[a_{ij}(t)]$, $b_i(t)$ and $c(t)$ be uniformly Lipschitz and that f be Lipschitz in each place.

We place our problem in a Banach space as follows: The analysis for the linear portion may be found in J. Goldstein [13]. Let X be the Banach space of bounded continuous functions which vanish at ∞ and are equipped with the supremum norm. We formally define,

$$(L(t)u(x) = \sum_{i,j=1}^{n} a_{ij}(t)\partial^2 u(x)/\partial x_i \partial x_j + \sum_{i=1}^{n} b(t)\partial u\ (x)/\partial x_i + c(t)u(x). \tag{1.18}$$

It can be shown [13] that there exists a strongly continuous ω-contractive semigroup having infinitesimal generator $A(t)$ in X such that $A(t)v = L(t)v$ for all $v \in \{v \in V | \partial v/\partial x_i, \partial^2 v/\partial x_i \partial x_j \in X,\ 1 \leq i,j \leq n$ and v has compact support in $R^n\}$. It is readily apparent that $\{A(t) | t \in [0,T]\}$ satisfies (A.1) through (A.3) and consequently generates a linear evolution operator on X.

We let $C = C([-r,0],X)$ be the space of continuous functions from, $[-r,0]$ to X with the supremum norm. If $\phi \in C$ we define $F(t,\phi) = f(t,\phi(-r))$. It is immediate that F so defined satisfies requirements (F.1) and (F.2). We thus rewrite (4.5) as a delay equation in X,

$$\dot{x}(\phi)(t) = A(t)x(\phi)(t) + F(t,x_t(\phi)). \tag{1.19}$$

Our theory guarantees the existence of mild to (4.6) and provides their product integral. Furthermore if $a_{ij}(k,t)$, $b_i(k,t)$, $c(k,t)$ and $f(k,t\bullet)$ satisfy the same conditions with the same constants as $a_{ij}(t)$, $b_i(t)$, $c(t)$ and $F(t,\bullet)$ and $\lim_{k\to\infty} a_{ij}(k,t) = a_{ij}(t)$, $\lim_{k\to\infty} b_i(k,t) = b_i(t)$, $\lim_{k\to\infty} c(k,t) = c(t)$ and $\lim_{k\to\infty} F(k,t,\phi) = F(t,\phi)$ then the mild solutions to the associated abstract problems

$$x(\phi,k)(\phi) = W_k(t,\tau)\phi(0) + \int_\tau^t W_k(t,s)F(s,x_x(\phi,k))ds$$

converge to $x(\phi)(t)$.

A more detailed version of this paper complete with proofs will appear at a later date, c.f. [11].

REFERENCES

[1] Brewer, D. W. (1975). "A nonlinear semigroup for a functional differential equation", Dissertation, University of Wisconsin.

[2] Brezis, H., and Pazy, A. (1972). "Convergence and approximation of semigroups of nonlinear operators in Banach spaces", *J. Functional Analysis*, 63-64.

[3] Browder, F. (1964). "Nonlinear equations of evolution", *Ann. Math. 80*, 485-523.

[4] Crandall, M., and Liggett, T. (1971). "Generation of semigroups of nonlinear transformations on general Banach spaces", *Amer. J. Math. 93*, 265-298.

[5] Crandall, M., and Pazy, A. (1972). "Nonlinear evolution equations in Banach spaces", *Israel J. Math. 11*, 57-94.

[6] DiToro, D. M. (1969). "Stream Equations and Method of Characteristics", Journal of the Sanitory Engineering Division, Amer. Soc. Civil Eng., *Proc. 95*, 699-703.

[7] Dyson, J., and Bressan, R. Villella. "Functional differential equations and nonlinear evolution operators", *Edinburgh J. Math.*, (to appear).

[8] Fitzgibbon, W. (1973). "Approximations of nonlinear evolution equations", *J. Math. Soc. Japan 25*, 211-221.

[9] Fitzgibbon, W. (1977). "Nonlinear Volterra equations with infinite delay", *Mönat für Math. 84*, 275-288.

[10] Fitzgibbon, W. (1978). "Semilinear functional differential equations", *J. Diff. Equations, 29*, 1-14.

[11] Fitzgibbon, W. "Representation and approximations of solutions to semilinear Volterra equations with delay", to appear in *J. Diff. Equations*.

[12] Fitzgibbon, W. (1977). "Stability for abstract nonlinear Volterra equations involving finite delay", *J. Math. Anal. Appl. 60*, 429-434.

[13] Flaschka, H., and Leitman, M. (1975). "On semigroups of nonlinear operators and the solution of the functional differential equation $\ddot{x}(t) = F(x_t)$", *J. Math. Anal. Appl. 49*, 649-658.

[14] Goldstein, J. (1969). "Abstract evolution equations", *Trans. Amer. Math. Soc. 141*, 158-188.

[15] Hale, J. (1971). "Functional differential equations", *Appl. Math. Series Vol. 3*, Springer-Verlag, New York.

[16] Kato, T. (1967). "Nonlinear semigroups and evolution equations", *J. Math. Soc. Japan 19*, 508-520.

[17] Pazy, A. (1974). "Semi-groups of linear operators and applications to partial differential equations", Lecture University of Maryland, College Park, Md.

[18] Plant, A. "Nonlinear semigroups of translations in Banach space generated by functional differential equations in Banach space", *J. Math. Anal. Appl.*, (to appear).

[19] Travis, C., and Webb, G. (1978). "Existence, stability and compactness in the α-norm for partial functional differential equations", *Trans. Amer. Math. Soc.*, *240*, 129-143.

[20] Travis, C., and Webb, G. (1974). "Existence and stability for partial functional differential equation", *Trans. Amer. Math. Soc. 200*, 395-418.

[21] Travis, C., and Webb, G. "Partial differential equations with deviating arguments", *J. Math. Anal. Appl.* (to appear).

[22] Webb, G. (1974). "Autononomous nonlinear functional differential equations", *J. Math. Anal. Appl. 46*, 1-12.

[23] Webb, G. "Asymptotic stability for abstract nonlinear functional differential equations", *Proc. Amer. Math. Soc.*, (to appear).

[24] Webb, G. "Functional differential equations and nonlinear semigroups in L^p spaces", *J. Diff. Equations*, (to appear).

[25] Yosida, K. (1968). "Functional Analysis", Springer-Verlag, New York.

[26] Crandall, M. (1973). "A generalized domain for semigroup generators", *Proc. Amer. Math. Soc. 37*, 434-441.

ANGLE-BOUNDED OPERATORS AND UNIQUENESS OF PERIODIC SOLUTIONS OF CERTAIN ORDINARY DIFFERENTIAL EQUATIONS

Chaitan P. Gupta

Department of Mathematical Sciences
Northern Illinois University
DeKalb, Illinois

INTRODUCTION

Let $f\colon [0,T] \times \mathbb{R} \to \mathbb{R}$ and $p\colon [0,T] \to \mathbb{R}$ be given continuous functions such that $\int_0^T p(t)dt = 0$. In this paper we are concerned with the question of uniqueness of periodic solutions of the second order equation

$$\left.\begin{aligned} &-x'' + cx' + f(t,x) = p(t), \quad t \in [0,T] \\ &x(0) = x(T) \\ &x'(0) = x'(T) \end{aligned}\right\} \tag{1}$$

We study this via the use of angle-bounded mappings. Note that we do not require the functions $f(t,x)$ and $p(t)$ to be periodic. Our results are similar to earlier results of Chang [4] and Leach [6] concerning uniqueness of periodic solutions of (1).

In section 1, we give an example of an angle-bounded mapping. As far as we know this is the first example of a non-symmetric angle-bounded mapping.

In section 2, we use the example of section 1 to obtain uniqueness of periodic solutions of equation (1). We may remark

ISBN 0-12-434180-2

that the aim of this paper is to illustrate the usefulness and simplicity of angle-bounded mappings in uniqueness problems.

SECTION I

Let X be a Banach space and X^* the dual Banach space of X. We denote the duality pairing between $x \in X$ and $w \in X^*$ by (w,x). Let Y be a closed subspace of X^*. A linear mapping $L\colon D(L) \subset X \to Y$ is said to be *monotone* if $(Lx,x) \geq 0$ for every $x \in D(L)$.

Definition 1. Let $L\colon D(L) \subset X \to Y$ be a monotone linear mapping. L is said to be *angle-bounded* if there exists a constant $\alpha \geq 0$ such that

$$|(Lx,y) - (Ly,x)| \leq 2\alpha\sqrt{(Lx,x)}\,\sqrt{(Ly,y)}, \quad x,y \in D(L) \tag{1.1}$$

α is called the constant of angle-boundedness for L. Notice that if L is symmetric i.e. $(Lx,y) = (Ly,x)$ for $x,y \in D(L)$ then L is angle-bounded with $\alpha = 0$. The concept of an angle-bounded linear mapping was first introduced by Amann in [1]. Angle-bounded linear mappings are very much like positive symmetric mappings in a Hilbert space in the sense that they admit a splitting much like the same way as the square root of a positive symmetric mapping in a Hilbert space. We shall need the following proposition essentially due to Browder-Gupta [3] (see also, Browder [2], Hess [7], Gupta-deFigueiredo [5]).

Proposition 2. Let X be a given Banach space and Y a closed subspace of X^*. Let $K\colon X \to Y$ be a bounded, linear, angle-bounded mapping with constant of angle-boundedness α. Then, there exist a Hilbert space H, a bounded linear mapping $S\colon X \to H$, a skew-symmetric mapping $B\colon H \to H$ such that

(i) $K = S^*(I+B)S$ where S^* is the adjoint of the mapping $S\colon X \to H$ with $R(S^*) \subset Y$.

(ii) $\|S\|^2 \leq \|K\|$, $\|B\| \leq \alpha$

(iii) $(Kx,x) \geq \dfrac{1}{(1+\alpha^2)\|K\|}\|Kx\|^2$ for $x \in X$.

We omit the proof of Proposition 1 and refer the reader to [2], [3], [5]. We also note that the inequality (iii) in Proposition 1 was first explicitly-noted by Hess [7].

We now give an example of a non-symmetric angle-bounded mapping. Let $X = L^\infty[0,T]$ and $Y = L^1[0,T]$. In the following c will be a fixed real number, $c \neq 0$. Define a linear mapping $L\colon D(L) \subset X \to Y$ by setting

$$\left.\begin{aligned} D(L) &= \{x \in X \mid x' \text{ absolutely continuous on } [0,T], \\ &\qquad x(0) = x(T),\quad x'(0) = x'(T)\} \\ Lx &= -x'' + cx' \quad \text{for } x \in D(L). \end{aligned}\right\} \tag{1.2}$$

We note that for $x \in D(L)$

$$\begin{aligned} (Lx,x) &= \int_0^T (-x''+cx')x\,dt = -\int_0^T x''x\,dt + c\int_0^T x'x\,dt \\ &= \int_0^T x'^2 dt \geq 0. \end{aligned}$$

So $L\colon D(L) \subset X \to Y$ is monotone.

Theorem 3. The monotone linear mapping $L\colon D(L) \subset X \to Y$ defined above is angle-bounded.

Proof: For $x \in D(L)$, $y \in D(L)$ we have using integration by parts,

$$\begin{aligned} (Lx,y) &= \int_0^T (-x''+cx')y\,dt = -\int_0^T x''y\,dt + c\int_0^T x'y\,dt \\ &= -x'(T)y(T) + x'(0)y(0) + \int_0^T x'y'dt + c\int_0^T x'y\,dt. \\ &= \int_0^T x'y'\,dt + c\int_0^T x'y\,dt \end{aligned}$$

and

$$\begin{aligned}(Ly,x) &= \int_0^T(-y''+cy')x\,dt = -\int_0^T y''x\,dt + c\int_0^T y'x\,dt\\ &= \int_0^T y'x'dt + cy(T)x(t) - cy(0)x(0) - c\int_0^T yx'\,dt\\ &= \int_0^T x'y'dt - c\int_0^T x'y\,dt.\end{aligned}$$

So, for $x,y \in D(L)$,

$$\begin{aligned}|(Lx,y) - (Ly,x)| &= \left|2c\int_0^T x'y\,dt\right| = 2|c|\left|\int_0^T x'(y-y(0))dt\right|\\ &\leq 2|c|\left(\int x'^2dt\right)^{1/2}\left(\int_0^T (y-y(0))\,dt\right)^{1/2}\\ &\leq 2|c|T\left(\int_0^T x'^2dt\right)^{1/2}\left(\int_0^T y'^2dt\right)^{1/2}\end{aligned}$$

So $|(Lx,y) - (Ly,x)| \leq 2|c|T\sqrt{(Lx,x)}\sqrt{(Ly,y)}$ for $x,y \in D(L)$, and accordingly L is angle-bounded with constant of angle-boundedness equal to $|c|T$. It is clear that L is nonsymmetric for $c \neq 0$. Hence the theorem. //

Let, now, $X_1 = \{u \in L^\infty[0,T] \mid \int_0^T u(t)dt = 0\}$ and $Y_1 = \{u \in L^1[0,T] \mid \int_0^T u(t)dt = 0\}$. Clearly X_1 is a closed subspace of $X = L^\infty[0,T]$ and Y_1 is a closed subspace of $Y = L^1[0,T]$.

Lemma 4. X_1 is a vector subspace of the dual space Y_1^* of Y and for $u \in X_1$

$$\tfrac{1}{2}\|u\|_X \leq \|u\|_{Y_1^*} \leq \|u\|_X \tag{1.3}$$

Proof: Clearly $X_1 \subset L^\infty[0,T] = (L^1[0,T])^* = Y^* \subset Y_1^*$ algebraically since $Y_1 \subset Y$. Also for $u \in X_1$ it is clear from the definition of the dual norm in Y_1^* that

$$\|u\|_{Y_1^*} \leq \|u\|_X.$$

Now, for $v \in L^1[0,T]$ we have, using $\int_0^T u(t)dt = 0$, that

$$\left|\int_0^T uv\,dt\right| = \left|\int_0^T u(v - \frac{1}{T}\int_0^T v)dt + \int_0^T u(\int_0^T \frac{1}{T}\, v\, ds)\, dt\right|$$
$$= \left|\int_0^T u(v - \frac{1}{T}\int_0^T v\, ds)dt\right|$$
$$\leq \|u\|_{Y_1^*}\|v - \frac{1}{T}\int_0^T v\, ds\|_{Y_1} \leq 2 \cdot \|u\|_{Y_1^*}\|v\|_Y$$

It is now clear from $X = L^\infty[0,T] = (L^1[0,T])^* = Y$ that

$$\|u\|_X \leq 2 \cdot \|u\|_{Y_1^*}$$

Hence the Lemma.//

Definition 5. For $v \in Y = L^1[0,T]$ define $x \in L^\infty[0,T]$ by

$$x(t) = -\int_0^T\int_0^\xi e^{c(\xi-s)}u(s)ds d\xi - \frac{e^{ct} - 1}{c(e^{-cT}-1)}\int_0^T e^{-cs}u(s)ds. \quad (1.4)$$

We define a bounded linear integral operator $K: Y \to L^\infty[0,T] = X$ by

$$(Ku)(t) = x(t) - \frac{1}{T}\int_0^T x(t)dt, \quad t \in [0,T]. \quad (1.5)$$

Clearly for $u \in Y$, $Ku \in X_1$ since $\int_0^T (Ku)(t)dt = 0$. The following theorem shows that K/Y_1 is right-inverse to the linear mapping L and is an angle-bounded mapping. We shall need this theorem in section 2. We shall denote K/Y_1 by K itself.

Theorem 5. (i) For $u \in Y_1$, $(Ku)(t) \in D(L)$ and $LKu = u$.

(ii) The bounded linear mapping $K: Y_1 \to X_1 \subset Y_1^*$ is an angle-bounded mapping and for $u \in Y_1$

$$(Ku,u) \geq \frac{1}{4(1+c^2T^2)\|K\|}\|Ku\|_X^2 \quad (1.6)$$

Proof: (i) follows easily from (1.4), (1.5) and the definition of L as given in (1.2). We omit the technical details for the reader.

For (ii) let $u,v \in Y_1$ and $x = Ku$, $y = Kv \in D(L)$ so that $LKu = Lx = u$ and $LKv = Ly = v$. The monotonicity of K is immediate since

$$(Ku,u) = (Lx,x) \geq 0$$

and the angle-boundedness of K follows from

$$\begin{aligned}|(Ku,v) - (Kv,u)| &= |(Lx,y) - (Ly,x)| \\ &\leq 2|c|T\sqrt{(Lx,x)}\sqrt{(Ly,y)} \\ &= 2|c|T\sqrt{(Kx,x)}\sqrt{(Ky,y)}\end{aligned}$$

The inequality (1.6) is now immediate from Proposition 2 and lemma 4. //

SECTION II

In this section we apply the results of section 1 to get the uniqueness of solutions of the equation:

$$\left.\begin{aligned}&-x'' + cx' + f(t,x) = p(t), \quad t \in [0,T] \\ &x(0) = x(T) \\ &x'(0) = x'(T)\end{aligned}\right\} \tag{2.1}$$

As in section 1, let $X = L^{\infty}[0,T]$, $Y = L^1[0,T]$, $X_1 = \{u \in L^{\infty}[0,T] \mid \int_0^T u(t)dt = 0\}$ and $Y_1 = \{u \in L^1[0,T] \mid \int_0^T u(t)dt = 0\}$. Let $P: Y \to Y_1$ defined by $Pu = u - \frac{1}{T}\int_0^T u(t)dt$ denote the canonical projection of Y onto Y_1. Notice that P maps X onto X_1 and $P/X: X \to X_1$ is the canonical projection of X onto X_1. In the following we shall not distinguish between P and P/X and shall depend on the context for appropriate meaning. Let $L: D(L) \subset X \to Y$ be the linear mapping as defined in section 1 by (1.2) and $K: Y_1 \to X_1$ be the integral operator as defined by (1.5) in section 1 so that for $u \in Y_1$ we have

$$(Ku,u) \geq \frac{1}{4(1+c^2T^2)\|K\|} \|Ku\|_X^2 \quad \text{for } u \in Y_1$$

in view of theorem 5.

Lemma 6. For $v \in D(L)$, $(Lv,v) \geq \dfrac{1}{4(1+c^2T^2)\|K\|} \|Pv\|_X^2$.

Proof: For $v \in D(L)$ let $Lv = u$. Clearly $u \in Y_1$ and $Ku = v - \frac{1}{T}\int_0^T v(t)dt = Pv$. It then follows

$$(Lv,v) = (Ku,u) \geq \frac{1}{4(1+c^2T^2)\|K\|} \|Ku\|_X^2$$

$$= \frac{1}{4(1+c^2T^2)\|K\|} \|Pv\|_X^2,$$

using Theorem 5 and the fact that $u \in Y_1$ i.e. $\int_0^T u(t)dt = 0$. Hence the lemma. //

Theorem 7. Let $f\colon [0,T] \times \mathbb{R} \to \mathbb{R}$ be a continuous function such that there is a constant $\alpha \geq 0$ satisfying the following:

(i) $[f(t,x_1) - f(t,x_2)](x_1 - x_2) \geq -\alpha(x_1 - x_2)^2$ for $t \in T$, $x_1,x_2 \in \mathbb{R}$;

(ii) $\alpha T \leq \dfrac{1}{4(1+c^2T^2)\|K\|} \equiv \beta$ (say).

Let, now, u_1, u_2 be two solutions of (2.1) with $\int_0^T u_1 = \int_0^T u_2$. Then $u_1(t) = u_2(t)$ for $t \in [0,T]$.

Proof: Since u_1, u_2 are two solutions of (2.1) with $\int_0^T u_1 = \int_0^T u_2$ we have

$$Pu_1 - Pu_2 = u_1 - u_2 \quad \text{and}$$

$$Lu_1 + f(t,u_1(t)) = p(t)$$

$$Lu_2 + f(t,u_2(t)) = p(t)$$

for $t \in [0,T]$. It then follows from Lemma 5 and our assumption on f that

$$0 = (Lu_1-Lu_2,u_1-u_2) + \int_0^T (f(t,u_1(t))$$
$$- f(t,u_2(t))(u_1(t)-u_2(t))dt$$
$$\geq \beta\|Pu_1 - Pu_2\|_X^2 - \alpha\int_0^T (u_1(t)-u_2(t))^2dt$$
$$\geq (\beta-\alpha T)\|u_1 - u_2\|_X^2 \geq 0.$$

Hence $u_1(t) = u_2(t)$ for $t \in [0,T]$ and the proof of the theorem is complete. //

Corollary 8. Let $f\colon [0,T]\times\mathbb{R}\to\mathbb{R}$ be a continuous function such that $\frac{\partial f}{\partial x}(t,x)$ exists for every $(t,x)\in[0,T]\times\mathbb{R}$. Suppose that (i) $\frac{\partial f}{\partial x}(t,x) \geq -\alpha$ for $(t,x)\in[0,T]$ with $0 \leq \alpha T \leq \frac{1}{4(1+c^2T^2)\|K\|}$. Then, if u_1, u_2 are two solutions of (2.1)

$$\text{with } \int_0^T u_1 = \int_0^T u_2 \text{ we must have } u_1(t) \equiv u_2(t) \text{ for } t\in[0,T].$$

Proof: It is easy to see using mean-value theorem that f satisfies the conditions of Theorem 7. //

Remark 9. If we define $Ku(t) = x(t)$ where $x(t)$ is as defined by (1.4) of Section 1, it follows easily (essentially as above) that if u_1, u_2 are two solutions of (2.1) with $u_1(0) = u_2(0)$ then $u_1(t) \equiv u_2(t)$.

REFERENCES

[1] Amann, H. (1969). "Ein Existenz-und Eindeutigkeit fur die Hammersteinsche Gleichung in Banach raumen", *Math. Zeit. 111*, 175-190.

[2] Browder, F. E. (1971). "Nonlinear functional Analysis and Nonlinear Integral Equations of Hammerstein and Urysohn Type", Contributions to Nonlinear Functional Analysis, *Pub. No. 27, MRC*, Univ. of Wisconsin, 425-501.

[3] Browder, F. E., and Gupta, C. P. (1969). "Nonlinear monotone operators and integral equations of Hammerstein type", *Bull. Amer. Math. Soc. 75*, 1347-1353.

[4] Chang, S. H. (1976). "Periodic Solutions of Certain Differential Equations With Quasibounded Nonlinearities", *Jour. Math. Anal. Appl. 56*, 165-171.

[5] DeFigueiredo, D. G., and Gupta, C. P. (1973). "On the variational method for the existence of solutions of nonlinear equations of Hammerstein type", *Proc. Amer. Math. Soc. 40*, 470-476.

[6] Leach, D. E. (1970). "On Poincare's perturbation theorem of W. S. Lond", *J. Differential Equations 7*, 34-50.

[7] Hess, P. (1971). "On nonlinear equations of Hammerstein type in Banach spaces", *Proc. Amer. Math. Soc. 30*, 308-312.

COMPARTMENTAL MODELS OF BIOLOGICAL SYSTEMS: LINEAR AND NONLINEAR

John A. Jacquez

Department of Physiology
The University of Michigan
Ann Arbor, Michigan

I. INTRODUCTION

Richard Bellman has often said that mathematics is constantly renewed by the challenge of application to the solution of real problems and that the corollary to that is that mathematics would become a dull game without the input from real world problems. I suspect that is in part why I am here: I am a physiologist and a dabbler in mathematics and have been concerned with the problems of modeling biological processes and systems.

In this century the biological sciences have been, and continue to be, a growing source of a remarkable variety of mathematical problems. I want to tell you something about the modeling process in the biological sciences and then go on to talk about an area of modeling that has come to be called compartmental analysis. Because nonlinear analysis is the major subject of this meeting, I shall try to emphasize problems with nonlinear systems as I go along.

ISBN 0-12-434180-2

II. MODELS AND THE MODELING PROCESS IN BIOLOGY

To understand some of the problems and the unique features of modeling in the biological sciences it will help to compare how mathematical models are used in the physical and in the biological sciences.

The great flowering of physics and chemistry in the latter part of the last century and the first part of this century gave rise to problems that added impetus to the development of differential equations, ordinary and partial, and the statistical mechanics. These arose at the two ends of a spectrum of size of a system. At one end of that spectrum modeling involved the application of a few basic laws to systems that were not only relatively simple in structure but the systems were well defined in the sense that their structures were exactly specifiable. At the other end of the spectrum, statistical mechanics developed as a method for handling systems that were large but were complex only in the sense that they were made up of large numbers of identical units and the interactions between the units were given by a few simple laws. Classical mathematical physics grew rapidly because it dealt primarily with problems at these two ends of the spectrum. The problems in between, involving fairly complex systems but not made up of so large a number of units as to allow of easy statistical treatment or if made up of a sufficiently large number of units, having complex interactions between the units, these problems have been difficult to handle.

For the most part biological systems fall in that in between group and the development of good mathematical models has been a slow and difficult business. What are the problems? Besides complexity in structure and in interactions there is considerable variability from species to species and even from individual to individual within the species. To give an example, suppose we are interested in the mechanisms by which kidneys produce a concentrated urine. The kidneys of different species differ considerably in size and structure and it took years of experimental work before we understood the major processes involved. We think that we have unraveled that knot far enough so that we now have some not too complex models that show many of the features of the concentrating function of real kidneys. But these models are not models of specific kidneys and in fact they do not include all of the complexity of any real kidney. This illustrates one of the significant differences between modeling in the physical and in the biological sciences. In the physical sciences mathematical models are often such good mappings of structure and process that we can expect that simulated experiments run on the model will give results that match those from the same experiment run on the real system to many significant figures. We often cannot expect that for models of biological systems where the model abstracts only the major features of the system and neglects many small features. The problem of evaluating such models is more difficult. We cannot expect a close match between the results of an experiment simulated on the model and the same experiment run on the real system so we have to depend on more general criteria such as similar patterns of behavior for experiments involving a wide variety of initial and boundary conditions.

These differences have some fairly obvious implications for modeling in the biological sciences. For one, it puts a premium on detailed knowledge and understanding of the properties of the biological system. As a consequence biologists have played the major role in model specification. But once a model is specified the professional mathematician plays the major role for he is the one that is best equipped to attack questions such as the following. Is the model well posed? Can it be imbedded in a more general class of problems? What are the general properties of solutions?

Mathematical modeling of biological systems has advanced to the point where it is relatively easy to exhibit a catalog of examples of problems which lead to, say, systems of nonlinear differential equations of initial value or boundary value type. If you want some interesting examples see the recent book by J. D. Murray [1] and issues of journals such as the Journal of Theoretical Biology and Mathematical Biosciences. Rather than hop from one example to another I would like to concentrate on a system or methodology of modeling that has turned out to be very useful and which has come to be called compartmental analysis. More importantly, for the biologist and particularly the physiologist compartmental modeling is a natural way for him to think about his problem; it allows him to build models in terms of constructs that are familiar and close to the substantive content of his field.

III. COMPARTMENTAL SYSTEMS

First I would like to spend a little time talking aobut compartmental systems. Some of you no doubt know all about compartmental systems and for you I regret this repetition but I feel I must set the stage for what is to follow. Moreover, it is important: Compartmental modeling has come to be used as a systematic method of modeling for a variety of problems in the biomedical area.

A. *Definitions*

A compartment is an amount of some material which acts kinetically like a distinct, homogeneous, well-mixed amount of material. A compartmental system consists of one or more compartments which interact by exchanging the material. There may be inputs into one or more compartments from outside the system and there may be excretions from the compartments of the system.

In the definition of a compartment I emphasize the idea of kinetic homogeneity. In real problems this has to be considered in terms of the rates of the processes involved. There is no such thing as instantaneous mixing but if mixing within each of various amounts of material is rapid in relation to the transfers between them a compartmental representation is appropriate. Another point that needs to be emphasized is that a compartment is not a volume. It may be the amount of material in some well defined volume. But note that an element present in two chemically distinct forms with a reaction transforming one compound into the other and with both of the compounds uniformly distributed in the same volume is representable as a two compartment system.

B. *Equations and Connectivity Diagrams*

We often represent a compartment, this kinetically homogeneous amount of material, by a box, as in Figure 1. In the box representation for the ith compartment of such a system, q_i is the size of the compartment, i.e., the ith amount of this material which is kinetically homogeneous. Note one of the inherent restrictions of compartmental systems; all q_i are non-negative quantities. The arrows to and from the jth compartment are labeled with fractional transfer coefficients, which may or may not be constants. The excretion to the environment is indicated by fractional excretion coefficient f_{0i} and the input from outside

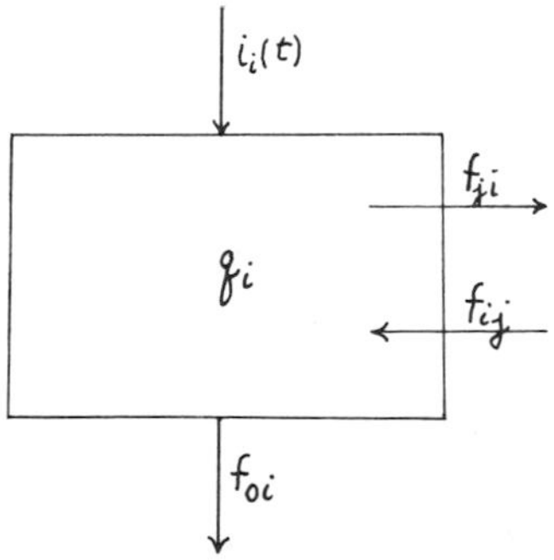

FIGURE 1

the system is $i_i(t)$. Thus, the equation for the ith compartment is (1).

$$\dot{q}_i = -(f_{0i} + \sum_{j \neq i} f_{ji})q_i + \sum_{j \neq i} f_{ij}q_j + i_i(t) \tag{1}$$

The coefficients of q_i are usually lumped into one, giving $f_{ii} = -(f_{0i} + \sum_{j \neq i} f_{ji})$

$$\dot{q}_i = f_{ii}q_i + \sum_{j \neq i} f_{ij}q_j + i_i(t) \tag{2}$$

Thus, in vector-matrix form:

$$\dot{q} = fq + i(t) \tag{3}$$

in which f is the matrix of coefficients. Frequently we have systems for which all f_{ij} are constants, giving us the familiar linear systems with constant coefficients. Occasionally some f_{ij} are time dependent, usually periodic, giving us linear systems with time dependent coefficients. More often the f_{ij} are functions of some of the q_j but not explicit functions of time, giving us a type of nonlinear system. Sometimes the transfer from one compartment to another takes a finite time giving us differential-difference equations such as the following for constant transfer coefficients.

$$\dot{q}_i = f_{ii}q_i(t) + \sum_{j \neq i} f_{ij}q_j(t - \tau_{ij}) + i_i(t) \tag{4}$$

We diagram compartmental systems by directed line segments. Such a diagram with inputs and excretions is a connectivity diagram as shown in Figure 2. The properties of the homogeneous system that depend only on the non-zero flows are exhibited by converting the connectivity diagram to a directed graph by dropping the inputs and defining a terminal compartment that receives all excretions. Thus Figure 2 becomes the digraph of Figure 3.

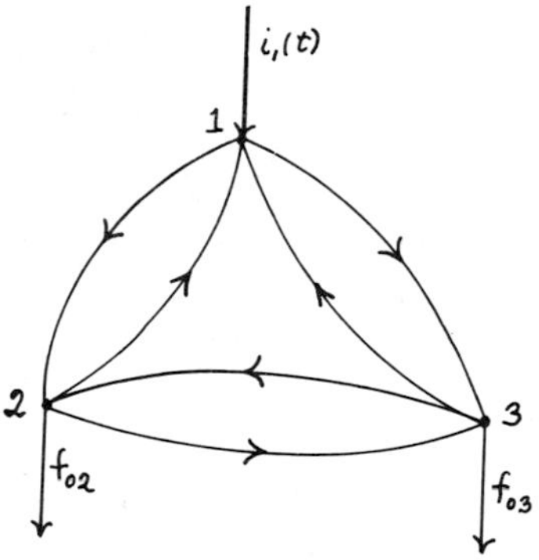

FIGURE 2

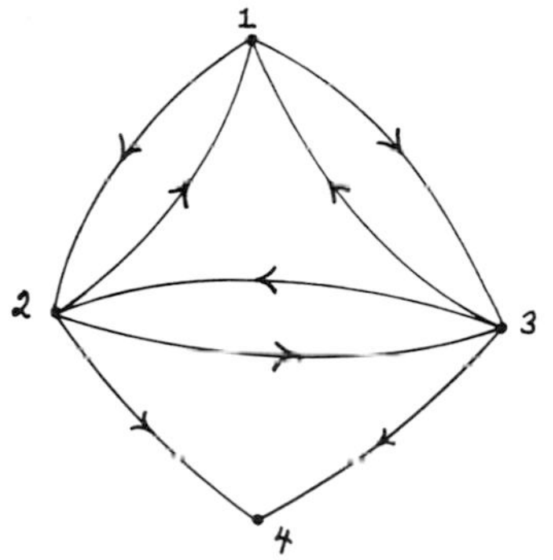

FIGURE 3

C. *Applications*

Let me give you some illustrations to show you the variety of processes that can be modeled with use of compartmental models.

The distribution of many materials in the body, normal constituents of the body or drugs, is often well modeled by compartmental models. Even though mixing in the circulation takes 2 - 3 circulation times it is surprising how well compartmental models represent the distribution of an injected compound, once the initial mixing transient is gone. Often nonlinear models are required and then a standard experimental technique is to put the system in a steady state and to follow the distribution of a small amount of material labeled with radioactive tracer. For a steady state system, linear or nonlinear, the distribution of the tracer follows linear kinetics.

Another set of phenomena that are well modeled by compartmental models are the growth and maturation of various cell types in the body. Cells such as the red blood cells which pass through a series of distinct and easily recognized stages in their maturation can be thought of as passing from one compartment to another during maturation.

As a final example, I would point to ecology and the modeling of food chains and competitive interactions between species as another area in which compartmental models have been used.

In what follows, I would like to show you some of the interesting general problems that have arisen in compartmental systems and will try to emphasize the nonlinear systems.

IV. NONLINEAR COMPARTMENTAL SYSTEMS

It turns out that the nonlinear systems we meet are for the most part of a special type. For example, in models of distribution of a compound in the body the transfers between compartments often represent passage of the compound across cell membranes. This process not infrequently occurs by a mechanism called active transport or facilitated diffusion. In these, the chemical compound binds to specific components of the cell membrane and then is released unchanged to the other side of the membrane. The number of binding sites to which compounds can bind is fixed so the rate of transfer saturates, i.e. it approaches a maximum rate asymptotically as the concentration of compound increases. If time lags are not significant the equations can be written as in (5).

$$\dot{q}_i = f_{ii}q_i + \sum_{j\neq i} f_{ij}q_j + i_i(t) \tag{5}$$

For a system of the type just described:

(1) The f_{ij} are functions of q and a vector of parameters but are not explicit functions of time. Usually f_{ij} is a function of only q_i and q_j and of some parameters.

(2) f_{ij} is a bounded function in each of the q_k with the following properties. As a function of q_k, f_{ij} is either a non-decreasing function of q_i or a non-increasing function. Since the f_{ij}, $i \neq j$, are non-negative, this means that for all other variables constant, as q_i increases, f_{ij} either increases to a positive limit value or it decreases to a non-negative limit value.

Given that the fractional transfer coefficients are constrained in this way it seems to me that nonlinear systems of this type must have many of the properties of linear systems. For the analyst, perhaps general properties of such restricted classes of nonlinear systems would be much easier to obtain than for general nonlinear systems. This would be of interest to those of us concerned with applications.

V. IDENTIFICATION AND THE INVERSE PROBLEM

First I want to concentrate on linear systems with constant coefficients. After defining the identification problem in this context the difficulties involved in the corresponding problem with nonlinear systems will be obvious.

A. *The Inverse Problem*

For compartmental systems it is useful to divide the problem into three stages.

1. *Order of System and Connectivities.* One of the first problems is to determine the number of compartments and their connectivities, i.e., the non-zero fractional transfer coefficients. The first thing to do is to follow any compartment and

determine the number of exponential components in the decay curve. This gives a lower bound on the number of compartments. For biological systems it is usually not possible to observe all compartments. Consequently the compartmental model is seldom determined solely by observing compartmental decay curves. Information on anatomical structure, bio-chemistry and physiology is used in conjunction with observations on one or more compartments to arrive at a model which must generally be viewed as a minimal model. One problem which I cannot discuss here is the effect of the choice of sampling times on the detectability of exponential decay components and hence on the theoretical countability of the number of compartments.

2. *Identifiability.* Suppose we know the number of compartments and their connectivities, i.e., we know which fractional transfer coefficients are non-zero, though not their values. The problem for the experimentalist is to take samples from some compartments and estimate the values of the fractional transfer coefficients. However, since the experimentalist does not have access to all compartments of this system he cannot even be sure that all of the parameters are fully defined from the data set he can gather, let alone carry out the statistical estimation. For this reason we separate the statistical estimation problem from the question of the theoretical determinability of the parameters. Let us define the latter more precisely. Given the system order and connectivities we specify that only the compartments of a given subset can be followed and that the measurements are error free, do these measurements uniquely determine all of the non-zero fractional transfer coefficients? This is the identification problem or what is also called the problem of structural identifiability.

3. *Statistical Estimation.* If the system is identifiable one can proceed to actually estimate the parameters. And, because the measurements are in fact not error free the problem is one of

statistical estimation of nonlinear parameters because the fractional exchange coefficients do not all appear linearly in the parameters directly estimated by the time courses of different compartments. However, it should be noted that when the system is not identifiable it is possible to fix some of the fractional transfer coefficients from other considerations to make the remaining problem identifiable and then carry out a statistical estimation on the remaining transfer coefficients, conditional on the values chosen for the fixed set.

As you can see, the inverse problem encompasses a number of interesting and difficult subproblems. Of these I would like to concentrate on the identification problem.

B. *General Linear Systems Theory and the Identification Problem*

Before I look at linear compartmental systems let me remind you of some material from the theory of general linear systems, material that is very closely related. In this I follow the presentation of Kalman, [2].

1. The State Variable Description of Linear Systems. The states of a linear dynamical system are defined by the values of a set of variables called the state variables. Suppose $x(t)$ is a vector of state variables $(n \times 1)$ and that $y(t)$ is a vector $(p \times 1)$ of the observed quantities which may be linear combinations of the components of $x(t)$. Let $i(t)$ be a vector of nonzero inputs $(m \times 1)$. Then a general linear system, starting from rest, has the equations (6), in which $A(t)$ is $(n \times n)$, $B(t)$ is $(n \times m)$, and $C(t)$ is $(p \times n)$.

$$\dot{x}(t) = A(t)x + B(t)i(t); \quad x(0) = 0 \tag{6}$$

$$y(t) = C(t)x$$

I shall be concerned only with time-invariant linear systems, i.e. systems for which A, B and C are constant matrices.

2. *The Input-Output Description of Linear Systems.* Another way to characterize a linear system is in terms of its response to unit impulse inputs. Consider a system at rest at $t = t_0$. Let $h_{ij}(t,t_0)$ be the output, $y_i(t)$ due to unit impulse input into x_j. The matrix $h(t,t_0)$ is the impulse response matrix or transfer function in the time domain. The output for arbitrary input is given by equation (7). We generally let $t_0 = 0$. For a time invariant system $h(t,\tau) = h(t - \tau)$ and equation (7),

$$y(t) = \int_{t_0}^{t} h(t,\tau)B(t)i(\tau)d\tau \tag{7}$$

can be written in the form of (8).

$$y(t) = \int_{0}^{t} h(t - \tau)Bi(\tau)d\tau \tag{8}$$

We frequently refer to $h(t)$ in terms of its Laplace transform, the transfer function.

3. *Relation Between Input-Output and State Equation Descriptions.* The two descriptions of dynamical systems are related but not equivalent. Given the dynamical equations, the impulse response matrix is fully determined. The reverse is not true. Given the impulse response matrix the dynamical equations may or may not be uniquely determined. This provides us with a particularly sharp statement of the identification problem. The transfer function or impulse response matrix for the input-output experiment can be considered to be what is observed. The identification problem is then: under what conditions does the impulse response matrix uniquely identify the system?

Given the impulse response matrix it is generally possible to find many state equation descriptions that generate the impulse response matrix. Each such description is called a realization of $h(t,\tau)$. The number of components of the state vector of a realization is its dimensionality. Among the realizations is a subset of the smallest dimensionality, say n_0. Thus one answer at the moment is that the impulse response matrix determines a subset of irreducible representations. The problem of uniqueness we consider specifically in the context of compartmental systems.

C. *The Identification Problem for Linear Compartmental Systems With Constant Coefficients*

1. *A Necessary Condition.* It is instructive to consider the identification problem in terms of the properties of the directed graph of a system. I shall develop a necessary condition rather heuristically, in terms of the ideas of input and output reachability. First, let me remind you of two simple ideas from graph theory. The reachable set, $R(x)$, of a point x of a graph is the set of all points reachable from x. The antecedent set, $Q(x)$, of a point x is the set of all points of the graph from which x is reachable.

We define an identification experiment as follows. Given a compartmental system starting at rest, i.e. $q(0) = 0$, we inject inputs into a subset of the compartments of a system and the time courses of some combinations of another subset of compartments are followed. Are all of the non-zero f_{ij} determinable? To have even a possibility of determining a particular f_{ij}, the transfer which has fractional transfer coefficient f_{ij} must affect the observed compartments. To make the argument concrete consider the system shown in Figure 4. For the identification experiment, material (usually radioactively labeled) is injected into compartment 1 and compartment 8 is observed. All transfers that affect compartments on paths from 1 to 8 can influence the time course of compartment 8. Thus f_{21} affects compartment 1 but f_{32} does not; f_{41} and f_{56} affect compartments 1 and 6. respectively, which are on a path from 1 to 8, but f_{45} and f_{54} do not. Thus the transfer coefficients f_{32}, f_{45} and f_{54} cannot possibly affect the observations. The transfer of

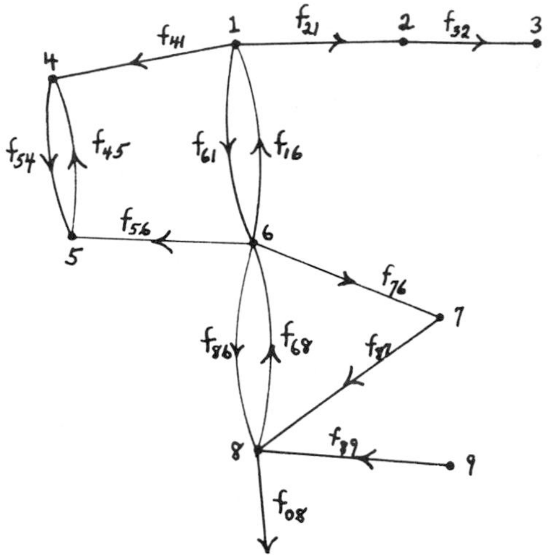

FIGURE 4

coefficient f_{89} is of a different type. It can potentially influence 8 but because there is no path for labeled material to pass from 1 to 9, none can reach 9 to transfer to 8 and so f_{89} cannot influence the observations for this experiment. Because f_{32}, f_{45}, f_{54} and f_{89} cannot possible appear in the observations of compartment 8 this system is structurally non-identifiable for this experiment.

a. Input reachability. Let S_i be the set of compartments that receive inputs in the identification experiment and let V be the set of all compartments. Then we must have $R(S_i) = V$ where $R(S_i)$ is the set of all compartments reachable from S_i. Isolated compartments are excluded from consideration.

b. Output reachability. For all compartments which have at least one path leaving them there must be a path to an observed compartment. Let S_0 be the set of observed compartments and let T be the set of terminal compartments. Then the antecedent set of S_0, $Q(S_0)$, which is the set of all compartments from which S_0 can be reached, must be $V - T$.

2. *The Transfer Function for an Identification Experiment and Sufficient Conditions.* Assuming the necessary conditions are satisfied, some of the f_{ij} may still not be identifiable. In order to determine identifiability we must examine the transfer function for the identification experiment to see whether all of the f_{ij} appear in the transfer function and if so, whether all are uniquely determined. The material which follows owes much to the papers of Cobelli and coworkers [3 - 5].

Consider an identification experiment with a linear compartmental system with constant coefficients in which there are r distinct inputs, into r different compartments. The equations for the system for this particular experiment are (9) and (10). Note that $i_d(t)$ represents the inputs of the identification

$$\dot{q} = fq + i_d(t) \tag{9}$$

$$y = Cq \tag{10}$$

experiment and is an n-vector with r non-zero components. We rewrite it in the form (11) in which $i_r(t)$ is an r-vector of non-zero elements and B is an $n \times r$ matrix with elements 0 or 1 so that (11) holds; every

$$i_d(t) = Bi_r(t) \tag{11}$$

column of B has just one 1 in it. If there are p observations, C is a $p \times n$ matrix; the observations may be individual compartments or linear combinations of the compartment values. A washout curve is an example of the latter. We assume *all* compartments are initially zero, $q(0) = 0$. Taking Laplace transforms gives us (12) and (13). Note that $H(s) = (sI - F)^{-1}$ is the general transfer function for the system. That is to say $h_{ij}(s)$

$$Q = (sI - F)^{-1}Bd_r(s) \tag{12}$$

$$Y = C(sI - f)^{-1}Bd_r(s) \tag{13}$$

is the transfer function for the response in compartment i due to unit impulsive input into j. Consequently equation (14) gives the transfer function

$$G(s) = CH(s)B = C(sI - f)^{-1}B \tag{14}$$

for this particular identification experiment on this system. In effect it gives us the combinations of the $h_{ij}(s)$ which form the elements of $G(s)$, i.e., which are available from the observations in this experiment. Recall that the inverse $(sI - F)^{-1}$ is given by (15).

$$H(s) = (sI - f)^{-1} = \frac{\text{adj}(sI - f)}{|sI - f|} \tag{15}$$

The determinant can be written in expanded form, the characteristic polynomial, as in (16).

$$|sI - f| = s^n + \alpha_1 s^{n-1} + \ldots + \alpha_n \tag{16}$$

The coefficients, α_i, are functions of the roots of the characteristic polynomial, the eigenvalues, since $|sI - f|$ can also be written in factored form as in (17).

$$|sI - f| = (s - \lambda_1)(s - \lambda_2)\ldots(s - \lambda_n) \tag{17}$$

The adjoint matrix can also be expanded in a matrix polynomial as in (18),

$$\text{adj}(sI - f) = R_0 s^{n-1} + R_1 s^{n-2} + \ldots + R_{n-1} \tag{18}$$

in which $R_0, R_1, \ldots, R_{n-1}$ are matrices, given by equations (19). Substituting

$$\begin{aligned} R_0 &= I \\ R_1 &= f + \alpha_1 I \\ R_2 &= f^2 + \alpha_1 f + \alpha_2 I \\ &\vdots \\ R_{n-1} &= f^{n-1} + \alpha f^{n-2} + \ldots + \alpha_{n-1} I \end{aligned} \tag{19}$$

in equation (14) gives us (20) for the transfer function for the identification

$$G(s) = \frac{1}{|sI - f|}\Big\{CB(s^{n-1} + \alpha_1 s^{n-2} + \ldots + \alpha_{n-1} \tag{20}$$

$$+ CfB(s^{n-2} + \alpha_1 s^{n-3} + \ldots + \alpha_{n-2})$$

$$\vdots$$

$$+ Cf^{n-2}B(s + \alpha_1) + Cf^{n-1}B\Big\}$$

experiment. The denominator is of degree n in s but all terms in the numerator are of lower degree. The problem now reduces to an examination of all of the coefficients of the terms in the numerator and denominator to see of all f_{ij} are represented and estimable from coefficients. Though easy to state in theory this may be a difficult task as soon as the system is larger than 3 - 4 compartments. The maximum number of parameters estimable is determined by the number of non-zero coefficients in the numerator and denominator of (20). If $\alpha_n \neq 0$ the denominator can determine no more than n parameters. For the coefficients in the numerator the terms of CB, CfB, ... $Cf^{n-1}B$ have to be examined to see how many non-zero terms are present. Assuming there are enough non-zero coefficients, the identification problem for linear compartmental systems reduces to a classical problem in nonlinear algebra. Given N nonlinear algebraic equations in N unknowns, when is there a unique solution for the N unknowns?

D. *Nonlinear Systems and the Identification Problem*

The identification problem is much more difficult for nonlinear compartmental systems. In what follows I again assume that the f_{ij} are functions of q and a vector of parameters but that they are not explicit functions of time. The problem can be posed at really two levels of difficulty. If the functional forms of the f_{ij} are unknown we indeed have a difficult problem. However, if the functional form is known but the parameters that appear in the f_{ij} are unknown the problem is generally a nonlinear parameter estimation problem.

1. *Identification From Steady State Studies.* For small systems physiologist have developed a direct experimantal attack on the problem. Put the system in a steady state; then the vector q is constant and hence the f_{ij} are constant. It is easy to show that if one then introduces a small amount of radioactive tracer labeled material, it distributes with linear kinetics with the steady state values of the f_{ij}. Thus one can determine the values of the f_{ij} for this particular steady state from the theory for linear compartmental systems. This process is then repeated for a number of different steady states of the system. Thus one obtains values of q and the corresponding f_{ij} as functions of q. Obviously the work involved increases exponentially in the number of compartments so that it is practical only for small systems.

2. *MacLaurin Series.* Another possibility is to use a power series expansion.

$$\dot{q}(t) = f(q, i(t), t, \theta), \quad t \in [0, T] \tag{21}$$

$$y(t) = g(q, \theta) \tag{22}$$

In equations (21) and (22) θ is a vector of unknown parameters and f and g are analytic at $t = 0$. Let $a_k(0)$ be the coefficient of the MacLaurin series for $y(t)$. Then the system is identifiable if the set of equations (23) have a unique solution for θ [6].

$$g^k(q(0), \theta) = a_k(0) \tag{23}$$

VI. COMPLEXITY AND STABILTY

Finally let me bring up a problem that occurs in the theory of compartmental systems but is actually more general, applying to many systems that are in some respects similar to compartmental systems in that the system is representable as a digraph but there may not be a conservation law such as the conservation of mass which imposes such stringent constraints on compartmental systems. The problem has to do with the relation between complexity and stability of systems. It has to do with large systems such as socio-economic systems, ecosystems, the systems that constitute a particular organism. The problem has been raised by intuitive judgements to the effect that complexity leads to stability in some large systems and fragility in others. Attempts to model complex economic systems or ecosystems has stimulated efforts to define the problem more exactly and to derive rigorous results.

The definition of this problem hinges on the meaning of the terms *complex systems* and *stability*. The first is not well defined and the second is not used in the sense in which it is used in the stability theory of differential equations.

1. *Complex System*. The idea of a complex system incorporates two main concepts. One is that the system is made up of a large number of subunits, the other is that there are many *interactions* between subunits. The meaning of "complexity" in this context is carried as much or more by the interactions as by the number of units in the system.

2. *Stability*. The term "stability" is used in a way that is different from the usual definitions. Let me remind you of some of the definitions of stability.

a. Differential equations. The classical idea of stability is that if a steady state system is subjected to perturbations in the state variables it returns to the steady state.

b. Parametric stability. This is similar to the previous. A small shift in parameters of a steady state system leads to a smooth transition to a new, nearby steady state. This is a definition of importance for biological systems.

c. Structural stability. Consider an ecosystem represented by a digraph, the nodes being the species. Suppose that as a result of some catastrophe one or more species disappear, i.e. one or a few nodes are deleted. Does the system shift smoothly to a new steady state without violent oscillations or is there a break down of the system with loss of more nodes?

d. Connective stability. This refers to systems such as economic systems where the transfers between nodes represent shifts between various components of an economy. Suppose some transfers are deleted. Stability in this sense implies a smooth shift to a new steady state without large oscillations or breakdown of the system into smaller digraphs due to disappearance of more connections.

The sorts of stability implied in the complexity-stability problem are of the type of structural stability and connective stability. The last few years have seen the appearance of work on this problem [7 - 9]. More can be expected as the formulation of the problem becomes more exact.

REFERENCES

[1] Murray, J. D. (1977). "Lectures on Nonlinear Differential-Equation Models in Biology", Clarendon Press, Oxford.

[2] Kalman, R. E. (1963). "Mathematical description of linear dynamical systems", *SIAM J. Control 1*:152.

[3] Cobelli, C., and Romanin-Jacur, G. (1975). "Structural identifiability of strongly connected biological compartmental systems", *Med. Biol. Eng.* :831.

[4] Cobelli, C., and Romanin-Jacur, G. (1976). "Controllability, observability and structural identifiability of multi-input and multi-output biological compartmental systems", *IEEE Trans. Biomed. Eng. BME-23:*93.

[5] Cobelli, C., and Romanin-Jacur, G. (1976). "On the structural identifiability of biological compartmental systems in a general input-output configuration", *Math. Biosci. 30:* 139.

[6] Pohjanpalo, H. "System identifiability based on the power series expansion of the solution", *Math. Biosci.* (In press).

[7] Ladde, G. S. (1976). "Cellular systems I. Stability of chemical systems", *29:* 309.

[8] Ladde, G. S. (1976). "Cellular systems II. Stability of compartmental systems", *30:*1.

[9] Siljak, D. D. (1978). "Large-Scale Dynamic Systems", Elsevier North-Holland.

NEW OPTIMIZATION PROBLEMS FOR DYNAMIC MULTICONTROLLER DECISION THEORY

*Harriet Kagiwada**

HFS Associates
Los Angeles, California

I. INTRODUCTION

There are numerous applications of the theory of optimal control and decision-making[1-4]. They are as varied as the mechanical controllers in airplanes and spacecraft, the neurophysiological controllers in the human body, and human beings themselves as controllers and decision-makers in complex situations. In order to treat such problems, there have been developed dynamic programming, statistical decision theory, nonlinear filtering theory, team theory, and other theories and methods.

Recently, my colleagues and I made a fundamental advance. We are now able to formulate and solve, in many cases, dynamic multicontroller decision problems[5,6]. These problems go beyond the classical ones named above in that they combine the time varying aspects of dynamic programming with the multicontroller aspects of team decision theory. And, since most of these problems lead

**The research was supported by AFOSR Grant No. 77-3383, by NIH Grant GM23732, and by a grant-in-aid from Sigma Delta Epsilon/ Graduate Women in Science.*

ISBN 0-12-434180-2

to integral equations, we make full use of modern methods for the solution of such equations[7].

Why study dynamic, multicontroller decision problems? They are important in many areas. These include (1) the military problem of the determination of optimal command decisions to be made by subordinate commanders who must coordinate their activities among themselves and over the course of time in order to optimally achieve an objective set down by superordinate headquarters. (2) In business organizations, there are various specialists in different areas of financial operations who try to cooperate to achieve maximum profit for their firms. (3) In the human body, there are many decentralized controllers constantly at work to keep the body functioning well, even in so simple a task as standing upright.

In modeling organizations, we must realize that they are complex systems[2]. There are individual differences in the amount, accuracy and timeliness of information presented to each decisionmaker. Yet, the whole concept of a team means that these separate controllers must cooperate, and coordinate their actions, in order to optimally achieve the common objective of the team. This will sometimes be done at some expense to the individuals. Such decisions are affected by uncertainty about the environment, and still further complicated by any lack of communication among the team members.

Yet it is possible to formulate such optimization problems under certain circumstances. One of the key assumptions is that all team decision-makers carry out their optimal policies. The optimal control policies are, as one would expect, feedback decision rules. For time-dependent problems, these optimal policies can be determined through use of dynamic programming.

II. PROFIT MAXIMIZATION OF A FIRM — AN EXAMPLE

To illustrate, let us discuss a simplified version of the problem of maximizing the profit of a firm over a specified time period[5]. We have two decision-makers: the president of the firm, who determines the investment policy, and the foreman, who determines the hiring policy. The president has knowledge of the interest rate for the coming time period. The foreman's information variable is the wage rate for the period. They do not communicate these information variables to each other.

Let us define the variables. The state variable is k, the firm's capital at time t, where t takes on discrete values $t = 0,1,2,\ldots,N$, and the meaning of t is the number of time units (say, weeks) to go in the overall period. The information variables are the interest rate r and the wage rate w, of course at the time when there are t units to go. The amount of investment to be made by the president is denoted I, and the amount of labor to be hired by the foreman is denoted L. These are the dependent variables - the optimal policies, or decision rules - which we wish to determine.

The equations in the model of the firm are assumed to be the following. The capital accumulation is given by the equation for the current capital as

$$K = k + I, \tag{1}$$

i.e., the new capital amount is the sum of the capital already on hand and the investment currently (and immediately) being made. The immediate gain in this unit of time is

$$G = G(K,L,w,r) = pf(K,L) - wL - rK, \tag{2}$$

where p is the unit price and $f = f(K,L)$ is the production function which gives the number of units of the product produced using capital K and labor L. The initial capital is given.

The random occurrence of the pair of information variables w and r is expressed by the probability density function $g(w,r)$.

The objective is to maximize the expected profit of the firm over N units of time by making the sequence of optimal decisions for investment of capital,

$$I = I_t(k,r), \tag{3}$$

and for hiring of labor,

$$L = L_t(k,w), \tag{4}$$

for $t = 0,1,2,\ldots,N$.

We define an important auxiliary function:

$P_t(k)$ = the expected profit with t periods to go, beginning with capital k on hand, and following a sequence of optimal investment and hiring policies. (5)

We then apply Bellman's Principle of Optimality[1], which relates the expected profit with t periods to go, to that with $t - 1$ periods to go. This yields

$$P_t(k) = \max_{I,L} \iint [G(k+I,L,w,r) + P_{t-1}(k+I)]\, g(r,w)\,dr\,dw, \tag{6}$$

and leads to the recurrence relation,

$$P_t(k) = \max_{I,L} \iint [p\, f(k+I,L) - wL - r(k+I) + P_{t-1}(k+I)]\, g(r,w)\,dr\,dw, \tag{7}$$

with initial condition,

$$P_0(k) = 0. \tag{8}$$

The person-by-person optimality conditions are

$$0 = \int \frac{\partial}{\partial I} [p\, f(k+I,L) - wL - r(k+I) + P_{t-1}(k+I)]\, g(r,w)\,dw, \quad \text{for all } r, \tag{9}$$

$$0 = \int \frac{\partial}{\partial L} [p\, f(k+I,L) - wL - r(k+I) + P_{t-1}(k+I)]\, g(r,w)\,dr, \quad \text{for all } w. \tag{10}$$

We denote the partial derivative of $f = f(k+I,L)$ with respect to the first argument by $f_1(k+I,L)$, and that with respect to the second argument by $f_2(k+I,L)$. The derivative of P is denoted P'. We introduce the conditional probability functions $g(w\backslash r)$ and $g(r\backslash w)$. We also write $I(r) = I_t(k,r)$, and $L(w) = L_t(k,w)$. Then the optimality conditions become

$$0 = \int [p\, f_1(k+I(r),L(w)) - r + P'_{t-1}(k+I(r))]\, g(w\backslash r)dw, \quad \text{for all } r, \tag{11}$$

$$0 = \int [p\, f_2(k+I(r),L(w)) - w]\, g(r\backslash w)dr, \quad \text{for all } w. \tag{12}$$

These equations may be interpreted to mean that the decision makers' optimal policies are such that the marginal conditional expected profit is made equal to zero.

These equations form a system of nonlinear integral equations, in general. In particular, with a quadratic production function, they reduce to a pair of linear Fredholm integral equations. Much can be done in this case both analytically and computationally. It is known from the theory that the optimal policies are linear in the information variables and the state k, and that the expected profits are quadratic in k.

These optimal policies are to be determined for $t = 0,1,2,\ldots,N$, and for all allowed values of k.

III. DISCUSSION

It is easy to see that this example generalizes to more decision makers, more state variables, more information variables, different communication patterns, time dependent dynamics and gains, and so forth. The general problem reduces to that of solving a sequence of systems of nonlinear integral equations. Some imbedding methods have already been found to be applicable[3,7]. Search methods are of possible utility in special circumstances, e.g., when no information variables are known.

Much remains to be done in the area of numerical solution of these dynamic multicontroller decision problems. Much also needs to be done on the analytical side. Progress along these lines and many interesting applications can be expected in the coming years.

ACKNOWLEDGMENTS

The author wishes to acknowledge the contributions of Professors Robert Kalaba and James Hess to the development of this work.

REFERENCES

[1] Bellman, R., and Kalaba, R. (1966). "Dynamic Programming and Modern Control Theory", Academic Press, New York.

[2] Marschak, J., and Radner, R. (1972). "Economic Theory of Teams", Yale University Press, New Haven.

[3] Ider, A., Hess, J., Kagiwada, H., and Kalaba, R. (June 1977). "Team Decision Theory and Integral Equations", *J. Optimization Theory and Applications,* 251-264.

[4] Hess, J., Kagiwada, H., and Kalaba, R. (1977). "Command, Control, Communications and Team Decision Theory", Decision Information for Tactical Command and Control, Ed. by R. Thrall, C. Tsokos and J. Turner, R. M. Thrall and Associates, Houston, 231-245.

[5] Hess, J., Kagiwada, H., Kalaba, R., and Tsokos, C. "Cooperative Dynamic Programming", *Appl. Math. and Comp.,* (to appear).

[6] Kagiwada, H., and Kalaba, R. (Feb. 23-25, 1978). "C^3 Modeling", Proceedings of First Annual Workshop on the Information Linkages Between Applied Mathematics and Industry, Naval Postgraduate School, Monterey, California.

[7] Kagiwada, H., and Kalaba, R. (1974). "Integral Equations via Imbedding Methods", Addison-Wesley, Reading.

STABILITY TECHNIQUE AND THOUGHT PROVOCATIVE DYNAMICAL SYSTEMS II*

G. S. Ladde

Mathematics Department
The State University of New York at Potsdam
Potsdam, New York

SECTION I

Recently [2], an attempt has been made to study stability analysis of deterministic nonlinear systems in the context of logarithmic norm. In addition, it has been shown that logarithmic norm provides a suitable mechanism for unification, comparison and applications of stability conditions

In the present work, we extend the deterministic results [2] to random thought provocative nonlinear dynamical systems. Furthermore, the presented stability conditions demonstrate partial solutions to several problems that are of practical interest, for example, measurability of complexity and random disturbances, deterministic vs. stochastic, and sensitivity of stability. In short, the work provides variety of Lyapunov functions together with stability conditions that exhibit statistical properties and also show the usefulness to the stability analysis of dynamical

**The research reported herein was supported by the U.S. Army under Grant Number: DAAG29-77-G0062.*

ISBN 0-12-434180-2

processes in biological, physical and social sciences. However, details are reported in [4].

In Section 2, we present certain preliminary material. In Section 3, we give a basic stability result that unifies the earlier results in a systematic and unified way. In addition, we state some observations that illustrate the practical significance of the stability conditions.

SECTION II

Let us consider a system of random differential equations

$$x'(t,w) = A(t,x,w)f(x,w), \quad x(t_0,w) = x_0(w) \tag{2.1}$$

where $x \in R^n$; $A(t,x,w) = (a_{ij}(t,x,w))$ is an $n \times n$ random matrix function whose elements $a_{ij} \in M[R_+ \times B(0,\rho),R[(\Omega,F,P),R]]$, and $a_{ij}(t,x,w)$ are almost-surely sample continuous in x for fixed t; $f \in R[B(0,\rho),R[(\Omega,F,P),R^n]]$, $f(x,w) = (f_1(x,w), f_2(x,w),\ldots,f_n(x,w))^T$ and its sample derivative $f_x(x,w) = \frac{\partial f}{\partial x}(x,w)$ exists; $f_x(x,w)$ is almost-surely sample continuous on $B(0,\rho)$; $B(0,\rho) = \{x \in R^n\colon\ |x_i| < \rho_i,\ i \in I = \{1,2,\ldots,n\}, \rho_i > 0\}$. Further assume that the random functions $A(t,x,w)$ and $f(x,w)$ are smooth enough to assure the existence of sample solution process $x(t,w)$ of (2.1). Furthermore, without loss in generality, we assume that $x \equiv 0$ is unique solution of (2.1). Let us assume that

$$\mu(f_x(x,w)A(t,x,w)) = \lim_{h\to 0^+} \frac{1}{h}\Big[\|I + hf_x(x,w)A(t,x,w)\| - 1\Big]$$

exists and satisfies the relation

$$P\left\{w\colon \limsup_{t\to\infty}\left[\frac{1}{t-t_0}\int_{t_0}^{t} \sup_{x\in B(0,\gamma)}\Big[\mu(f_x(x,w)A(s,x,w))\Big]ds\right]\le -a\right\} = 1 \tag{2.2}$$

for some $a = a(t_0) > 0$ and $0 < \gamma < \rho$.

SECTION III

In this section, we shall formulate a basic stability result which plays an important role in unifying and systemizing the stability analysis of (2.1).

Theorem 3.1. Assume that $f(x,w)$ and $A(t,x,w)$ in (2.1) satisfies (2.2). Then, the trivial solution process $x \equiv 0$ of (2.1) is exponentially stable with probability one.

Proof: Define $v(x,w) = \|f(x,w)\|$, where $f(x,w)$ is as defined in (2.1). From the conditions on $f(x,w)$, it is obvious that $v(x,w)$ is positive definite and locally Lipschitzian in x. For small $h > 0$, we have

$$v(x+hA(t,x,w)f(x,w)) = \|f(x+hA(t,x,w)f(x,w)\|$$
$$= \|f(x)+hf_x(x,w)A(t,x,w)f(x,w)+o(t,x,h,w)\|$$

This together with the definition of $\mu(f_x(x,w)A(t,x,w))$ and the properties of norm, yields

$$D^+_{(2.1)}v(x,w) \le \mu(f_x(x,w)A(t,x,w)) \quad \text{w.p.1.} \tag{3.1}$$

for $(t,x) \in B(0,\rho)$. From (2.2), (3.1) and applying results in [1], we have

$$v(x(t),w) \le u_0(w)\exp\left[\int_{t_0}^{t} \sup_{x \in \overline{B}(0,\gamma)} \mu(f_x(x,w)A(s,x,w))ds\right] \tag{3.2}$$

whenever $v(x_0(w),w) \le u_0(w)$. The rest of the proof follows by following the standard argument that is used to study stability properties of deterministic differential systems.

A remark similar to Remarks 3.1 and 3.2 [2] can be formulated, similarly. Further, we remark that the stability condition (2.2) can be reformulated in the context of laws of large numbers with regard to elements of random rate matrix $A(t,x,w)$. For details, see [3].

Finally, remarks concerning the usefulness of the stability analysis can be made similar to deterministic study [2]. However, further details are reported in [4].

REFERENCES

[1] Ladde, G. S. (1976). "Systems of differential inequalities and stochastic differential equations III", *J. of Mathematical Physics, Vol. 17,* 2113-2119.

[2] Ladde, G. S. (1977). "Stability Technique and Thought Provocation Dynamical Systems", Nonlinear Systems and Applications (Ed. by V. Lakshmikantham), Academic Press, 211-218.

[3] Ladde, G. S. (1977). "Logarithmic norm and stability of linear systems with random parameters", *Int. J. Systems Sci., Vol. 8,* 1057-1066.

[4] Ladde, G. S. (1977). "Competitive Processes II. Stability of Random Systems", *J. of Theoretical Biology, Vol. 68,* 331-354.

REACTION-DIFFUSION EQUATIONS IN ABSTRACT CONES*

V. Lakshmikantham
A. S. Vatsala

Department of Mathematics
The University of Texas at Arlington
Arlington, Texas

S. Leela

Department of Mathematics
State University of New York
Geneseo, New York

I. INTRODUCTION

Let T be the temperature and n the concentration of a combustible substance. A simple model governing the combustion of the material is given by

$$\left.\begin{aligned} \frac{\partial T}{\partial t} &= K_1 \Delta T + Qn \exp(-E/RT), \\ \frac{\partial n}{\partial t} &= K_2 \Delta n - n \exp(-E/RT), \end{aligned}\right] \tag{1.1}$$

where the constant Q is the heat of reaction; the constants K_1, K_2 are thermal, material diffusion coefficients; the term $\exp(-E/RT)$ is the Arrenhius rate factor; E is the activation

**Research partially supported by U.S. Army Research Grant DAAG29-77-G0062.*

ISBN 0-12-434180-2

energy; and R is the universal gas constant. Equations (1.1) are considered on a bounded domain Ω with the boundary conditions

$$T = T_0, \quad \frac{\partial n}{\partial \tau} = 0 \quad \text{on} \quad \partial\Omega, \tag{1.2}$$

together with initial conditions

$$T(x,0) = T_0(x), \quad n(x,0) = n_0(x), \tag{1.3}$$

under the assumption $T_0(x) \geq T_0$. Here τ denotes an outward normal. A discussion of the derivation of the general equations of chemical kinetics may be found in the books by Gavalas [14] and Frank-Kamenetzky [12]. The particular model (1.1) to (1.3) is discussed by Gelfand [13] and Sattinger [30]. See also Hlavacek and Hofmann [15].

Let u, v denote certain measures of total population such as number of individuals, mass, area of shade cast by plants, etc. A simple model describing populations which are diffusing through Ω as well as interacting with each other is

$$\left.\begin{aligned} u_t &= a_1 \Delta u + uM(u,v), \\ v_t &= a_2 \Delta v + vN(u,v), \end{aligned}\right] \tag{1.4}$$

where $a_1, a_2 \geq 0$, Δ is the Laplace operator in x and M, N are suitable functions. Equations (1.4) are considered with the boundary conditions

$$\frac{\partial u}{\partial \tau} = 0, \quad \frac{\partial v}{\partial \tau} = 0 \quad \text{on} \quad R_+ \times \partial\Omega \tag{1.5}$$

and the initial conditions

$$u(0,x) = u_0(x), \quad v(0,x) = v_0(x) \quad \text{on} \quad \Omega. \tag{1.6}$$

Models of the type (1.4) to (1.6) occur in studies of population genetics [2,34], conduction of nerve impulses [2,10], chemical reactions [1,11] and several other biological questions [1,33]. See also [5,6,7,9,23,24,25,29,31,33,36].

If we confine our attention to equations which are independent of space variable x, the equations (1.4) become the Kolmogorov form of ordinary differential equations describing interactive growth of two populations. This approach has a long history dating from the pioneering work of Lotka and Volterra and still occupies a central portion in mathematical ecology [24,25].

It is therefore clear that many models in chemical, biological and ecological processes involve a system of parabolic differential equations of the form

$$u_t^i = a^i \Delta u^i + f_i(t,x,u_1,\ldots,u_N,u_{x_1}^i,\ldots,u_{x_n}^i) \tag{1.7}$$

subject to appropriate initial boundary conditions. Such systems are called weakly coupled systems.

No model of the form (1.7) can be relied on to be in quantitative agreement with real world systems. Volterra himself recognized this fact. He emphasized consistently that differential equations are, at best, only rough approximations of actual ecological systems. They would apply only to animals without age or memory, which eat all the food they encounter and immediately convert it into offspring. Nevertheless, system (1.7) is worth studying since it might offer better qualitative understanding of real-world problems and since such study might help in constructint and analyzing more complex models.

One of the importnat and effective techniques in the qualitative analysis of dynamical systems is the comparison method or theory of differential inequalities [20,34]. This method is also a useful tool for the qualitative analysis of reaction-diffusion systems [3,4,8,16,17,19,22]. Unfortunately, comparison theorems for systems impose monotonicity requirements on the reaction terms which are physically unreasonable and thus not suited to the problems at hand. To circumvent this monotonic restrictions, one has two alternatives:

(i) since this difficulty is essentially due to the choice of the cone relative to the system, namely, the cone of nonnegative elements of R^n, a possible answer lies in choosing an appropriate cone other than R^n_+ to work in a given situation [18,21],

(ii) extend suitably the important classical result of Müller [26] which leads to the new notion of quasi-solutions that have computational advantage.

In this paper, we shall develop these ideas. Throughout, we consider systems in a Banach space and work with arbitrary cones. Section 2 deals with some auxiliary results. Initial boundary value problems of reaction-diffusion type are investigated in Section 3. Finally, in Section 4, we apply the results to some concrete situations and demonstrate the advantages.

II. PRELIMINARY RESULTS

Let E be a real Banach space with $\|\cdot\|$ and let E^* denote the set of continuous linear functionals. A proper subset K of E is said to be a cone if

$$\lambda K \subset K, \quad \lambda \geq 0, \quad K + K \subset K, \quad K = \overline{K} \text{ and } K \cap (-K) = \{0\}.$$

Here $\overline{K}$ denotes the closure of K. Let us denote by K^0 the interior of K and assume that K^0 is nonempty. The cone K induces a partial ordering on E defined by

$$u \leq v \text{ iff } v - u \in K \text{ and } u < v \text{ iff } v - u \in K^0.$$

A linear functional $\phi \in E^*$ is said to be a positive linear functional if $\phi(x) \geq 0$ whenever $x \in K$. Let K^* denote the set of positive linear functional. It then follows that K is contained in the closed half space

$$C_\phi = [x \in E: \ \phi(x) \geq 0, \ \phi \text{ a positive linear functional}].$$

Thus the positive linear functionals are support functionals and since K is a cone in E, K is the intersection of all the closed half spaces which support it. If $S \subset K^*$ and $K = \cap[C_\phi: \phi \in S]$ then S is said to generate the cone K. Let $S_0 = [\phi \in S: \|\phi\| \leq \gamma_0]$ and $\overline{S}_0$ be the closure of S_0 in the weak star topology.

We need the following lemma.

Lemma 2.1. Let $S \subset U = [\phi \in E^*: \|\phi\| \leq \gamma_0]$. (i) If $v(x) = \sup[\phi(u(x)): \phi \in S]$, where $u \in C[R^n, E]$, then $v(x)$ is continuous in R^n; (ii) if $u \in E$ and $d = \sup[\phi(u): \phi \in S]$ then there exists a $\psi \in \overline{S}$ such that $\psi(u) = d$.

The proof of this lemma for the case $n = 1$ and $v(x) = \inf[\phi(u(x): \phi \in S]$ is given in [32]. The proof of the general case as stated follows similar arguments.

We shall assume throughout that S_0 generates the cone K.

Let Ω be a bounded domain in R^n and let $H = (t_0, \infty) \times \Omega$, $t_0 \geq 0$. Suppose that the boundary ∂H of H is split into two parts ∂H_0, ∂H_1 such that $\partial H = \partial H_0 \cup \partial H_1$ and $\partial H_0 \cap \partial H_1$ is empty.

A vector τ is said to be an outer normal at $(t,x) \in \partial H_1$, if $(t, x-h\tau) \in \Omega$ for small $h > 0$. The outer normal derivative is then given by $\frac{\partial u}{\partial \tau}(t,x) = \lim_{h \to 0} \frac{u(t,x)-u(t, x-h\tau)}{h}$. We shall always assume that an outer normal exists on ∂H_1 and the functions in question have outernormal derivatives on ∂H_1.

If $u \in C[\overline{H}, E]$ and if the partial derivatives u_t, u_x, u_{xx} exist and are continuous in H, then we shall say that $u \in C$. As before, E is a real Banach space and we shall assume K is a cone in E generated by S_0.

The following lemmas are basic to our discussions.

Lemma 2.2. Assume that

(i) $m, n \in C$ and $m(t,x) < 0 < n(t,x)$ on ∂H_0;

(ii) $\frac{\partial m}{\partial \tau}(t,x) < 0 < \frac{\partial n}{\partial \tau}(t,x)$ on ∂H_1, where τ is the outer normal on ∂H_1;

(iii) if $(t_1,x_1) \in H$ and $\phi \in \overline{S}_0$ such that $m(t_1,x_1) \leq 0 \leq n(t_1,x_1)$ and either $\phi(m(t_1,x_1)) = 0$, $\phi(m_{x_i}(t_1,x_1)) = 0$, $i = 1,2,\ldots,n$ and the quadratic form $\sum_{i,j=1}^{n} \lambda_i\lambda_j\phi(m_{x_ix_j}(t_1,x_1)) \leq 0$, $\lambda \in R^n$, or $\phi(n(t_1,x_1)) = 0$, $\phi(n_{x_i}(t_1,x_1)) = 0$, $i = 1,2,\ldots,n$, and the quadratic form $\sum_{i,j=1}^{n} \lambda_i\lambda_j\phi(n_{x_ix_j}(t_1,x_1)) \geq 0$, $\lambda \in R^n$,

then either $\phi(m_t(t_1,x_1)) < 0$ or $\phi(n_t(t_1,x_1)) > 0$ respectively. Under these conditions, we have

$$m(t,x) < 0 < n(t,x) \quad \text{on } \overline{H}.$$

Proof. We set $p(t,x) = \sup[\phi(m(t,x)): \phi \in S_0]$ and $q(t,x) = \inf[\phi(n(t,x)): \phi \in S_0]$. By Lemma 2.1, p, q are continuous on $\overline{H}$. To prove Lemma 2.2, it is enough to prove $p(t,x) < 0 < q(t,x)$ on $\overline{H}$. Suppose that this is not true. Then there exists $t_1 > t_0$ such that $p(t,x) < 0 < q(t,x)$ on $[t_0,t_1) \times \overline{\Omega}$ and either $p(t_1,x) \leq 0$ on $\overline{\Omega}$ or $q(t_1,x) \geq 0$ on $\overline{\Omega}$. Consider first the case $p(t_1,x) \leq 0$ on $\overline{\Omega}$. Since p is continuous, there exists an $x_1 \in \overline{\Omega}$ such that $p(t_1,x_1) = 0$ is the maximum of $p(t_1,x)$. Hence by Lemma 2.1 there is a $\psi \in \overline{S}_0$ such that $p(t_1,x_1) = \psi(m(t_1,x_1))$. Clearly $(t_1,x_1) \notin \partial H_0$ by (i). Also $(t_1,x_1) \notin \partial H_1$. For if $(t_1,x_1) \in \partial H_1$, then $\frac{\partial p}{\partial \tau}(t_1,x_1) = \lim_{h \to 0} \frac{p(t_1,x_1)-p(t_1,x_1-h\tau)}{h} \geq 0$. On the other hand, by (ii), $\frac{\partial p(t_1,x_1)}{\partial \tau} \equiv \psi(\frac{\partial m}{\partial \tau}(t_1,x_1)) < 0$ which is a contradiction. Thus $(t_1,x_1) \in H$ and $p(t_1,x)$ attains an interior maximum at $(t_1,x_1) \in H$. Consequently, if we define $r(t,x) = \psi(m(t,x))$ and note that $r(t_1,x_1) = p(t_1,x_1) = 0$, we have $r(t_1 - h, x_1) < 0$ for small $h > 0$. It therefore follows that $r_t(t_1,x_1) \geq 0$ which implies that $\psi(m_t(t_1,x_1)) \geq 0$. Since $p(t_1,x)$ attains an interior maximum at (t_1,x_1), we also have $\psi(m_{x_i}(t_1,x_1)) = 0$,

$i = 1,2,\ldots,n$ and $\sum_{i,j=1}^{n} \lambda_i\lambda_j\psi(m_{x_ix_j}(t_1,x_1)) \le 0$, $\lambda \in R^n$. Hence by (iii), $\psi(m_t(t_1,x_1)) < 0$ leading to a contradiction. The lemma is therefore proved.

The following special case of Lemma 2.2 will be stated separately for later use.

Lemma 2.3. Assume that

(i) $m \in C$ and $m(t,x) < 0$ on ∂H_0 ($n \in C$ and $n(t,x) > 0$ on ∂H_0);

(ii) $\frac{\partial m}{\partial \tau}(t,x) < 0$ on ∂H_1 ($\frac{\partial n}{\partial \tau}(t,x) > 0$ on ∂H_1);

(iii) if $(t_1,x_1) \in H$ and $\phi \in \overline{S}_0$ such that $m(t_1,x_1) \le 0$, $\phi(m(t_1,x_1)) = 0$, $\phi(m_{x_i}(t_1,x_1)) = 0$, $i = 1,2,\ldots,n$ and $\sum_{i,j=1}^{n} \lambda_i\lambda_j\phi(m_{x_ix_j}(t_1,x_1)) \le 0$, $\lambda \in R^n$, then $\phi(m_t(t_1,x_1)) < 0$ ($n(t_1,x_1) \ge 0$, $\phi(n(t_1,x_1)) = 0$, $\phi(n_{x_i}(t_1,x_1)) = 0$, $i = 1,2,\ldots,n$ and $\sum_{i,j=1}^{n} \lambda_i\lambda_j\phi(n_{x_ix_j}(t_1,x_1)) \ge 0$, then $\phi(n_t(t_1,x_1)) > 0$).

Under these conditions, we have

$$m(t,x) < 0 \text{ on } \overline{H} \quad (n(t,x) > 0 \text{ on } \overline{H}).$$

III. MAIN RESULTS

We shall first discuss results concerning differential inequalities.

Let $f \in C[\overline{H} \times E \times E^n \times E^{n^2}, E]$, where E^i stands for $E \times E \times E \ldots$, i times. A function f is said to be quasimonotone nondecreasing in u relative to K for each $(t,x) \in \overline{H}$, if for any $u,v \in E$, $u_x, v_x \in E^n$, $u_{xx}, v_{xx} \in E^{n^2}$ and $\phi \in \overline{S}_0$ such that $u \le v$, $\phi(u) = \phi(v)$, $\phi(u_{x_i}) = \phi(v_{x_i})$, $i = 1,2,\ldots,n$ and $\sum_{i,j=1}^{n} \lambda_i\lambda_j\phi(u_{x_ix_j} - v_{x_ix_j}) \le 0$, $\lambda \in R^n$, we have

$$\phi(f(t,x,u,u_x,u_{xx})) \le \phi(f(t,x,v,v_x,v_{xx})).$$

For the case $E = R^N$ and $K = R^N_+$, the quasimonotone condition on f implies that $f_i(t,x,u,u_x,u_{xx}) = f_i(t,x,u,u^i_x,u^i_{xx})$, for each i, $1 \leq i \leq N$.

We can now prove the following comparison result.

Theorem 3.1. Suppose that

(i) $v,w \in C$, $f \in C[\overline{H} \times E \times E^n \times E^{n^2}, E]$, f is quasimonotone nondecreasing in u relative to K and

$$v_t \leq f(t,x,v,v_x,v_{xx}), \quad w_t \geq f(t,x,w,w_x,w_{xx}) \quad \text{on } H,$$

where $v = v(t,x)$, $w = w(t,x)$;

(ii) (a) $v(t,x) < w(t,x)$ on ∂H_0;

(b) $\dfrac{\partial v(t,x)}{\partial \tau} < \dfrac{\partial w}{\partial \tau}(t,x)$ on ∂H_1;

Then $v(t,x) < w(t,x)$ on $\overline{H}$, if one of the inequalities in (i) is strict.

Proof. Assume that one of the inequalities in (i) is strict. Then it is enough to show that $m(t,x) = v(t,x) - w(t,x)$ satisfies the conditions of Lemma 2.3. It is easy to see that the conditions (i) and (ii) of Lemma 2.3 hold. We therefore need to verify the condition (iii). Let $(t_1,x_1) \in H$ and $\phi \in \overline{S}_0$ be such that $m(t_1,x_1) \leq 0$, $\phi(m(t_1,x_1)) = 0$, $\phi(m_{x_i}(t_1,x_1)) = 0$ and $\sum_{i,j=1}^{n} \lambda_i\lambda_j\phi(m_{x_ix_j}(t_1,x_1)) \leq 0$, $\lambda \in R^n$. This implies that $v(t_1,x_1) \leq w(t_1,x_1)$, $\phi(v(t_1,x_1)) = \phi(w(t_1,x_1))$, $\phi(v_{x_i}(t_1,x_1)) = \phi(w_{x_i}(t_1,x_1))$, and $\sum_{i,j=1}^{n} \lambda_i\lambda_j\phi(v_{x_ix_j}(t_1,x_1) - w_{x_ix_j}(t_1,x_1)) \leq 0$, $\lambda \in R^n$. Thus, in view of (i), it follows that

$$\begin{aligned}\phi(m_t(t_1,x_1)) &= \phi(v_t(t_1,x_1) - w_t(t_1,x_1)) \\ &< \phi(f(t_1,x_1,v(t_1,x_1),v_x(t_1,x_1),v_{xx}(t_1,x_1)) \\ &\quad - f(t_1,x_1,w(t_1,x_1),w_x(t_1,x_1),w_{xx}(t_1,x_1))) \leq 0,\end{aligned}$$

which proves the theorem.

We can dispense with the strict inequality needed in Theorem 3.1 if f satisfies the following condition:

(C_0) $z \in C$, $z > 0$ on $\overline{H}$, $\frac{\partial z}{\partial \tau}(t,x) \geq \gamma > 0$ on ∂H_1 and for sufficiently small $\varepsilon > 0$, either

(a) $\varepsilon z_t > f(t,x,v,v_x,v_{xx}) - f(t,x, v-\varepsilon z, v_x-\varepsilon z_x, v_{xx}-\varepsilon z_{xx})$.

or

(b) $\varepsilon z_t > f(t,x, w+\varepsilon z, w_x+\varepsilon z_x, w_{xx}+\varepsilon z_{xx}) - f(t,x,w,w_x,w_{xx})$ on H, where $v,w \in C$.

Theorem 3.2. Let the assumption (i) of Theorem 3.1 hold. Suppose further that the condition (C_0) is satisfied. Then the relations

(ii) (a) $v(t,x) \leq w(t,x)$ on ∂H_0,

(b) $\frac{\partial v}{\partial \tau}(t,x) \leq \frac{\partial w}{\partial \tau}(t,x)$ on ∂H_1,

imply $v(t,x) \leq w(t,x)$ on H.

Proof. Assume that the condition (b) of (C_0) holds. Consider $\tilde{w} = w + \varepsilon z$. We have

$$\begin{aligned}\tilde{w}_t &= w_t + \varepsilon z_t \geq f(t,x,w,w_x,w_{xx}) + \varepsilon z_t \\ &> f(t,x, w+\varepsilon z, w_x+\varepsilon z_x, w_{xx}+\varepsilon z_{xx}) \\ &= f(t,x,\tilde{w},\tilde{w}_x,\tilde{w}_{xx}) \quad \text{on } H.\end{aligned}$$

Also $v(t,x) < \tilde{w}(t,x)$ on ∂H_0 and on ∂H_1, $\frac{\partial \tilde{w}}{\partial \tau}(t,x) = \frac{\partial w}{\partial \tau}(t,x) + \varepsilon\frac{\partial z}{\partial \tau}(t,x) \geq \frac{\partial v}{\partial \tau}(t,x) + \varepsilon\gamma > \frac{\partial v}{\partial \tau}(t,x)$. Thus the functions, v, $\tilde{w}$ satisfy the assumptions of Theorem 3.1 and hence $v(t,x) < \tilde{w}(t,x)$ on $\overline{H}$. Taking limit as $\varepsilon \to 0$ yields the desired result and the proof is complete.

As an example, consider the case when $E = R$, $f(t,x,u,u_x,u_{xx}) = au_{xx} + bu_x + F(t,x,u)$ where $au_{xx} = \sum_{i,j=1}^{n} a_{ij}u_{x_i x_j}$, $bu_x = \sum_{j=1}^{n} b_j u_{x_j}$, and suppose that F if Lipschitzian in u for a constant $L > 0$. Suppose that the boundary ∂H_1 is regular, that is, there exists a function $h \in C$ such that $h(x) \geq 0$, $\frac{\partial h(x)}{\partial \tau} \geq 1$ on ∂H_1 and h_x, h_{xx} are bounded. Let $M > 1$, $H(x) = e^{LMh(x)} \geq 1$, $z(t,x) = e^{Nt}H(x)$, where $N = LM + A$, $|aH_{xx} + bH_x| \leq A$. Then

$$\frac{\partial z}{\partial \tau}(t,x) = LM\frac{\partial h(x)}{\partial \tau} \cdot z(t,x) \geq LM = \gamma > 0 \quad \text{on} \quad \partial H_1 \quad \text{and}$$

$$z_t - az_{xx} - bz_x \geq [N - A]z = LMz > Lz.$$

Consequently, using Lipschitz condition of F, we arrive at

$$\varepsilon z_t > \varepsilon[az_{xx} + bz_x] + F(t,x, w+\varepsilon z) - F(t,x,w),$$

which is exactly the condition (b) of (C_0).

Remark. If ∂H_1 is empty so that $\partial H_0 = \partial H$, the assumption (C_0) in Theorem 3.2 can be replaced by a weaker hypothesis, namely a one sided Lipschitz condition of the form

(C_1) $f(t,x,u,P,Q) - f(t,x,v,P,Q) \leq L(u - v)$, $u > v$.

In this case, it is enough to set $\tilde{w} = w + \varepsilon e^{2Lt}y_0$, $y_0 \in K^0$ so that

$$v(t,x) < \tilde{w}(t,x) \quad \text{on} \quad \partial H$$

and

$$\begin{aligned} \tilde{w}_t &\geq f(t,x,w,w_x,w_{xx}) + 2\varepsilon Le^{2Lt}y_0 \\ &\geq f(t,x,\tilde{w},\tilde{w}_x,\tilde{w}_{xx}) + \varepsilon Le^{2Lt}y_0 \\ &> f(t,x,\tilde{w},\tilde{w}_x,\tilde{w}_{xx}) \quad \text{on} \quad H. \end{aligned}$$

Even when ∂H_1 is not empty, the condition (C_1) is enough provided (ii)(b) is strengthened to $\frac{\partial v}{\partial \tau} + Q(t,x,v) \leq \frac{\partial w}{\partial \tau} + Q(t,x,w)$ on ∂H_1, where $Q \in C[\overline{H} \times E,E]$ and $Q(t,x,u)$ is strictly increasing in u. To see this observe that $\tilde{w} > w$ and hence $Q(t,x,w) < Q(t,x,\tilde{w})$ which gives the desired strict inequality needed in the proof. Of course, if Q is not strictly increasing or $Q \equiv 0$, then the condition (C_0) become essential.

Let us next consider the mixed problem

$$u_t = f(t,x,u,u_x,u_{xx}), \tag{3.1}$$

$$u(t,x) = u_0(t,x) \text{ on } \partial H_0 \text{ and } \frac{\partial u}{\partial \tau}(t,x) = 0 \text{ on } \partial H_1, \tag{3.2}$$

and assume that the solutions of (3.1) and (3.2) exist on $\overline{H}$.

A closed set $F \subset E$ is said to be flow-invariant relative to the system (3.1), (3.2) if for every solution $u(t,x)$ of (3.1), (3.2), we have

$$u_0(t,x) \in F \text{ on } \partial H_0 \text{ implies } u(t,x) \in F \text{ on } \overline{H}.$$

The function $f(t,x,u,u_x,u_{xx})$ is said to be quasi-nonpositive (quasi-nonnegative) if $u \leq 0$ $(u \geq 0)$, $\phi(u) = 0$, $\phi(u_{x_i}) = 0$, $i = 1,2,\ldots,n$ and $\sum_{i,j=1}^{n} \lambda_i\lambda_j\phi(u_{x_ix_j}) \leq 0$ $\left(\sum_{i,j=1}^{n} \lambda_i\lambda_j\phi(u_{x_ix_j}) \geq 0\right)$, $\lambda \in R^n$ for some $\phi \in \overline{S}_0$, then $\phi(f(t,x,u,u_x,u_{xx})) \leq 0$ $(\phi(f(t,x,u,u_x,u_{xx})) \geq 0)$.

The following results on flow-invariance are useful in obtaining bounds on solutions of (3.1), (3.2).

Theorem 3.3. Assume that f is quasi-nonpositive and that the condition (C_0)(a) holds with $v = u$, where $u = u(t,x)$ is any solution of (3.1), (3,2). Then the closed set $\overline{Q}$ is flow-invariant relative to the system (3.1), (3.2), where $Q = [u \in E: u < 0]$.

Proof. We set $m(t,x) = u(t,x) - \varepsilon z(t,x)$, where $u(t,x)$ is any solution of (3.1), (3.2) such that $u_0(t,x) \in \overline{Q}$ on ∂H_0 and $\varepsilon > 0$ sufficiently small. We wish to show that this $m(t,x)$ satisfies the conditions of Lemma 2.3. In view of (C_0), it is clear that $m(t,x) < 0$ on ∂H_0 and $\frac{\partial m}{\partial \tau}(t,x) < 0$ on ∂H_1. Let $(t_1,x_1) \in H$ and $\phi \in \overline{S}_0$ be such that $m(t_1,x_1) \leq 0$, $\phi(m(t_1,x_1)) = 0$, $\phi(m_{x_i}(t_1,x_1)) = 0$, $i = 1,2,\ldots,n$ and $\sum_{i,j=1}^{n} \lambda_i\lambda_j\phi(m_{x_ix_j}(t_1,x_1)) \leq 0$, $\lambda \in R^n$. Then by (C_0) and the fact f is quasi-nonpositive, we obtain

$$\begin{aligned}\phi(m_t(t_1,x_1)) &= \phi(u_t(t_1,x_1) - \varepsilon z_t(t_1,x_1)) \\ &\leq \phi(f(t_1,x_1,u(t_1,x_1),u_x(t_1,x_1),u_{xx}(t_1,x_1)) \\ &\quad - \varepsilon z_t(t_1,x_1)) \\ &< \phi(f(t_1,x_1,m(t_1,x_1),m_x(t_1,x_1), m_{xx}(t_1,x_1)) \\ &\leq 0\end{aligned}$$

Hence by Lemma 2.3, $m(t,x) < 0$ on $\overline{H}$ which implies as $\varepsilon \to 0$, the flow-invariance of $\overline{Q}$, proving the theorem.

Remark. Theorem 3.2 can be derived as a consequence of Theorem 3.3. For this purpose, we set $d = v - w$, so that

$$\begin{aligned}d_t = F(t,x,d,d_x,d_{xx}) &\equiv f(t,x,w+d,(w+d)_x,(w+d)_{xx}) \\ &\quad - f(t,x,w,w_x,w_{xx}) + P(t,x,v,w)\end{aligned}$$

where $P(t,x,v,w) = v_t - f(t,x,v,v_x,v_{xx}) - w_t + f(t,x,w,w_x,w_{xx})$. If $d \leq 0$, $\phi(d) = 0$, $\phi(d_{x_i}) = 0$, $i = 1,2,\ldots,n$ and $\sum_{i,j=1}^{n} \lambda_i\lambda_j\phi(d_{x_ix_j}) \leq 0$, for some $\phi \in \overline{S}_0$, then using quasi-monotonicity of f, we get,

$$\begin{aligned}\phi(F(t,x,d,d_x,d_{xx})) &\equiv \phi[f(t,x,w+d,(w+d)_x,(w+d)_{xx}) \\ &\quad - f(t,x,w,w_x,w_{xx}) + P(t,x,v,w) \\ &\leq \phi[f(t,x,w+d,(w+d)_x,(w+d)_{xx} \\ &\quad - f(t,x,w,w_x,w_{xx})] \leq 0.\end{aligned}$$

Hence F is quasi-nonpositive. Also, since

$$\begin{aligned}\varepsilon z_t &> f(t,x,v,v_x,v_{xx}) - f(t,x, v-\varepsilon z, v_x-\varepsilon z_x, v_{xx}-\varepsilon z_{xx}) \\ &= f(t,x,w+d,(w+d)_x,(w+d)_{xx}) \\ &\quad - f(t,x, w+d-\varepsilon z, (w+d-\varepsilon z)_x, (w+d-\varepsilon z)_{xx}) \\ &= F(t,x,d,d_x,d_{xx}) - F(t,x,d-\varepsilon z,(d-\varepsilon z)_x,(d-\varepsilon z)_{xx}).\end{aligned}$$

The claim now follows from Theorem 3.3.

The following corollaries are useful in some situations, whose proofs also we omit.

Corollary 3.1. Assume that f is quasi-nonnegative and that the condition (C_0)(b) holds with $w = u(t,x)$. Then the closed set $\bar{Q}$ is flow-invariant relative to (3.1), (3.2) where $Q = [u \in E: u > 0]$.

Corollary 3.2. Suppose that the condition (C_0) holds with $v = w = u$. Assume also that the following condition holds:

if $u \leq b$, $\phi(u) = \phi(b)$, $\phi(u_{x_i}) = 0$, $i = 1,2,\ldots,n$ and $\sum_{i,j=1}^{n} \lambda_i\lambda_j\phi(u_{x_ix_j}) \leq 0$, $\lambda \in R^n$ for some $\phi \in \bar{S}_0$, then $\phi(f(t,x,u,u_x,u_{xx})) \leq 0$, and

if $a \leq u$, $\phi(u) = \phi(a)$, $\phi(u_{x_i}) = 0$, $i = 1,2,\ldots,n$ and $\sum_{i,j=1}^{n} \lambda_i\lambda_j\phi(u_{x_ix_j})) \geq 0$, $\lambda \in R^n$ for some $\phi \in \bar{S}_0$, then $\phi(f(t,x,u,u_x,u_{xx})) \geq 0$.

Then the closed set $\bar{w}$, where $w = [u \in E: a < u < b, \; a,b \in E]$ is flow-invariant relative to (3.1), (3.2).

We shall next consider a comparison result which yields upper and lower bounds for solutions of (3.1), (3.2) in terms of solutions of ordinary differential equations.

Theorem 3.4. Assume

(i) $u = u(t,x)$ is any solution of (3.1), (3.2) and the condition (C_0) holds with $v = w = u$;

(ii) $g_1, g_2 \in C[R_+ \times E, E]$, $g_1(t,r)$, $g_2(t,r)$ are quasi-monotone nondecreasing in r relative to K and for $(t,x,u) \in \bar{H} \times E$, $\phi \in \bar{S}_0$ if $\phi(u_{x_i}) = 0$, $i = 1,2,\ldots,n$, $\sum_{i,j=1}^{n} \lambda_i\lambda_j\phi(u_{x_ix_j}) \leq 0$, $\lambda \in R^n$,

$$\phi(f(t,x,u,u_x,u_{xx})) \leq \phi(g_1(t,u)),$$

and if $\phi(u_{x_i}) = 0$, $i = 1,2,\ldots,n$, $\sum_{i,j=1}^{n} \lambda_i\lambda_j\phi(u_{x_ix_j}) \geq 0$, $\lambda \in R^n$,

$$\phi(g_2(t,u)) \leq \phi(f(t,x,u,u_x,u_{xx})).$$

(iii) $r(t)$, $\rho(t)$ are solutions of

$$r' = g_1(t,r), \quad r(t_0) = r_0, \quad \rho' = g_2(t,\rho), \quad \rho(t_0) = \rho_0,$$

respectively existing on $[t_0,\infty)$ such that

$$\rho(t) \leq u_0(t,x) \leq r(t), \quad \text{on } \partial H_0.$$

Then $\rho(t) \leq u(t,x) \leq r(t)$ on H.

Proof. Setting $m(t,x) = u(t,x) - r(t)$, we see that m satisfies

$$m_t = F(t,x,m,m_x,m_{xx}),$$

$$m(t,x) = u_0(t,x) - r(t) \quad \text{on } \partial H_0, \quad \text{and} \tag{3.4}$$

$$\frac{\partial m}{\partial \tau}(t,x) = \frac{\partial u}{\partial \tau}(t,x) = 0 \quad \text{on } \partial H_1,$$

where $F(t,x,m,m_x,m_{xx}) = f(t,x,m+r,u_x,u_{xx}) - g_1(t,r)$

We shall show that (3.3), (3.4) satisfies the assumptions of Theorem 3.3. Let $m \leq 0$, $\phi(m) = 0$, $\phi(m_{x_i}) = 0$, $i = 1,2,\ldots,n$ and $\sum_{i,j=1}^{n} \lambda_i\lambda_j\phi(m_{x_ix_j}) \leq 0$, $\lambda \in R^n$, for some $\phi \in \overline{S}_0$. This implies that $u \leq r$ and $\phi(u) = \phi(r)$ and consequently, the quasimonotonicity of g_1 yields $\phi(g_1(t,u)) \leq \phi(g_1(t,r))$. It now follows from (ii) and the fact $m_x = u_x$, $m_{xx} = u_{xx}$,

$$\phi(F(t,x,m,m_x,m_{xx})) \leq \phi(f(t,x,u,u_x,u_{xx}) - g_1(t,u)) \leq 0,$$

proving F is quasi-nonpositive.

We have

$$\varepsilon z_t > f(t,x,u,u_x,u_{xx}) - f(t,x, u-\varepsilon z, u_x-\varepsilon z_x, u_{xx}-\varepsilon z_{xx})$$
$$= f(t,x,m+r,m_x,m_{xx}) - f(t,x, m+r-\varepsilon z, m_x-\varepsilon z_x, m_{xx}-\varepsilon z_{xx})$$
$$= F(t,x,m,m_x,m_{xx}) - F(t,x, m-\varepsilon z, m_x-\varepsilon z_x, m_{xx}-\varepsilon z_{xx}).$$

This proves that F satisfies the condition (C_0)(a) with $v = m$. Thus, by Theorem 3.3, it follows that $m(t,x) \leq 0$ on $\overline{H}$ which proves that $u(t,x) \leq r(t)$ on $\overline{H}$.

By setting $m = u - \rho$ and showing that the assumptions of Corollary 3.1 hold, we can similarly show $\rho(t) \leq u(t,x)$ on $\overline{H}$. This proves the theorem.

Corollary 3.3. If $\overline{w}$ is flow-invariant relative to the system (3.1), (3.2), there exist functions g_1, g_2 satisfying the assumptions of Theorem 3.4.

Proof. We construct g_1, g_2 as follows: for $\phi \in \overline{S}_0$,

$$\phi(g_1(t,u)) = \sup[\phi(f(t,x,v,v_x,v_{xx})): \ x \in \overline{\Omega}, \ a \leq v \leq u,$$
$$\phi(v) = \phi(u), \ \phi(v_{x_i}) = 0, \ i = 1,2,\ldots,n$$
$$\text{and} \ \sum_{i,j=1}^{n} \lambda_i\lambda_j\phi(v_{x_ix_j}) \leq 0, \ \lambda \in R^n],$$

and

$$\phi(g_2(t,u)) = \inf[\phi(f(t,x,v,v_x,v_{xx})): \ x \in \overline{\Omega}, \ u \leq v \leq b,$$
$$\phi(v) = \phi(u), \ \phi(v_{x_i}) = 0, \ i = 1,2,\ldots,n$$
$$\text{and} \ \sum_{i,j=1}^{n} \lambda_i\lambda_j\phi(v_{x_ix_j}) \geq 0, \ \lambda \in R^n].$$

Although Theorem 3.4 and Corollary 3.2 provide upper and lower bounds for solutions (3.1)-(3.2) whenever f is not quasimonotone the bounds that result may not be sharper in view of the construction of g_1, g_2. We shall now discuss a comparision result which offers better bounds under much weaker assumptions. This result is based on the classical result of Müller [26].

Theorem 3.5. Assume that

(i) for $\phi \in \bar{S}_0$, $v, w \in C$, $\phi(v_t) \leq \phi(f(t,x,\sigma,\sigma_x,\sigma_{xx})$ for all $\sigma \in E$, such that $v \leq \sigma \leq w$, $\phi(v) = \phi(\sigma)$, $\phi(v_{x_i}) = \phi(\sigma_{x_i})$, $i = 1,2,\ldots,n$ and $\sum_{i,j=1}^{n} \lambda_i\lambda_j\phi(\sigma_{x_ix_j} - v_{x_ix_j}) \geq 0$, $\lambda \in R^n$, and,

$\phi(w_t) \geq \phi(f(t,x,\sigma,\sigma_x,\sigma_{xx})$ for all $\sigma \in E$ such that $v \leq \sigma \leq w$,

$\phi(w) = \phi(\sigma)$, $\phi(w_{x_i}) = \phi(\sigma_{x_i})$, $i = 1,2,\ldots,n$ and $\sum_{i,j=1}^{n} \lambda_i\lambda_j\phi(\sigma_{x_ix_j} - w_{x_ix_j}) \leq 0$;

(ii) the condition (C_0) holds with (a), (b) of (C_0) replaced by the weaker conditions

(a*) $\phi(\varepsilon z_t) > \phi(f(t,x,\sigma,\sigma_x,\sigma xx) - f(t,x, \sigma-\varepsilon z, \sigma_x-\varepsilon z_x, \sigma_{xx}-\varepsilon z_{xx}))$ whenever $v \leq \sigma \leq w$, $\phi(v) = \phi(\sigma)$, $\phi(v_{x_i}) = \phi(\sigma_{x_i})$, $i = 1,2,\ldots,n$ and $\sum_{i,j=1}^{n} \lambda_i\lambda_j\phi(\sigma_{x_ix_j} - v_{x_ix_j}) \geq 0$, $\lambda \in R^n$ and

(b*) $\phi(\varepsilon z_t) > \phi(f(t,x, \sigma+\varepsilon z, \sigma_x+\varepsilon z_x, \sigma_{xx}+\varepsilon z_{xx}) - f(t,x,\sigma,\sigma_x,\sigma_{xx}))$ whenever $v \leq \sigma \leq w$, $\phi(w) = \phi(\sigma)$, $\phi(w_{x_i}) = \phi(\sigma_{x_i})$, $i = 1,2,\ldots,n$ and $\sum_{i,j=1}^{n} \lambda_i\lambda_j\phi(\sigma_{x_ix_j} - w_{x_ix_j}) \leq 0$.

(iii) $u(t,x)$ is any solution of (3.1), (3.2) such that $v \leq u_0 \leq w$ on ∂H_0 and $\frac{\partial v}{\partial \tau} \leq \frac{\partial u}{\partial \tau} \leq \frac{\partial w}{\partial \tau}$ on ∂H_1.

Then $v(t,x) \leq u(t,x) \leq w(t,x)$ on $\bar{H}$.

Proof. We shall first assume that v, w satisfy strict inequalities and prove the conclusion of the theorem for strict inequalities. We let $m = u - w$ and $n = u - v$ on $\bar{H}$. Then m, n verify the conditions (i), (ii) of Lemma 2.2. To check (iii), let $(t_1,x_1) \in H$ and $\phi \in S_0$ be such that $m(t_1,x_1) \leq 0 \leq n(t_1,x_1)$ and either $\phi(m(t_1,x_1)) = 0$, $\phi(m_{x_i}(t_1,x_1)) = 0$, $i = 1,2,\ldots,n$, and $\sum_{i,j=1}^{n} \lambda_i\lambda_j\phi(m_{x_ix_j}(t_1,x_1)) \leq 0$, $\lambda \in R^n$ or $\phi(n(t_1,x_1) = 0$, $\phi(n_{x_i}(t_1,x_1)) = 0$, $i = 1,2,\ldots,n$ and

$\sum_{i,j=1}^{n} \lambda_i\lambda_j\phi(n_{x_ix_j}(t_1,x_1)) \geq 0$. Suppose that the first alternative holds. Then it implies that, at (t_1,x_1), we have $v \leq u \leq w$, $\phi(u) = \phi(w)$, $\phi(u_{x_i}) = \phi(w_{x_i})$, $i = 1,2,\ldots,n$ and $\sum_{i,j=1}^{n} \lambda_i\lambda_j\phi(u_{x_ix_j} - w_{x_ix_j}) \leq 0$, $\lambda \in R^n$. Hence $\phi(m_t(t_1,x_1))$ $= \phi(u_t - w_t) < \phi(f(t_1,x_1,u,u_x,u_{xx}) - f(t_1,x_1,u,u_x,u_{xx})) = 0$, which proves (iii). The proof of the second case is similar. Thus by Lemma 2.2 we get $m(t,x) < 0 < n(t,x)$ on $\overline{H}$ and this proves the claim of the theorem for strict inequalities.

Consider now $\tilde{w} = w + \varepsilon z$, $\tilde{v} = v - \varepsilon z$. Observe that if $v \leq \sigma \leq w$ then $\tilde{v} \leq \tilde{\sigma} \leq \tilde{w}$ where $\tilde{\sigma} = \sigma + \varepsilon z$. Hence from (b*) of (C_0), and (i), it follows that

$$\begin{aligned}\phi(\tilde{w}_t) &= \phi(w_t + \varepsilon z_t) > \phi(f(t,x,\sigma,\sigma_x,\sigma_{xx}) \\ &\quad + f(t,x,\ \sigma+\varepsilon z,\ \sigma_x+\varepsilon z_x,\ \sigma_{xx}+\varepsilon z_{xx}) - f(t,x,\sigma,\sigma_x,\sigma_{xx})) \\ &= \phi(f(t,x,\tilde{\sigma},\tilde{\sigma}_x,\tilde{\sigma}_{xx}),\end{aligned}$$

whenever $\tilde{v} \leq \tilde{\sigma} \leq \tilde{w}$, $\phi(\tilde{\sigma}) = \phi(\tilde{w})$, $\phi(\tilde{\sigma}_{x_i}) = \phi(\tilde{w}_{x_i})$, $i = 1,2,\ldots,n$ and $\sum_{i,j=1}^{n} \lambda_i\lambda_j\phi(\tilde{\sigma}_{x_ix_j} - \tilde{w}_{x_ix_j}) \leq 0$, $\lambda \in R^n$. One can similarly show that

$$\phi(\tilde{v}_t) < \phi(f(t,x,\tilde{\sigma},\tilde{\sigma}_x,\tilde{\sigma}_{xx})), \quad \tilde{\sigma} = \tilde{v} - \varepsilon z,$$

whenever $\tilde{v} \leq \tilde{\sigma} \leq \tilde{w}$, $\phi(\tilde{\sigma}) = \phi(\tilde{v})$, $\phi(\tilde{\sigma}_{x_i}) = \phi(\tilde{v}_{x_i})$, $i = 1,2,\ldots,n$ and $\sum_{i,j=1}^{n} \lambda_i\lambda_j\phi(\tilde{\sigma}_{x_ix_j} - \tilde{w}_{x_ix_j}) \geq 0$, $\lambda \in R^n$. Since $\tilde{v} < u < \tilde{w}$ on ∂H_0 and $\frac{\partial\tilde{v}}{\partial\tau} < \frac{\partial u}{\partial\tau} < \frac{\partial\tilde{w}}{\partial\tau}$ on ∂H_1, we get immediately

$$v(t,x) - \varepsilon z(t,x) < u(t,x) < w(t,x) + \varepsilon z(t,x) \quad \text{on } \overline{H}$$

for arbitrary $\varepsilon > 0$. Letting $\varepsilon \to 0$, we obtain the stated result completing the proof.

Corollary 3.4. Let the assumptions (i), (ii) of Theorem 3.4 hold without g_1, g_2 being quasimonotone nondecreasing. Suppose that the conditions (ii), (iii) of Theorem (3.5) are satisfied. Assume further that

$$\phi(w'(t)) \geq \phi(g_1(t,\sigma)) \quad \text{for all} \quad v \leq \sigma \leq w \quad \text{and} \quad \phi(\sigma) = \phi(w),$$

and

$$\phi(v'(t)) \leq \phi(g_2(t,\sigma)) \quad \text{for all} \quad v \leq \sigma \leq w \quad \text{and} \quad \phi(\sigma) = \phi(v).$$

Then $v(t) \leq u(t,x) \leq w(t)$ on $\overline{H}$.

It is important to note that the functions v, w do not depend on the space variable x in the foregoing corollary.

IV. APPLICATIONS

A significant amount of attention has been given to systems of nonlinear reaction-diffusion equations of the form

$$u_t = A\Delta u + F(t,u) \tag{4.1}$$

in $R_+ \times \Omega$ with the initial condition

$$u(0,x) = u_0(x) \quad \text{on} \quad \overline{\Omega} \tag{4.2}$$

and the Neuman boundary condition

$$\frac{\partial u}{\partial \tau}(t,x) = 0 \quad \text{on} \quad (0,\infty) \times \partial\Omega. \tag{4.3}$$

See for example [2,7,22,27,28,36]. In (4.1) Δ denotes the Laplace operator in $x \in R^n$, $u,F,u_0 \in R^N$ and A is a diagonal matrix. Equations of the type (4.1) have been very important in the modeling and study of population dynamics, nuclear and chemical reactions, and several models of nerve conduction [1,2,10,11, 33]. For example, $u(t,x)$ may represent the concentration of the chemical at time t and location x. The boundary condition (4.3) implies that no chemical flows in or out of the boundary $\partial\Omega$. In population growth, this boundary condition means that there is no migration across the boundary.

Consider the standard cone in R^N, namely, $K = R^N_+ = [u \in R^n: u_i \geq 0,\ i = 1,2,\ldots,N]$. Clearly the set $S = [\phi \in K^*: \phi(u) = u_i,\ i = 1,2,\ldots,N] \subset K^*$ generates the cone K. Let us note that the weak cou;ling of the system (4.1) suggests the choice of this special cone. Thus the inequality $u \leq v$ implies the component-wise inequalities $u_i \leq v_i$, $i = 1,2,\ldots,N$.

One of the simplest results that can be deduced from the results of Section 4, concerning the problem (4.1) to (4.3), is the following:

Theorem 4.1. Assume that

(i) $A \geq 0$ and $u(t,x)$ is any solution of (5.1) to (5.3) existing on $R_+ \times \bar{\Omega}$;

(ii) $F(t,x) - F(t,y) \leq B(x - y)$, $x \geq y$, where B is an $n \times n$, nonnegative matrix;

(iii) the boundary $\partial\Omega$ is regular, i.e., there exists a $h \in C$ such that $h(x) \geq 0$ on $\bar{\Omega}$, $\dfrac{\partial h(x)}{\partial \tau} \geq \gamma > 0$ on $\partial\Omega$ and h_x, h_{xx} are bounded.

Then the following conclusions are valid:

(a) If $u_i = 0$, $u_j \geq 0$, $j \neq i$, $j = 1,2,\ldots,N$, implies $F_i(t,u) \geq 0$, then $u(t,x) \geq 0$ on $R_+ \times \bar{\Omega}$ provided $u_0(x) \geq 0$ on $\bar{\Omega}$;

(b) If $F(t,u)$ is quasimonotone nondecreasing in u relative to R^N_+, that is, for each i, $1 \leq i \leq N$, $F_i(t,u)$ is nondecreasing in u_j, $j \neq i$ and if the solutions $r(t)$, $\rho(t)$ of $y' = F(t,y)$ with $r(0) = \bar{y}_0$, $\rho(0) = \underline{y}_0$ exist on R_+, then

$$\rho(t) \leq u(t,x) \leq r(t) \quad \text{on } R_+ \times \bar{\Omega}, \tag{4.4}$$

provided that $\underline{y}_0 \leq u_0(x) \leq \bar{y}_0$ on $\bar{\Omega}$;

(c) If $F(t,u)$ is quasimonotone nondecreasing in u, $F(t,0) \equiv 0$, then

$0 \le u_0(x)$ on $\bar{\Omega}$ implies that $u(t,x) \ge 0$ on $R_+ \times \bar{\Omega}$,

$0 < u_0(x)$ on $\bar{\Omega}$ implies that $u(t,x) > 0$ on $R_+ \times \bar{\Omega}$, and

$0 \le u_0(x) \le y_0$ on $\bar{\Omega}$ implies that $0 \le u(t,x) \le r(t)$ on $R_+ \times \bar{\Omega}$, where $r(t)$ is the same function assumed in (b);

(d) If $F(t,u)$ is not quasimonotone and if the closet set $\bar{\omega} = [u \in R^N: \ a \le u \le b]$ is flow invariant relative to (4.1) to (4.3), then the estimate (4.4) holds $r(t)$, $\rho(t)$ are now being the solutions of

$$r' = g_1(t,r), \quad r(0) = \bar{y}_0, \quad \rho' = g_2(t,\rho), \quad \rho(0) = \underline{y}_0,$$

where $g_{1i}(t,u) = \max[F_i(t,v): \ a \le v \le u, \ v_i = u_i]$, $g_{2i}(t,u) = \min[F_i(t,v): \ u \le v \le b, \ v_i = u_i]$, $1 \le i \le N$;

(e) If $F(t,u)$ is not quasimonotone and $\bar{\omega}$ is not known to be flow invariant, then (4.4) holds if $r(t)$, $\rho(t)$ satisfy the relations

$$r_i' \ge F_i(t,\sigma) \quad \text{for all } \sigma \text{ such that } \rho \le \sigma \le r \text{ and } \sigma_i = r_i,$$

$$\rho_i' \le F_i(t,\sigma) \quad \text{for all } \sigma \text{ such that } \rho \le \sigma \le r \text{ and } \sigma_i = \rho_i,$$

$1 \le i \le N$.

Proof. The conclusion (a) follows from Corollary 3.1. Theorem 3.4 yields (b) with the choice $F = g_1 = g_2$. Uniqueness of solutions of $y' = F(t,y)$ together with the fact that $F(t,0) \equiv 0$, implies (c). Corollary 3.3 gives the conclusion (d) whereas (e) follows from Corollary 3.4.

As an example of Theorem 4.1, consider $E = R^2$ so that $K = R_+^2$. Suppose that $F_1(t,u_1,u_2)$ is nondecreasing in u_2 and $F_2(t,u_1,u_2)$ is nonincreasing in u_1. Then the functions r and ρ have to verify

$$r_1' \ge F_1(t,r_1,r_2), \quad \rho_1' \le F_1(t,\rho_1,\rho_2)$$

$$r_2' \ge F_2(t,\rho_1,r_2), \quad \rho_2' \le F_2(t,r_1,\rho_2)$$

Of course, the functions F_1, F_2 have to satisfy a Lipschitz condition.

Let us next demonstrate the advantage of employing a suitable cone other than R_+^N in the study of reaction-diffusion equations. Consider the system

$$\left.\begin{aligned}\frac{\partial u_1}{\partial t} &= a_{11}\Delta u_1 + a_{12}\Delta u_2 + F_1(t,u_1,u_2),\\ \frac{\partial u_2}{\partial t} &= a_{21}\Delta u_1 + a_{22}\Delta u_2 + F_2(t,u_1,u_2),\end{aligned}\right] \tag{4.5}$$

with (4.2) and (4.3) where $u = (u_1,u_2)$. Clearly this system is not weakly coupled in the sense of system (4.1). Consequently, if we choose to work relative to the cone R_+^2, we can not draw any conclusions concerning (4.5) as in Theorem 4.1. However, suppose we notice that for some $\alpha,\beta > 0$ with $\beta > \alpha$, we have the relations

$$b_1 \equiv \alpha a_{22} - \alpha^2 a_{12} = \alpha a_{11} - a_{21} \geq 0,$$

$$b_2 \equiv \beta a_{22} - \beta^2 a_{12} = \beta a_{11} - a_{21} \geq 0.$$

Then, considering the cone $K = [u \in R^2\colon u_2 \leq \beta u_1, \text{ and } u_2 \geq \alpha u_1]$ and noting that $S = [\phi\colon\ \phi_1(u) = u_2 - \alpha u_1,\ \ \phi_2(u) = \beta u_1 - u_2]$ generates the cone K, we can write (5.5) as

$$\left.\begin{aligned}\frac{\partial}{\partial t}(u_2 - \alpha u_1) &= b_1\Delta(u_2 - \alpha u_2) + \tilde{F}_1(t,u_1,u_2)\\ \frac{\partial u}{\partial t}(\beta u_1 - u_2) &= b_2\Delta(\beta u_1 - u_2) + \tilde{F}_2(t,u_1,u_2)\end{aligned}\right] \tag{4.6}$$

where $\tilde{F}_1 = F_2 - \alpha F_1$ and $\tilde{F}_2 = \beta F_1 - F_2$. It is easy to see that (4.6) is weakly coupled relative to K. We therefore have the following result observing that $K \subset R_+^2$.

Theorem 4.2. Assume that $\tilde{F} = (\tilde{F}_1,\tilde{F}_2)$ is quasimonotone nondecreasing relative to K and $\tilde{F}$ satisfies a uniqueness condition as in (ii) of Theorem 4.1. Suppose that (iii) of Theorem 4.1 holds. Then

$$\underline{y}_{20} - \alpha\underline{y}_{10} \leq u_{20}(x) - \alpha u_{10}(x) \leq \bar{y}_{20} - \alpha\bar{y}_{10},$$

$$\beta\underline{y}_{10} - \underline{y}_{20} \leq \beta u_{10}(x) - u_{20}(x) \leq \beta\bar{y}_{10} - \bar{y}_{20}, \quad \text{on } \bar{\Omega}$$

implies

$$\rho_1(t) \leq u_1(t,x) \leq r_1(t),$$

$$\rho_2(t) \leq u_2(t,x) \leq r_2(t) \quad \text{on } R_+ \times \bar{\Omega}.$$

REFERENCES

[1] Aris, R. (1975). "The mathematical theory of diffusion and reaction in permeable catalysts", Clarendon Press, Oxford, England.

[2] Aronson, D. G., and Weinberger, H. F. (1975). "Nonlinear diffusion in population genetics, combustion and nerve propagation", Lecture Notes, Vol. 446, Springer Verlag, New York.

[3] Chandra, J., and Davis, P. W. "Comparison theorems for systems of reaction diffusion equations", to appear in Proc. Int. Conf. on Applied Nonlinear Analysis.

[4] Chandra, J. (Aug., 1978). "Some comparison theorems for reaction diffusion equations", to appear in Proc. Int. Conf. on "Recent trends in differential equation", Trieste.

[5] Cohen, D. S. (1971). "Multiple stable solutions of non-linear boundary value problems arising in chemical reactor theory", *SIAM Appl. Math.* *20*, 1-13.

[6] Cohen, D. S., and Laetsch, T. W. (1970). "Nonlinear boundary value problems suggested in chemical reactor theory", *J. Diff. Eqs.* *7*, 217-226.

[7] Conway, E. D., and Smoller, J. A. (1977). "Diffusion and the predator-prey interaction", *SIAM J. Appl. Math. 33*, 673-686.

[8] Conway, E. D., and Smoller, J. A. (1977). "A comparison technique for systems of reaction diffusion equations", *Comm. Par. Diff. Eq., 2,* 679-697.

[9] Crank, J. (1955). "The Mathematics of Diffusion", Chap. VIII. Clarendon Press, Oxford.

[10] Fitzhugh, R. (1969). "Mathematical models of excitation and propagation in nerves", Biological Engineering, H. P. Schwann, Ed., McGraw-Hill, New York.

[11] Fife, P. C. (1976). "Pattern formation in reacting and diffusing systems", *J. Chem. Phys. 64,* 554-564.

[12] Frank-Kamenetzky, D. A. (1955). "Diffusion and heat exchange in chemical kinetics", translated by N. Thon, Princeton Univ. Press, Princeton, NJ.

[13] Gelfand, I. M. (1963). "Some problems in the theory of quasilinear equations", *AHS Trans. Ser. 2, 29,* 295-381.

[14] Gavalas, G. R. (1968). "Nonlinear Differential Equations of Chemically Reacting Systems", Springer, New York.

[15] Hlavacek, V., and Hofmann, H., (1970). "Modeling of chemical reactors. XVI. Steady state axial heat and mass transfer in tubular reactors", *Chem. Eng. Sci. 25,* 173-185.

[16] Lakshmikantham, V., and Vaughn, R. L. "Reaction-diffusion inequalities in cones", to appear in Appl. Anal.

[17] Lakshmikantham, V., and Vaughn, R. L. "Parabolic differential inequalities in cones", to appear in J.M.A.A.

[18] Lakshmikantham, V., and Leela, S. (1977). "Cone valued Lyapunov functions", *Nonlinear Anal. 1,* 215-222.

[19] Lakshmikantham, V., (Aug., 1978). "Comparison theorems for reaction-diffusion equations in abstract cones", to appear in Proc. Int. Conf. on "Recent trends in differential equations", Trieste.

[20] Lakshmikantham, V., and Leela, S. (1969). "Differential and Integral Inequalities, Vols. I and II", Academic Press.

[21] Lakshmikantham, V. (1974). "On the method of vector Lyapunov functions", Proc. Twelfth Alberton Conf. on Circuit and System Theory, 71-76.

[22] McNabb, A. (1961). "Comparison and existence theorems for multicomponent diffusion systems", *J. Math. Anal. Appl. 3*, 133-144.

[23] Maynard-Smith, J. (1968). "Mathematical Ideas in Biology", Cambridge Univ. Press, London/New York.

[24] Maynard-Smith, J. (1974). "Models in Ecology", Cambridge Univ. Press, Cambridge.

[25] May, R. (1973). "Stability and complexity in model eco-systems", Princeton Univ. Press, Princeton.

[26] Müller, M. (1926). "Über das Fundamentaltheorem in der Theorie der gewohnlichen Differentialgleichungen", *Math. Z. 26*, 619-645.

[27] Pao, C. V. (1974). "Positive solution of a nonlinear diffusion system arising in chemical reactors", *J. Math. Anal. Appl. 46*, 820-835.

[28] Rosen, G. (1975). "Solutions to systems of nonlinear reaction diffusion equations", *Bull. Math. Biol. 37*, 277-289.

[29] Rauch, J., and Smoller, J. A. (1978). "Qualitative theory of the Fitzhugh-Nagumo equation", *Advances in Math, 27*, 12-44.

[30] Sattinger, D. H. (1975). "A nonlinear parabolic system in the theory of combusion", *Quart. Appl. Math.*, 47-61.

[31] Samuelson, P. (1971). "Generalized predator-prey oscillations in ecological and economic equilibrium", *Proc. Nat. Acad. Sci. U.S.A. 68*, 980-983.

[32] Thompson, R. (1977). "An invariance property of solutions to second order differential inequalities in ordered Banach spaces", *SIAM J. Math. Anal. 8*, 592-603.

[33] Turing, A. M. (1952). "On the chemical basis of morphogenesis", *Phil. Trans. Roy. Soc. London Ser. B, 237,* 37-52.

[34] Waltman, P. E. (1964). "The equations of growth", *Bull. Math. Biophys. 26,* 39-43.

[35] Walter, W. (1970). "Differential and Integral Inequalities", Springer-Verlag.

[36] Williams, S. A., and Chow, P. Z. (1978). "Nonlinear reaction diffusion models for interacting populations", *J.M.A.A. 62,* 157-169.

NUMERICAL SOLUTION OF NEURO-MUSCULAR SYSTEMS*

K. V. Leung

Department of Computer Science
Concordia University
Montreal, Quebec, Canada

M. N. Oğuztöreli

Department of Mathematics
University of Alberta
Edmonton, Alberta, Canada

R. B. Stein

Department of Physiology
University of Alberta
Edmonton, Alberta, Canada

I. INTRODUCTION

In a recent series of papers a mathematical model has been developed to study the oscillations in a neuro-muscular system (cf. [6],[7],[8],[9],[10],[11]). These oscillations arise from the interaction of a skeletal muscle with its load, due to

**This work was partially supported by the National Research Council of Canada (Grant NRC A-4342 and Grant NRC A-4345) and the Medical Research Council of Canada (Grant NRC MT-3307).*

ISBN 0-12-434180-2

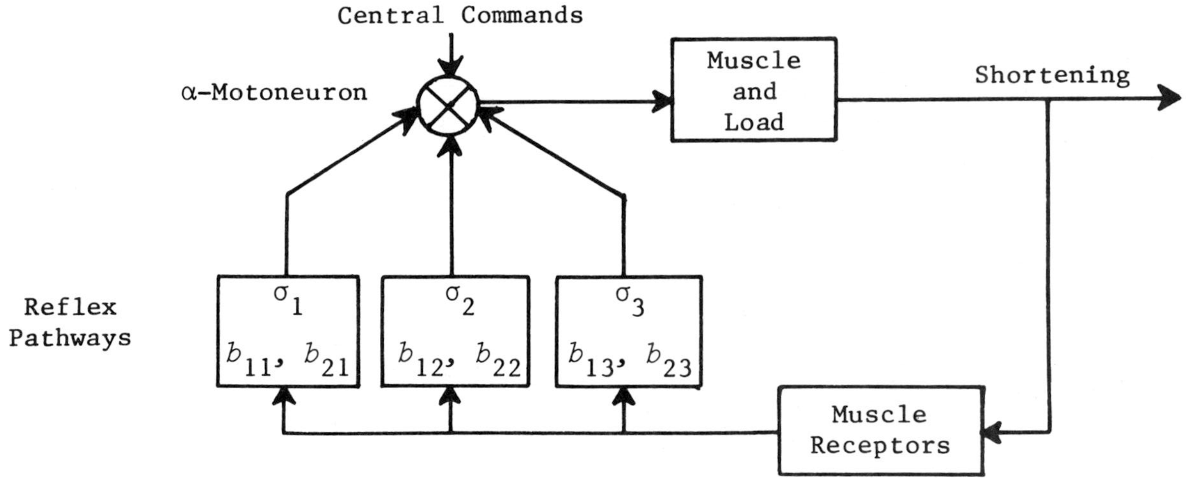
Central Commands
α-Motoneuron
Muscle and Load
Shortening
Reflex Pathways
σ_1 b_{11}, b_{21}
σ_2 b_{12}, b_{22}
σ_3 b_{13}, b_{23}
Muscle Receptors

FIGURE 1

instabilities in the neural feedback pathways involved in the control of muscles, and/or are generated by the central nervous system. The mathematical model is in agreement with a consider-albe amount of collected data related to animal and human studies (cf. [1],[2],[4]). The basic model of a reflex system is shown in Figure 1.

Let $\ell(t)$ be the shortening in the muscle length at time t, supposing that $\ell = 0$ in the resting position. It has been shown in the above mentioned references that $\ell(t)$ satisfies a differential difference equation of the form

$$\ell^{IV}(t) = c_4\dddot{\ell}(t) + c_3\ddot{\ell}(t) + c_2\dot{\ell}(t) + c_1\ell(t) + ag(t) + bf(u(t)) \quad (1.1)$$

where $\dot{\ell}(t) = \frac{d\ell(t)}{dt}$, etc., and

$$u(t) = \sum_{k=1}^{3} [b_{1k}\ell(t-\sigma_k) + b_{2k}\dot{\ell}(t-\sigma_k)], \quad (1.2)$$

$$f(t) = \begin{cases} u & \text{in the linear regime,} \\ \tanh \frac{u}{\beta} & \text{in the nonlinear regime,} \end{cases} \quad (1.3)$$

and

$$g(t) = \sin(\Omega t + \theta), \quad (1.4)$$

where a (≥ 0), b (<0), c_i (<0), σ_k (>0), b_{jk} (>0), Ω (>0), and θ are certain real constants whose values can be determined experimentally. Let T_0 be a positive number. Put

$$\sigma = \max\{\sigma_1, \sigma_2, \sigma_3\}. \quad (1.5)$$

The main initial value problem we are interested in in the present work can be formulated as follows:

Find a solution $\ell(t)$ to equation (1.1) for $0 \leq t \leq T_0$ subjected to the initial conditions

$$\ell(t) = z_1(t),\ \dot{\ell}(t) = z_2(t),\ \ddot{\ell}(t) = z_3(t),\ \dddot{\ell}(t) = z_4(t) \quad (1.6)$$

for $-\sigma \leq t \leq 0$, where $z_1(t)$, $z_2(t)$, $z_3(t)$ and $z_4(t)$ are four given sufficiently smooth functions in $-\sigma \leq t \leq 0$ such that

$$z_{k+1}(t) = \dot{z}_k(t) \quad (k = 1,2,3). \tag{1.7}$$

The degree of smoothness of the function $z_1(t)$ will be discussed in Section V.

Note that this initial value problem has a unique solution which depends continuously on all the parameters appearing in the equation [5]. Further, the solution is always bounded in the nonlinear regime [6]. Several physiologically significant properties of the solution in the linear regime have already been established in the references mentioned above. In the nonlinear regime, however, the available analytical methods are not powerful enough to produce satisfactory quantitative results which can be used in the experimental studies.

The objective of the present work is the construction numerically of the solution of the initial value problem with any prescribed accuracy. The full implementation of the numerical analysis given here and related computer simulations will be presented elsewhere.

The required numerical solution will be achieved by the use of the finite element method (cf. [3],[12]) in three steps:

Step No. 1: Reformulation of the problem by transforming Equation (1.1) into a system of four differential difference equations of the first order for four functions $x_1(t)$, $x_2(t)$, $x_3(t)$ and $x_4(t)$ of time t.

Step No. 2: Derive a recursive formula to compute higher order derivatives $x_s^{(q)}(t)$, $s = 1,2,3,4$ and $q = 1,2,\dots,Q$.

Step No. 3: Derive a single step method with variable step size to compute $x_s(t)$ at a set of discrete gridpoints $t_1 = 0$, $t_2 = h_1 + t_1$, $t_3 = h_2 + t_2,\dots,t_{r+1} = h_r + t_r,\dots$, where h_r is the size of the r-th step.

II. REFORMULATION OF THE PROBLEM

We begin with Step No. 1 with the introduction of the new functions $x_1(t)$, $x_2(t)$, $x_3(t)$ and $x_4(t)$ by the classical relationships:

$$\ell(t) = x_1(t),\ \dot{\ell}(t) = x_2(t),\ \ddot{\ell}(t) = x_3(t),\ \dddot{\ell}(t) = x_4(t). \tag{2.1}$$

Further, let the shift operator Δ_k $(k = 1,2,3)$ be defined as follows:

$$\Delta_k \phi(t) = \phi(t - \sigma_k). \tag{2.2}$$

Put

$$\Delta = [\Delta_1, \Delta_2, \Delta_3]. \tag{2.3}$$

Consider the vectors

$$\left.\begin{aligned} X(t) &= [x_1(t), x_2(t), x_3(t), x_4(t)]^T, \\ X_{12}(t) &= [x_1(t), x_2(t)]^T \\ F(t) &= [0,\ 0,\ 0,\ f(u(t))]^T, \\ G(t) &= [0,\ 0,\ 0,\ g(t)]^T, \\ Z(t) &= [z_1(t), z_2(t), z_3(t), z_4(t)]^T, \end{aligned}\right\} \tag{2.4}$$

and the matrices

$$A = \begin{pmatrix} 0 & 1 & 0 & 0 \\ 0 & 0 & 1 & 0 \\ 0 & 0 & 0 & 1 \\ c_1 & c_2 & c_3 & c_4 \end{pmatrix}, \qquad B = \begin{pmatrix} b_{11} & b_{21} \\ b_{12} & b_{22} \\ b_{13} & b_{23} \end{pmatrix} \tag{2.5}$$

where the superscript T denotes the transpose of the vector or matrix associated with. Then the function $u(t)$ defined by Equation (1.2) can be written in the form

$$u(t) = \sum_{k=1}^{3} [b_{1k}x_1(t-\sigma_1) + b_{2k}x_2(t-\sigma_k)] = \Delta BX_{12}(t) \qquad (2.6)$$

and Equations (1.1) and (1.6) can be rewritten respectively in the vector-matrix form

$$\frac{d}{dt} X(t) = AX(t) + bF(t) + aG(t) \quad (0 \leq t \leq T_0) \qquad (2.7)$$

and

$$X(t) = Z(t) \quad (-\sigma \leq t \leq 0). \qquad (2.8)$$

Hence, the initial value problem formulated in Section I can be reformulated as follows:

Find the solution of the system (2.7) for $0 \leq t \leq T_0$ which satisfies the initial condition (2.8).

This completes the Step No. 1.

III. COMPUTATION OF HIGHER ORDER DERIVATIVES

Consider Equation (2.7). By successive differentiation of Equation (2.7) we find the recursive formula

$$X^{(q+1)}(t) = AX^{(q)}(t) + bF^{(q)}(t) + aG^{q}(t) \qquad (3.1)$$

$$(q = 0,1,2,\ldots,Q-1)$$

where

$$X^{(q)} = \frac{d^qX}{dt^q}, \quad F^{(q)} = \frac{d^qF}{dt^q}, \quad G^{(q)} = \frac{d^qG}{dt^q}. \qquad (3.2)$$

We have three sections on the right-hand side of Equation (3.1):

Section I. $X^{(q)}$. The values of $X^{(q)}$ have been computed by the same recursive formula (3.1) in the previous step.

Section II: $F^{(q)}$. Clearly

$$F^{(q)}(t) = \left[0,\ 0,\ 0,\ \frac{d^q f(u)}{dt^q}\right]^T. \tag{3.1}$$

Two algorithms have to be implemented depending on whether $f(u) = u$ (linear regime) or $f(u) = \tanh \frac{u}{2}$ (nonlinear regime):

(a) *Linear regime:* In this case we have

$$\frac{d^q f(u)}{du^q} = u^{(q)} = \Delta B[x_1^{(q)}, x_2^{(q)}]^T \tag{3.4}$$

$$= [b_{11}x_1^{(q)} + b_{21}x_2^{(q)}]_{\bar{t}_1=t-\sigma_1} + [b_{12}x_1^{(q)} + b_{22}x_2^{(q)}]_{\bar{t}_2=t-\sigma_2}$$

$$+ [b_{13}x_1^{(q)} + b_{23}x_2^{(q)}]_{\bar{t}_3=t-\sigma_3}$$

The computation of the higher order derivatives of $x_1(t)$ and $x_2(t)$ at time $\bar{t}_i = t - \sigma_k$ $(k = 1,2,3)$ is shown in Section V.

(b) *Nonlinear regime:* In this case we have

$$f(u) = 1 - 2v$$

where

$$v = \frac{1}{1+e^u}, \quad u = \Delta B X_{12}. \tag{3.6}$$

Hence

$$\frac{d^q}{dt^q} f(u) = -2 \frac{d^q}{dt^q} v(u(t)) \tag{3.7}$$

and

$$\frac{d^q F(t)}{dt^q} = -2\left[0,\ 0,\ 0,\ \frac{d^q}{dt^q} v(u(t))\right]^T. \tag{3.8}$$

The computation of $\frac{d^q}{dt^q} v(u(t))$ will be shown in Section VI.

Section III: $G^{(q)}$. Clearly we have

$$G^{(q)}(t) = \left[0,\ 0,\ 0,\ \frac{d^q}{dt^q}\, g(t)\right]^T \tag{3.9}$$

and

$$\frac{d^q}{dt^q}\, g(t) = \Omega^q \sin(\Omega t + \theta + q\tfrac{\pi}{2}). \tag{3.10}$$

This completes the Step No. 2.

IV. COMPUTATION OF $x_s(t)$ AT GRIDPOINTS

Let $t_1 = 0$, $t_2 = t_1 + h_1$, $t_3 = t_2 + h_2, \ldots,$ $t_{r+1} = t_r + h_r, \ldots$ be a set of successive gridpoints. Let

$$x_{s,r} = x_s(t_r) \quad (s = 1,2,3,4; \quad r = 1,2,3,\ldots). \tag{4.1}$$

Consider $x_{s,r+1} = x_s(t_{r+1})$ with $t_{r+1} = t_r + h_r$ where h_r is the size of the r-th step. Using Taylor's expansion we find

$$x_{s,r+1} = x_{s,r} + h_r x^{(1)}_{s,r} + \frac{h_r^2}{2!}\, x^{(2)}_{s,r} + \ldots + \frac{h_r^q}{q!}\, x^{(q)}_{s,r} + \ldots + \frac{h_r^Q}{Q!}\, x^{(Q)}_{x,r} \tag{4.2}$$

where

$$x^{(q)}_{s,r} = \frac{d^q}{dt^q}\, x_s \bigg|_{t=t_r} .$$

The step size h_r of the r-th step is determined by the relationship

$$\max_s \left| \frac{h_r^Q}{Q!}\, x^{(Q)}_{s,r} \right| < \varepsilon \tag{4.4}$$

where ε is a prescribed error bound. This completes the Step No. 3.

V. COMPUTATION OF THE DERIVATIVES OF $x_s(t)$ AT $\bar{t}_k$

To compute the derivatives of the functions $x_s(t)$, $s = 1,2,3,4$, at time $\bar{t}_k = t_r - \sigma_k$ we shall distinguish five cases depending on the location of $\bar{t}_k$. To simplify our presentation we shall denote $\bar{t}_k$ by simply $\bar{t}$.

1°) $-\sigma \leq \bar{t} \leq 0$. Since $x_s(t)$ is given in the interval $-\sigma \leq t \leq 0$,

$$x_s(t) = \frac{d^{s-1}}{dt^{s-1}} z_1(t) \quad (-\sigma \leq t \leq 0), \tag{5.1}$$

by Equations (1.6)-(1.7), we have

$$x_s^{(q)}(\bar{t}) = z_1^{(s+q-1)}(t) = \left.\frac{d^{s+q-1}}{dt^{s+q-1}} z_1(t)\right|_{t=\bar{t}} \tag{5.2}$$

for $s = 1,2,3,4$ and $q = 1,2,\ldots,Q$. We assume that the initial function $z_1(t)$ is $(Q+5)$-times continuously differentiable in the interval $-\sigma \leq t \leq 0$.

2°) $0 = t_1 = t_\lambda < \bar{t} < t_{\lambda+1} = t_2 \leq t_r$ (Figure 2). In this case $x_s^{(q)}(\bar{t})$ will be approximated by a cubic polynomial $\tilde{x}_s^{(q)}(\bar{t})$ subject to the conditions that

(i) $x_s^{(q)}(t)$ and $\tilde{x}_s^{(q)}(t)$ are numerically equal at the gridpoints t_λ $(= t_1)$ and $t_{\lambda+1}$ $(= t_2)$:

$$\left.\begin{aligned} \tilde{x}_s^{(q)}(t_1) &= x^{(q)}(t_1) = x_{s,0}^{(q)}, \\ \tilde{x}_s^{(q)}(t_2) &= x^{(q)}(t_2) = x_{s,1}^{(q)}, \end{aligned}\right\} \tag{5.3}$$

and that

(ii) the first and second derivatives of $\tilde{x}_s^{(q)}(t)$ and $x_s^{(q)}(t)$ are numerically equal at the gridpoint t_λ $(= t_1)$:

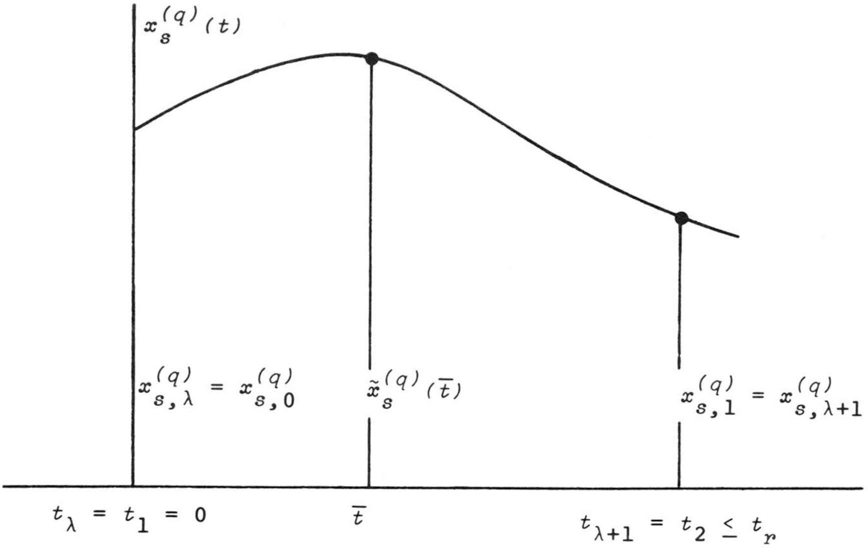

FIGURE 2

$$\left.\begin{aligned} \tilde{x}_s^{(q+1)}(t_1) &= x_s^{(q+1)}(t_1) = z_1^{(s+q)}(0), \\ \tilde{x}_s^{(q+2)}(t_1) &= x_s^{(q+2)}(t_1) = z_1^{(s+q+1)}(0), \end{aligned}\right\} \tag{5.4}$$

Accordingly, the expression

$$\tilde{x}_s^{(q)}(t) = x_{s,0}^{(q)} + t\, x_{s,0}^{(q+1)} + \frac{1}{2} t^{-2} x_{s,0}^{(q+2)} + \frac{x_{s,1}^{(q)} - x_{s,0}^{(q)} - t_2 x_{s,0}^{(q+1)} - \frac{1}{2} t_2^2 x_{s,0}^{(q+2)}}{t_2^3} t^{-3} \tag{5.5}$$

will be used as the approximate value of $x_s^{(q)}(\bar{t})$.

<u>3°) $0 = t_1 = t_{\lambda-1} < t_\lambda < \bar{t} < t_{\lambda+1} = t_3 = t_r$</u> (Figure 3). In this case $t_2 < \bar{t} < t_3 = t_r$ and $x_s^{(q)}(t)$ will be approximated by a cubic polynomial $\tilde{x}_s^{(q)}(t)$ subject to the conditions that

(i) $x_s^{(q)}(t)$ and $\tilde{x}_s^{(q)}(t)$ are numerically equal at the three successive points $t_{\lambda-1}$, t_λ and $t_{\lambda+1}$ (i.e., t_1, t_2 and t_3, respectively):

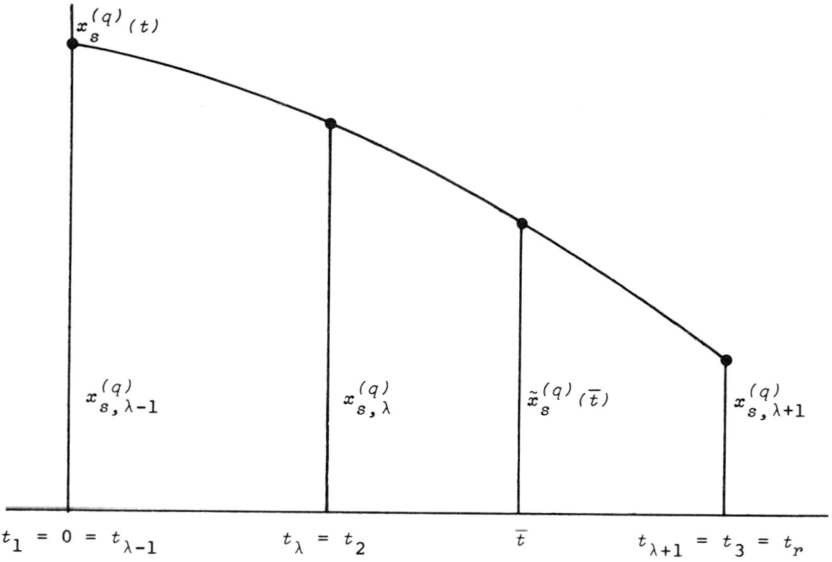

FIGURE 3

$$\left.\begin{aligned} \tilde{x}_s^{(q)}(t_1) &= x_s^{(q)}(t_1) = x_{s,0}^{(q)}, \\ \tilde{x}_s^{(q)}(t_2) &= x_s^{(q)}(t_2) = x_{s,1}^{(q)}, \\ \tilde{x}_s^{(q)}(t_3) &= x_s^{(q)}(t_3) = x_{s,2}^{(q)}, \end{aligned}\right\} \tag{5.6}$$

and that

(ii) the first derivatives of $x_s^{(q)}(t)$ and $\tilde{x}_s^{(q)}(t)$ are also numerically equal at the gridpoint $t = t_1 = 0$:

$$\tilde{x}_s^{(q+1)}(t_1) = x_s^{(q+1)}(t_1) = x_{s,0}^{(q+1)}. \tag{5.7}$$

Accordingly, the expression

$$\tilde{x}_s^{(q)}(\bar{t}) = x_{s,0}^{(q)} + x_{s,0}^{(1+1)}\bar{t} + \frac{\Delta_1}{\Delta}\bar{t}^2 + \frac{\Delta_2}{\Delta}\bar{t}^3 \tag{5.8}$$

where

$$\Delta = \begin{vmatrix} t_1^2 & t_1^3 \\ t_2^2 & t_2^3 \end{vmatrix} . \tag{5.9}$$

$$\Delta_1 = \begin{vmatrix} x_{s,1}^{(q)} - x_{s,0}^{(q)} - t_1 x_{s,0}^{(q+1)} & t_1^3 \\ x_{s,2}^{(q)} - x_{s,0}^{(q)} - t_2 x_{s,0}^{(q+1)} & t_2^3 \end{vmatrix} , \tag{5.10}$$

and

$$\Delta_2 = \begin{vmatrix} t_1^2 & x_{s,1}^{(q)} - x_{s,0}^{(q)} - t_1 x_{s,0}^{(q+1)} \\ t_2^2 & x_{s,2}^{(q)} - x_{s,0}^{(q)} - t_2 x_{s,0}^{(q+1)} \end{vmatrix} , \tag{5.11}$$

will be used as the approximate value of $x_s^{(q)}(\bar{t})$.

<u>4°) $0 \le t_{\lambda-1} < t_\lambda < \bar{t} < t_{\lambda+1} < t_{\lambda+2} \le t_r$</u> (Figure 4). In this case $x_s^{(q)}(t)$ will be approximated at $t = \bar{t}$ $(t_\lambda < \bar{t} < t_{\lambda+1})$

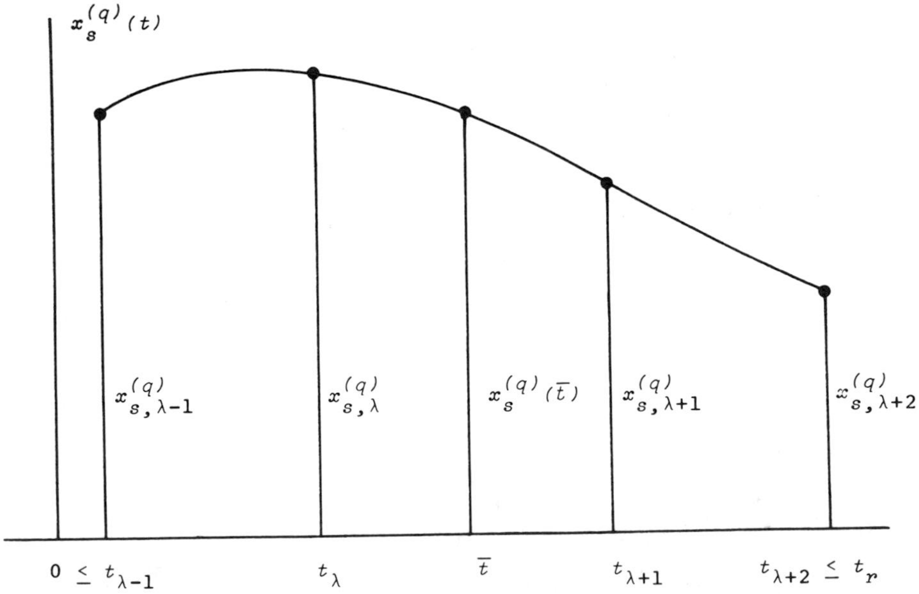

FIGURE 4

by a cubic Lagrange interpolating polynomial $\tilde{x}_s^{(q)}(t)$ subject to the conditions that $\tilde{x}_s^{(q)}(t)$ is numerically equal to $x_s^{(q)}(t)$ at the four successive gridpoints $t_{\lambda-1}$, t_λ, $t_{\lambda+1}$ and $t_{\lambda+2}$:

$$\tilde{x}_s^{(q)}(t_{\lambda+\nu}) = x_s^{(q)}(t_{\lambda+\nu}) = x_{s,\lambda+\nu}^{(q)} \quad (\nu = -1,0,1,2). \tag{5.12}$$

Hence

$$\begin{aligned} x_s^{(q)}(\bar{t}) \sim \tilde{x}_s^{(q)}(\bar{t}) = {} & \frac{(\bar{t}-t_\lambda)(\bar{t}-t_{\lambda+1})(\bar{t}-t_{\lambda+2})}{(t_{\lambda-1}-t_\lambda)(t_{\lambda-1}-t_{\lambda+1})(t_{\lambda-1}-t_{\lambda+2})} x_{s,\lambda+1}^{(q)} \\ & + \frac{(t-t_{\lambda-1})(t-t_{\lambda+1})(t-t_{\lambda+2})}{(t_\lambda-t_{\lambda-1})(t_\lambda-t_{\lambda+1})(t_\lambda-t_{\lambda+2})} x_{s,\lambda}^{(q)} \\ & + \frac{(\bar{t}-t_{\lambda-1})(\bar{t}-t_\lambda)(\bar{t}-t_{\lambda+2})}{(t_{\lambda+1}-t_{\lambda-1})(t_{\lambda+1}-t_\lambda)(t_{\lambda+1}-t_{\lambda+2})} x_{s,\lambda+1}^{(q)} \\ & + \frac{(\bar{t}-t_{\lambda-1})(\bar{t}-t_\lambda)(\bar{t}-t_{\lambda+1})}{(t_{\lambda+2}-t_{\lambda-1})(t_{\lambda+2}-t_\lambda)(t_{\lambda+2}-t_{\lambda+1})} x_{s,\lambda+2}^{(q)}. \end{aligned} \tag{5.13}$$

And, finally, we have the following case:

<u>5°) $0 \leq t_{\lambda-2} < t_{\lambda-1} < t_\lambda < \bar{t} < t_{\lambda+1} = t_r$</u> (Figure 5). In this case $x_s^{(q)}(\bar{t})$ will be approximated by a cubic Lagrange interpolating polynomial $\tilde{x}_s^{(q)}(t)$ subject to the conditions that $\tilde{x}_s^{(q)}(t)$ is numerically equal to $x_s^{(q)}(t)$ at the four consecutive gridpoints $t_{\lambda-2}$, $t_{\lambda-1}$, t_λ, $t_{\lambda+1}$:

$$\tilde{x}_s^{(q)}(t_{\lambda+\nu}) = x_s^{(q)}(t_{\lambda+\nu}) = x_{s,\lambda+\nu}^{(q)} \quad (\nu = -2,-1,0,1).$$

Accordingly we have the approximation formula:

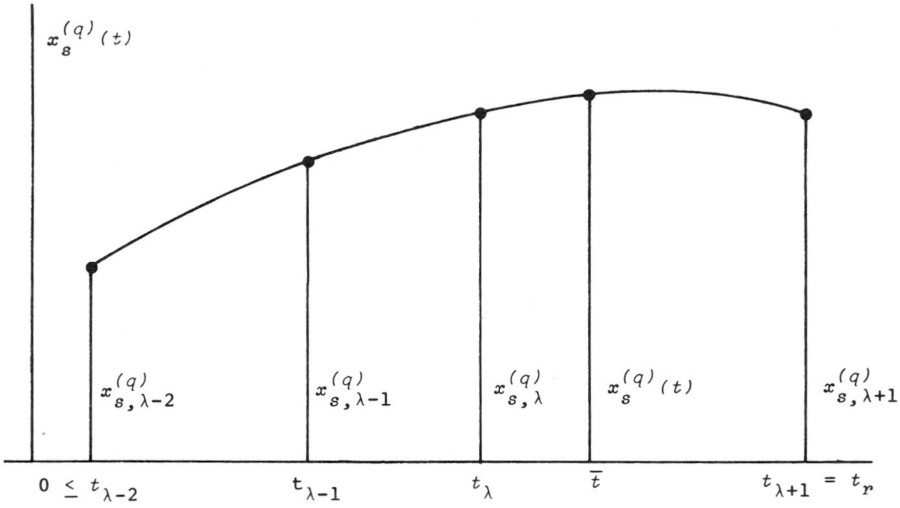

FIGURE 5

$$x_s^{(q)}(\bar{t}) \sim \tilde{x}_s^{(q)}(\bar{t}) = \frac{(\bar{t}-t_{\lambda-1})(\bar{t}-t_{\lambda})(\bar{t}-t_{\lambda+1})}{(t_{\lambda-2}-t_{\lambda-1})(t_{\lambda-2}-t_{\lambda})(t_{\lambda-2}-t_{\lambda+1})} x^{(q)}_{s,\lambda-2} \tag{5.15}$$

$$+ \frac{(\bar{t}-t_{\lambda-2})(\bar{t}-t_{\lambda})(\bar{t}-t_{\lambda+1})}{(t_{\lambda-1}-t_{\lambda-2})(t_{\lambda-1}-t_{\lambda})(t_{\lambda-1}-t_{\lambda+1})} x^{(q)}_{s,\lambda-1}$$

$$+ \frac{(\bar{t}-t_{\lambda-2})(\bar{t}-t_{\lambda-1})(t-t_{\lambda+1})}{(t_{\lambda}-t_{\lambda-2})(t_{\lambda}-t_{\lambda-1})(t_{\lambda}-t_{\lambda+1})} x^{(q)}_{s,\lambda}$$

$$+ \frac{(\bar{t}-t_{\lambda-2})(\bar{t}-t_{\lambda-1})(\bar{t}-t_{\lambda})}{(t_{\lambda+1}-t_{\lambda-2})(t_{\lambda+1}-t_{\lambda-1})(t_{\lambda+1}-t_{\lambda})} x^{(q)}_{s,\lambda+1}.$$

This completes the computation of $x_s^{(q)}(\bar{t})$, where $\bar{t} = t - \sigma_k$.

VI. COMPUTATION OF SUCCESSIVE DERIVATIVES OF $v(u(t))$

Consider the function $v(u)$ defined by Equation (3.6)

$$v = (1 + e^u)^{-1}, \quad u = \Delta BX_{12}. \tag{6.1}$$

Put

$$[n] = \frac{d^n v}{du^n}, \quad (m) = \frac{d^m u}{dt^m}. \tag{6.2}$$

Then, applying the chain rule, we obtain successive derivatives of v with respect to t:

$$\frac{dv}{dt} = [1](1) \tag{6.3}$$

$$\frac{d^2v}{dt^2} = [2](1)^2 + [1](2)$$

$$\frac{d^3v}{dt^3} = [3](1)^3 + 3[2](1)(2) + [1](3)$$

$$\frac{d^4v}{dt^4} = [4](1)^4 + 6[3](1)^2(2) + 4[2](1)(3) + 3[2](2)^2 + [1](4)$$

$$\frac{d^5v}{dt^5} = [5](1)^5 + 10[4](1)^3(2) + 10[3](1)^2(3) + 15[3](1)(2)^2 + 5[2](1)(4) + 10[2](2)(3) + [1](5)$$

$$\frac{d^6v}{dt^6} = [6](1)^6 + 15[5](1)^4(2) + 20[4](1)^3(3) + 45[4](1)^2(2)^2 + 15[3](1)^2(4) + 60[3](1)(2)(3) + 15[3](2)^3 + 6[2](1)(5) + 15[2](2)(4) + 10[2](3)^2 + [1](6).$$

. . . .

The computation of $(m) = \frac{d^n u}{dt^n}$ is shown in Section III(a). In the following we shall deal with the computation of $[n] = \frac{d^n v}{du^n}$. For this purpose, put

$$w = ve^u \tag{6.4}$$

and

$$P_n(w) = \sum_{\lambda=0}^{n-1} a_{n,\lambda} w^\lambda \tag{6.5}$$

with

$$\left.\begin{aligned} &a_{n,-1} = 0, \quad a_{n,n} = 0, \quad a_{1,0} = -1, \\ &a_{n,\lambda} = (\lambda+1)[a_{n-1,\lambda} - a_{n-1,\lambda-1}]. \end{aligned}\right\} \tag{6.6}$$

Hence we have

$$\begin{aligned} P_1(w) &= -1 \\ P_2(w) &= -1 + 2w \\ P_3(w) &= -1 + 6w - 6w^2 \\ P_4(w) &= -1 + 14w - 36w^2 + 24w^3 \\ P_5(w) &= -1 + 30w - 150w^2 + 240w^3 - 120w^4 \\ P_6(w) &= -1 + 62w - 540w^2 + 1560w^3 - 1800w^4 + 720w^5 \\ &\cdots\cdots\cdots\cdots\cdots\cdots \end{aligned} \tag{6.7}$$

It can be easily verified, by comparing the coefficients of w^λ on both sides, that

$$P_n(w) = (1-2w)P_{n-1}(w) + w(1-w)P'_{n-1}(w). \tag{6.8}$$

Further, we have the formula

$$\frac{d^n v}{du^n} = v\, w\, P_n(w) \tag{6.9}$$

for any n, which can be proved by the mathematical induction. Indeed, we can easily verify that formula (6.9) is true for $n = 1, 2, 3 \ldots$ If the formula is true for $n - 1$, then we have the identity

$$\begin{aligned} \frac{d^{n-1} v}{du^{n-1}} &= v\, w\, P_{n-1}(w) \\ &= v\, w \sum_{\lambda=0}^{n-2} a_{n-1,\lambda} w^\lambda, \end{aligned} \tag{6.10}$$

Then, differentiating Equation (6.10) with respect to u and employing the formula (6.8), we find

$$\frac{d^n v}{du^n} = \frac{d}{du}(vw)P_{n-1}(w) + v\,w\,P'_{n-1}(w)\,\frac{dw}{du} \tag{6.11}$$

$$= v\,w\,(1-2w)P_{n-1}(w) + v\,w^2\,(1-w)P'_{n-1}(w)$$

$$= v\,w\,[(1-2w)P_{n-1}(w) + w(1-w)P'_{n-1}(w)]$$

$$= v\,w\,P_n(w),$$

as required. This completes the computation of $(n) = \dfrac{d^n v}{du^n}$.

VI. IMPLEMENTATION OF THE NUMERICAL ANALYSIS

The implementation of the analysis presented in the foregoing sections has been carried out according to the following chart:

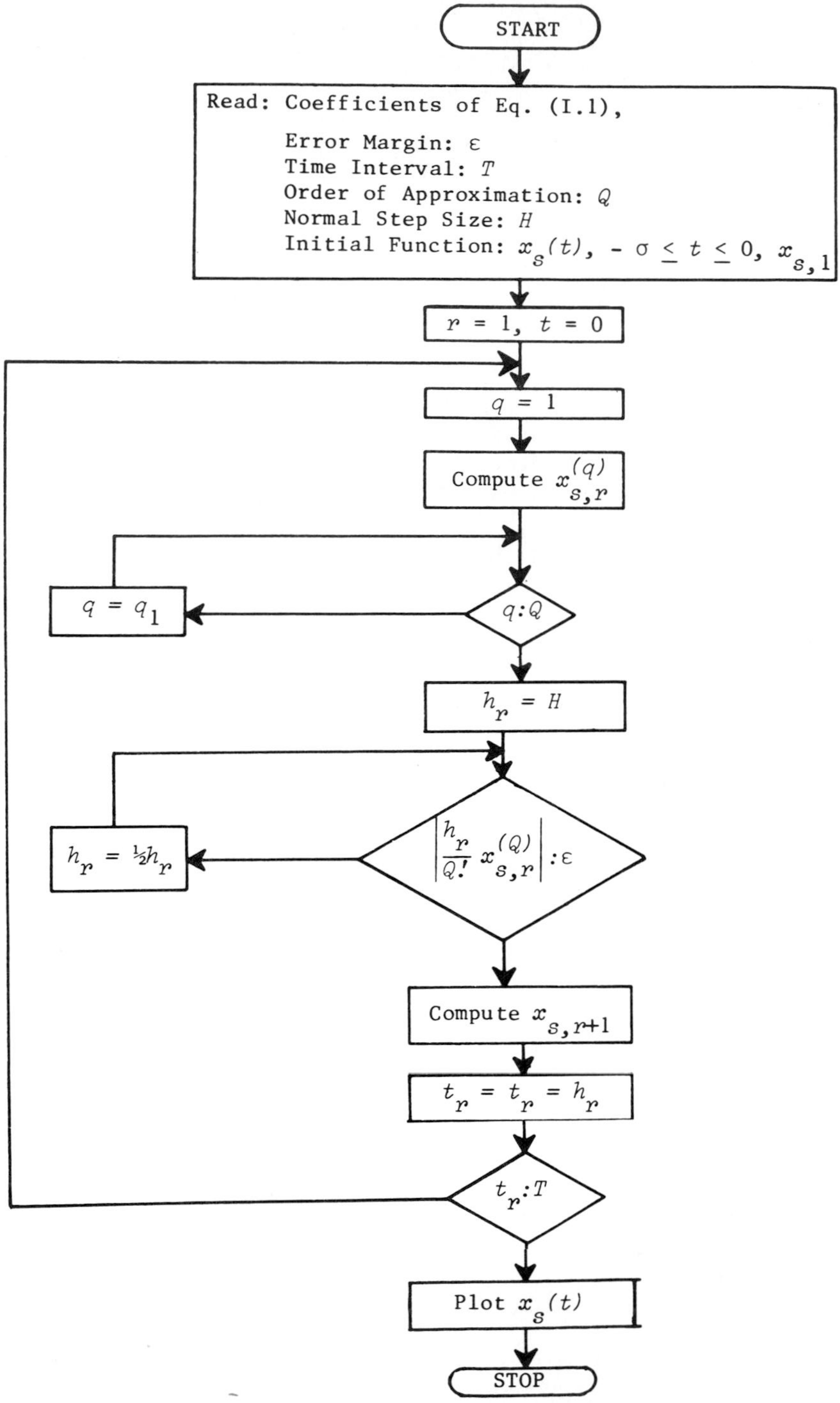
START
Read: Coefficients of Eq. (I.1),
Error Margin: ε
Time Interval: T
Order of Approximation: Q
Normal Step Size: H
Initial Function: $x_s(t)$, $-\sigma \le t \le 0$, $x_{s,1}$
$r = 1$, $t = 0$
$q = 1$
Compute $x_{s,r}^{(q)}$
$q = q_1$
$q:Q$
$h_r = H$
$h_r = \frac{1}{2}h_r$
$\left|\frac{h_r}{Q!} x_{s,r}^{(Q)}\right| : \varepsilon$
Compute $x_{s,r+1}$
$t_r = t_r = h_r$
$t_r:T$
Plot $x_s(t)$
STOP

VIII. PROGRAM DESCRIPTION

A computer program in the *FORTRAN* language has been prepared according to the numerical analysis presented in this paper. This program consists of a main program and of seven subroutines. In the program b is denoted by *BB*, Ω is denoted by *W*, and θ is called *THETA*. Further, σ_k are denoted by *SIG(k)*, c_k are denoted by *-C(k)* and b_{jk} are denoted by *B(j,k)*, respectively.

The subroutines are:

DFF, GRIDD, GT, LAGG, TANHP, TAYLOR and *Z1*.

All arrays and vectors are specified in *dimension* statement which should be altered if necessary.

Parameters (constants) are inputs from the *INPUT*, results are on the *OUTPUT* and saved on *TAPE 1* for eventual plotting with *FORMAT* (*F6.3*, *4F10.5*); t, $x(t)$, $x'(t)$, $x''(t)$ and $x'''(t)$ are printed and saved.

All parameters and dimension requirements are in the comments part of the main program. Some of the parameters are:

- *LIMIT:* Let h_0 be the user supplied maximum step size. Then *LIMIT* is such that

$$LIMIT \geq 8\,\frac{\sigma}{h_0}\,; \tag{8.1}$$

  Arrays *XX* and *TT* are then adjusted.

- *KDMAX* should be left to *5*.

- *ITERMAX* is the required number of iterations.

- $KDM \leq KDMAX$ is the requested order of Taylor expansion.

- $LINEAR = \begin{cases} 0 & \text{Nonlinear regime is chosen,} \\ 1 & \text{Linear regime is chosen.} \end{cases}$

User supplied function *Z1* is build as follows:

```
      FUNCTION Z1(K,T)

C     K is the order of differentiation
C     T is the real argument of Z1
C     Z1 returns the K-th order derivative of Z1 at T
C     This is an example with z1(t) = t
      Z1 = 0.
      IF (K.GE.2) RETURN
      Z1 = 1.
      IF (K.EQ.1) RETURN
      Z1 = T
      RETURN
      END.
```

Data set up is arranged as follows:

A, BB, W, THETA	(Card 1)
B	(Card 2) in the order *B(1,1), B(2,1), B(1,2),...*
C	(Card 3) in the order *C(1), C(2), C(3), C(4)*
SIG	(Card 4) in the order σ_1, σ_2, σ_3
ITERMAX, KDM, HO, LINEAR	(Card 5)

A series of physiologically significant simulations have been done in the Sir George Williams Campus of Concordia University in Montreal with *CDC 640* Computer System. The listings of the program can be obtained from the first author.

REFERENCES

[1] Bawa, P., Mannard, A., and Stein, R. B. (1976). "Effects of elastic loads on contractions of cat muscles", *Biol. Cybern.* *22*, 129-137.

[2] Bawa, P., Mannard, A., and Stein, R. B. (1976). "Predictions and experimental tests of a visco-elastic muscle model using elastic and inertial loads", *Biol. Cybern.* *22*, 139-145.

[3] Fix, G., and Strand, G. (1973). "An Analysis of the Finite Element Method", Prentice-Hall.

[4] Mannard, A., and Stein, R. B. (1973). "Determination of the frequency response of isometric soleus muscle in the cat using random nerve stimulation", *J. Physiol., London* *229*, 275-296.

[5] Oǧuztöreli, M. N. (1966). "Time-Lag Control Systems", Academic Press.

[6] Oǧuztöreli, M. N., and Stein, R. B. (1975). "An analysis of oscillations in neuro-muscular systems", *J. Math. Biol.* *2*, 87-105.

[7] Oǧuztöreli, M. N., and Stein, R. B. (1976). "The effects of multiple reflex pathways on the oscillations in neuro-muscular systems", *J. Math. Biol.* *3*, 87-101.

[8] Oǧuztöreli, M. N., and Stein, R. B. (1977). "Some recent progresses in neuro-muscular systems", in Nonlinear Systems and Applications - An International Conference, Academic Press, Ed. V. Lakshmikantham, 257-293.

[9] Oǧuztöreli, M. N., and Stein, R. B. "Interactions between centrally and peripherally generated neuro-muscular systems, *J. Math. Biol.*, (in press).

[10] Stein, R. B., and Oǧuztöreli, M. N. (1976). "Tremor and other oscillations in neuro-muscular systems", *Biol. Cybern.* *22*, 147-157.

[11] Stein, R. B., and Oǧuztöreli, M. N. (1976). "Does the velocity sensitivity of muscle spindles stabilize the stretch reflex?", *Biol. Cybern.* *23*, 219-228.

[12] Zienkiewicz, O. C. (1977). "The Finite Element Method", 3rd Edition, McGraw-Hill.

SEPARATRICES FOR DYNAMICAL SYSTEMS

Roger C. McCann

Department of Mathematics
Mississippi State University
Mississippi State, Mississippi

The concept of separtrix is alluded to in several texts on differential equations, e.g., [2, page 119], [3, page 250], [4, page 24], [6, page 223], [7, page 170], [12, page 46], [13, page 237], [14, page 312], [15, page 18]. In all but [4], [6], and [7] the concept is left undefined. In [4] separatrix is defined only in the special case that the flow contains a saddle point. In [7] the separatrix must be contained in a limit set. Only in [6] is a general definition attempted (for two dimensional systems) and it applies only to special situations (see Figure 4 below).

Intuitively, the trajectories indicated in the following diagrams are separatrices. The indicated trajectories, which are a portion of a limit set, do not satisfy the definition of separatrix in [7].

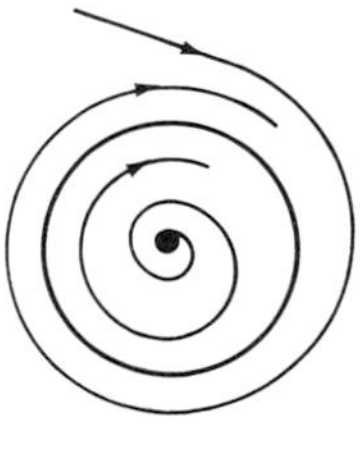

FIGURE 1

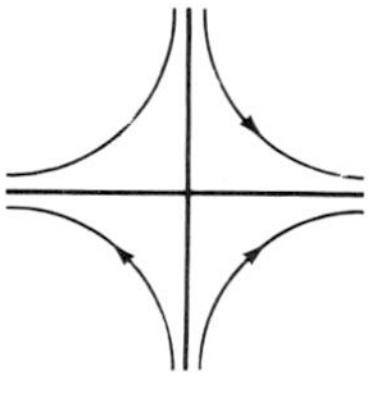

FIGURE 2

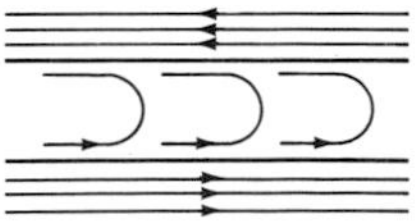

FIGURE 3

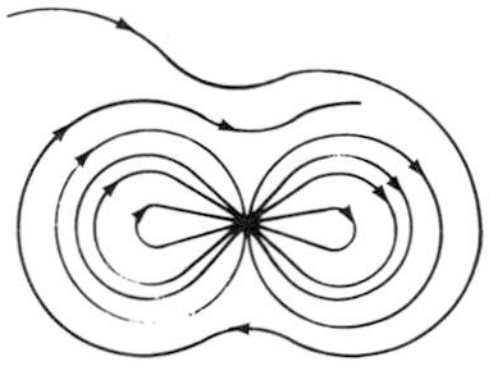

FIGURE 4

ISBN 0-12-434180-2

In [7, page 223] Lefschetz states "Roughly speaking a separatrix is a true path (not a critical point) behaving topologically abnormally in comparison with neighboring paths." If this statement is regarded as describing a local condition, then the trajectory T in Figure 3 is not a separatrix. However, if the condition is considered globally, then T is a separatrix. The intuitive notion described by Lefschetz will be formalized to a precise definition of separatrix.

It appears that the entire trajectorial structure of a planar flow is determined, at least partially, by the structure of the separatrices: [5], [8], [10], [11]. In [9] Markus defines separatrix for unstable flows on an n-manifold M and shows, among other things, that if the flow is completely unstable and there are no separatrices, then the orbit space is Hausdorff and the natural mapping of M onto the orbit space is the projection of a fibre bundle. Hence, in such a setting the absence of separatrices assures the existence of a subset N of M such that each trajectory intersects N in exactly one point and this intersection occurs in a continuous manner.

The trajectories of the system on a space X may be classified as follows:

(i) singular, if the trajectory is a single point,

(ii) periodic, if the trajectory is homeomorphic with a circle,

(iii) ordinary, if the trajectory is homeomorphic with the real line,

(iv) recurrent, if the closure of the trajectory coincides with the positive and negative limit sets of the trajectory.

The sets of all points which lie on trajectories which are singular, periodic, ordinary, or recurrent will be denoted by S, P, O, and R respectively.

Let $\sim$ denote the equivalence relation which identifies points on the same trajectory.

Proposition 1. If int $P \neq \phi$, then there is an open, dense, invariant subset P_1 of P such that $P_1/\sim$ is Hausdorff in the induced topology.

Proposition 2. If int $O \neq \phi$, then there is an open, dense, invariant subset O_1 of O such that $O_1/\sim$ is Hausdorff in the induced topology.

Proposition 3. If int$(R - (S \cup \text{int } P) = \phi$, then there is a maximal open, dense, invariant subset U of $X - S$ such that $U/\sim$ is Hausdorff in the induced topology.

Definition 4. A trajectory T is called a separatrix if there is an open, dense, invariant subset U of $X - S$ such that $U/\sim$ is Hausdorff and $T \in \partial U$.

This definition generalizes that of Markus for unstable flows and those found in [4], [5], [6], [7] for planar flows. There is another concept of separatrix found in [1] which is not equivalent to Markus' and, hence, not to ours, for unstable flows.

Theorem 5. A flow contains no separatrices if and only if $(X - S)/\sim$ is Hausdorff.

Thus, we see that this concept of separatrix corresponds with the intuitive notion expressed by Lefschetz.

Being separatrix free does not mean that the flow is parallelizible as it does in Markus' paper [9]. For example

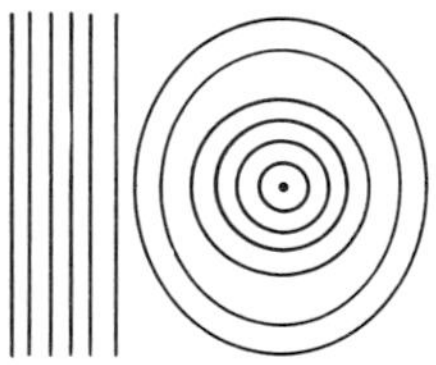

Corollary. If the system contains no separatrices, then there is a subset Y of $X - S$ such that each nonperiodic trajectory in $X - S$ intersects Y in exactly one point and each periodic

trajectory intersects Y in a finite number of points. Moreover, the intersection is continuous in the Hausdorff topology.

Clearly this corollary generalizes the aforementioned result by Markus for completely unstable flows. If the assumption $\text{int}(R - S \cup \text{int } P) = \phi$ is made, then there are relatively few trajectories which are separatrices.

Theorem 6. If $\text{int}(R - (S \cup \text{int } P)) = \phi$, then the set of points which lie on separatrices is nowhere dense in $X - S$.

REFERENCES

[1] Bhatia, N., and Franklin, L. (1972). "Dynamical systems without separatrices", *Funkcialij Ekvacioj 15*, 1-12.

[2] Birkhoff, G., and Rota, G. (1969). "Ordinary Differential Equations", Second Edition, Blaisdell, Watham (Mass.).

[3] Braver, F., and Nohel, J. (1969). "Qualitative Theory of Ordinary Differential Equations", W. A. Benjamin, New York.

[4] Davis, T., and James, Eleanor (1966). "Nonlinear Differential Equations", Addisoin-Wesley, Reading (Mass.).

[5] Kaplan, W. (1940). "Regular families of curves filling the plane, I and II", *Duke Math. J. 7(1940)*, and *8(194)*, 11-46.

[6] Lefschetz, S. (1957). "Differential Equations: Geometric Theory", Second Edition, Interscience, New York.

[7] Lefschetz, S. (1948). "Lectures on Differential Equations", Princeton Univ. Press, Princeton.

[8] Markus, L. (1954). "Global structure of ordinary differential equations in the plane", *Trans. Amer. Math. Soc., 76*, 127-148.

[9] Markus, L. (1964). "Parallel dynamical systems", *Topology 8*, 47-57.

[10] McCann, R. (1970). "Planar dynamical systems without critical points", *Funkcialiaj Ekvacioj, 13*, 67-95.

[11] McCann, R. (1970). "A classification of center-foci", *Pacific J. Math.*, *32*, 467-78.

[12] Minorsky, N. (1962). "Nonlinear Oscillations", Van Nostrand, Princeton.

[13] Reissig, R., Sansone, G., and Conti, R. (1974). "Nonlinear Differential Equations of Higher Order", Noordhoff, Leyden.

[14] Ritger, P., and Rose, N. (1968). "Differential Equations with Applications", McGraw-Hill, New York.

[15] Struble, R. (1962). "Nonlinear Differential Equations", McGraw-Hill, New York.

STABILITY PROBLEMS FOR HOPF BIFURCATION*

P. Negrini

Istituto di Matematica
Università di Camerino
Camerino, Italy

L. Salvadori[**]

Istituto di Matematica
Università di Roma
Roma, Italy

I. INTRODUCTION

Consider the one-parameter family of differential equations

$$\dot{z} = f(\mu, z), \quad (\cdot) = d/dt \tag{1.1}$$

where $f \in C^{k+1}[(-\bar{\mu}, \bar{\mu}) \times D_a, \mathbb{R}^2]$, and $f(\mu, 0) \equiv 0$. Here $\bar{\mu}$, $a > 0$, $D_a = \{z \in \mathbb{R}^2 : \|z\| < a\}$, and $k \geq 3$. Denoting by $\alpha(\mu) \pm i\beta(\mu)$ the eigenvalues of $D_z f(\mu, 0)$ we shall suppose throughout the paper that $\alpha(0) = 0$, $\alpha'(0) \neq 0$, and $\beta(0) > 0$.

We are concerned with the general problem of asymptotic stability of the periodic orbits arising in the Hopf bifurcation for

*The details will appear in the paper "Attractivity and Hopf Bifurcation", Nonlinear Analysis, T.M.A., Pergamon Press.

**Visiting Professor at the University of Texas at Arlington, Department of Mathematics, Arlington, Texas 76019, U.S.A.

ISBN 0-12-434180-2

(1.1). Such property is related to the asymptotic behavior of the flow relative to $\mu = 0$ (the critical value of the parameter) near the origin 0 of $\mathbb{R}^2$. Actually the bifurcating periodic orbits are found to be attracting under the general assumption that 0 is asymptotically stable for $\mu = 0$, and there exists an odd integer $h \in \{3,\ldots,k\}$ such that the above character of 0 is recognizable in a suitable sense by the terms of $f(0,\cdot)$ of degree $\leq h$ (h-asymptotic stability). We denote this property by $P(h)$, and we point out (Sect. 3) that:

(i) The occurrence of $P(h)$ can be recognized by using a classical procedure of Poincaré [1], and this procedure is reduced to the analysis of linear algebraic systems.

(ii) $P(h)$ is equivalent to the condition given by Marsden and McCracken ([2], Th. 3B.3) upon the first h derivatives with respect to c, of the displacement function $V(\mu,c)$, computed at $c = 0$. (In particular, $P(3)$ is equivalent to the condition that 0 is a "vague attractor" for $\mu = 0$, in the sense of a definition of Ruelle and Takens [3]).

(iii) $P(h)$ is found to be equivalent to the condition for $f(0,\cdot)$ to have at $z = 0$ a $(s,-)$ singularity in the sense of Takens' definition [4], with $h = 2s + 1$.

With respect to the problem of asymptotic stability of bifurcating periodic orbits, we emphasize that $P(h)$ is not only sufficient but also necessary for the bifurcating periodic orbits to be attracting and for the attractivity to have an appropriate structural character (Th. 4.3). Thus the quoted theorem (3B.3) in [2] is reinterpreted and enriched. Moreover, the assumption that 0 is asymptotically stable (but not h-asymptotically stable) for $\mu = 0$ is not in general sufficient for the bifurcating periodic orbits to be attracting: this is shown by a counter-example (see Remark 4.5) and contradicts Theorem (3B.4) in [2].

Finally, the significance of the present approach is illustrated by a discussion, in Section 5, of a problem already treated in [5,6].

II. PRELIMINARIES

Let $\psi: (-\bar{\mu},\mu) \times D_a \to \mathbb{R}$, $(\mu,x,y) \to \psi(\mu,x,y)$, be an application. We shall say that ψ is of class C_s^r, $s \le r$, if $\psi \in C^r$ and the MacLaurin expansion in (x,y) of ψ begins with terms of degree $\ge s$. A suitable linear transformation $\tau(\mu)$: $\mathbb{R}^2 \to \mathbb{R}^2$ can be found so that (1.1) takes the form:

$$\begin{aligned} \dot{x} &= \alpha(\mu)x - \beta(\mu)y + p(\mu,x,y) \\ \dot{y} &= \alpha(\mu)y + \beta(\mu)x + q(\mu,x,y), \end{aligned} \tag{2.1}$$

where $p,q \in C_2^{k+1}[(-\bar{\mu},\bar{\mu}) \times D_a, \mathbb{R}]$. Letting $x = r\cos\theta$, $y = r\sin\theta$ in (2.1), and observing that $\beta(0) > 0$, we can find $\bar{\mu}$, $a > 0$, such that for all $\mu \in (-\bar{\mu},\bar{\mu})$, the differential equation of the orbits of (2.1), in the polar coordinates (r,θ), is

$$\frac{dr}{d\theta} = R(\mu,r,\theta), \tag{2.2}$$

where $R \in C^k[(-\bar{\mu},\bar{\mu}) \times D_a \times \mathbb{R}, \mathbb{R}]$.

In addition, for every $\mu \in (-\bar{\mu},\bar{\mu})$ the following three stability properties concerning the origin 0 of $\mathbb{R}^2$ as solution of (2.1) are found to be equivalent: asymptotic stability (resp. complete instability) attractivity (resp. repulsivity), 0 is an attracting (resp. repulsing) focus. We denote by $r(\theta,\mu,c)$ the noncontinuable solution of (2.2) through $\theta = 0$, $r = c$. It is easily seen that if $\bar{a} \in (0,a)$ and $\bar{\mu}$ are sufficiently small, for every $\mu \in (-\bar{\mu},\bar{\mu})$ and $c \in [0,\bar{a})$, $r(\theta,\mu,c)$ exists on $[0,2\pi]$. We set $V(\mu,c) = r(2\pi,\mu,c) - c$. V is the so-called displacement function for (2.1); obviously $V \in C^k$.

We recall the C^{k+1} version in $\mathbb{R}^2$ of the well-known bifurcation theorem of Hopf ([2], Th. (3.1), (A) and (B)).

2.1. Theorem. There exists a number $\varepsilon \in (0,\bar{a})$ and a function $\mu \in C^{k-1}[(0,\varepsilon), \mathbb{R}]$, with $\mu(0) = \mu'(0) = 0$, and $\sup\{|\mu(c)|, c \in [0,\varepsilon)\} \equiv \sigma < \bar{\mu}$, such that given any $c \in (0,\varepsilon)$ and

$\mu \in (-\sigma,\sigma)$, the orbit of (2.1) passing through $(c,0)$ is closed if and only if $\mu = \mu(c)$.

III. ASYMPTOTIC STABILITY AND COMPLETE INSTABILITY OF THE ORIGIN FOR $\mu = 0$, RECOGNIZED BY MEANS OF REDUCED SYSTEMS

Setting $\lambda = \beta(0)$, $X(x,y) = p(0,x,y)$ and $Y(x,y) = q(0,x,y)$, consider the system (2.1) for $\mu = 0$:

$$\begin{aligned}\dot{x} &= -\lambda y + X(x,y)\\ \dot{y} &= \lambda x + Y(x,y).\end{aligned} \tag{3.1}$$

We shall denote by X_i, Y_i the homogeneous polynomials of degree $i \in \{2,\ldots,k+1\}$ in the MacLaurin expansion of X,Y.

(A) Suppose that X,Y are analytic. Then the displacement $V(0,c)$ is also analytic; we denote it by $v(c)$ and by v_i the coefficients in its MacLaurin expansion. Set for $(x,y) \in D_a$:

$$F(x,y) = x^2 + y^2 + \sum_{i=3}^{m} F_i(x,y), \tag{3.2}$$

where F_i is a homogeneous polynomial of degree i, and $m \geq 3$. The derivative, $\dot{F}_{(3.1)}$, of F along the solutions of (3.1) is analytic on D_a and its power series in (x,y) starts with terms of degree ≥ 3. The classical procedure of Poincaré mentioned in Section 1, consists either in the construction of power series satisfying formally the condition of being first integrals of (3.1), or in the determination of m and $F_3,\ldots,F_m$ in a manner that $\dot{F}_{(3.1)}$ be a sign definite function. We recall this procedure and also its connections with the stability problem of the null solution of (3.1).

We shall denote by $\left[\dot{F}_{(3.1)}\right]_j$, $j \geq 3$, the homogeneous polynomical of degree j in the expansion of $\dot{F}_{(3.1)}$. Assume $F_3,\ldots,F_{m-1}$ are such that $\left[\dot{F}_{(3.1)}\right]_3 = \ldots = \left[\dot{F}_{(3.1)}\right]_{m-1} = 0$; then one has [1]:

(1) if m is odd, then F_m can be determined in one and only one way in order to obtain $\left[\dot{F}_{(3.1)}\right]_m = 0$;

(2) if m is even, then F_m can be chosen in an infinite number of ways such that $\left[\dot{F}_{(3.1)}\right]_m = G_m(x^2 + y^2)^{m/2}$, where G_m is a well determined constant.

In both cases the determination of F_m is obtained by solving an algebraic system of linear equations. The right-hand sides of these equations involve λ and the coefficients of F_i, X_i, Y_i, $i \in \{2,\ldots,m-1\}$, and, if m is even, the unknown constant G_m. We set:

$$M = \sup\{m \geq 3 \mid \text{there exist } F_3,\ldots,F_{m-1} \text{ such that } \left[\dot{F}_{(3.1)}\right]_3 = \cdots = \left[\dot{F}_{(3.1)}\right]_{m-1} = 0\}.$$

M will be called the index of the system (3.1). Moreover, the constant G, defined by $G = 0$ or $G = G_M$, according to $M = \infty$ or $M \in \mathbb{N}$ will be called the Poincaré constant of (3.1).

Case (a): $M = \infty$. Then there exist series of the kind

$$F = x^2 + y^2 + F_3 + \ldots \tag{3.3}$$

satisfying formally the condition of being integrals of (3.1). It is known [1] that actually (3.1) admits in this case an analytic integral of the form (3.3). By virtue of the Lyapunov stability theorem, one concludes that $x \equiv y \equiv 0$ is a (non asymptotically) stable solution of (3.1). Moreover $v \equiv 0$ and therefore, for every $c \in [0,\bar{a})$, the orbit of (3.1) through $(c,0)$ is closed.

Case (b): $M \in \mathbb{N}$. Then there exists a polynomial of the form (3.2) with $m = M$ such that

$$\dot{F}_{(3.1)}(x,y) = G(x^2 + y^2)^{M/2} + \eta(x,y), \tag{3.4}$$

where $G \neq 0$ and η is an analytic function whose expansion in x,y begins with terms of degree $\geq M + 1$. Therefore the solution $x \equiv y \equiv 0$ of (3.1) is asymptotically stable if $G < 0$ and completely unstable if $G > 0$. Moreover,

$$v_i = 0, \quad i \in \{2,\dots,M-2\}, \quad v_{M-1} \neq 0, \tag{3.5}$$

and v_{M-1} has the same sign as G ([7], p. 361).

(B) Now we consider the system (3.1) in the general case, that is $X,Y \in C^{k+1}$. We need the following definition.

3.1. Definition. Let h be an integer $\in \{2,\dots,k\}$. The null solution $x \equiv y \equiv 0$ of (3.1) is said to be h-asymptotically stable (resp. h-completely unstable) if

(1) for every ξ,τ $C[D_a, \mathbb{R}]$ of order $> h$, the solution $x \equiv y \equiv 0$ of the system

$$\begin{aligned} \dot{x} &= -\lambda y + X_2(x,y) + \dots + X_h(x,y) + \xi(x,y) \\ \dot{y} &= \lambda x + Y_2(x,y) + \dots + Y_h(x,y) + \tau(x,y) \end{aligned} \tag{3.6}$$

is asymptotically stable (resp. completely unstable);

(2) property (1) is not satisfied when h is replaced by any $m \in \{2,\dots,h-1\}$.

Now the following theorem involving the displacement function V relative to (2.1) can be proved.

3.2. Theorem. Let h be an integer $\in \{2,\dots,k\}$. Then, the following propositions are equivalent.

(u) The solution $x \equiv y \equiv 0$ of (3.1) is h-asymptotically stable (resp. h-completely unstable).

(v) The index of the system

$$\begin{aligned} \dot{x} &= -\lambda y + X_2(x,y) + \dots + X_h(x,y) \\ \dot{y} &= \lambda x + Y_2(x,y) + \dots + Y_h(x,y) \end{aligned} \tag{3.7}$$

is equal to $h + 1$ and the relative Poincaré constant is < 0 (resp. > 0).

(w) One has

$$\frac{\partial^i V}{\partial c^i}(0,0) = 0 \quad \text{for} \quad i \in \{1,2,\dots,h-1\}, \quad \text{and} \quad \frac{\partial^h V}{\partial c^h}(0,0) < 0$$

(resp. > 0).

In addition, if any of propositions (u), (v), (w) holds, then h is odd.

To give an idea of the proof, we limit ourselves to prove that (u) $\models\!\Rightarrow$ (v). Assume that the origin O of (3.1) is h-asymptotically stable; then, setting $\xi = \tau = 0$ in (3.6), we recognize that $x \equiv y \equiv 0$ is an asymptotically stable solution of (3.7). Therefore the index of (3.7) is an even number $M \in \mathbb{N}$ and the relative Poincaré constant is a number $G < 0$. Hence there exists a polynomial F of the form (3.2) with $m = M$ such that

$$\dot{F}_{(3.7)}(x,y) = G(x^2 + y^2)^{M/2} + o(r^M), \quad r = \sqrt{x^2 + y^2}\,.$$

The condition (2) of Definition 3.1 obviously requires that $M \geq h + 1$. Suppose $M > h + 1$ and assume in (3.6) $\xi = ax(x^2 + y^2)^{\frac{M}{2}-1}$, $\tau = ay(x^2 + y^2)^{\frac{M}{2}-1}$ where $a > -\frac{G}{\Sigma}$. We get

$$\dot{F}_{(3.6)}(x,y) = (2a + G)(x^2 + y^2)^{M/2} + o(r^M)$$

and the solution $x \equiv y \equiv 0$ of the corresponding system (3.6) is completely unstable. Then, a contradiction.

(C) The concepts of h-asymptotic stability (resp. h-completed instability) and that of $(s,-)$ (resp. $(s,+)$) singularity, $h = 2s + 1$, introduced by Takens in [4], are equivalent. Precisely, consider the vector field ν defined by the r.h.s. of (3.1):

$$\nu: \quad (x,y) \to (-\lambda y + X(x,y),\ \lambda x + Y(x,y)).$$

and assume that X, Y are C_2^∞ functions.

3.3. *Definition.* (Takens). Let $s \in \mathbb{N}$. The vector field ν is said to have at the origin O a $(s,-)$ (resp. a $(s,+)$) singularity if a C^∞ diffeomorphism Φ: $(x,y) \to (x_1,x_2)$ of D_a in $\mathbb{R}^2$ can be found such that, if $\Phi_*(\nu)$ is the transformed of ν under Φ, we have:

$$\Phi_*(\nu) = \phi(x_1,x_2)[-x_2+x_1\delta(x_1^2+x_2^2)^s + \overline{X}_1,\ x_1+x_2\delta(x_1^2+x_2^2)^s + \overline{X}_2],$$

where $\phi \in C^\infty$ is a positive function, $\delta = -1$ (resp. $\delta = 1$), and $\overline{X}_1, \overline{X}_2 \in C_{s+1}^\infty$.

The following theorem can be proved, again by using the arguments of the Poincaré procedure.

3.4. *Theorem.* Suppose that $X, Y \in C_2^\infty$ and $s \in \mathbb{N}$. The vector field ν has at the origin O a $(s,-)$ (resp. a $(s,+)$) singularity if and only if O is $(2s+1)$-asymptotically stable (resp. $(2s+1)$-completely unstable).

IV. ATTRACTING AND REPULSING CLOSED ORBITS

The present section is devoted to an analysis of the relationship for (2.1) between attractivity properties of the origin O of $\mathbb{R}^2$ for $\mu = 0$ and the family of periodic bifurcating orbits. This family will be denoted by $\{(c,\mu(c)\ c \in (0,\varepsilon)\}$, where the function $\mu(c)$ introduced by Theorem 2.1 will be called the bifurcation function for (2.1). We emphasize that whenever we use the expression "the bifurcating periodic orbits are attracting", we mean that attractivity is actually occurring for all periodic orbits for which c is sufficiently small. Finally, we notice that for periodic orbits there is an equivalence between the concepts of asymptotic stability (resp. complete instability) and attractivity (resp. repulsivity).

4.1. Lemma. Suppose that 0 is asymptotically stable (resp. completely unstable) for $\mu = 0$. Then a necessary and sufficient condition for the bifurcating periodic orbits to be attracting (resp. repulsing) is that there exists an $\varepsilon^* \in (0,\varepsilon)$ such that: (1) the restriction of the bifurcating function to $[0,\varepsilon^*)$ is 1-1; (2) $\mu(c)\alpha'(0) > 0$ (resp. $\mu(c)\alpha'(0) < 0$)) on $(0,\varepsilon^*)$.

The proof of the sufficiency is a consequence of the existence of a family of attracting (resp. repulsing) invariant compact sets, homeomorphic to closed annulus centered at the origin [8,9]. Necessity follows by arguments involving the well-known Bendixon theorem on the limit sets of orbits in $\mathbb{R}^2$.

Now, given any odd $h \in \{3,\ldots,k\}$ we want to consider the case of bifurcating attracting (resp. repulsing) periodic orbits in which this structure is preserved under modifications of the right hand sides of (2.1) that do not change the functions α, β and those terms of X and Y having degree $\leq h$. With this in mind, denote by S_h the set of couples (P,Q) of functions $\in C_2^{k+1}[(-\bar{\mu},\bar{\mu}) \times D_a, \mathbb{R}]$ such that $[P(0,x,y)]_i = X_i(x,y)$ and $[Q(0,x,y)]_i = Y_i(x,y)$, $i \in \{2,\ldots,h\}$. For $(P,Q) \in S_h$, let $V_{P,Q}$ and $\mu_{P,Q}$ be the displacement and the bifurcation functions respectively for the one-parameter family of differential systems

$$\begin{aligned} \dot{x} &= \alpha(\mu)x - \beta(\mu)y + P(\mu,x,y) \\ \dot{y} &= \alpha(\mu)y + \beta(\mu)x + Q(\mu,x,y). \end{aligned} \tag{4.1}$$

4.2. Definition. Let $h \in \{3,\ldots,k\}$. The bifurcating periodic orbits of (2.1) are said to be h-attracting (resp. h-repulsing) if:

(a) for any $(P,Q) \in S_h$ the periodic orbits of (4.1) are attracting (resp. repulsing).

(b) condition (a) is not satisfied when h is replaced by any odd $m \in \{3,\ldots,h-2\}$.

The properties of the periodic orbits given in Definition 4.2 are completely characterized by the following theorem.

4.3. Theorem. The bifurcating periodic orbits of (2.1) are h-attracting (resp. h-repulsing), if and only if 0 is h-asymptotically stable (resp. h-completely unstable) for $\mu = 0$.

Outline of the Proof. Suppose for instance that $\alpha'(0) > 0$. By using Lemma 4.1, sufficiency can be proved by showing that the bifurcation function $\mu_{P,Q}$ is 1-1 and > 0 on an interval $(0,\varepsilon^*)$. To prove this we notice that from the identity $V_{P,Q}(c, \mu_{P,Q}(c)) \equiv 0$, it follows

$$\frac{\partial^{s+1}}{\partial c^{s+1}} V_{P,Q}(0,0) = -(s+1)\frac{\partial^2}{\partial c \partial \mu} V_{P,Q}(0,0)\mu_{P,Q}^{(s)}(0), \tag{4.2}$$

for every $s \leq k-1$ such that $\mu_{P,Q}^{(1)}(0) = \mu_{P,Q}^{(2)}(0) = \ldots = \mu_{P,Q}^{(s-1)} = 0$. In addition, we can prove $\frac{\partial^2 V_{P,Q}}{\partial c \partial \mu}(0,0) > 0$. Since 0 is h-asymptotically stable (resp. h-completely unstable) for

$$\begin{aligned} \dot{x} &= -\lambda y + P(0,x,y) \\ \dot{y} &= \lambda x + Q(0,x,y), \end{aligned} \tag{4.3}$$

Theorem (3.2) implies that:

$$\mu_{P,Q}^{(i)}(0) = 0 \quad \text{for } i \in \{1,\ldots,h-2\} \quad \text{and} \quad \mu_{P,Q}^{(h-1)}(0) > 0$$

$$(\text{resp. } < 0)$$

so that $\mu_{P,Q}$ is strictly increasing in an interval $[0,\varepsilon^*)$. To complete the proof of sufficiency, we observe that for every odd $m \in \{3,\ldots,h-2\}$ we can choose $(P,Q) \in S_m$ in order to obtain that 0 is m-completely unstable (resp. m-asymptotically stable) for (4.3).

To prove the necessity part we observe that if the index of (3.7) is $> h+1$, then by a suitable $(P,Q) \in S_h$, we could obtain by means of the sufficiency part of the theorem, that 0 is h-completely unstable (resp. h-asymptotically stable) for (4.3). Then $M = h + 1$ and obviously, using again the argument of sufficiency, we have $G < 0$ (resp. $G > 0$).

4.4. Corollary. Suppose that the right hand sides of (2.1) are C^∞ functions of (μ,x,y) and for $\mu = 0$ they are analytic in (x,y). If $\alpha'(0) > 0$ (resp. < 0), then exactly one of the following possibilities holds:

(a) 0 is asymptotically stable for $\mu = 0$; the bifurcating periodic orbits are attracting and occur only for $\mu > 0$ (resp. < 0);

(b) 0 is completely unstable for $\mu = 0$; the bifurcating periodic orbits are repulsing and occur only for $\mu < 0$ (resp. > 0);

(c) 0 is stable but not attracting for $\mu = 0$; the bifurcation periodic orbits have this same property and occur only for $\mu = 0$.

In particular, the bifurcating periodic orbits are attracting if and only if 0 is asymptotically stable for $\mu = 0$. A similar statement holds for repulsivity.

4.5. Remark. The condition that the origin is asymptotically stable for $\mu = 0$ (but not h-asymptotically stable) is not sufficient in general to guarantee that the bifurcating periodic orbits are attracting. Indeed consider the system:

$$\dot{x} = \mu x - y - xf(x,y) - x(x^2 + y^2)^{s+1}$$
$$\dot{y} = \mu y + x - yf(x,y) - y(x^2 + y^2)^{s+1} \tag{4.4}$$

where $s \in \mathbb{N}$ is ≥ 3 and $f(x,y) = (x^2 + y^2)^s \sin^2(x^2 + y^2)^{-1}$ for $(x,y) \neq (0,0)$, $f(0,0) = 0$. One can prove that the null solution of (4.4) for $\mu = 0$ is asymptotically stable. Moreover, setting

$$\mu(c) = c^{2s}\sin^2 c^{-2} + c^{2(s+1)} \quad \text{for} \quad c > 0, \quad \text{and} \quad \mu(0) = 0,$$

we recognize that given any $c > 0$ and $\mu \in \mathbb{R}$, the orbit of (4.5) relative to the couple (c,μ) is closed if and only if $\mu = \mu(c)$. It is seen that given any $\varepsilon > 0$, the function $\mu(c)$ is not 1-1 on $[0,\varepsilon)$. Then the condition (1) in Lemma (4.1) is not satisfied and our remark is proved. We notice that this is a counterexample for Theorem (3B.4) in [2].

V. APPLICATION TO THE FITZHUGH NERVE CONDUCTION EQUATIONS

Consider the Fitzhugh differential system

$$\begin{aligned} \dot{x}_1 &= \eta + x_2 + x_1 - \frac{1}{3}x_1^3 \\ \dot{x}_2 &= \rho(a - x_1 - bx_2), \end{aligned} \tag{5.1}$$

where $\eta \in \mathbb{R}$ is a parameter and a,b,ρ are real constants with $b,\rho \in (0,1)$. We need some known properties of (5.1) [5,6,10,11].

(i) For every $\eta \in (-\infty,\infty)$ there exists a unique equilibrium point $(x_1(\eta),x_2(\eta))$ for (5.1) and there are two values of η namely η_1 and η_2 such that $x_1(\eta_1) = -(1 - \rho b)^{1/2}$, $x_1(\eta_2) = (1 - \rho b)^{1/2}$.

(ii) Denote by S_j $(j = 1,2)$ the system obtained setting in (5.1) $\eta = \eta_j + \mu$ and $x_1 = y_1 + x_1(\eta_j + \mu)$, $x_2 = y_2 + x_2(\eta_j + \mu)$. Moreover, let $\alpha_j(\mu) \pm i\beta_j(\mu)$ be the eigenvalues of the linear part of S_j. Then one has:

$$\alpha_1(0) = \alpha_2(0) = 0, \quad \beta_1(0) = \beta_2(0) = \lambda,$$

$$\alpha_1'(0) > 0, \quad \alpha_2'(0) < 0,$$

where $\lambda = (\rho - \rho^2 b^2)^{1/2}$.

(iii) By means of a suitable linear transformation $(y_1,y_2) \mapsto (x,y)$, S_j takes the form:

$$\dot{x} = \frac{\rho}{\lambda}[b(1 - x_1^2(\eta_j+\mu)) - 1]y - \frac{2\rho b}{\lambda} x_1(\eta_j+\mu)y^2 - \frac{4\rho b}{3\lambda} y^3$$
$$\dot{y} = [1 - x_1^2(\eta_j+\mu) - \rho b]y + \lambda x - 2x_1(\eta_j+\mu)y^2 - \frac{4}{3} y^3. \tag{5.2}$$

Thus, the analysis of the periodic orbits of (5.1) in a neighborhood of $(\eta_j, x_1(\eta_i), x_2(\eta_j))$ is reduced to the analysis of the periodic orbits of (5.2) arising in a neighborhood of the origin 0 of $\mathbb{R}^2$ and of $\mu = 0$. We notice that the right hand sides of (5.2) satisfy the hypotheses assumed for (1.1) and in addition they are analytic in (μ, x, y). Consequently, the question of attractivity properties of the above periodic orbits is completely resolved by Corollary 4.4 through the stability properties of the solution $x \equiv y \equiv 0$ of (5.2) when $\mu = 0$. (Actually system (5.2) does not have the form (2.1), but this form can be assumed by a linear transformation (Sect. 2)). Using the Poincaré procedure we obtain the following theorem:

5.1. Theorem. Given any $\rho, b \in (0,1)$ and $a \in \mathbb{R}$, one has: (1) if $1 + \rho b^2 - 2b \geq 0$, then the bifurcating periodic orbits of (5.2) are attracting and occur for $\mu > 0$ or $\mu < 0$ according to $j = 1$ or $j = 2$ respectively; (2) if $1 + \rho b^2 - 2b < 0$, then the bifurcating periodic orbits of (5.2) are repulsing and occur for $\mu < 0$ or $\mu > 0$ according to $j = 1$ or $j = 2$ respectively.

We observe that our results are the same as those found in [5] when $1 + \rho b^2 - 2b > 0$ and $1 + \rho b^2 - 2b < 0$. But our result concerning the case $1 + \rho b^2 - 2b = 0$ does not agree with that found in [6](°). Finally our computations seem to be simpler and thus more accessible than those in [5,6].

(°) *After this report was completed we have come to learn that an error in [6] has been corrected by the author and the revised result, which agrees with ours, will appear in J. Math. Anal. Appl.*

REFERENCES

[1] Sansone, G., and Conti, R. (1969). "Non-linear Differential Equations", Mac-Millan.

[2] Marsden, J. E., and McCraken, M. (1976). "The Hopf bifurcation and its applications", *Appl. Math. Sciences 19*, Springer-Verlag.

[3] Ruelle, D., and Takens, F. (1971). "On the nature of turbulence", *Comm. Math. Phys. 20*, 167-192.

[4] Takens, F. (1973). "Unfolding of certain singularities of vector fields: Generalized Hopf bifurcations", *J. Diff. Eq. 14*, 476-493.

[5] Hsü, In-Ding, and Kazarinoff, N. D. (1976). "An applicable Hopf bifurcation formula and instability of small periodic solutions of the Field-Noyes model", *J. Math. Anal. Appl.*, *55*, *(1)*, 62-89.

[6] Hsü, In-Ding. (1977). "A higher order Hopf bifurcation formula and its application to Fitzhugh's nerve conduction equations", *J. Math. Anal. Appl. 60*, 47-57.

[7] Lyapunov, M. A. (1969). "Problème général de la stabilité du mouvement", *Ann. of Math. Studies*, Princ. Univ. Press.

[8] Marchetti, F., Negrini, P., Salvadori, L., and Scalia, M. (1976). "Liapunov direct method in approaching bifurcation problems", *Ann. Mat. Pur. Appl. (iv) cviii*, 211-225.

[9] Chafee, N. (1968). "The bifurcation of one or more closed orbit from an equilibrium point of an autonomous differential system", *J. Diff. Eq. 4*, 661-679.

[10] Fitzhugh, R. (1961). "Impulses and physiological states in theoretical models of nerve membrane, *Biophys. J.*, 445-466.

[11] Troy, W. C. (1974). "Oscillation Phenomena in Nerve Conduction Equations", Ph.D. Dissertation, SUNY at Buffalo.

AN ITERATIVE METHOD FOR APPROXIMATING SOLUTIONS TO NONLINEAR PARTIAL DIFFERENTIAL EQUATIONS

*J. W. Neuberger**

Mathematics Department
North Texas State University
Denton, Texas

INTRODUCTION

In this note an iterative method is given for approximating solutions to boundary value problems for a class of partial differential equations. The method has been used in a number of FORTRAN programs. It is related to but is much more efficient than methods given in [3], [4].

For simplicity of exposition we discuss here a single second order quasi-linear equation on R^2:

$$r(u,u_1,u_2)u_{11} + 2s(u,u_1,u_2)u_{12} + t(u,u_1,u_2)u_{22} = 0 \qquad (1)$$

where r,s,t are given continuous functions on all of R^3 so that $r^2 + 2s^2 + t^2 > 0$. Systems of equations on spaces of higher dimension could have been considered just as well.

$(u_1 = \partial u/\partial x, \quad u_{12} = \partial^2 u/\partial y \partial x$ etc.$)$

*Supported in part by an NSF Grant.

ISBN 0-12-434180-2

I. FINITE DIFFERENCE APPROXIMATION

For further simplicity only rectangular grids with even spacing on $[0,1] \times [0,1]$ are considered. Irregularly shaped regions (even regions with holes) may be treated with only slight extra complication.

Fix a positive integer n and denote by G the grid $\{(i/n,\ j/n)\}_{i,j,=1}^{n}$. Denote by K the linear space of real-valued functions on the grid G. From among a number of interesting possibilities we use the following approximation to differentiation:

$$(D_i u)(p) = \begin{cases} (u(p+\delta e_i)-u(p-\delta e_i))/(2\delta) \text{ if } p+\delta e_i, p-\delta e_i \in G \\ (u(p+\delta e_i)-u(p))/\delta \text{ if } p-\delta e_i \notin G \\ (u(p)-u(p-\delta e_i))/\delta \text{ if } p+\delta e_i \notin G \end{cases}$$

$i = 1,2$ where $e_1 = \binom{1}{0}$, $e_2 = \binom{0}{1}$, $\delta = 1/n$. Then D_1^2, D_1D_2, D_2^2 approximate second differentiation.

With this notation a difference equation approximation to (1) is the problem of finding $u \in K$ such that

$$r(u,D_1u,D_2u)D_1^2u+2s(u,D_1u,D_2u)D_1D_2u+t(u,D_1u,D_2u)D_2^2u = 0. \quad (2)$$

It is usually too much to ask that such an equation be exactly satisfied on all of G. In the next section we introduce the weaker idea of quasi-solution. It is quasi-solutions that are then found.

To introduce boundary conditions, designate by G' a proper subset of G. Consider G' as 'boundary' for G. If $w \in K$ one has the boundary value problem of finding $u \in K$ which is a solution or a quasi-solution to (2) so that

$$u(p) = w(p), \quad p \in G'.$$

Denote by K_1 the set of all $u \in K$ such that $u(p) = 0$, $p \in G'$ and by K_0 the set of all $u \in K$ such that $u(p) = 0$, $p \in G/G'$.

Two equations on which extensive computations have been made using the method to be outlined in section 3 are

(i) the minimal surface equation

$$(1 + u_2^2)u_{11} - 2u_1u_2u_{12} + (1 + u_1^2)u_{22} = 0$$

and the sonic flow equation

(ii) $u_{11} + u_2u_{22} = 0.$

These were studied at the suggestion of Peter Lax, whose criticism of this entire development has been of great value.

II. QUASI-SOLUTIONS

Denote K^6 by H and denote by A the transformation from K to $L(H,K)$ defined by

$$A(u)\begin{pmatrix} x \\ \begin{pmatrix} p \\ q \end{pmatrix} \\ \begin{pmatrix} f \\ g \\ h \end{pmatrix} \end{pmatrix} = r(u,D_1u,D_2u)f+2s(u,D_1u,D_2u)g+t(u,D_1u,D_2u)h \tag{3}$$

if $u \in K$, $(x,p,q,f,g,h,) \in H$. From here on it is assumed that equations have been normalized so that $r^2 + 2s^2 + t^2 = 1$.

For $u \in K$, define

$$\|u\| = (\sum_{p \in G} u(p)^2)^{1/2}.$$

For a norm in H, take the inner product norm induced by K.

Denote by π_1 the orthogonal projection of K onto K_1 and by π_0 the orthogonal projection of K onto K_0 (assuming a fixed boundary set G' as in section 1).

Define D: $K \to H$ by

$$Du = \begin{pmatrix} u \\ \begin{pmatrix} D_1 u \\ D_2 u \end{pmatrix} \\ \begin{pmatrix} D_1^2 u \\ D_1 D_2 u \\ D_2^2 u \end{pmatrix} \end{pmatrix}, \quad u \in K. \tag{4}$$

Using (3) and (4), rewrite (2) as the problem of finding $u \in K$ such that

$$A(u)Du = 0. \tag{5}$$

Note that $A(u)A(u)^* = I$, the identity on K and that consequently, $A(u)^*A(u)$ is an orthogonal projection on H (the orthogonal projection onto the orthogonal complement of the null-space of $A(u)$) if $u \in K$.

Note that $u \in K$ satisfies (5) if and only if

$$(A(u)D)^*(A(u)D)u = 0. \tag{6}$$

Definition. $u \in K$ is a quasi-solution (relative to the boundary set G') of (5) provided

$$\pi_1(A(u)D)^*(A(u)D)u = 0. \tag{7}$$

We attempt here an indication of conditions under which a quasi-solution might be a reasonable approximation to a solution to (1).

For K one has the natural basis $\{e_p\}p \in G$ where

$$e_p(q) = \begin{cases} 1 & \text{if } q = p \\ 0 & \text{if } q \neq p, \quad p,q \in G. \end{cases}$$

Regard, for u satisfying (7), $A(u)D$ as an $(n+1) \times (n+1)$ matrix relative to this basis. Call a column of $A(u)D$ a boundary column if it equals $A(u)De_p$ for some $p \in G'$. Observe that (7) implies that $A(u)D\,u$ is orthogonal to all non-boundary columns of $A(u)D$. For well chosen boundary conditions and u

satisfying (7) any boundary column of $A(u)D$ is 'nearly' equal to its orthogonal projection onto the span of the non-boundary columns of $A(u)D$. This would mean $\|A(u)Du\|$ is small and so u is a reasonable *finite* approximation to a solution to (1).

Much more remains to be done concerning analysis of quasi-solutions.

III. DESCRIPTION OF ITERATION PROCESS

The transformation D^*D serves as a Laplacian for a Dirichlet space with Dirichlet norm $\|u\| \equiv \|Du\|_H$, $u \in K_G$ (See [1]) even though not all of the axioms of [1] are satisfied here. Crucial to our iteration process is the following transformation C defined in terms of this Laplacian:

$$C \equiv \pi_1 D^*D|_{K_1} \tag{8}$$

Observe (see [4]) that C is symmetric and positive definite and hence C^{-1} exists.

For fixed r,s,t (and hence A) as well as G' consider the following iteration process starting with a choice of $w \in K$:

$$w_0 = w,\ w_{k+1} = w_k - C^{-1}\pi_1(A(w_k)D)^*(A(w_k)D)w_k,\ k = 0,1,2,\ldots . \tag{9}$$

We are interested in conditions under which this process converges to a quasi-solution of (5). Note that for $\{w_k\}_{k=0}^{\infty}$ satisfying (9), $w_k(p) = w(p)$, $p \in G'$ since the range of C^{-1} is K_1 and all members z of K_1 satisfy $z(p) = 0$, $p \in G'$. Hence $w_{k+1}(p) = w_k(p)$, $p \in G'$, $k = 0,1,\ldots$. Therefore, if $\{w_k\}_{k=0}^{\infty}$ converges, its limit z it agrees with the 'initial estimate' w on G'.

For a first theorem we consider a linear problem. Suppose $A(u)$ is independent of u, $u \in K$, and define $F = A(u)$, any $u \in K$.

Theorem 1. Suppose the problem

$$FDy = 0, \quad y \in K_1, \tag{10}$$

has only the solution $y = 0$. If $w \in K$, then the sequence $\{w_k\}_{k=0}^{\infty}$ defined in (9) converges to $z \in K$ such that

(i) $z(p) = w(p), \quad P \in G'$ and

(ii) $\pi_1(FD)^*(FD)z = 0$.

Suppose $J(u) = A(u)^*A(u)$ for all $u \in K$. For the next theorem we make the following assumption:

$$J \text{ is locally Lipschitz.} \tag{11}$$

Define $L: K \to L(K_1, K)$ such that if $u \in K$ then

$$L(u) = A(u)D|_{K_1}$$

The next theorem gives conditions under which (5) has a quasi-solution for "small" initial estimate w.

Theorem 2. Suppose that $L(0)$ is invertible. There is $r_1 > 0$ such that if $w \in K$, $\|w\| < r_1$, then the iteration (9) converges to $z \in K$ such that

(i) $z(p) = w(p), \quad P \in G'$,

(ii) $\pi_1(A(z)D)^*(A(z)D)z = 0$.

In computational practive the process (9) seems to converge without assumption (11) or the invertibility assumption on $L(0)$. Moreover the local character of the conclusion does not seem necessary. The purpose of this note is to present the background for working computer codes and to present theorems which at least point in the direction of an explanation of why the codes work. Much work remains to be done in this direction.

IV. PROOFS

For a given choice of A, G' and w consider the sequence $\{w_k\}_{k=0}^{\infty}$ generated by (9). Define $\{z_n\}_{n=0}^{\infty}$ by $z_k = C^{1/2}\pi_1 w_k$, $k = 0,1,\ldots$. Denote $\pi_0 w$ by x. Then $\pi_1 w_k = C^{-1/2}z_k$ and so $w_k = \pi_0 w_k + \pi_1 w_k = x + C^{-1/2}z_k$, $k = 0,1,\ldots$. Hence $\{w_k\}_{k=0}^{\infty}$ converges if and only if $\{z_k\}_{k=0}^{\infty}$ converges.

For $v \in K_0$ define $\alpha_v\colon K_0 \to K_0$, $\beta_v\colon K_1 \to K_0$, $\gamma_v\colon K_1 \to K_1$ so that

$$\alpha_v(y) = \pi_0(A(v+y)D)^*(A(v+y)D)|_{K_0}$$

$$\beta_v(y) = \pi_0(A(v+y)D)^*(A(v+y)D)|_{K_1}$$

$$\gamma_v(y) = \pi_1(A(v+y)D)^*(A(v+y)D)|_{K_1}$$

for all $y \in K_1$. Note that

$$\beta_v(y)^*\colon K_0 \to K_1$$

In terms of this notation the iteration (9) becomes

$$z_{k+1} = z_k - C^{-1/2}\gamma_x(C^{-1/2}z_k)C^{-1/2}z_k - C^{-1/2}\beta_x(C^{-1/2}z_k)^*x, \qquad (12)$$

$$k = 0,1,2,\ldots ,$$

since

$$\begin{aligned} &C^{-1/2}\pi_1(A(w_k)D)^*(A(w_k)D)w_k \\ &= C^{-1/2}\pi_1(A(w_k)D)^*(A(w_k)D)(\pi_1 w_k + \pi_0 w_k) \\ &= C^{-1/2}\gamma_x(C^{-1/2}z_k)\pi_1 w_k + C^{-1/2}\beta_x(C^{-1/2}z_k)^*x \\ &= C^{-1/2}\gamma_x(C^{-1/2}z_k)C^{-1/2}z_k + C^{-1/2}\beta_x(C^{-1/2}z_k)^*x \end{aligned}$$

using $\pi_0 w_k = x$, $k = 0,1,2,\ldots$.

Proof of Theorem 1. Define $\gamma\colon K_1 \to K_1$ so that $\gamma = \pi_1(FS)^*(FD)|_{K_1}$ and $\beta\colon K_1 \to K_0$ so that $\beta = \pi_0(FD)^*(FD)|_{K_1}$. Then the iteration (12) is here

$$z_{k+1} = z_k - C^{-1/2}\gamma C^{-1/2}z_k - C^{-1/2}\beta^*x, \quad k = 0,1,2,\ldots . \qquad (13)$$

Suppose $y \in K_1$, $y \neq 0$. Then $\langle C^{-1/2}\gamma C^{-1/2}y,y\rangle$ $= \langle C^{-1/2}\pi_1(FD)^*(FD)|_K\, C^{-1/2}y,y\rangle = \langle FDC^{-1/2}y,FDC^{-1/2}y\rangle > 0$ since by hypothesis $FD(C^{-1/2})y \neq 0$. Therefore $C^{-1/2}\gamma C^{-1/2}$ is a positive symmetric transformation from $K_1 \to K_1$.

But for $z \in K_1$, starting as above $\langle C^{-1/2}\gamma C^{-1/2}z,z\rangle$ $= \langle FDC^{-1/2}z,FDC^{-1/2}z\rangle = \langle F^*FDC^{-1/2}z,DC^{-1/2}z\rangle \leq \langle DC^{-1/2}z,DC^{-1/2}z\rangle$ $= \langle C^{-1/2}\pi_1 D^*DC^{-1/2}z,z\rangle = \langle C^{-1/2}CC^{-1/2}z,z\rangle = \langle z,z\rangle$. Therefore $C^{-1/2}\gamma C^{-1/2} \leq I$. Since $C^{-1/2}\gamma C^{-1/2} > 0$, it follows that $I - C^{-1/2}\gamma C^{-1/2} < I$ and hence there is $\varepsilon > 0$ such that

$$|I - C^{-1/2}\; C^{-1/2}| \leq 1 - \varepsilon.$$

Hence since $z_{k+2} = z_{k+1} - C^{-1/2}\gamma C^{-1/2}z_{k+1} - C^{-1/2}\beta^* x$ it follows that $z_{k+2} - z_{k+1} = (I - C^{-1/2}\gamma C^{-1/2})(z_{k+1} - z_k)$ and so $\|z_{k+2} - z_{k+1}\| \leq (1-\varepsilon)\|z_{k+1} - z_k\|$, $k = 0,1,2,\ldots$. This assures the convergence of $\{z_k\}_{k=0}^{\infty}$ to some $u \in K_1$ and hence also one has the convergence of the related sequence $\{w_k\}_{k=0}^{\infty}$ to $z \equiv x + C^{-1/2}u$. From (13) it is clear that $C^{-1/2}\gamma C^{-1/2}u$ $+ C^{-1/2}\beta^*x = 0$, i.e., $\gamma C^{-1/2}u + \beta^*x = 0$. $\gamma C^{-1/2}u + \beta^*x = 0$, i.e.

$$\pi_1(FD)^*(FD)z = 0 \text{ since } \pi_1(FD)^*(FD)z = \pi_1(FD)y + \pi_1(FD)^*(FD)x$$
$$= \gamma_y + \beta^*x.$$

Now $z(p) = y(p) + x(p) = x(p) = w(p)$, $p \in G'$. Hence z is a quasi-solution to (5) which satisfies the required boundary conditions.

Proof of Theorem 2. Note that L is continuous and hence there are $r_1,b > 0$ so that $b < 1$ and if $u \in K$, $\|u\| < r_1$ and $y \in K_1$, then $\|L(u)y\| \leq b\|y\|$. Denote by b_1 a number so that $0 < b_1 < 1$ and $\|C^{-1/2}y\| \geq b_1\|y\|$ for all $y \in K_1$.

From (11) it follows that there is $r_2,M_3 > 0$ such that the sets $\{\gamma_v\}_{\|v\|\leq r_2}$, $v \in K_0$ and $\{\beta^*_v\}_{\|v\|\leq r_2}$, $v \in K_0$ are uniformly locally Lipschitz at 0 with Lipschitz constant M_3. Note that since r, s and t are bounded functions there is $M_1 > 0$ so that $|\gamma_v(y)| \leq M_1$, $|\beta_v(y)| \leq M_1$ if $v \in K_0$, $y \in K_1$.

Lemma. Suppose $v \in K_0$, $y \in K_1$, $0 < r_3 < r_1$, $\|y\| \leq r_3$, $\|v\| \leq M_2 \equiv r_3(1 - (1 - b^2b_1^2)^{1/2})/(|C^{-1/2}|M_1)$ and $z = (I - C^{-1/2}\gamma_v(C^{-1/2}y)C^{-1/2})y - C^{-1/2}\beta_v(C^{-1/2}y)^*v$. Then $\|z\| \leq r_3$.

Proof of Lemma. Note that $\|(I - C^{-1/2}\gamma_v(C^{-1/2}y)C^{-1/2})y\|^2$ $= \|y\|^2 - 2\langle C^{-1/2}\gamma_v(C^{-1/2}y)C^{-1/2}y,y\rangle + \langle C^{-1/2}\gamma_v(C^{-1/2}y)C^{-1/2}y,$ $C^{-1/2}\gamma_v(C^{-1/2}y)C^{-1/2}y\rangle$, $\langle C^{-1/2}\gamma_v(C^{-1/2}y)C^{-1/2}y,y\rangle$ $= \langle \pi_1(A(v + C^{-1/2}y)D)^*(A(v + C^{-1/2}y)D)|_{K_1}C^{-1/2}y,C^{-1/2}y\rangle$ $= \|A(v + C^{-1/2}y)DC^{-1/2}y\|^2 \geq b^2\|C^{-1/2}y\|^2 \geq b^2b_1^2\|y\|^2$ and $\langle C^{-1/2}\gamma_v(C^{-1/2}y)C^{-1/2}y,C^{-1/2}\gamma_v(C^{-1/2}y)C^{-1/2}y\rangle$ $\leq \langle C^{-1/2}\gamma_v(C^{-1/2}y)C^{-1/2}y,y\rangle$ since $C^{-1/2}\gamma_v(C^{-1/2}y)C^{-1/2}$ is symmetric, non-negative and $\leq I$ on K_1.

Hence $\|(I - C^{-1/2}\gamma_v(C^{-1/2}y)C^{-1/2})y\|^2 \leq \|y\|^2 - b^2b_1^2\|y\|^2$ and so $\|y - C^{-1/2}\gamma_v(C^{-1/2}y)C^{-1/2}y\| \leq \|y\|(1 - b^2b_1^2)^{1/2}$. (Note in passing that $|I - C^{-1/2}\gamma_v(C^{-1/2}y)C^{-1/2}| \leq (1 - b^2b_1^2)^{1/2}$).

Therefore $\|z\| \leq \|y - C^{-1/2}\gamma_v(C^{-1/2}y)C^{-1/2}y\|$ $+ \|C^{-1/2}\beta_v(C^{-1/2}y)^*v\| \leq \|y\|(1 - b^2b_1^2)^{1/2} + |C^{-1/2}|M_1\|v\| \leq r_3$, using the inequality for $\|v\|$ in the hypothesis. This concludes a proof of the lemma.

By induction one has, using this lemma, that if $z_0 \in K_1$, $x \in K_0$, $\|z_0\| \leq r_3 \leq r_1$, $\|x\| \leq M_2$ (M_2 as in the lemma), then $\|z_k\| \leq r_3$, $k = 0,1,2,\ldots$.

Choose r_3 so that $0 < r_3 < r_1$ and $|C^{-1/2}|M_3r_3$ $\leq (1 - (1 - b^2b_1^2)^{1/2})/3$ and suppose $y \in K_1$, $\|y\| \leq r_4$ $\equiv \sup\{r_2,r_3\}$, $x \in K_0$, $\|x\| \leq r_4(1 - b^2b_1^2)^{1/2})|(|C^{-1/2}|M_1)$.

Take $z_0 = y$ and define $\{z_k\}_{k=0}^{\infty}$ so that (12) holds. By the lemma, $\|z_k\| \leq r_4$, $k = 0,1,2,\ldots$. It is an easy calculation that

$$\begin{aligned} z_{k+2}-z_{k+1} &= (I-C^{-1/2}\gamma_x(C^{-1/2}z_k)C^{-1/2})(z_{k+1}-z_k) \\ &-[C^{-1/2}\gamma_x(C^{-1/2}z_{k+1})C^{-1/2}-C^{-1/2}\gamma_x(C^{-1/2}z_k)C^{-1/2}]z_{k+1} \\ &-[C^{-1/2}\beta_x(C^{-1/2}z_{k+1})^*-C^{-1/2}\beta_x(C^{-1/2}z_k)^*]x, \end{aligned}$$

$k = 0,1,2,\ldots$.

But then for $k = 0,1,2,\ldots$,

$$\begin{aligned} \|z_{k+2}-z_{k+1}\| &\leq |I-C^{-1/2}\gamma_x(C^{-1/2}z_k)C^{-1/2}|\ \|z_{k+1}-z_k\| \\ &+2|C^{-1/2}|M_3r_4\|C^{-1/2}z_{k+1}-C^{-1/2}z_k\| \\ &\leq[(1-b^2b_1^2)^{1/2}+(2(1-(1-b^2b_1^2)^{1/2})/3)|C^{-1/2}|]\|z_{k+1}-z_k\| \\ &\leq[2/3+(1-b^2b_1^2)^{1/2}/3]\|z_{k+1}-z_k\| \quad \text{since} \quad |C^{-1/2}| \leq 1. \end{aligned}$$

But $2/3 + (1 - b^2b_1^2)^{1/2}/3 < 1.$

Hence $\{z_k\}_{k=0}^{\infty}$ converges to some $u \in K_1$ and so the related sequence $\{w_k\}_{k=0}^{\infty}$ of (9) converges to $z = x + C^{-1/2}u$ $(w_0 = w = x + C^{-1/2}y)$. That $\pi_1(A(z)D)^*(A(z)D)z = 0$ and $z(p) = w(p)$, $p \in G'$ follows essentially as in the argument for Theorem 1.

REFERENCES

[1] Beurling, A., et Deny, J. (1958). "Espaces de Dirichlet I. Le Cas Elementaire", *Acta Math.* *99*, 203,224.

[2] Neuberger, J. W. (1976). "Projection Methods for Linear and Nonlinear Systems of Partial Differential Equations", Springer-Verlag, Lecture Notes, 564.

[3] Neuberger, J. W. "Boundary Value Problems for Systems of Nonlinear Partial Differential Equations", Springer-Verlag, Lecture Notes, (to appear).

[4] Neuberger, J. W. (1977). "Iteration for Systems of Nonlinear Partial Differential Equations", Proceedings of Conference on Nonlinear Equations in Abstract Spaces, University of Texas at Arlington, Academic Press.

ON THE EXISTENCE OF INVARIANT MEASURES

Giulio Pianigiani

Institute for Physical Science and Technology
University of Maryland
College Park, Maryland

and

Istituto Matematico "U. Dini",
University of Florence,
Florence, Italy

I. INTRODUCTION

Let (X,β,μ) be a probability space and let T be a mapping from X into itself such that $T^{-1}(E) \in \beta$ iff $E \in \beta$. For all $f \in L^1(X,\mu)$ we are interested in the mean value of T along the trajectory $\gamma(x) = (x,T(x),T^2(x),\ldots)$, that is we are interested in the existence of the following limit:

$$\lim \frac{1}{n} \sum_{i=1}^{n} f(T^i(x)) \overset{\text{def}}{=} g_f(x) \tag{1.1}$$

The Birkoff ergodic theorem guarantees the existence of the limit (1.1) for μ-almost all $x \in X$, provided the measure μ is invariant under T ($\mu(T^{-1}E) = \mu(E)$ for all $E \in \beta$). If μ is ergodic ($T^{-1}(E) = E$ iff $\mu(E)$ is either 0 or 1) then $g_f(x) = \int_X f d\mu$ for μ-almost all $x \in X$.

The problem of the existence of the mean value (1.1) may be, therefore, reduced to the existence of an invariant measure.

ISBN 0-12-434180-2

II. KRILOFF-BOGOLIUBOFF THEOREM

In what follows a measure means any probability measure defined on the σ-field of the Borel subsets of X.

Theorem 1 (Kriloff-Bogoliuboff). Let X be a compact Housdorff topological space and let T be a continuous map from X into itself. Then there exists a measure μ invariant under T.

Sketch of the Proof. Fix $x_0 \in X$. For all $f \in C(X)$ define:

$$Af = \operatorname*{Lim}_{n} \frac{1}{n} \sum_{i=1}^{n} f(T^{i}(x_0)) \tag{2.1}$$

where Lim is the Banach limit (see [2], p. 73). A is a positive linear functional on the space $C(X)$ of all continuous real functions on X. By Riesz representation theorem there exists a measure μ such that:

$$Af = \int_X f d\mu \tag{2.2}$$

On the other hand, $Af\circ T = \operatorname*{Lim}_{n} \frac{1}{n} \sum_{i=1}^{n} f(T^{i+1}(x_0))$ and from the properties of Lim it follows $Af\circ T = Af$ which implies μ is invariant under T.

Unfortunately the measure μ constructed above can be, sometimes, rather trivial. Suppose, for instance, that there exists a point x_0 fixed under T. The measure μ supported on x_0 is surely invariant (and ergodic). The Birkoff theorem holds and the limit (1.1) exists for μ-almost all x; but μ-almost all means in this case "only for x_0". In order to get more information we need the invariant measure μ to be "more regular", that is, it must be supported in a "large" set. A natural question is to ask if there are continuous measures μ, i.e. $\mu\{x\} = 0$ for each singleton $\{x\}$. It is easily seen that if the map T does not have some "expansivity" property then a continuous measure need not exist. Let, for example, X be $[0,1]$ and $T(x) = x/2$. The only invariant measure is the measure supported on the fixed point $x = 0$. A sufficient condition for the existence of a continuous invariant measure is presented in the following theorem.

Theorem 2 [6]. Let X be an Housdorff topological space and let $T: X \to X$ be continuous. If there exists a finite collection of compact sets A_i, $i = 1,\ldots,k$ such that for some iterate T^n of T $(n \geq 1)$;

$$\bigcap_{i=1}^{k} T^n(A_i) \supset \bigcup_{i=1}^{k} A_i \quad \text{and} \quad \bigcap_{i=1}^{k} A_i = \emptyset \tag{2.2}$$

then there exists a continuous measure invariant under T.

When $X = [0,1]$, a condition which guarantees the existence of a continuous invariant measure is the existence of a point of period $2^n.3$ for some $n \geq 0$. (See [1], [8]).

III. ABSOLUTELY CONTINUOUS INVARIANT MEASURES

A number of interesting problems arise when the space X is the unit interval $[0,1]$. In $[0,1]$ there is a "natural measure" m (the Lebesgue measure) and it is far more important to look for absolutely continuous invariant measures, i.e., measures μ for which $d\mu = fdm$ where f is summable.

Rényi was studying the maps T_r: $[0,1] \to [0,1]$ defined by $T_r(x) = rx$ (mod 1). If r is an integer, then the Lebesgue measure is invariant. If r is not an integer the question is more complicated and it was unsolved until 1957. Rényi [13] proved that for all $r > 1$ there exists exactly one absolutely continuous invariant measure.

The problem of the existence of absolutely continuous invariand measures can be reformulated in the following way: suppose μ is invariant, that is $\mu = \mu \circ T^{-1}$. If f is the density $d\mu/dm$ of μ, the density of $\mu \circ T^{-1}$ is given by $\sum_{i=1}^{n} |\phi_i'(x)| f(\phi_i(x))$ where ϕ_i are the local inverses of T. The operator P: $L^1[0,1] \to L^1[0,1]$ defined by

$$P(f(x)) = \sum_i |\phi_i'(x)\ f(\phi_i(x))$$

is called "the Frobenius Perron operator relative to T" and it has the following properties:

(i) P is positive, i.e., $f(x) \geq 0$ implies $Pf(x) \geq 0$;

(ii) $\|Pf\| = \|f\|$;

(iii) $Pf = f$ if and only if the measure μ defined by $d\mu = fdm$ is invariant.

The existence of an absolutely continuous invariant measure is equivalent to the existence of a fixed point of P in $L^1[0,1]$. Lasota and Yorke [7] proved the following theorem. (Here "piecewise" means "finite number of intervals").

Theorem 3 [7]. Let $T: [0,1] \to [0,1]$ be a piecewise C^2 map and let

$$|T'(x)| \geq \lambda > 1 \quad \text{whenever } T' \text{ is defined.} \tag{3.1}$$

Then there exists an absolutely continuous invariant measure.

In order to prove this theorem they consider the sequence $\{P^n(f)\}_{n=1}^{\infty}$ for any function $f \in L^1[0,1]$ and they prove that under the hypotheses of theorem the sequence of the averages $\{\frac{1}{n}\sum_{i=1}^{n} P^i(f)\}$ converges strongly to some function $g \in L^1[0,1]$ which is fixed under P. Once more the crucial condition is the "expansivity condition" (3.1), and it can be shown that this condition cannot be weakened. In fact if in Theorem 3 we substitute (3.1) by $|T'(x)| > 1$ the existence of an absolutely continuous invariant measure can fail.

However there are in the literature examples of transofrmations for which there exists absolutely continuous invariant measures but the condition (3.1) is not satisfied. Consider, for instance, the map $T(x) = 4x(1-x)$. This map is conjugate to the map $T_1(x) = 2x$ for $x \leq \frac{1}{2}$ and $2 - 2x$ for $x \geq \frac{1}{2}$ via the homeomorphism $F(x) = \frac{2}{\pi} \arcsin \sqrt{x}$, that is $T_1 = F \circ T \circ F^{-1}$. Therefore there exists an absolutely continuous measure invariant under T even if the map T does not satisfy condition (3.1) since $T'(\frac{1}{2}) = 0$.

The following theorem holds.

Theorem 4 [11]. Let $T: [0,1] \to [0,1]$ be a piecewise monotonic map. If there exists a positive real function $h \in L^1[0,1]$ such that:

(i) $|T'(x)|h(T(x))/h(x) \geq \lambda > 1$ a.e.;

(ii) $h(x)/(T'(x)h(T(x)))$ is piecewise absolutely continuous;

then there exists an absolutely continuous invariant measure.

To understand the meaning of the function h in Theorem 4, suppose T is piecewise monotonic and there exists an absolutely continuous measure μ invariant under T. Suppose f is the density of μ and $f > 0$ a.e. By writing $Pf = f$ we obtain

$$\sum_i |\phi_i'(x)|f(\phi_i(x)) = f(x) \tag{3.2}$$

where ϕ_i are the local inverses of T. From (3.2) it follows $|\phi_i'(x)|f(\phi_i(x)) \leq f(x)$ which by substituting $T(x)$ yields

$$|T'(x)|f(T(x))/f(x) \geq 1. \tag{3.3}$$

The last inequality shows that the condition (i) of Theorem 4 is "almost" necessary.

In order to find the function h in Theorem 4 consider the following: it can be proved that if such a function h exists then the sequence $\{\frac{1}{n}\sum_{i=1}^{n-1} P^i(1)(x)\}$ converges strongly to some fixed point of P.

Then we can assume

$$h(x) = \frac{1}{n}\sum_{i=1}^{n-1} P^i(1)(x) \quad \text{for some convenient } n$$

and with such an h try to prove (i) and (ii) of Theorem 4. With this approach is actually possible to give much simpler proofs of known results [11] as well as prove new results [11,12].

IV. THE QUADRATIC MAP

There has been a lot of interest in the behavior of the discrete process described by the map

$$T(x) = Ax(1-x) \quad x \in [0,1]$$

when the parameter A varies between 0 and 4. Several authors studied this problem in particular: Smale and Williams [15], Guckeneheimer [3], Li and Yorke [8], May [10]. This "simple" dynamical process exhibits a very complicated behavior. It is known that for $0 \le A < 1$ the origin is attracting and all points tend asymptotically to it. For $A = 1$, the origin becomes unstable and there appears another fixed point $x_0 = 1-1/A$. For $1 < A < 3$ this point is attracting all $(0,1)$. For $A = 3$ x_0 becomes unstable and bifurcates into a periodic orbit of period 2. These two points are stable and attracting for $3 \le A < A_1$. For $A = A_1$ the orbit of period 2 becomes unstable and bifurcates into a periodic orbit of period 4 and so on for $8, 16, \ldots, 2^n \ldots$. For $A = A_c = 3.57 \ldots$ we have an irreversible bifurcation in the sense that for $A < A_c$ we have only orbits of period 2^n for $n \le n(A)$ while for $A > A_c$ there are infinitely many periodic orbits with different periods. It is not completely known what the behavior is for values of A between A_c and 4. It is known, however, that there are infinitely many values of $A \in (A_c, 4)$ for which there exists an attracting period orbit which attracts almost all points in $(0,1)$ and there exists infinitely many values of $A \in (A_c, 4)$ for which all periodic orbits are repelling.

As we said in the previous section, for $A = 4$ there exists an absolutely continuous measure which is invariant under T_A. Recently Ruelle [14] found another value of the parameter A

($A = 3.67\ldots$) for which there exists an absolutely continuous invariant measure. Let $S \subset [0,4]$ be the set of those values of the parameter A for which there exists an absolutely continuous invariant measure. We conjecture that S is the complement of an open dense set, and S is uncountable (possibly of positive Lebesgue measure). What we can prove is that S is at least countably infinite.

Theorem 5 [12]. Let $T_A(x) = Ax(1-x)$ and let $A \in [0,4]$ be such that:

(i) there exists an $n > 1$ such that $T_A^n(\frac{1}{2}) = 1 - T_A^{2n}(\frac{1}{2})$ and $T_A^n(\frac{1}{2}) < \frac{1}{2}$.

(ii) $T_A^n(I) = I$ and $T_A^i(I) \cap (\frac{1}{A}, 1-\frac{1}{A})$ is empty $i = 1, \ldots, n-1$, where $I = (T_A^n(\frac{1}{2}), T_A^{2n}(\frac{1}{2}))$.

Then there exists an absolutely continuous measure invariant under T_A. Moreover there exist infinitely many $A \in [0,4]$ satisfying (i) and (ii).

In order to prove the above theorem we consider T_A^n restricted to the interval I. By applying Theorem 4 we can show there exists an absolutely continuous measure μ invariant under T_A^n. Of course the measure $\overline{\mu} = (\mu + \mu\circ T^{-1} + \ldots + \mu\circ T^{-n+1})/n$ is absolutely continuous and invariant under T_A.

V. APPLICATIONS

To conclude this paper we want to mention the following: in [9] Lorenz studies a system of 3 ordinary differential equations. He finds an ellipsoid E such that all the trajectories eventually fall in and once inside they remain inside. In order to study the asymptotical behavior, he integrates numerically the system and then he considers the successive local maxima $z(t_n)$ for the z variable. He finds that $z(t_{n+1}) = T(z(t_n))$ where T maps an interval into itself and is a two-branched curve with $|T'(x)| \geq \lambda > 1$. Therefore there exists an absolutely continuous measure invariant under T. Then the behavior of the system can be studied statistically via this measure.

REFERENCES

[1] Butler, J., and Pianigiani, G. (1978). "Periodic points and chaotic functions", *Bull. Austral. Math. Soc. 18*, 255-265.

[2] Dunford, N., and Schwartz, J. T. (1958). "Linear Operator" I. General Theory, *Pure and Appl. Math., Vol. 7, Interscience*, N.Y.

[3] Guckenheimer, J. (1977). "On the bifurcation of maps in the interval", *Inventiones Math., 39*, 165-178.

[4] Guckenheimer, J., Oster, G., and Ipaktchi, A. (1977). "The dynamics of density dependent population model", *J. Math. Biology, 4*, 101-147.

[5] Hoppensteadt, F., and Hyman, J. (1977). "Periodic solutions of a logistic difference equation", *SIAM J. Appl. Math. 12*, 78-81.

[6] Lasota, A., and Pianigiani, G. (1977). "Invariant measures on topological spaces", *Boll. Un. Matem. Ital. 14-B*, 592-603.

[7] Lasota, A., and Yorke, J. (1973). "On the existence of invariant measures for piecewise monotonic transformations", *Trans. Amer. Math. Soc. 186*, 481-488.

[8] Li, T. Y., and Yorke, J. (1975). "Period 3 implies chaos", *Amer. Math. Monthly, 82*, 985-992.

[9] Lorenz, E. (1963). "Deterministic nonperiodic flow", *J. Atmos. Sci., 20*, 130-141.

[10] May, R. (1974). "Biological population with non-overlapping generations: stable points, stable cycles and chaos", *Science 186*, 695-697.

[11] Pianigiani, G. "Existence of invariant measures for piecewise continuous transformations", *Annals Polon. Math.*, to appear.

[12] Pianigiani, G. "Absolutely continuous invariant measures for the process $x_{n+1} = Ax_n(1-x_n)$", to appear in *Boll. Un. Matem Ital.*

[13] Renyi, A. (1957). "Representation for real numbers and their ergodic properties", *Acta. Math. Acad. Sci. Hungary, 9*, 477-493.

[14] Ruelle, D. (1977). "Applications conservant une mesure absolutment continue per rapport a dx sur [0,1]", *Comm. Math. Phys., 55*, 477-493.

[15] Smale, S., and Williams, R. (1976). "The qualitative analysis of a difference equation of population growth", *J. Math. Biology 3*, 1-5.

THE ROLE OF DIRECT FEEDBACK IN THE CARDIAC PACEMAKER

Richard E. Plant

Department of Mathematics
University of California
Davis, California

I. INTRODUCTION

The human cardiovascular system has been the study of intense theoretical research. This research has considered all aspects of the system, including pacemaker activity, muscular contraction, ventricular geometry, venous and arterial flow, and so forth (e.g., Noordergraff, 1967). While some workers have developed large scale models incorporating several components of the circulatory system, the complexity of this system has resulted in models which are tractable only to simulation studies, and whose accuracy is difficult to determine in any case. More recently it has been recognized that by studying the circulatory systems of simpler animals, one may construct models which are of a relatively simple nature. Moreover, these models may incorporate several components of the system and still provide the expectation of reasonable accuracy.

Such a model has been constructed for the heart of the matis shrimp *Squilla* (Plant, 1976a). The anatomy and physiology of

*This research was supported by NIH Grant Number NS-13777-01.

ISBN 0-12-434180-2

particular animal are such that many of the usual simplifying assumptions made in the construction of circulatory system models (such as the law of Laplace, neglection of fluid inertia, cylindrical configuration of the heart, the "windkessel" model of the circulatory system, etc.) are approximately true.

Figure 1 provides a schematic diagram of the model using an electrical analogy in which voltage represents pressure and current represents fluid flow. Under the influence of an external neural pacemaker which is coupled in a feedback loop to the heart, systolic contraction is introduced, forcing the blood from the heart through a one way valve, represented by a diode. The blood passes through the circulatory system, represented by a capacitor in series with two resistors, r_c and r_p. The blood then flows into the pericardial sinus, a large chamber surrounding the heart; this sinus is represented in the model by a ground. Diastolic expansion is induced by contraction of the suspensory ligaments, represented in the model by springs (Fig. 1). These ligaments, which are fixed at one end to the exoskeleton, serve to hold the heart in place. The blood then flows through a second set of

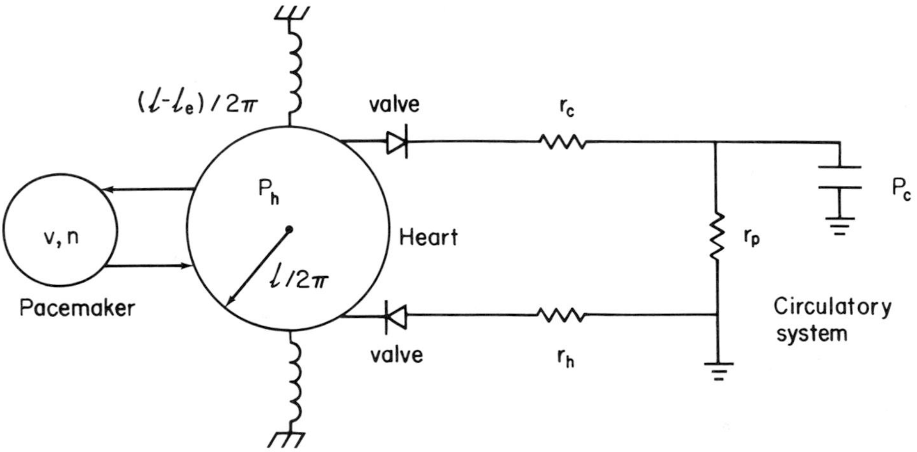

FIGURE 1

valves and thus returns to the heart. The heart wall is modeling using a standard Hill three-element muscle model (Fung, 1970).

In our discussion of the model, we will introduce the system by considering the "free-running" model in which the pacemaker of Figure 1 receives no feedback, in other words, in which the dynamics of the pacemaker are independent of the rest of the variables. In fact, the pacemaker is thought to receive feedback information as to the state of the heart. Specifically, the nerve cells of the pacemaker system send dendritic processes into the cardiac muscle wall, and these processes are thought to be stretch receptors which measure the length of the cardiac muscle. We shall subsequently incorporate a representation of this feedback into the model. It will be shown that this inclusion enables the model heart to simulate one particular aspect of the real heart which is not simulated by the "free running" model. By this means we shall be able to infer the function and some information on the underlying form of the stretch receptor feedback.

The equations of the free running model are (Plant, 1976a)

$$\dot{v} = g_N[s_I(v)[1-v] - nv] \tag{1a}$$

$$\dot{n} = \tau_K^{-1}[s_K(v) - n] \tag{1b}$$

$$\dot{x} = -\frac{f_0(v,x) - k_s f_s(\ell-\omega x)}{a(x) + k_s f_s(\ell-\omega x)} \tag{1c}$$

$$\dot{\ell} = \ell^{-1}[-\sigma(p_h-p_c) + \sigma(-p_h/r_c)] \tag{1d}$$

$$\dot{p}_c = \sigma(p_h-p_c) - p_c/r_p \tag{1e}$$

together with the algebraic equation

$$p_h = \ell^{-1}[\rho_r f_r(\ell) + \rho_s f_s(\ell-\omega x) - f_e[\ell_e-\ell]]. \tag{1f}$$

Here v represents pacemaker membrane potential, n is pacemaker potassium activation, x is muscle contractile element length, ℓ is heart circumference, p_c is circulatory system pressure, and p_h is cardiac pressure. The meanings of the various functions and constants are described in the original paper (Plant, 1976a). The factor g_N in equation (1a) multiplied only the term $s_I(v)[1-v]$ in the original model. This change has been made in order to facilitate comparison of the model to be developed in this paper with the original. In addition, a rudimentary form of feedback was included in the original model but will be abandoned in this paper.

The pacemaker is described by equations (1a,b), which represent a simple analog for excitable membrane equations (Plant, 1977a). In fact, there is evidence (I. M. Cooke, personal communication) that the physiological process represented by the parameter n is governed by internal calcium ion concentration and not membrane voltage as in equation (1b). However, the dynamics of models for these two types of process are qualitatively similar (Plant, 1978), and we will therefore retain the simpler formulation of equation (1b).

Figure 2 shows the phase plane of equations (1a,b). This represents a relaxation oscillation in which v is the "rapidly

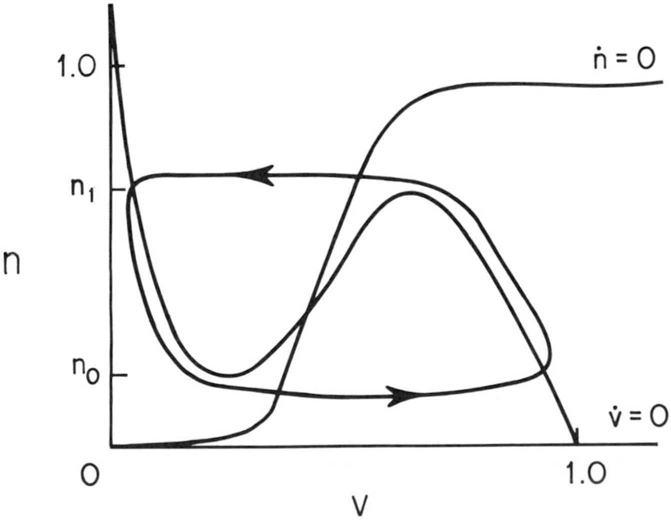

FIGURE 2

varying" term. In the free-running model this oscillator serves as a periodic forcing term to the passive (x, ℓ, p_c) system to produce a periodic, cycling solution. Numerical solution of the equations (Plant, 1976b) indicates that the free running model simulates many of the experimentally observed properties of the real system. One noteworthy example is that the free running model satisfies Starling's law of the heart, which states that stroke work is an increasing function of end diastolic volume. The review of Noordergraaf (1967) may be consulted for a discussion of this "law".

One aspect of the behavior of the real system which is not simulated by the original model is shown in Figure 3, which is a sketch based on results published by Maynard (1960) and by Izquierdo (1932). The significance of this figure is that as the heart is stretched, the frequency as well as the amplitude of contraction is increased. A stretching of the heart is equivalent to an increase in the equilibrium length ℓ_e, and in the original model the oscillation frequency is relatively independent of ℓ_e. Thus we may expect that dendrietic stretch feedback plays a role in modulating the frequency of oscillation of the system.

In order to study this possibility we shall make use of the formalism of the theory of discontinuous oscillations as described in an earlier paper (Plant, 1977b). We will study the reduced system obtained (Plant, 1976b) by making a "quasi steady-state"

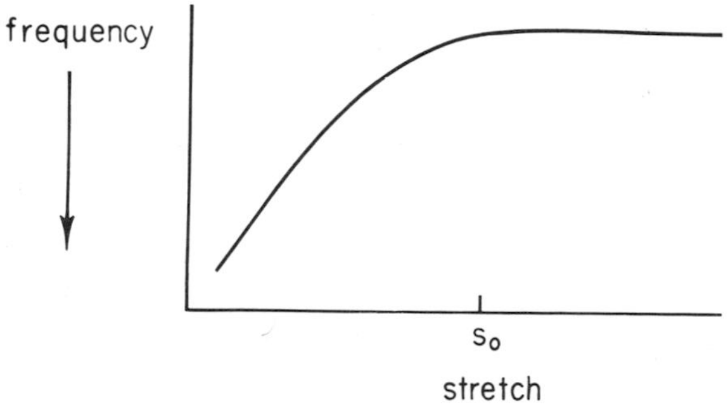

FIGURE 3

hypothesis on the variables x and p_c. This results in the following reduced system

$$\dot{v} = g_N[s_I(v)[1-v] - nv] \tag{2a}$$

$$\dot{n} = \tau_K^{-1}[s_K(v) - n] \tag{2b}$$

$$\dot{\ell} = \ell^{-1}[-\sigma(\frac{p_h}{1+r_p}) + \sigma(-p_h/r_h)] \tag{2c}$$

where

$$p_h = \ell^{-1}[\rho_r f_{rs}(\ell) + \rho_s f_{rs}(\ell-\omega x) - f_e[\ell_e-\ell]] \tag{2d}$$

and $x(\ell,v)$ satisfies

$$f_0(v,\ell,x) = k_s f_{rs}(\ell-\omega x). \tag{2e}$$

We make use of the fact that v is "repidly varying" to replace equation (2a) by

$$0 = g_N[s_I(v)[1-v] - nv]. \tag{2a'}$$

The system (2a', 2b-c) is referred to as the singular approximation of the system (2a-e), and the solution of such systems has been formally defined (Plant, 1977b). Basically, the solution moves along the surface in (v,n,ℓ) space defined by equation (2a'), called the "slow manifold", until it reaches a point at which it cannot remain on that surface, at which point it "jumps" discontinuously to another portion of the surface.

Figure 2 then represents a projection of this surface onto the v,n plane (note that equation (2a') is independent of ℓ). The values n_0 and n_1 are the values of n at which the jump from one portion of the surface to another occurs. These points, which are called "boundary points", are defined by the equation $\frac{\partial G}{\partial v} = 0$ where $G(v,\ell)$ is the function defined by solving equation (2a') for n in terms of v and ℓ.

It follows from equation (2a') that for the free running model $G_0(v,\ell)$ is given by

$$G_0(v,\ell) = \frac{g_N}{v} s_I(v)[1-v] \tag{3}$$

so that

$$\frac{\partial G_0}{\partial v} = \frac{g_N}{v^2}[s_I(v)[1-v]v - s_I(v)]. \tag{4}$$

Since both s_I and s_I are positive it follows that $\partial G_0/\partial v$ will be negative unless $s_I(v) < s_I(v)[1-v]v$. This inequality holds only for values of v not to close to zero or one, as is reflected in the shape of the curve in Figure 2.

Figure 4a represents a plot of the curves of boundary points in the surface defined by equation (2a') as seen from along the v axis. The oscillation moves from a point on the n_0 curve to a point on the n_1 curve and back again. The fact that these "curves" are actually parallel straight lines is a result of the lack of dependence of equation (2a') on ℓ.

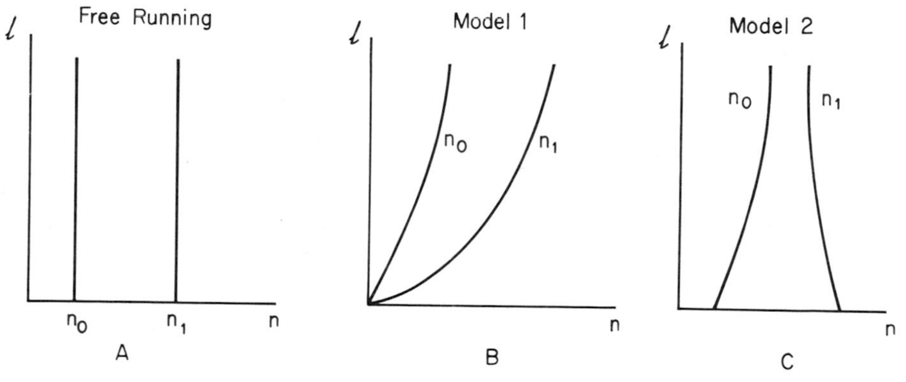

FIGURE 4

We are now in a position to introduce stretch receptor feed-fack into the model. We will actually consider two alternative models, which will be called Model 1 and Model 2. These will be compared as to their ability to reproduce the data represented in Figure 3.

In Model 1 we assume that increased stretch results in the activation of a voltage dependent inward current conductance. This assumption may be represented mathematically by the multiplication of the term $s_I(v)[1-v]$ in equation (1a) by a monotonic, non-decreasing, non-negative function $g_1(\ell)$, yielding

$$\dot{v} = g_N[g_1(\ell)s_I(t)[1-v] - nv] \tag{5}$$

as a replacement for equations (1a) and (2a) in the full and reduced models, and to the replacement of equation (2a) with the equation

$$0 = g_N[g_1(\ell)s_I(v)[1-v] - nv] \tag{5'}$$

in the singular approximation. The function $G_1(v,\ell)$ obtained by solving equation (5') for n is

$$G_1(v,\ell) = \frac{g_N}{v} g_1(\ell)s_I(v)[1-v] \tag{6}$$

so that we have

$$\frac{\partial G_1}{\partial v} = \frac{g_N}{v^2} g_1(\ell)[s_I(v)[1-v]v - s_I(v)]. \tag{7}$$

Therefore the values of v such that $\partial G_1/\partial v > 0$ are unchanged, but the values n_0 and n_1 become proportional to the value of $g_1(\ell)$ as sketched in Figure 4b.

In Model 2 we assume that increased stretch results in the activation of a "leakage" conductance which passes both inward and outward current. This system resembles models for the functioning of the synapse proposed, for example, by Lewis (1968).

We may incorporate this assumption into the model by adding a monotonic, non-decreasing, non-negative function $g_2(\ell)$ as follows:

$$\dot{v} = g_N[[s_I(v) + g_2(\ell)][1-v] - [n + g_2(\ell)]v]. \tag{8}$$

The full and reduced models are formed by replacing equations (1a) and (2a) with equation (8), and the singular approximation is formed by replacing equation (2a') with the equation

$$0 = g_N[[s_I(v) + g_2(\ell)][1-v] - [n + g_2(\ell)]v]. \tag{8'}$$

The function $G_2(v,\ell)$ defined by solving equation (8') for n is given by

$$G_2(v,\ell) = \frac{g_N}{v}[s_I(v)[1-v] + g_2(\ell)[1-2v]] \tag{9}$$

resulting in

$$\frac{\partial G_2}{\partial v} = \frac{g_N}{v^2}[s_I(v)[1-v]v - s_I(v) - g_2(\ell)]. \tag{10}$$

Since the non-negative quantity $g_2(\ell)$ is subtracted from the original difference $s_I(v)[1-v]v - s_I(v)$, the range of v such that $\partial G_2/\partial v$ is positive will be reduced as $g_2(\ell)$ is increased. Moreover, equation (9) indicates that the effect of $g_2(\ell)$ on $G_2(v,\ell)$ is null when $v = 1/2$. Therefore if the parameters of the model are such that $\partial G_2/\partial v \geq 0$ for $v = 1/2$, then the value of n_0 will be increasing function of $g_2(\ell)$ while n_1 will be a decreasing function of $g_2(\ell)$. If $g_2'(\ell) > 0$ then the projection of the boundary curves onto the (n,ℓ) plane for this system will resemble those of Figure 4c.

Let us now obtain an estimate of the effect of increased ℓ on the period T of oscillation for each of the three cases of Figure 4. To do this we will estimate the integral

$$T = \tau_K \left[\int_{n_0}^{n_1} \frac{dn}{\phi_1(n)-n} + \int_{n_1}^{n_0} \frac{dn}{\phi_0(n)-n} \right] \tag{11}$$

where

$$\phi_i(n) = s_K(G_i^{-1}(n,\ell(n))), \quad i = 0,1 \tag{12}$$

the functions G_i^{-1} being defined along the trajectories in the slow manifold ending in n_i, for $i = 0$ and 1. To obtain estimate of T we note from Figure 2 that when the solution of the singular approximation is on that portion of the slow manifold with the higher value of v (i.e., that portion defining G^{-1}), we have $s_K(v) \simeq 1$; similarly, on the other portion of the slow manifold we have $s_K(v) \simeq 0$. We therefore make the approximations

$$\phi_1(n) \simeq 1, \quad \phi_0(n) \simeq 0$$

which give

$$T \simeq \tau_K \left[\int_{n_0}^{n_1} \frac{dn}{1-n} + \int_{n_0}^{n_1} \frac{dn}{n} \right]$$

or

$$T \simeq \ln \left(\frac{[1-n_0]n_1}{[1-n_1]n_0} \right) . \tag{13}$$

For the free running model, Figure 4a then implies that T is approximately independent of ℓ. For Model 1, Figure 4b implies that T is an increasing function of ℓ. To see this, note that if we write $n_0(\ell) = \hat{n}_0 g_1(\ell)$ and $n_1(\ell) = \hat{n}_1 g_1(\ell)$ then the argument of approximation (13) is given by $[\hat{n}_1/\hat{n}_0][1-\hat{n}_0 g_1(\ell)]/[1-\hat{n}_1 g_1(\ell)]$ and we have

$$\frac{d}{d\ell}\left(\frac{1-\hat{n}_0 g_1(\ell)}{1-\hat{n}_1 g_1(\ell)} \right) = g_1'(\ell)[\hat{n}_1-\hat{n}_0][1-\hat{n}_1 g_1(\ell)]^{-2} \tag{14}$$

which is non-negative wherever it is defined.

For Model 2, Figure 4c implies that T is a decreasing function of ℓ since as n_0 and n_1 approach each other the argument of approximation (13) approaches unity. These estimations of the effect of increased stretch on period were tested by numerical solution of the equations (1a-e). The values of the functions and parameters were the same as those used in the original model. The new parameters were given by

$$g_N = 25.0, \quad g_1(\ell) = \begin{cases} 1.2\ell: & \ell \geq 0 \\ 0: & \ell < 0 \end{cases}, \quad g_2(\ell) = \begin{cases} -.08+.1\ell: & \ell \geq .8 \\ 0: & \ell < .8 \end{cases}$$

The equations were solved for $\ell_e = 4$ and $\ell_e = 6$. The results of this computation are given in Table I. As shown, the estimations obtained earlier were confirmed.

The conclusion is that in the free running model the frequency is independent of stretch, Model 1 has a frequency which decreases with increased stretch, and Model 2 has a frequency which increases with increased stretch. A comparison of this result with Figure 3 indicates that Model 2 is more likely than Model 1 to be the correct description. We may therefore interpret the results of this modelling study as follows. We propose that the function of the dendritic stretch receptors of the

TABLE I

Model	ℓ_e	*period, T*
free running	4	5.6
	6	5.6
1	4	4.4
	6	4.5
2	4	5.0
	6	3.9

system is to provide an increased frequency with increased stretch. Moreover, these stretch receptors are hypothesized to function by activating leakage channels through which ions may flow down their concentration gradients, similarly to the process thought to occur in the synapse. Finally, beyond a certain level of stretch, denoted by the letter s_0 in Figure 3, all of the channels are activated and the system becomes free running, with a pacemaker which is independent of the heart conditions.

There is an experimental test of the model which is quite simple, at least conceptually. This is to isolate, by either physical or pharmacological means, the dendritic stretch receptors from the pacemaker cell bodies. The model predicts that such an isolated pacemaker would be free running, and therefore have a frequency which is independent of stretch.

ACKNOWLEDGMENTS

I am grateful to Professor I. M. Cooke and Professor E. C. Zeeman for some useful and stimulating discussions.

REFERENCES

[1] Fung, Y. C. (1970). "Mathematical Representation of the Mechanical Properties of the Heart Muscle", *J. Biomech. 3*, 381-404.

[2] Izquierdo, J. J. (1932). "A study of the Crustacean Heart Muscle", *Proc. Roy. Soc. Land. Ser. B, 109*, 229-250.

[3] Lewis, E. R. (1968). "Using Electronic Circuits to Model Simple Neuroelectric Interactions", *Proc. IEEE, 56*, 931-949.

[4] Maynard, D. M. (1960). "Circulation and Heart Function", Chapter 5 in The Physiology of Crustacea, T. H. Waterman, ed., New York: Academic Press.

[5] Noordergraaf, A. (1969). "Hemodynamics", Chapter t of Biological Engineering, H. P. Schwann, ed., New York: McGraw-Hill.

[6] Plant, R. E. (1976a). "A Simple Model for the Crustacean Cardiac Pacemaker Control System", *Math. Biosci. 32*, 275-290.

[7] Plant, R. E. (1976b). "Analysis of a Model for the Crustacean Cardiac Pacemaker", *Math. Biosci. 32*, 291-305.

[8] Plant, R. E. (1977a). "Simple Analogs for Nerve Membrane Equations", Nonlinear Systems and Applications, V. Lakshmikantham, ed., 647-655.

[9] Plant, R. E. (1977b). "Crustacean Cardiac Pacemaker Model - An Analysis of the Singular Approximation", *Math. Biosci. 36*, 149-171.

[10] Plant, R. E. (1978). "The Effects of Calcium^{++} on Bursting Neurons: A Modelling Study", *Biophys. J. 21*, 217-237.

THE CURRENT STATE OF THE N-BODY PROBLEM

Harry Pollard

Department of Mathematics
Purdue University
West Lafayette, Indiana

ABSTRACT

Through the year 1966 the most comprehensive reports on the state of the n-body problem can be found in Chapter V of A. Wintner's book "The Analytical Foundations of Celestial Mechanics" (Princeton, 1941) and the author's "Mathematical Introduction to Celestial Mechanics" (Prentice-Hall, 1966). It is the purpose of this survey lecture to review some of the progress made since that time. There are two major questions: (a) What physical conditions correspond to the occurrence of a singularity in the (necessarily) analytic solution of the problem? In particular what is the role of collision? (b) If no singularities occur, what is the behavior of the system as $t \to \infty$?

We introduce our notation: U is the self-potential (= negative of potential energy). T the kinetic energy, $h = T - U$ the total energy, I one half the moment of inertia related to the preceding quantities by the Lagrange-Jacobi identity $\ddot{I} = 2T - U$. Let r denote the minimum distance between pairs of particles, R the maximum. Then positive constants A, B exist such that $B \leq rU \leq A$, $BR^2 \leq I \leq AR^2$. In other

ISBN 0-12-434180-2

words, U^{-1} is a measure of how close together particles get, and $\sqrt{I}$ how far apart.

The major classical result concerning problem (a) asserts that a singularity occurs at time t_1 if and only if $r(t) \to 0$ as $t \to t_1$. This does *not* assert the occurrence of a collision. In 1970 Sperling proved an important conjecture of von Zeipel: *if the system remains bounded, then the sigularity is due to physical collisions*. In 1973 Saari drew the same conclusion provided the system does not expand too rapidly. The latter depends on theorem of Pollard and Saari (1968) that a singularity is due to collisions if and only if $U(t) \sim C(t_1 - t)^{-2/3}$ as $t \to t_1$ for some constant C. It also follows from this last theorem (Saari, 1972-1973) that the set of initial conditions leading to collision singularities is of Lebesgue measure zero and of first Baire category.

We can only sample some of the progress made in problem (b).

A standard theorem (the Virial Theorem) states that if I and T remain bounded for all time, then the time averages $\hat{T}$ and U exist and $2\hat{T} = \hat{U}$ which holds if and only if R grows less rapidly than t, that is $R/t \to 0$, $t \to \infty$.

Sometimes it is stated that a particle must escape from a system of positive energy. The (erroneous) proof starts with $\ddot{I} = T + h$. If $h > 0$ then $I \to \infty$. Therefore $\sum m_k r_k^2 \to \infty$. But it cannot be concluded as is sometimes done, that some $r_k \to \infty$. Pollard (1967) proves that if U is square-integrable than a particle must escape.

STABILITY OF McSHANE SYSTEMS[1,2]

A. N. V. Rao
C. P. Tsokos

Department of Mathematics
University of South Florida
Tampa, Florida

I. INTRODUCTION

In this paper we shall investigate the stability of McShane systems of the form

$$dx^i(t;\omega) = \sum_{\rho=1}^{m} f_\rho^i(t,x(t;\omega);\omega)dZ_\rho(t;\omega)$$

$$+ \sum_{\rho,\sigma=1}^{m} h_{\rho,\sigma}^i(t,x(t;\omega);\omega)dZ_\rho(t;\omega)dZ_\sigma(t;\omega)$$

$$x^i(0:\omega) \equiv x_0^i(\omega), \quad i = 1,2,\ldots,n \tag{1.1}$$

where

[1]*These results were presented at the 84th Annual Meeting of the American Mathematical Society, Atlanta, Georgia, January 3-8, 1978.*

[2]*This research was supported by the United States Air Force, .ir Force Office of Scientific Research, under Grant No. AFOSR-'6-2711.*

ISBN 0-12-434180-2

(i) $t \in R_+ \equiv [0,\infty)$ and $\omega \in \Omega$, the supporting set of a complete probability measure space (Ω, A, P);

(ii) f^i_ρ, $h^i_{\rho,\sigma}$: $R_+ \times R^n \times \Omega \to R$;

(iii) $Z(t:\omega)$ with subscript is a real-valued stochastic process;

(iv) $x(t;\omega)$ is a n-vector process with components $x^i(t;\omega)$ $(i = 1,2,\ldots,n)$.

The integrals appearing in equation (1.1) are to be understood as McShane integrals [4].

The stability behavior of the system (1.1) will be investigated using the theory of differential inequalities in a stochastic setting. The role of differential inequalities in the study of the qualitative aspects of differential equations is well documented [see [3]]. More recently these techniques have been extended to investigate the stability properties of random differential equations, [2], [6], Ito-type equations [1], [5], and stochastic functional equations [2]. In the present study, we extend these techniques to systems involving McShane integrals.

II. PRELIMINARIES

Let (Ω, A, P) be a complete probability measure space. We shall assume that there is a family of sub-σ-algebras A_t, $t \in R_+$, in A, such that for $s < t$, $A_s \subset A_t$. We shall also assume with respect to equation (1.1) that

(H_1) every process denoted by $Z(t;\omega)$ with subscript, is a real valued stochastic process adapted to A_t with almost sure (a.s.) continuous sample paths and satisfying the condition $E[|Z(t;\omega)-Z(s;\omega)|^r|A_s] \le K(t-s)$ where $0 \le s \le t < \infty$, $r = 1,2,4$ and K is some constant; also $Z(t;\omega)$ is independent of $x_0(\omega)$;

(H_2) For $x,y \in R^n$, the functions f^i_ρ, $h^i_{\rho,\sigma}$ satisfy the following conditions:

$$\|f^i_\rho(t,x;\omega) - f^i_\rho(t,y;\omega)\| \leq L(\omega)\|x - y\|$$

$$\|h^i_{\rho,\sigma}(t,x;\omega) - h^i_{\rho,\sigma}(t,y;\omega)\| \leq L(\omega)\|x - y\|$$

$f^i_\rho(t,0;\omega)$, $h^i_{\rho,\sigma}(t,0;\omega) \equiv 0$, where $L(\omega)$ is a.s. finite;

(H_3) For each n-vector mean square continuous process $x(t;\omega)$ that is adapted to A_t, the processes $f^i_\rho(t,x(t;\omega);\omega)$ and $h^i_{\rho,\sigma}(t,x(t;\omega);\omega)$ are mean square continuous and adapted to A_t.

Under the conditions (H_1)-(H_3), it is known [4] that, if $E[x_0(\omega)]^2 < \infty$, then there exists a process $x(t;\omega)$, which is mean square continuous, separable, adapted to A_t and which satisfies equation (1.1). We shall now state four basic definitions of stochastic stability with respect to the system (1.1). It is possible to define other notions of stability [3], [1] and obtain suitable results. We shall denote $c[[0,t],R^+]$ to be the set of all continuous non-negative real functions on the interval $[0,t]$.

Definition 2.1. The system (1.1) is stable in the mean, if for every $\varepsilon > 0$ $\exists$ $a\delta = \delta(\varepsilon) > 0$, such that $E\|x(t;\omega)\| < \varepsilon$, for all solutions $x(t;\omega)$ of (1.1) with $\|x(0,\omega)\| < \delta$.

Definition 2.2. The system (1.1) is asymptotically stable in the mean if it is stable in the mean and if, in addition, $E\|x(t;\omega)\| \to 0$ as $t \to \infty$ for all solutions $x(t;\omega)$ with $\|x(0;\omega)\| < \delta$.

Definition 2.3. The system (1.1) is stable with probability one (w.p.1) if for every p, $\lambda > 0$, $\exists\delta = \delta(p,\lambda) > 0$ such that

$$P[\sup_t \|x(t;\omega)\| > \lambda] \leq p, \quad t > 0$$

for all solutions $x(t;\omega)$ with $\|x(0;\omega)\| < \delta$.

Definition 2.4. The system (1.1) is asymptotically stable w.p.1 if it is stable w.p.1 and in addition $\|x(t;\omega)\| \to 0$ as $t \to \infty$ w.p.1 for all solutions $x(t;\omega)$ with $\|x(0;\omega)\| < \delta$.

Definition 2.5. A function b is said to belong to class K if $b \in C[R_+,R_+]$, $b(0) = 0$ and $b(u)$ is strictly increasing in u.

Definition 2.6. A function a is said to belong to class K^* if $a \in C[R_+ \times R_+,R^+]$, $a(t,0) \equiv 0$ and $a(t,u)$ is increasing in u for each $t \in R_+$.

Corresponding to the stochastic system (1.1), we shall consider a deterministic comparision system described by the scalar differential equation

$$u' = g(t,u) \tag{2.1}$$

$$u(0) = u_0$$

We shall assume that $g(t,0) \equiv 0$ and that equation (2.1) has global solutions for all u_0. With respect to the system (2.1), we shall define the following notions of stability [3].

Definition 2.7. The system (2.1) is stable if for any $\varepsilon > 0$, there is a $\delta = \delta(\varepsilon)$ such that

$|u(t)| < \varepsilon$, for all solutions $u(t)$ of (2.1) with $|u(0)| < \delta$.

Definition 2.8. The system (2.1) is asymptotically stable if it is stable and in addition

$|u(t)| \to 0$ as $t \to \infty$, for all solutions $u(t)$ with $|u(0)| < \delta$.

Throughout the paper, we consider sums involving subscripts or superscripts i, ρ, σ. For notational simplicity we let

$$\sum_i \equiv \sum_{i=1}^{n}, \quad \sum_{\rho,\sigma} \equiv \sum_{\rho,\sigma=1}^{m}, \quad \text{unless otherwise stated.}$$

III. A COMPARISON THEOREM

In this section, we shall develop a comparison theorem for the stochastic system (1.1), which will prove to be a basic result for the stability theorems. We shall first state a theorem analogous to the Ito formula for the differentiation of composite functions, the proof which can be found in [4].

Theorem 3.1. Let the hypothesis (H_1)-(H_3) be satisfied. Let $V(t,x)\colon R_+ \times R^n \to R_+$, have continuous partial derivatives $V_0 \equiv \frac{\partial v}{\partial t}$, $V_i \equiv \frac{\partial v}{\partial x^i}$, $V_{ij} \equiv \frac{\partial^2 v}{\partial x^i \partial x^j}$. Let $x(t;\omega)$ be a solution of (1.1). Then the composite function $V(t,x(t;\omega))$ satisfies the equation.

$$dV = V_0 dt + \sum_{i,\rho} V_i f^i_\rho dZ_\rho + \sum_{i,\rho,\sigma} \left\{ V_i h^i_{\rho,\sigma} + \frac{1}{2} \sum_{j=1}^{n} V_{ij} f^i_\rho f^j_\rho \right\} dZ_\rho dZ_\sigma \tag{3.1}$$

where V_0, V_i, V_{ij} are evaluated at $(t,x^1(t;\omega),\ldots,x^n(t;\omega))$. For notational convenience, we shall define

$$LV \equiv V_0 + \sum_{i,\rho} KV_i f^i_\rho + \sum_{i,\rho,\sigma} K\left\{ V_i h^i_{\rho,\sigma} + \frac{1}{2} \sum_{j=}^{n} V_{ij} f^i_\rho f^j_\sigma \right\}$$

where K is the Lipschitz constant in (H_1) associated with the process $Z(t;\omega)$.

We shall now state the comparision result.

Theorem 3.2. Suppose there exist function $g(t,u)$ and $V(t,x)$ such that

(i) $g(t,u) \in c[R_+ \times R,R]$ and is concave in u for each $t \in R_+$;

(ii) $V(t,x) \in c[R_+ \times R^n,R_+]$, is continuously differentiable in t, twice continuously differentiable in x such that

$$|V_0| + |V_i| \; \|x\| + |V_{ij}| \; \|x\|^2 \le k_1 + k_2\|x\|^2$$

for some positive constants k_1, k_2;

(iii) $LV \le g(t,V)$.

Then for any solution $x(t;\omega)$ of (1.1) with $V(0,x_0) \le U_0$, we have

$$E[V(t,x(t;\omega)) \mid x_0] \le r(t)$$

where $r(t)$ is the maximal solution of the scalar comparison system (2.1).

Proof. Set

$m(t) = E[V(t,x(t;\omega))]$. Then for small $h > 0$, we have

$$m(t+h) - m(t) = E[V(t+h,\ x(t+h;\omega)) - V(t,x(t;\omega))]$$

$$= E[E[V(t+h,\ x(t+h;\omega)) - V(t,x(t;\omega)) \mid A_t]] \quad (3.2)$$

It is not difficult to show (see [4] pp. 115-119) that

$$E[V(t+h,\ x(t+h;\omega) - V(t,x(t;\omega)) \mid A_t]$$
$$= hLV(t,x(t;\omega)) + O(h). \quad (3.3)$$

Hence

$$D^+m(t) \equiv \limsup_{h\to 0^+} \frac{m(t+h) - m(t)}{h}$$

$$= E[LV(t,x(t;\omega)) \mid A_t] \le E[g(t,V(t,x(t;\omega))]$$

$$\le g(t,m(t)). \quad (3.4)$$

The assertion of the theorem now follows from the well known result in the theory of differential inequalities [3].

IV. STABILITY RESULTS

We shall now prove some stability theorems for the system (1.1).

Theorem 4.1. Assume that the hypotheses of theorem (3.2) hold. Assume further that there exist functions $b \in K$, $a \in K^*$ such that b is convex and

$$b(\|x\|) \leq V(t,x) \leq a(t, \|x\|).$$

Then

(i) The stability of the comparison system (2.1) implies the stability in the mean of the stochastic system (1.1);

(ii) The asymptotic stability of the comparison system (2.1) implies the asymptotic stability in the mean of the stochastic system (1.1).

Proof. Suppose the comparison system is stable. Then for every $\varepsilon > 0$, there is a $\delta_1 = \delta_1(\varepsilon) > 0$ such that if $|u_0| < \delta_1$, then $r(t,u_0) < b(\varepsilon)$. Choose $\rho_1 > 0$ such that $a(0,\rho_1) < u_0$. Also, since $a \in K^*$, we can find a $\delta^1 = \delta^1(\varepsilon) > 0$ such that $\|x_0\| < \delta^1$ implies that $a(0, \|x_0\|) \leqq \delta_1$. Let $\delta = \min(\delta^1, \rho_1)$. Then for any x_0 with $\|x_0\| < \delta$ we have

$$V(0,x_0) \leq a(0, \|x_0\| \leq a(0,\rho_1) < u_0.$$

Hence by theorem (3.1), we obtain

$$E[V(t,x(t;\omega))] \leq r(t,u_0) < b(\varepsilon). \tag{4.1}$$

This together with the assumption $b(\|x\|) \leq V(t,x)$, yields the deisred result.

The asymptotic stability of the comparison system implies that in equation (4.1).

$$\lim_{t\to\infty} r(t,u_0) = 0.$$

Hence, it follows that

$\lim_{t\to\infty} b(E\|x(t;\omega)\|) = 0$ which in turn implies that $E\|x(t;\omega)\| \to 0$ as $t \to \infty$.

Theorem 4.2. Assume that the conditions of theorem (4.1) hold. Suppose the system (1.1) has bounded solutions. Then

(i) The stability of the comparison system implies the stability w.p.1 of the stochastic system.

(ii) The asymptotic stability of the comparison system implies the asymptotic stability w.p.1 of the stochastic system.

Proof. (i) Suppose the comparison system is stable. Then from theorem (4.1), we have for every $\varepsilon_1 > 0$, a $\delta > 0$, such that

$$E\|x(t;\omega)\| < \varepsilon_1 \tag{4.2}$$

provided $\|x_0\| < \delta$. The separability of the process $x(t;\omega)$ implies that there is an enumerable subset S of R^+ with the property (a.s.) that a sample function has the same bound on R^+ as on S. Hence we shall consider $x(t;\omega)$ for $t = t_0, t_1, \ldots, t_n$ $(t_0 < t_1 < t_2 \ldots)$. Define

$$A_n \equiv \left\{\omega: \ \|x(t_n;\omega)\| > \lambda\right\}$$

$$A \equiv \bigcup_{n=1}^{\infty} A_n = \left\{\omega: \ \sup_n \|x_n(t;\omega)\| > \lambda\right\}$$

and

$$B = A_1, \quad B_n = A_1^c \cap A_2^c \ldots \cap A_{n-1}^c \cap A_n \qquad (n > 1).$$

We note that the sets $B_1, B_2, \ldots$ are disjoint and $\cup B_n = A$. Also from the definition of B-sets, we have, for any $\lambda > 0$

$$\lambda\, P[\sup_n \|x(t;\omega)\| > \lambda] = \lambda\, P[A]$$

$$\leq \sum_{n=1}^{\infty} \int_{B_n} \|x(t_n;\omega)\|\, d\,P. \tag{4.3}$$

The series

$$\sum_{n=1}^{\infty} [\int_{B_n} \|x(t_n;\omega)\| dP] \text{ is convergent as} \tag{4.4}$$

$\|x(t_n;\omega)\|$ is bounded by assumption. Hence, given any $\varepsilon_2 > 0$ we can find a $N + N(\varepsilon_2)$ such that

$$\sum_{n=1}^{n} \int_{B_n} \|x_n(t;\omega)\| \, d\, p \leq N\varepsilon \tag{4.5}$$

Using (4.5) and (4.4) in (4.3), we get

$$\lambda P [\sup \|x(t_n;\omega)\| > \lambda] \leq \varepsilon_2 + N\varepsilon_1.$$

Since ε_1, ε_2 are arbitrary, the result follows.

The proof of (ii) is similar and therefore not given.

REFERENCES

[1] Ladde, G. S., Lakshmikantham, V., Lin, P. T. (Feb., 1973). "Differential Inequalities and Ito Type Stochastic Differential Equations", Tech. Report No. 29, University of Rhode Island, Kingston, R.I.

[2] Ladde, G. S. (June, 1974). "Differential Inequalities and Stochastic Functional Equations", *J. Math. Phys., Vol. 15, No. 6.*

[3] Lakshmikantham, V., and Leela, S. (1969). "Differential and Integral Inequalities", Academic Press.

[4] McShane, E. J. (1974). "Stochastic Calculus and Stochastic Models, Academic Press.

[5] Rao, A. N. V., and Tsokos, C. P. (1977). "On the Behavior of Stochastic Differential Systems", *Mathematische Nachrishten,* 167-175.

[6] Wonham, W. M. (1970). "Random Differential Equations in Control Theory, Probabilistic Methods in Applied Mathematics", Ed. Bharucha-Reid, A. T., Vol. 2, Academic Press.

CONSTRUCTIVE TECHNIQUES FOR ACCRETIVE AND MONOTONE OPERATORS*

Simeon Reich

Department of Mathematics
University of Southern California
Los Angeles, California

Let E^* be the dual of a real Banach space E, and let $J\colon E \to 2^{E^*}$ be the (normalized) duality mapping: For $x \in E$, $Jx = \{x^* \in E^*\colon (x,x^*) = |x|^2 = |x^*|^2\}$. Recall that a subset A of $E \times E$ is called accretive if for each $y_i \in Ax_i$, $i = 1,2$, there exists j in $J(x_1 - x_2)$ such that $(y_1 - y_2, j) \geq 0$. A subset M of $E \times E^*$ is said to be monotone if for each $y_i \in Mx_i$, $i = 1,2$, $(x_1 - x_2, y_1 - y_2) \geq 0$. The purpose of this paper is to discuss certain aspects of the constructive solvability of equations involving accretive and monotone operators. We intend to present new and recent convergence results, error estimates, applications, and open problems.

We begin by presenting a new result for strongly accretive operators, namely those A for which there exists a positive α such that $A - \alpha I$ is also accretive. The result is concerned with the scheme

*Partially supported by the National Science Foundation under Grant MCS 78-02305.

ISBN 0-12-434180-2

$$x_{n+1} \in x_n - \lambda_n A x_n, \quad n \geq 0 \tag{1}$$

where $x_0 \in E$ and $\{\lambda_n\}$ is a positive sequence.

Theorem 1. Let A be a strongly accretive operator with a zero, and let E^* be uniformly convex. Suppose that a sequence $\{x_n\}$ can be defined by (1) and that $\{(x_n - x_{n+1})/\lambda_n\}$ is bounded. If $\lim_{n\to\infty} \lambda_n = 0$ and $\sum_{n=0}^{\infty} \lambda_n = \infty$, then $\{x_n\}$ converges strongly to the zero of A.

Proof. Since E^* is uniformly convex, there is [14, p. 89] a continuous nondecreasing function $b\colon [0,\infty) \to [0,\infty)$ such that $b(0) = 0$, $b(ct) \leq cb(t)$ for $c \geq 1$, and $|x+y|^2 \leq |x|^2 + 2(y, Jx) + \max\{|x|, 1\}|y|b(|y|)$ for all x and y in E. Let z be the zero of A, and denote $(x_n - x_{n+1})/\lambda_n$ by $y_n \in Ax_n$. We have $|x_{n+1}-z|^2 \leq |x_n - z|^2 - 2\lambda_n(y_n, J(x_n - z)) + \max\{|x_n - z|, 1\} \cdot \max\{|y_n|, 1\}|y_n|\lambda_n b(\lambda_n)$, and $(y_n, J(x_n - z)) \geq \alpha|x_n - z|^2$. It follows that $\{x_n\}$ is bounded and that $|x_{n+1} - z|^2 \leq (1 - 2\lambda_n\alpha)|x_n - z|^2 + M\lambda_n b(\lambda_n)$ for some constant M. This last inequality combined with $\{\lambda_n\} \in c_0 \setminus \ell^1$ implies that $\lim_{n\to\infty} x_n = z$.

The following result (cf. [10]) provides a sufficient condition for the construction of the sequence $\{x_n\}$ with bounded $\{y_n\}$.

Proposition 2. An accretive $A \subset E \times E$ is locally bounded in the interior of its domain if either

(a) E^* is uniformly convex, or

(b) E is uniformly convex and smooth.

Proof. Let the ball $B(x_0, r)$ be contained in $D(A)$, and suppose that $y_n \in Ax_n$ with $x_n \to x_0$. If $|u| \leq r$, then $(y - y_n, J(u + x_0 - x_n)) \geq 0$ for $y \in A(u + x_0)$. If (a) holds, then $(y_n, J(u)) \leq (y, J(u + x_0 - x_n)) - (y_n, J(u + x_0 - x_n) - J(u)) \leq C(u) + \alpha_n|y_n|$ with $\alpha_n \to 0$. If (b) holds, then $J^{-1}(J(x_0 - x_n) + B(0, r^*)) - J^{-1}(J(x_0 - x_n)) \subset B(0, r)$ for some positive r^*. Therefore $J(x_0 - x_n) + B(0, r^*) \subset J(x_0 - x_n + B(0, r))$, and

$(y-y_n, J(x_0-x_n)+v^*) \geq 0$ for all $|v^*| \leq r^*$. Consequently, $(y_n,v^*) \leq (y,J(x_0-x_n)+v^*) - (y_n, J(x_0-x_n)) \leq C(v^*) + \beta_n|y_n|$ with $\beta_n \to 0$. Thus in both cases the result follows from (an extension of) the uniform boundedness principle.

Proposition 2 implies that in the setting of Theorem 1, if z is in the interior of the domain of A, and x_0 is close enough to z, then $\{x_n\}$ can indeed be defined by (1) with bounded $\{y_n\}$ provided $\sup\{\lambda_n\}$ is small enough. We do not know if Theorem 1 can be extended to other Banach spaces the duals of which are not necessarily uniformly convex.

The next proposition provides us with information concerning the function b which was used in the proof of Theorem 1. We denote the modulus of convexity of a Banach space E by δ_E.

Proposition 3. If $\delta_{E^*}(\varepsilon) \geq K\varepsilon^r$ for some $K > 0$ and $r \geq 2$, then for $t \leq M$, $b(t) \leq ct^{s-1}$ with $s = r/(r-1)$.

Proof. Let F be a Banach space with $\delta_F(\varepsilon) \geq K_1\varepsilon^r$. Since for $j_1 \in J_F x$ and $j_2 \in J_F y$, $|(x+y)/2|^2 \geq |x|^2 + (y-x,j_1)$ and $|(x+y)/2|^2 \geq |y|^2 + (x-y,j_2)$, we obtain $(x-y,j_1-j_2) \geq |x|^2 + |y|^2 - \frac{1}{2}|x+y|^2 \geq K_2|x-y|^r$ for some positive K_2 and bounded x and y. Therefore J_{F^*} is Hölder continuous with exponent $1/(r-1) = s-1$ on bounded sets. Taking $F = E^*$, we see that the result follows from the definition of b in [14].

If $E = L^p$, $1 < p < \infty$, then $s = p$ if $1 < p \leq 2$, and $s = 2$ if $2 \leq p < \infty$.

Proposition 3 enables us to obtain a convergence rate in the setting of Theorem 1. We claim that if $\lambda_n = \frac{s}{2\alpha(n+1)}$, then $|x_n-z| = O(n^{-\frac{s-1}{2}})$. Indeed we have $|x_{n+1}-z|^2 \leq (1-\frac{s}{n+1})|x_n-z|^2 + M_1\frac{1}{(n+1)^s}$. Let $M_2 = \max\{|x_1-z|^2,M_1\}$. Then $|x_1-z|^2 \leq M_2$, and $|x_n-z|^2 \leq M_2\frac{1}{n^{s-1}}$ implies that $|x_{n+1}-z|^2 \leq (1-\frac{s}{n+1})\frac{M_2}{n^{s-1}} + \frac{M_2}{(n+1)^s} \leq (1-\frac{1}{n+1})^s\frac{M_2}{n^{s-1}} + \frac{M_2}{(n+1)^s} = \frac{M_2}{(n+1)^{s-1}}$

This rate of convergence agrees with the one obtained in [6] (see also [9]) for the Hilbert space case ($s = 2$). More convergence rate results can be found in [12].

When A is not strongly accretive, it is natural to consider the following scheme

$$x_{n+1} \in x_n - \lambda_n(Ax_n + \theta_n x_n), \quad n \geq 0, \tag{2}$$

where $x_0 \in E$ and $\{\lambda_n\}$ and $\{\theta_n\}$ are positive sequences. In the Hilbert space case this scheme has been studied in [2] and [7]. Here we present a quick proof of the following result, which can also be deduced from [14, Theorem 3.2]. We denote $((1-\lambda_n\theta_n)x_n - x_{n+1})/\lambda_n$ by $y_n \in Ax_n$. A is called m-accretive if $R(I+A) = E$.

Theorem 4. Let E be a uniformly convex Banach space with a uniformly convex dual and a duality mapping that is weakly sequentially continuous at zero, and let $A \subset E \times E$ be an m-accretive operator with a zero. Suppose that a sequence $\{x_n\}$ can be defined by (2) such that $\{x_n\}$ and $\{y_n\}$ are bounded. If $\{\theta_n\}$ is decreasing, $\lim_{n\to\infty} \theta_n = 0$, $\sum_{n=0}^{\infty} \lambda_n\theta_n = \infty$, $\lim_{n\to\infty} b(\lambda_n)/\theta_n = 0$, and $\lim_{n\to\infty} \frac{(\theta_{n-1}/\theta_n - 1)}{\lambda_n\theta_n} = 0$, then $\{x_n\}$ converges strongly to a zero of A.

Proof. Let J_r be the resolvent of A, and denote J_{1/θ_n} by u_n. We have $|u_n - u_{n-1}| \leq M_1(\theta_{n-1}/\theta_n - 1)$, and $|x_{n+1} - u_n|^2$
$\leq |x_n - u_n|^2 - 2\lambda_n(y_n + \theta_n x_n, J(x_n - u_n)) + M_2\lambda_n b(\lambda_n)$
$\leq (1-2\lambda_n\theta_n)|x_n - u_n|^2 + M_2\lambda_n b(\lambda_n) \leq (1-2\lambda_n\theta_n)|x_n - u_{n-1}|^2$
$+ M_3(\theta_{n-1}/\theta_n - 1) + M_2\lambda_n b(\lambda_n)$. Therefore $x_{n+1} - u_n \to 0$. Since $\{u_n\}$ converges strongly to a zero of A, this completes the proof.

In ℓ^p, $1 < p < \infty$, $\lambda_n = (n+1)^{-a}$ and $\theta_n = (n+1)^{-b}$ satisfy the hypotheses of Theorem 4 if $0 < b < (s-1)a$ and $a + b < 1$. (We do not know if Theorem 4 is true in L^p, $p \neq 2$.)

It is also possible to establish (weak)convergence results for (1) even if A is not strongly accretive. Here we consider the case $A = I - T$ where T is nonexpansive $(|Tx-Ty| \leq |x-y|)$. (1) takes the form

$$x_{n+1} = (1-\lambda_n)x_n + \lambda_n Tx_n, \quad n \geq 0, \tag{3}$$

where $0 \leq \lambda_n \leq 1$, and $x_0 \in C$, a closed convex subset of E. The following result extends [17, Theorem 1]. We use Lorentz's concept of almost convergence.

Theorem 5. Let C be a closed convex subset of a uniformly convex Banach space E with a Fréchet differentiable norm and let $T\colon C \to C$ be a nonexpansive mapping with a fixed point. Let $\{x_n\}$ be defined by (3). If $\lim_{n\to\infty} \lambda_n = 1$, then $\{x_n\}$ is weakly almost convergent to a fixed point of T.

We omit the proof of this theorem. Note that if $\sum_{n=0}^{\infty} \lambda_n = \infty$ and $\{\lambda_n\}$ does not tend to 1, then $\{x_n\}$ converges weakly to a fixed point of T by [17, Theorem 2].

In addition to the explicit schemes (1), (2) and (3) one can also study the implicit schemes

$$x_n \in x_{n+1} + \lambda_{n+1} Ax_{n+1}, \quad n \geq 0, \tag{4}$$

$$x_n \in x_{n+1} + \lambda_{n+1}(Ax_{n+1} + \theta_{n+1}x_{n+1}), \quad n \geq 0, \tag{5}$$

and

$$x_{n+1} = \frac{1}{1+\lambda_{n+1}} x_n + \frac{\lambda_{n+1}}{1+\lambda_{n+1}} Tx_{n+1}, \quad n \geq 0. \tag{6}$$

Method (5) was considered in [15] and [16]. If A is continuous, results concerning this implicit method (with errors) can be applied to the explicit method (2).

Now let E be a uniformly convex Banach space with a Fréchet differentiable norm, $A \subset E \times E$ an m-accretive operator with a zero, and $\{\lambda_n\}$ a positive sequence. Let $\{x_n\}$ be defined by

(4) and denote $(x_n - x_{n+1})/\lambda_{n+1}$ by $y_{n+1} \in Ax_{n+1}$. If $\delta_E(\varepsilon) \geq K\varepsilon^r$ for some $K > 0$ and $r \geq 2$, and $\sum_{n=1}^{\infty} \lambda_n^r = \infty$, then $|y_n| \leq \dfrac{d(x_0, A^{-1}0)}{(K \sum_{j=1}^{n} \lambda_j^r)^{1/r}}$ and $\{x_n\}$ converges weakly to a zero of A [17, Theorem 3]. This result can be improved for (6), where $A = I - T$. ($d(x,D)$ is the distance between a point $x \in E$ and a subset $D \subset E$.)

Theorem 6. Let C be a closed convex subset of a uniformly convex Banach space E with a Fréchet differentiable norm, and $T\colon C \to C$ a nonexpansive mapping with a fixed point. Let $\{x_n\}$ be defined by (6). If $\delta_E(\varepsilon) \geq K\varepsilon^r$ ($K > 0$, $r \geq 2$) and $\sum_{n=1}^{\infty} \lambda_n = \infty$, then $|Ax_n| \leq \dfrac{d(x_0, A^{-1}0)}{(4K \sum_{j=1}^{n} \lambda_j)^{1/r}}$ and $\{x_n\}$ converges weakly to a fixed point of T.

Proof. If $|v| \leq |u|$, then $\frac{1}{2}(u+v, Ju) \leq |u|^2 \left(1 - \delta\left(\frac{|u-v|}{|u|}\right)\right)$, and $(u-v, Ju) \geq 2|u|^2 \delta\left(\frac{|u-v|}{|u|}\right)$. Therefore $(Ax-Ay, J(x-y)) \geq 2|x-y|^2 \delta\left(\frac{|Ax-Ay|}{|x-y|}\right)$. Let $\{x_n\}$ and $\{y_n\}$ be defined by (6) with starting points x_0 and y_0. We obtain $2\sum_{j=1}^{n} \lambda_j (Ax_j - Ay_j, J(x_j - y_j)) \leq |x_0 - y_0|$. The result now follows by letting y_0 be a fixed point of T.

Since we also have $(Ax_j - Ay_j), J(x_j - y_j)) \geq 2|x_j - y_j|\,|x_0 - y_0| \cdot \delta\left(\frac{|Ax_j - Ay_j|}{|x_0 - y_0|}\right)$ for all $j \geq 0$, we see that in the setting of Theorem 6, $\{x_n\}$ converges weakly to a fixed point of T even if the modulus of convexity of E is not of power type. It can also be shown that if $\{x_n\}$ is defined by (4) and $\sum_{n=1}^{\infty} \lambda_n = \infty$, then the sequence $\{\sum_{j=1}^{n} \lambda_j x_j / \sum_{j=1}^{n} \lambda_j\}$ converges weakly to a zero of A provided E is uniformly convex with a Fréchet differentiable norm. (See [8] for a more general result.)

We remark in passing that the beginning of the proof of Theorem 6 can be used to show that if E is uniformly convex and smooth, and A satisfies the convergence condition in the sense of [12], then so does its Yosida approximation A_r $(r > 0)$.

We also mention the following result which is of interest when T is fixed point free. Let $cl(R(A))$ denote the closure of the range of $I - T$.

<u>*Proposition 7*</u>. Let E be a uniformly convex Banach space with a uniformly Gâteaux differentiable norm, $T: E \to E$ a nonexpansive mapping, v the point of least norm in $cl(R(A))$, and $\{x_n\}$ a sequence defined by either (3) or (6). If $\sum_{n=0}^{\infty} \lambda_n(1-\lambda_n) = \infty$ in the first case and $\sum_{n=1}^{\infty} \lambda_n = \infty$ in the second, then $\lim_{n\to\infty} Ax_n = v$ and $\lim_{n\to\infty}|x_n| = \infty$ if and only if T is fixed point free.

This proposition follows from the ideas of [18].

We are not aware of results for monotone operators (outside Hilbert space) that are analogous to Theorems 1 and 4, except in rather special cases (cf. [19]). The situation changes if an auxiliary Hilbert space can be found. Here is an example (cf. [11]).

Let $M: E^* \to E$ be a (nonlinear) operator such that $M + \beta I$ is monotone for some positive β, and $K: E \to E^*$ a bounded linear operator. Suppose that there exist a Hilbert space and bounded linear operators $A: E \to H$, $C: H \to E^*$ and $D: H \to H$ such that D is one-to-one and onto, $(D^{-1}h,h) \geq \gamma|h|^2$ for some $\gamma > 0$ and all $h \in H$, $K = CDA$, and $(Ax,h) = (x,Ch)$ for $x \in E$ and $h \in H$. Then $(D^{-1}h_1 + AMC(h_1) - D^{-1}h_2 - AMC(h_2), h_1-h_2) \geq (\gamma-\beta\|C\|^2)|h_1-h_2|^2$. Also, if $D^{-1}v + AMC(v) = 0$, then $u = Cv$ is a solution of the Hammerstein equation $u + KM(u) = 0$. Thus we can use Theorem 1 (if $\gamma > \beta\|C\|^2$) and Theorem 4 (if $\gamma \geq \beta\|C\|^2$) to construct such a solution. The assumptions on K are satisfied if it is angle-bounded [5].

We do not know if Theorem 4 is true in L^p, $p \neq 2$, because we do not know if $\lim_{r\to\infty} J_r z$ always exists. The analogous problem for monotone operators has an affirmative answer (cf. [4]). Indeed let E^* be strictly convex with a Fréchet differentiable norm, $M \subset E \times E^*$ a maximal monotone operator with a zero, and $z \in E^*$. For each $r > 0$ there is a unique $x_r \in E$ such that $Jx_r + rMx_r \ni z$. $\{x_r\}$ is bounded. Suppose that as $r_n \to \infty$, $\{x_{r_n}\}$ converges weakly to u, and denote x_{r_n} by x_n and $(z-Jx_n)/r_n$ by $y_n \in Mx_n$. u belongs to $M^{-1}0$. Since $(Jx_n-Ju,x_n-u) + r_n(y_n,x_n-u) = (z-Ju,x_n-u)$, $\lim_{n\to\infty}(Jx_n-Ju,x_n-u) = 0$ and $x_n \to u$. Since $(z-Ju,u-y) \geq 0$ for all $y \in M^{-1}0$, u is unique, and the strong $\lim_{r\to\infty} x_r$ exists and belongs to $M^{-1}0$. Another approximation theorem can be found in [3].

Zeros of certain monotone operators correspond to solutions of certain partial differential equations. (Zeros of accretive operators correspond to equilibrium points of the semigroups they generate.) Convergent schemes may be used to prove existence. In probabilistic analysis they may be used to establish existence of random solutions to random equations. There are also applications to optimization theory.

We conclude with another result on accretive operators. Let A be an m-accretive operator in a reflexive E with $0 \in R(A)$, $g\colon [0,\infty) \to [0,\infty)$ a nonincreasing function of class C^1 such that $\lim_{t\to\infty} g(t) = 0$ and $\int_0^\infty g(t)dt = \infty$, $x \in E$, $x_0 \in D(A)$, and $u\colon [0,\infty) \to E$ the strong solution of the initial value problem

$$\begin{cases} u'(t) + Au(t) + g(t)u(t) \ni g(t)x \\ u(0) = x_0. \end{cases}$$

Under certain conditions, the strong $\lim_{t\to\infty} u(t)$ exists and belongs to $A^{-1}0$.

This leads to a doubly iterative procedure for constructing zeros of A [14].

Here we remark that $\lim_{t\to\infty} u(t)$ exists when $A = I - P$ where $P: E \to C$ is a nonexpansive retraction onto a closed and convex $C \subset E$, and E has a uniformly Gâteaux differentiable norm. Indeed let $x_n = u(t_n)$ with $t_n \to \infty$. We will show that there is a subsequence of $\{x_n\}$ that converges strongly to a point v in C. We assume for simplicity that E is separable. Then we may also assume (by passing to a subsequence) that $f(z) = \lim_{n\to\infty}|x_n - z|$ exists for all $z \in E$ (cf. [1]). Since $Ax_n \to 0$, $f(z)$ attains its minimum over E at a point $v \in C$. We have $\limsup_{n\to\infty}(z-v, J(x_n - v)) \leq 0$ for all $z \in E$ and $\limsup_{n\to\infty}(x_n - x, J(x_n - y)) \leq 0$ for all $y \in C$. Taking $z = x$ and $y = v$, we see that $x_n \to v$. Since $(v-x, J(v-y)) \leq 0$ for all $y \in C$, the strong $\lim_{t\to\infty} u(t)$ exists.

This argument can also be applied to $\lim_{r\to\infty} J_r x$. It shows that C is, in fact, a sunny nonexpansive retract of E (cf. [13, p. 288]).

ACKNOWLEDGMENT

Part of this paper was prepared at the Mathematics Research Center, University of Wisconsin-Madison.

REFERENCES

[1] Baillon, J. B. "Générateurs et semi-groupes dans les espaces de Banach uniformément lisses", to appear.

[2] Bakušinskii, A. B., and Poljak, B. T. (1974). "On the solution of variational inequalities", *Soviet Math. Dokl. 15*, 1705-1710.

[3] Brézis, H., and Sibony, M. (1968). "Méthodes d'approximation et d'iteration pour les opérateurs monotones", *Arch. Rational Mech. Anal. 28*, 59-82.

[4] Browder, F. E. (1966). "Existence and approximation of solutions of nonlinear variational inequalities", *Proc. Nat. Acad. Sci. U.S.A. 56*, 1080-1086.

[5] Browder, F. E., and Gupta, C. P. (1969). "Monotone operators and nonlinear integral equations of Hammerstein type", *Bull. Amer. Math. Soc. 75*, 1347-1353.

[6] Bruck, R. E. (1973). "The iterative solution of the equation $y \in x + Tx$ for a monotone operator T in Hilbert space", *Bull. Amer. Math. Soc. 79*, 1258-1261.

[7] Bruck, R. E. (1974). "A strongly convergent iterative solution of the equation $0 \in U(x)$ for a maximal monotone operator U in Hilbert space", *J. Math. Anal. Appl. 48*, 114-126.

[8] Bruck, R. E., and Passty, G. B. "Almost convergence of the infinite product of resolvents in Banach spaces", to appear.

[9] Dunn, J. C. (1978). "Iterative construction of fixed points for multivalued operators of the monotone type", *J. Functional Analysis 27*, 38-50.

[10] Fitzpatrick, P. M., Hess, P., and Kato, T. (1972). "Local boundedness of monotone-type operators", *Proc. Japan Acad. 48*, 275-277.

[11] Fitzpatrick, P. M., and Petryshyn, W. V. "Galerkin method in the constructive solvability of nonlinear Hammerstein equations with applications to differential equations", to appear.

[12] Nevanlinna, O., and Reich, S. (1978). "Strong convergence of contraction semigroups and of iterative methods for accretive operators in Banach spaces", MRC report.

[13] Reich, S. (1976). "Asymptotic behavior of semigroups of nonlinear contractions in Banach spaces", *J. Math. Anal. Appl. 53*, 277-290.

[14] Reich, S. (1978). "An iterative procedure for constructing zeros of accretive sets in Banach spaces", *Nonlinear Analysis 2*, 85-92.

[15] Reich, S. "Iterative methods for accretive sets", *Proc. Conf. on Nonlinear Equations*, Academic Press, to appear.

[16] Reich, S. "Constructing zeros of accretive operators", *Applicable Analysis*, to appear.

[17] Reich, S. "Weak convergence theorems for nonexpansive mappings in Banach spaces", *J. Math. Anal. Appl.*, to appear.

[18] Reich, S. "On infinite products of resolvents", *Atti Accad. Naz. Lincei*, to appear.

[19] Vaĭnberg, M. M. (1961). "On the convergence of the method of steepest descent for nonlinear equations", *Sibirsk. Mat. Ž. 2*, 201-220. (Russian)

SUPOR Q: A BOUNDARY PROBLEM SOLVER FOR *ODE's*

M. R. Scott
H. A. Watts

Applied Mathematics Division
Sandia Laboratories
Albuquerque, New Mexico

ABSTRACT

For several years now we have been interested in the development of high quality software for the solution of two-point boundary value problems for ordinary differential equations. In references [1,2] Scott and Watts discussed algorithmic matters of a computer code called *SUPORT*, a solver for linear problems. Even before the original version was completed it was felt that various other capabilities would be beneficial. Indeed, feedback from users (*SUPORT* has been widely disseminated) has been gratifying and has reinforced our plans for modifying the code, striving for more versatility and improving the efficiency in appropriate circumstances. Some of the more recent changes have been reported in references [3,4]. Still others are being studied and contemplated at the present time.

The most important new development anticipated was the combination of the powerful techniques used by *SUPORT* with a linearization process to form a solver for nonlinear boundary value problems. Some preliminary results in this direction have been compiled with a code called *SUPOR Q* and reported in

ISBN 0-12-434180-2

references [5,6]. *SUPOR Q* is being designed to solve linear problems directly so that the *SUPORT* code can eventually be eliminated. However, because substantial improvements were still being contemplated for the working version of *SUPOR Q*, we undertook parallel developments of the *SUPORT* code, such as those mentioned above. These improvements are now being transferred to the *SUPOR Q* code. The subject of this paper, as described in more detail below, deals with the status of *SUPOR Q* as a general purpose two-point boundary problem solver, emphasizing aspects of efficiency in the integration processes.

The technique of interest for solving the linear two-point boundary value problem is to combine reduced superposition with an orthonormalization process. Briefly, we consider problems of the form

$$y'(x) = F(x)y(x) + g(x),$$

$$Ay(a) = \alpha, \quad By(b) = \beta,$$

a system of n equations with $n - k$ boundary conditions at a and k boundary conditions at b. For our initial value procedure we assume the integration proceeds from a towards b. To use superposition we compute k solutions to the homogeneous equation along with a particular solution of the inhomogeneous equation, yielding $k + 1$ independent solution vectors which are sufficient to form a basis for the problem solution space. (If $g(x) \equiv 0$ and $\alpha = 0$, only k solutions need to be computed.)

If the spectral width of the eigenvalue spectrum of F is extremely large, it will be impossible to maintain linear independence of the computed initial value solutions over the complete interval $[a,b]$ if this interval length is at all large. In this case, simple superposition will fail and it becomes necessary to use other techniques such as an orthonormalization procedure. By this we mean the following. Each time the solutions begin to lose their linear independence numerically, they are reorthonormalized before integratiion proceeds. Continuity of

y is achieved by matching the solutions over successive orthonormalization subintervals. Thus, the process consists of: (1) the integration sweep from a to b while storing the orthogonalization information at the points of orthonormalization and homogeneous and particular solution values at all the designated output points of interest. (2) the backward sweep of using the continuity conditions and stored information to calculate the solution values y at the output points.

We shall examine ways of ameliorating the cost of integrating the (possibly large) number of independent solutions which is necessary for the process. Essentially, we want to take advantage of the linearity of the equations being solved. We examine these possibilities and look at numerical integration schemes which are particularly efficient in such circumstances.

For nonlinear problems of the form

$$y'(x) = f(x,y),$$

$$\phi(y(a)) = 0, \quad \psi(y(b)) = 0,$$

we advocate the use of quasilinearization (Newton's Method) to generate a sequence of linear boundary value problems, thereby taking advantage of the features of the linear solver. An important aspect of this process is to provide an approximation to the previously computed (discrete) solution everywhere it is needed in the formulation of the equation at the current iteration. The effects of this on the integration algorithms will be addressed.

Typically, the code returns the solution of the boundary value problem at a predetermined set of output points. However, since it will usually be of interest to locate the places where the solution varies most rapidly and since we generally do not know the precise behavior of the solution a priori, we have included an algorithm for adding or deleting output points from one iteration to the next. This is particularly important for problems having a lot of structure in the solution, such as boundary layers. Our aim has been to provide this information in a way that adequately describes the solution character without undue penalities on the computational cost and storage.

REFERENCES

[1] Scott, M. R., and Watts, H. A. (1977). "Computational Solution of Linear Two-Point Boundary Value Problems via Orthonormalization", *SIAM J. Numer. Anal.*, *14*, 40-70.

[2] Scott, M. R., and Watts, H. A. (1975). "SUPORT - A Computer Code for Two-Point Boundary Value Problems via Orthonormalization", Sandia Laboratories Report, SAND75-019, Albuquerque, New Mexico.

[3] Darlow, B. L., Scott, M. R., and Watts, H. A. (1977). "Modifications of SUPORT, A Linear Boundary Value Problem Solver: Part I - Pre Assigning Orthonormalization Points, Auxiliary Initial Value Problem, Disk or Tape Storage", Sandia Laboratories Report, SAND77-1328, Albuquerque, New Mexico.

[4] Darlow, B. L., Scott, M. R., and Watts, H. A. (1977). "Modification of SUPORT, A Linear Boundary Value Problem Solver: Part II - Inclusion of an Adams Integrator", Sandia Laboratories Report, SAND77-1690, Albuquerque, New Mexico.

[5] Scott, M. R., and Watts, H. A. (1976). "A Systematized Collection of Codes for Solving Two-Point Boundary Value Problems", Numerical Methods for Differential Systems, L. Lapidus and W. Schiesser, eds., Academic Press, New York.

[6] Scott, M. R., and Watts, H. A. (1977). "Computational Solution of Nonlinear Two-Point Boundary Value Problems", Proceedings of the 5th Symposium of Computers in Chemical Engineering, Czechoslovakia, also published as a Sandia Laboratories report SAND77-0091, Albuquerque, New Mexico.

This work was supported by U.S. Department of Energy.

SOME RECENT DEVELOPMENTS IN STABILITY OF GENERAL SYSTEMS

Peter Seibert

Departamento de Matemáticas
Y Ciencia de la Computación
Universidad Simón Bolívar
Caracas, Venezuela

INTRODUCTION

In the almost 100 years since the publication of M. A. Liapunov's famous monograph [17], the method of the Liapunov function, or "second method", has become an ever more widely used tool in the study of stability properties of dynamical systems. It has been applied to new problems, such as stability under persistent perturbations, global asymptotic stability, etc., but until rather recently, the basic form in which it is presented and used has changed remarkably little. Trends towards more general and abstract theories started with the appearance of Zubov's book [39] in 1957, and new approaches, such as the use of vector Liapunov functions and the invariance principle, have emerged. Today one can speak of "two second methods", one using autonomous Liapunov functions, and the other (more classical one) using functions depending explicitly on the time. While the first is of greater formal simplicity, in the second one the question of existence is less problematic.

ISBN 0-12-434180-2

With the increasing ramification of the "second method", unified theories have become more of a necessity. Several efforts in this direction have already been undertaken, and the points of view adopted have varied widely. We may mention here (without claiming completeness) the papers by Bushaw [5,6], Dana [8], Lakshmikantham and his school, represented by the monograph [15] and a large number of papers, J. Auslander and the author [1], Habets and Peiffer [10], and Pelczar [20-24].

A difficulty which easily arises in the attempt to establish a unified theory is the great number of possible stability concepts with all the complexities this entails. (In Bushaw's classification [6], for instance, the number of stability types is in the millions). In order to avoid such excessive proliferation, the theory we present here starts with a bare minimum of structure and from a few extremely simple definitions. This is sufficient to construct a theory which encompasses all the usual concepts of Liapunov stability and some others. From these simple foundations the theory can be progressively refined. Our endeavor is to formulate a few very general theorems and to obtain, if possible, all known results as corollaries.

There is a particular emphasis on the question of existence of Liapunov-type functions. This problem is considered at various levels: First for abstract functions without any continuity requirements. Here the notion of admissibility is decisive. This relates the order structure of the collections of sets involved to the order structure of the real line. Then the existence of continuous and semicontinuous, Liapunov and para-Liapunov functions is discussed. The difference of behavior with respect to existence of autonomous and non-autonomous Liapunov functions also finds its natural explanation in this context.

The theory falls into two parts, one concerning stability, the other asymptoticity (including asymptotic stability). For reasons of limitations of space and time, I had to restrict myself to the first of the two, which is also the more firmly established. For the part concerning asymptoticity, I refer to the two technical reports [9] and [33].

A theory similar to ours, except that the algebraic structure of the time scale is preserved, has been developed by A. Pelczar, [20-24], for what he calls pseudo-dynamical systems.

I. SET-THEORETIC FRAMEWORK AND NOTATIONS

Let X denote a set, called the state space. By a *collection* on X we mean a nonempty set of subsets of X. If $\mathcal{A}$ and $\mathcal{B}$ are collections on X, we say

$\mathcal{A}$ *is coarser than* $\mathcal{B}$, in symbols $\mathcal{A} > \mathcal{B}$, if every set $A \in \mathcal{A}$ contains a set $B \in \mathcal{B}$.

The relation $>$ is obviously reflexive and transitive.

Functions Applied to Collections. Consider a function $\Phi\colon X \to 2^X$ (the collection of all subsets of X). For sets $A \subset X$ and collections $\mathcal{A}$ on X we define:

$$\Phi A = \{\Phi(x) \mid x \in A\};$$

$$\Phi\mathcal{A} = \{\Phi A \mid A \in \mathcal{A}\}.$$

Since Φ is monotone in the sense that $A \supset B$ implies $\Phi A \supset \Phi B$, the following implication holds:

$$\mathcal{A} > \mathcal{B} \Longrightarrow \Phi\mathcal{A} > \Phi\mathcal{B}. \tag{1.1}$$

We call ΦA the *orbit* of a set A.

II. STABILITY

By a *system* S we mean a quadruplet $(X, \Phi, \mathcal{D}, \mathcal{E})$, where X and Φ are the same as above, and $\mathcal{D}$ and $\mathcal{E}$ are collections on X. We define:

The system $S = (X, \Phi, \mathcal{D}, \mathcal{E})$ is *stable* if

$$\mathcal{E} > \Phi\mathcal{D}, \tag{2.1}$$

or explicitly, if every $E \in \mathcal{E}$ contains the orbit of some $D \in \mathcal{D}$.

Proposition 2.1. If $\mathcal{E}' > \mathcal{E}$ and $\mathcal{D} > \mathcal{D}'$, *and if* $(X,\Phi,\mathcal{D},\mathcal{E})$ is *stable, then* $(X,\Phi,\mathcal{D}',\mathcal{E}')$ *is also stable.*

This is an immediate consequence of (1.1).

Examples. 1. Consider a dynamical system[1] on a topological space X and let M be a compact subset of X. By $\mathcal{N}_M$ we denote the collection of all neighborhoods of M. Then we define $\mathcal{D} = \mathcal{E} = \mathcal{N}_M$, and Φ as the map which assigns to every point $x \in M$ its positive semiorbit. Now the condition (2.1) is equivalent to the stability of M, i.e., to the property that for any $U \in \mathcal{N}_M$ there exists a $V \in \mathcal{N}_M$ such that $\Phi V \subset U$.

2. We consider the same situation as in the preceding example, but assuming X to be metric. Furthermore, we assume only that M is closed, not necessarily compact. Beside the collection of topological neighborhoods, $\mathcal{N}_M$, (i.e., sets containing open sets containing M) we now have the collection of all metric neighborhoods (or ε-neighborhoods of M with $\varepsilon > 0$). This collection we denote by $\mathcal{M}_M$, and the open ε-neighborhoods of M by $N_\varepsilon(M)$. Then the following equivalences hold:

(a) If $\mathcal{D} = \mathcal{E} = \mathcal{M}_M$, (2.1) becomes *uniform stability* of M, i.e.

$$(\forall \varepsilon > 0)\ (\exists \delta > 0)\quad \Phi N_\delta(M) \subset N_\varepsilon(M).$$

(b) If $\mathcal{D} = \mathcal{N}_M$, $\mathcal{E} = \mathcal{M}_M$, (2.1) becomes *stability* of M:

$$(\forall\ \varepsilon > 0)\ (\exists V \in \mathcal{N}_M)\quad \Phi V \subset N_\varepsilon(M).$$

These definitions are easily seen to be equivalent to 4.1.3 and 4.1.1, respectively, of [4], Chapter V.

[1] *Our standard reference for the stability theory of dynamical systems is the monograph by N. P. Bhatia and G. P. Szegö,* [4].

(c) Finally, if $\mathcal{D} = \mathcal{E} = N_M$, (2.1) becomes *topological stability* as defined in [1-3,34].

Since $M_M > N_M$, we conclude from proposition 2.1 that uniform stability and topological stability both imply stability. Of uniform and topological stability, on the other hand, neither implies the other (cf. [1], p. 2).

3. Consider a nonautonomous differential equation, $\frac{dy}{dt} = f(t,y)$, where f is a function which satisfies conditions guaranteeing global existence and uniqueness of solutions. We will denote by $\sigma(t;y_0,t_0)$ the solution satisfying the initial condition $\sigma(t_0;y_0,t_0) = y_0$. We furthermore assume $f(t,0) \equiv 0$. According to the standard definition, the trivial solution $\sigma(t;0,t_0)$ is *stable* if for every $\varepsilon > 0$ there exists a $\delta = \delta(\varepsilon,t_0) > 0$ such that $\|y_0\| < \delta$ implies $\|\sigma(t;y_0,t_0)\| < \varepsilon$ for all $t \geq t_0$. Moreover, the rest position, $y = 0$, is called

(a) *stable* if the above condition holds for every t_0,

(b) *uniformly stable* if it is stable and δ is independent of t_0.

In order to express these conditions in our notations, put $X = R \times Y$, $\Phi(x_0) = \sigma([t_0,\infty);y_0,t_0)$ where $x_0 = (t_0,y_0)$, and define $\mathcal{E}$ as the collection of all ε-neighborhoods of the t-axis with $\varepsilon > 0$, $E_\varepsilon = \{(t,x) \mid t \in R, \|x\| < \varepsilon\}$. Then uniform stability of the rest position is obviously equivalent to the stability of $(X,\Phi,\mathcal{E},\mathcal{E})$, which is uniform stability of the t-axis in the sense of example 2.

Stability of the rest position, on the other hand, reduces to the following: Denote, for every t_0,δ, $D'_{t_0,\delta} = \{(t_0,x) \mid \|x\| < \delta\}$, and choose, for every ε and t_0, $\delta = \delta(\varepsilon,t_0)$ as in the definition above. Then define

$$D_\varepsilon = \cup\left\{D'_{t_0,\delta(\varepsilon,t_0)} \;\middle|\; t_0 \in R\right\}.$$

If, for fixed ε, $\delta(\varepsilon,t_0)$ can be chosen so as to be bounded away from 0 on finite intervals of t_0, D_ε is a neighborhood of the t-axis. It is easy to see that this is the case whenever f is bounded on a neighborhood of every point of the t-axis. Then, if we put $\mathcal{D} = \{D_\varepsilon \mid \varepsilon > 0\}$, stability of $y = o$ is equivalent to stability of $(X,\Phi,\mathcal{D},E)$, which is stability of the t-axis in the sense of the preceding example.

4. In the same context, the rest position is called *eventually stable* if for every given $\varepsilon > 0$ there exists a $\tau = \tau(\varepsilon) \in R$ and, for any $t_0 \geq \tau$, a $\delta = \delta(\varepsilon,t_0) > 0$ such that $\|y_0\| < \delta$ and $t_0 \geq \tau$ together imply $\|\sigma(t;y_0,t_0)\| < \varepsilon$ for all $t \geq t_0$. It is called *uniformly eventually stable* if δ does not depend on t_0. The notion of eventual stabilty is due to J. P. LaSalle and R. J. Rath [16], and was introduced independently by T. Yoshizawa in [37,38]. For more recent contributions see [15], sect. 3.14, and [12].

If we define X, Φ and E as in example 3 and denote by $\mathcal{D}_T$ the collection of all *half tubes* $D_{t_0,\delta} = \{(t,y) \mid t \geq t_0, \|y\| < \delta\}$, with $t_0 \in R$ and $\delta > 0$, stability of $(X,\Phi,\mathcal{D}_T,E)$ is equivalent to uniform eventual stability. Similarly, in order to rephrase the definition of eventual stability using our notations, we define topological half tubes, $D_{t_0,V}$, as intersections of neighborhoods V of the t-axis with half spaces X_{t_0}: $\{(t,y) \mid t \geq t_0, y \in Y\}$. We denote $\mathcal{D}_T^* = \{D_{t_0,V} \mid t_0 \in R, V \in N_{t\text{-axis}}\}$. Now, with the same proviso as in the preceding example, eventual stability can be expressed as stability of the system $(X,\Phi,\mathcal{D}_T^*,E)$.

Eventual stability can easily be generalized in such a way that to every stability type $(\mathcal{D},E)$, there corresponds an eventual stability type, provided the state space X is a product, $T \times Y$, where T (the generalized time scale) is endowed with a preorder (reflexive, transitive relation), $\geq$. To this end we associate to every set $D \subset X$ and time t the *t-tail* $D_t = \{(t',y) \in D \mid t' \geq t\}$, and to the collection $\mathcal{D}$ its corresponding *tail collection*, $\mathcal{D}_T = \{D_t \mid D \in \mathcal{D}, t \in T\}$. Then by *eventual stability* of $S = (X,\Phi,\mathcal{D},E)$ we mean that the system

$S_{\text{event}} := (X, \Phi, \mathcal{D}_T, E)$ is stable, i.e. that every E contains a ΦD_t. The definitions of uniform and non-uniform eventual stability given above are now special cases of this more general one.

III. LIAPUNOV-TYPE FUNCTIONS

3.1. Let $S = (X, \Phi, \mathcal{D}, E)$ be a system as defined in the preceding section, and consider functions

$$v: \quad X \to [0, \infty).$$

For every number $\beta > 0$ we define, with respect to v, the set

$$S_v^\beta = \{x \in X \mid v(x) < \beta\},$$

and denote the collection of all these sets by S_v.

We say v is a *para-SL-function* for the system S if it satisfies the conditions

$$E > S_v, \tag{S.1}$$

$$S_v > \mathcal{D}, \tag{S.2}$$

$$S_v > \Phi S_v. \tag{S.3}$$

Note that the last condition is satisfied whenever the sets S_v^α are invariant under Φ, i.e. $\Phi S_v^\alpha \subset S_v^\alpha$, or equivalently, if

$$y \in \Phi(x) \Longrightarrow v(y) \leq v(x); \tag{S.3*}$$

in other words, if v is *nonincreasing* with respect to Φ.[2]

If v satisfies the conditions (S.1), (S.2), and (S.3*), we will call it an *SL-function*.

[2]*A reasonably general condition for the monotonicity of* v *in terms of its generalized total derivative is given in lemma 9.2 of* [26].

Theorem 3.1. *The system* $S = (X, \Phi, \mathcal{D}, \mathcal{E})$ *is stable if it admits a para-SL-function.*

Proof. By applying, consecutively, (S.1), (S.3) and (S.2), and observing (1.1), we find $\mathcal{E} > S_v > \Phi S_v > \Phi\mathcal{D}$, hence, by transitivity of the relation $>$, stability of S.

3.2. Continuous functions v satisfying the condition (S.3) were called *para-Liapunov functions* by O. Hájek in [11]. He used them to characterize stability in certain cases where no continuous Liapunov function exists ([11], theorem 27).

The condition (S.3) is satisfied, in particular, if there exists a decreasing sequence of numbers $\beta_n \to 0$ such that the sets $S_v^{\beta_n}$ are invariant under Φ [whereas condition (S.3*) requires invariance of *all* S_v^{β}]. Functions with this property were used by J. A. Yorke in [36] to characterize stability of the origin for an autonomous differential system (theorem 4.6).

We will return to the question of existence of (semi-) continuous Liapunov-type functions in sections 3.5 and 3.6.

3.3. Existence of Para-SL-Functions. To assure the existence of a para-SL-function it is not enough to assume stability of the system. For instance, in the case of topological stability (defined in example 2.c, above) such a function need not exist (see [34], section 7, example IV). The formulation of the additional condition needed requires a few definitions.

First, throughout the remainder of this chapter, we assume that Φ defines a preorder on X.

If $\mathcal{A}$ is a collection on X, and $\mathcal{B} \subset \mathcal{A}$ satisfies the condition $\mathcal{A} > \mathcal{B}$, we say $\mathcal{B}$ is a *base* of $\mathcal{A}$. We call *admissible* any collection admitting a countable nested base. For example, let M be a closed set in a metric space X, and denote by $\mathcal{M}_M$ and $\mathcal{N}_M$ the collections of metric and topological neighborhoods of M, respectively. Then $\mathcal{M}_M$ is admissible [for instance, $\{N_{\varepsilon_n}(M)\}$, with $\varepsilon_n \to 0$, is a base]. $\mathcal{N}_M$, on the other hand, is in general not admissible (for instance, in the case $X = R^2$, $M = R^1$).

In particular, the collection S_v is always admissible.

If $(\mathcal{D},E)$ is a pair of collections, we say $(\mathcal{D},E)$ is *strongly admissible* if either $\mathcal{D}$ or E is admissible, *weakly admissible* if there exists an admissible collection S such that $E > S > \mathcal{D}$.

Since the relation $>$ is reflexive, strong admissibility implies weak admissibility.

In the above example, the pairs (M_M, M_M) and (N_M, M_M) are strongly admissible, while (N_M, N_M) is not even weakly admissible. (An example of a weakly admissible pair which is not strongly admissible was given in [31], section 3.)

Theorem 3.2. [31,34]. (a) *Necessary conditions for the existence of a para-SL-function for the system $S = (X,\Phi,\mathcal{D},E)$ are the following:*

(i) *S is stable,*

(ii) *The pair $(\mathcal{D},E)$ is weakly admissible.*

(b) *The conditions* (i) *and*

(ii) *$(\mathcal{D},E)$ is strongly admissible*

are sufficient for the existence of an SL-function.

Indeed, part (a), (i) is the contents of theorem 3.1; part (a), (ii) is an immediate consequence of the conditions (S.1) and (S.2), and of the admissibility of S_v; part b), on the other hand, is contained in the theorems 2 and 3 of [34].

3.4. Examples. 1. In the case of stability of a compact set M under a dynamical system (example 1 of sect. 2), condition (S.1) says that v does not take arbitrarily small values outside of any neighborhood of M, (S.2) requires that $v \to 0$ as $x \to M$, and (S.3) is satisfied if v is nonincreasing along the oriented orbits [condition (S.3*)]. These are the standard necessary and sufficient conditions for stability in this case (cf. [38], theorem 22.6).

2. Now let M be a closed set in a metric space endowed with a dynamical system (example 2 of sect. 2). Considering first the simple case of uniform stability, the interpretation of the condition (S.1) is the same as above, using metric neighborhoods, and (S.2) means that $v(x) \to 0$, uniformly as $x \to M$. Then the conditions become precisely those of Zubov, [39], theorem 12. (Note that Zubov's "stability" is Bhatia's uniform stability, the latter's being the terminology we have adopted here).

In the case of non-uniform stability, condition (S.1) means that v does not take arbitrarily small values outside of any metric neighborhood of M, and (S.2) says that $v \to 0$ along any sequence x_n tending to a point of M. These are exactly the conditions of theorem 4.5, chapt. V of [4].

3. In the case of stability of the rest position of a non-autonomous differential equation (example 3 of sect. 2), we consider the flow defined by the solutions on the product space $R \times Y$, and (uniform) stability of the origin of Y reduces to that of the set $M = R \times \{o\}$. The interpretation of the conditions (S.1) and (S.2) is then the same as in the preceding example. The conditions thus obtained are equivalent to those of Yoshizawa's theorem 18.5, [38], and of the following remark.

In all these examples the inverse theorem 3.2 can be applied, since the collections of metric neighborhoods are admissible.

4. An interesting application of our general theorem concerns the eventual properties discussed in example 4 of Section 2. In what follows, we assume that the time scale T is directed by the relation $\geq$, and that Φ is "directed into the future", i.e. $\Phi X_t \subset X_t$ for every t $(X_t := \{(t',y) \mid t' \geq t, y \in Y\})$. The following theorem gives a sufficient condition for the general notion of eventual stability defined in Section 2.

<u>Theorem 3.3.</u> *The system $S = (X,\Phi,\mathcal{D},\mathcal{E})$ is eventually stable if there exists a function v: $X \to R^+$ satisfying the following conditions:*

$$E > (S_v)_T; \tag{ES.1}$$

$$S_v > \mathcal{D}_T; \tag{ES.2}$$

$$S_v > (\Phi S_v)_T. \tag{ES.3}$$

The proof is straight forward, using the conditions (ES.1), (ES.3) and (ES.2), in this order, and the definitions and hypotheses involved.

[Note that the weaker theorem, obtained by replacing (ES.1) and (ES.3) by the stronger conditions (S.1) and (S.3), is a direct corollary of theorem 3.1. In this case, theorem 3.2 also immediately yields *necessary conditions* for eventual stability.]

In the case of *uniform eventual stability* the collections $\mathcal{D}$ and E are the metric neighborhood systems of the t-axis. Condition (ES.1) then means that, given $\varepsilon > 0$, there exists $\alpha > 0$ and $\tau \in T$ such that $v(t,y) \geq \alpha$ for all $t \geq \tau$ and $\|y\| \geq \varepsilon$. (ES.2), on the other hand, means that given $\beta > 0$, there exist $\delta > 0$ and τ such that $v(t,y) < \beta$ whenever $t \geq \tau$ and $\|y\| < \delta$. (ES.3) is, of course, fulfilled if v is nonincreasing under Φ. These conditions are all implied by the usual ones; cf., for instance, theorem 5.1, [12].

We now turn to Yoshizawa's sufficient conditions for uniform eventual stability ([38], theorem 17.4). His condition (i) is equivalent to our conditions (S.1) and (S.2), while his condition (ii), which may be written in the form $v'(t,x) = O(h)$, $\int_0^\infty |h(t)|dt < \infty$ [v': generalized total derivative], is easily seen to imply our condition (ES.3). Thus ours is a common generalization of both types of theorems on uniform eventual stability.

Taking as $\mathcal{D}$ the system of topological neighborhoods of the t-axis, one immediately obtains conditions for non-uniform eventual stability. These are usually omitted in the literature.

3.5. The Existence of Continuous and Semicontinuous Liapunov Functions. In his paper [29] (summarized in [30]), P. Salzberg studied the question under what conditions a continuous or semicontinuous SL-function exists. Let F denote a collection on a Hausdorff space X. A base B (in the sense defined in 3.3) of F is called *invariant* if each of its members is invariant under Φ. It is called *upper* [*lower*] *normal* if it can be written in the form $\{B_i \mid i \in I\}$, where I is a dense subset of R^+, and $B_i \subset \operatorname{int} B_{i'}$, $[\overline{B}_i \subset B_{i'}]$ whenever $i \leq i'$. It is called *normal* if it is both upper and lower normal.

Theorem 3.4. [29]. *The system $S = (X,\Phi,D,E)$ admits an upper [lower] semicontinuous SL-function if and only if there exists a collection F admitting an invariant upper [lower] normal base such that $E > F > D$. It admits a continuous SL-function if and only if there exists an invariant normal base satisfying the same relation.*

Example. Consider, on the real line, the flow with rest points at 0, $\pm 1/n$, $n \in N$, and orbits joining them, oriented away from the origin. [Analytic example: $\frac{dx}{dt} = |x| \sin^2 \frac{\pi}{x}$ for $x \neq 0$, 0 for $x = 0$]. We will denote by B_i $[i \in (0,\infty)]$ the sets $\Phi(-i,i)$. Then B_I, $I = (0,\infty)$, is an invariant upper normal base of N_0, and $\overline{B}_I$ is an invariant lower normal base. On the other hand, there does not exist any invariant normal base. Then according to the theorem, there exist upper and lower semicontinuous SL-functions. For instance, the function

$$v(t) = \begin{cases} 0 & \text{for } x = 0 \\ \frac{1}{n} & \text{for } \frac{1}{n+1} \leq x < \frac{1}{n} \end{cases}$$

is upper semicontinuous, and the one obtained by interchanging $\leq$ and $<$, is lower semicontinuous.

Remark. In the case of a dynamical system on a metric space, uniform stability of a closed set M always implies the existence of both upper and lower semicontinuous SL-functions. Indeed, it is easy to see that $v_1(x) = \inf_{t\leq 0} d(xt,M)$ is upper and $v_2(x) = \sup_{t\leq 0} d(xt,M)$ {with the notation of [4]} is lower semicontinuous.

3.6. On the Existence of Para-Liapunov Functions. A question which has been left open so far is the one concerning the existence of continuous para-SL-functions, or para-Liapunov functions.

Because of the limitation of space we will consider here only a rather special case, which, however, can easily be extended along the general lines indicated by the preceding result.

Theorem 3.5. Let M denote a closed, uniformly stable set under a preorder Φ in a metric space X. Then there exists a continuous para-SL-function v for the system (X,Φ,M_M,M_M) [or para-Liapunov function with $v^{-1}(0) = M$].

Proof. Uniform stability of M implies the existence of a decreasing sequence $\{\alpha_n\}$ tending to 0 such that the sets $E_n = N_{\alpha_n}(M)$ satisfy the condition $\Phi E_{n+1} \subset E_n$ $(n \in N)$, and consequently $\overline{\Phi E_{n+1}} \subset \overline{E_n} \subset \operatorname{int} E_{n-1} \subset \operatorname{int} \Phi E_{n-1}$ (putting $E_0 = X$). We define $B_k = \Phi E_{2k}$ $(k = 0,1,2,\ldots)$. Then $\overline{B_k} \subset \operatorname{int} B_{k-1}$, for all $k \in N$. Applying Urysohn's lemma to each of the sets $\overline{B_{k-1} - B_k}$, $k \in N$, we can construct a continuous function v from X into $[0,1]$ which is 0 on M and $\frac{1}{k} \leq v \leq \frac{1}{k-1}$ on $\overline{B_{k-1} - B_k}$. It is now easy to verify that v satisfies the conditions (S.1) through (S.3) for $\mathcal{D} = \mathcal{E} = M_M$.

In the example given above, it is easy to see that the function $v(x) = |x|$ is a para-SL-function.

The stability type studied by Hajek in [11], in the context of a dynamical system in a topological space, is our topological stability (apart from a concept called para-stability, too complicated to be discussed here). Only in the case of a set with compact boundary in a locally compact space (among other hypotheses) does he give a criterion involving a singel para-Liapunov function (theorem 27). In this case topological and uniform stability are equivalent, and his theorem coincides with ours. In the general case there cannot exist a para-SL-function (not even a discontinuous one) according to what was said in 3.3.

For the case of an autonomous differential equation, Yorke in [36] {which appeared before Hájek's paper} gives conditions (involving the generalized total derivative) for a function to be para-Liapunov, and uses this to establish a stability criterion (theorem 4.6).

IV. EXTENSIONS

4.1. Liapunov Families. The idea of using a family of Liapunov-type functions instead of a single one has been conceived independently by several authors. Certain types of such families were used to study asymptotic stability and attraction, for instance in theorem 17.1 of Krasovskiĭ's book [13]. These, however, remain outside the scope of this paper, and we will restrict ourselves to those families of functions which are used for studying Liapunov stability.

We give the following references (in chronological order): L. Salvadori, [27,28]; O. Hájek, [11] (in particular, theorem 28); J. Auslander, [2,3]. In spite of certain variations, the essential idea in all of these papers is the same. Auslander's results were subsequently extended by P. M. Salzberg and the author in [32], and similar families were recently applied to asymptoticity by G. Dankert and the author ([9]). The following exposition is based on the paper [32].

We consider a family V of functions $v\colon\ X \to R^+$, and define the collection $S(V)$ as the union of all collections S_v [defined in 3.1] with $v \in V$. We call V a *para-SL-family* for the system $S = (X,\Phi,\mathcal{D},E)$ if it satisfies the following conditions:

$$E > S(V); \tag{SF.1}$$

$$S(V) > \mathcal{D}; \tag{SF.2}$$

$$S(V) > \Phi S(V). \tag{SF.3}$$

{The first and third of these conditions are weaker than the corresponding ones, (S.1) and (S.3), applied to each function v individually; the second one is equivalent to (S.2) applied to each v.}

If V satisfies the conditions (SF.1) and (SF.2), and instead of (SF.3) the stronger one that v be non-increasing under Φ, we call V an SL-*family*.

Theorem 4.1 [32]. *If the system* S *admits a para-SL-family, it is stable; if it is stable, it admits an SL-family.*

Theorem 4.2 [32]. *If the system* S *admits a countable para-SL-family, and if* $\mathcal{D}$ *is closed under finite intersections, then* S *admits a para-SL-function.*

4.2. *Abstract Comparison Principle.* Comparison principles have been used in the theory of differential equations for a long time, and relatively recently they have also been applied to the study of stability properties [7]. First scalar, and then also vector differential equations and inequalities were used as comparison systems [10,14,15,18,19]. In the paper by Habets and Peiffer, a very general comparison principle is formulated which is applicable to whole classes of concepts rather than to individual ones.

The common feature of all comparison principles is the presence of a given system S (with an unknown property), a comparison system S_0 (with a known property), and a function v which relates one system to the other. The general problem then consists in finding conditions for v under which the known property of S_0 implies the desired property of S. We will see that very simple sufficient conditions can be formulated in the general framework of our unified theory.

Consider two systems $S = (X,\Phi,\mathcal{D},\mathcal{E})$ and $S_0 = (X_0,\Phi_0,\mathcal{D}_0,\mathcal{E}_0)$, S_0 being the comparison system, and a family V of mappings v from X into X_0. We call V a *para-SL-family of mappings* from S to S_0 (or *para-SL-mapping* if V is a single function) if it satisfies the following conditions:

$$\mathcal{E} > V^{-1}\mathcal{E}_0; \qquad \text{(SF-I)}$$

$$V^{-1}\mathcal{D}_0 > \mathcal{D}; \qquad \text{(SF-II)}$$

$$V^{-1}\Phi_0\mathcal{D}_0 > \Phi V^{-1}\mathcal{D}_0. \qquad \text{(SF-III)}$$

Here $V^{-1}\mathcal{E}_0$ (for instance) denotes the collection $\{v^{-1}E_0 \mid v \in V, E_0 \in \mathcal{E}_0\}$, and $v^{-1}E_0$ may be empty.

Theorem 4.3. *If S_0 is stable and there exists a para-SL-family of mappings from S to S_0, then S is stable.*

If V is a single mapping, this is essentially theorem 1 of [35].

Proof of the Theorem. One applied, successively, the conditions (SF-I), stability of S_0, (SF-III), and (SF-II).

In the special case where X_0 is R^+, Φ_0 is the identity mapping, and $\mathcal{D}_0$ and $\mathcal{E}_0$ are the collection $\{[0,\alpha), \alpha > 0\}$, theorem 4.3 reduces to theorem 4.1.

4.3. Conditions for Continuity. Since Liapunov stability is a special case of continuity, it is natural to ask for conditions in terms of auxiliary functions in order that a given function $f\colon X \to Y$ is continuous (in some sense). However, it will be necessary to use two functions, one defined on the domain X, the other on the range, Y. A theory along these lines has been developed by E. H. Rogers in [25]. Here we will briefly sketch the extension of the results of the preceding section to this case. The following results were obtained in collaboration with L. Mendoza.

Let X and Y be sets, f a function from X into Y, and $\mathcal{D}$ and $\mathcal{E}$ collections on X and Y, respectively. We say f is *$(\mathcal{D},\mathcal{E})$ - continuous* or *the system $S = (X,Y,f,\mathcal{D},\mathcal{E})$ is continuous*, if $\mathcal{E} > f\mathcal{D}$. {In particular, if $\mathcal{D}$ and $\mathcal{E}$ are the neighborhood filters of x and $f(x)$ $[x \in X]$, respectively, this becomes stability of f at x.}

Now, consider two such systems, S and $S_0 = (X_0,Y_0,f_0,\mathcal{D}_0,\mathcal{E}_0)$, and two families of functions, $U = \{u\}$, $v = \{v\}$, from X into X_0 and from Y into Y_0, respectively:

$$\begin{array}{ccc} X & \xrightarrow{f} & Y \\ \downarrow u & & \downarrow v \\ X_0 & \xrightarrow{f_0} & Y_0 \end{array}$$

We say (U,V) is a *para-CL-bifamily of mappings* from S to S_0 (or a *para-CL-pair of mappings* if U and V consist of a single function) if the following conditions are satisfied:

$$\mathcal{E} > V^{-1}\mathcal{E}_0; \qquad \text{(CF-I)}$$

$$U^{-1}\mathcal{D}_0 > \mathcal{D}; \qquad \text{(CF-II)}$$

$$V^{-1}f_0\mathcal{D}_0 > fU^{-1}\mathcal{D}_0. \qquad \text{(CF-III)}$$

In the special case $X_0 = Y_0 = R^+$, $f_0 =$ identity, $\mathcal{D}_0 = \mathcal{E}_0 = \{[0,\alpha) \mid \alpha > 0\}$, u and v are real-valued functions analogous to Liapunov functions and then we call (U,V) a *para-CL-bifamily of functions* for the system S.

Theorem 4.4. *If the system S_0 is continuous and there exists a para-CL-bifamily of mappings from S to S_0, then S is continuous.*

The proof is analogous to that of theorem 4.3, to which the present theorem reduces if one puts $X = Y$, $f = \Phi$, $f_0 = \Phi_0$, and $U = V$.

Theorem 4.5. The following condition is necessary and sufficient for the existence of a para-CL-pair of functions for the system $S = (X,Y,f,\mathcal{D},E)$:

There exists a pair of admissible collections $\mathcal{U}$ *and* $\mathcal{R}$ *on* X *and* Y *respectively, such that* $E > \mathcal{R}$, $\mathcal{U} > \mathcal{D}$, *and* $\mathcal{R} > f\mathcal{U}$.

These results can be applied to the problem of convergence of series and integrals (cf. [25]).

REFERENCES

[1] Auslander, J. (1970). "On stability of closed sets in dynamical systems", Sem. Diff. Eqs. Dynam. Systs., II, Univ. of Maryland, 1969, Lect. Notes Math., No. 144; Springer, 1-4.

[2] Auslander, J. (1973). "Non-compact dynamical systems", Recent Advances in Topol. Dynamics, Proc. Conf. Yale Univ., 1972, Lect. Notes Math., No. 318; Springer, 6-11.

[3] Auslander, J. (1977). "Filter stability in dynamical systems", *SIAM J. Math. Anal. 8*, 573-579.

[4] Bhatia, N. P., and Szegö, G. P. (1970). "Stability Theory of Dynamical Systems", Springer.

[5] Bushaw, D. (1967). "A stability criterion for general systems", *Math. Systems Theory 1*, 79-88.

[6] Bushaw, D. (1969). "Stabilities of Liapunov and Poisson types", *SIAM Rev. 11*, 214-225.

[7] Corduneanu, C. (1960). "Applications of differential inequalities to stabilty theory [Russian]", Anal. Stiintif. Univ. "A. I. Cuza", *Iasi 6*, 47-58.

[8] Dana, M. (1972). "Conditions for Liapunov stability", *J. Diff. Eqs. 12*, 596-609.

[9] Dankert, G., and Seibert, P. (1977). "Asymptoticity of general systems and Liapunov families", Techn. Rpt. DS 77-1, Univ. S. Bolivar (Publ. Nr. 21), Caracas, Venez.

[10] Habets, P., and Peiffer, K. (1973). "Classification of stability-like concepts and their study using vector Liapunov functions", *J. Math. Anal. Appl. 43*, 573-570.

[11] Hájek, O. (1972). "Ordinary and asymptotic stability of noncompact sets", *J. Diff. Eqs. 11*, 49-65.

[12] Kloeden, P. E. (1975). "Eventual stability in general control systems", *J. Diff. Eqs. 19*, 106-124.

[13] Krasovskiǐ, N. N. (1963). "Stability of motion", Stanford Univ. Press, [Russian original, Moscow, 1959].

[14] Lakshmikantham, V. (1965). "Vector Liapunov functions and conditional stability", *J. Math. Anal. Appl. 10*, 368-377.

[15] Lakshmikantham, V., and Leela, S. (1969). "Differential and Integral Inequalities", Academic Press.

[16] LaSalle, J. P., and Rath, R. J. (1964). "Eventual stability", Proc. 2nd IFAC Congr., Basle, 1963; Vol. 2; Butterworth, London, 556-560.

[17] Liapounoff, M. A. (1949). "Problème général de la stabilité du mouvement", Princeton Univ. Press, [Russian original: Kharcov, 1892].

[18] Matrosov, V. M. (1962). "On the theory of stability of motion [Russian]", *Prikl. Mat. Meh. 26*, 992-1002.

[19] Matrosov, V. M. (1968). "The comparison principle with vector-valued Liapunov function, I [Russian]", *Diff. Uravn. 4*, 1374-1386.

[20] Pelczar, A. (1971). "Stability of sets in pseudo-dynamical systems, I-IV", *Bull. Acad. Polon. Sci., Ser. Sci. Mat. Astr. Phys., 19*, 13-17; 951-957; *20* (1972), 673-677; *21* (1973), 911-916.

[21] Pelczar, A. (1973). "Stability questions in general processes and pseudo-dynamical systems", *Bull. Acad. Polon. Sci., Ser. Sci. Mat. Astr. Phys., 21*, 541-549.

[22] Pelczar, A. (1975). "Remarks on stability in local pseudo-dynamical systems", *Bull. Acad. Polon. Sci., Ser. Sci. Mat. Astr. Phys., 23*, 985-992.

[23] Pelczar, A. (1976). "Stability of motions in pseudo-dynamical systems", *Bull. Acad. Polon. Sci., Ser. Sci. Mat. Astr. Phys., 25,* 409-418.

[24] Pelczar, A. (1976). "Semistability in pseudo-dynamical systems", ibid., 419-428.

[25] Rogers, E. H. (1975). "Liapunov criteria for uniformity", *Math. Systems Theory 9,* 232-240.

[26] Roxin, E. O. (1965). "Stability in general control systems", *J. Diff. Eqs. 1,* 115-150.

[27] Salvadori, L. (1969). "Sulla stabilità del movimento", Le Matematiche (Sem. Mat. Univ. Catania) *24,* 218-239.

[28] Salvadori, L. (1971). "Ramiglie ad un parametro di funzioni di Liapunov nello studio della stabilità", *Sympos. Math.* (Ist. Naz. Alta. Mat.) *6,* 309-330.

[29] Salzberg, P. M. (1976). "On the existence of continuous and semicontinuous Liapunov functions", *Funkcial, Ekvac. 19,* 19-26.

[30] Salzberg, P. M. (1976). "Existence and continuity of Liapunov functions in general systems", Int. Symp., Providence, 1974, Vol. 2; Academic Press, 211-216.

[31] Salzberg, P. M., and Seibert, P. (1973). "A necessary and sufficient condition for the existence of a Liapunov function", *Funkcial. Ekvac. 16,* 97-101.

[32] Salzberg, P. M., and Seibert, P. (1975). "Remarks on a universal criterion for Liapunov stability", *Funkcial. Ekvac. 18,* 1-4.

[33] Salzberg, P. M., and Seibert, P. (1976). "A unified theory of attraction in general systems", Techn. Rpt. DS 76-1, Univ. S. Bolivar, Dpto. Mat., Publ. No. 11.

[34] Seibert, P. (1972). "A unified theory of Liapunov stability", *Funkcial. Ekvac. 15,* 139-147.

[35] Seibert, P. (1974). "Liapunov functions and the comparison principle", Dynamical Systems, Int. Symp., Providence, Vol. 2; Academic Press, 181-185.

[36] Yorke, J. A. (1967). "Invariance for ordinary differential equations", *Math. Systems Theory 1*, 353-372.

[37] Yoshizawa, T. (1966). "Eventual properties and quasi-asymptotic stability of a non-compact set", *Funkcial. Ekvac. 8*, 79-90.

[38] Yoshizawa, T. (1966). "Stability Theory by Liapunov's Second Method", *Publ. Math. Soc. Japan.*

[39] Zubov, V. I. (1964). "Methods of A. M. Liapunov and Their Application", Noordhoff, Groningen, [Russian original: Leningrad, 1957].

Added in proof, concerning footnote[2]: More useful criteria can be found in: Yorke, J. A. (1968). "Extending Liapunov's second method to non-Lipschitz Liapunov functions", Seminar on Differential Equations and Dynamical Systems, Lecture Notes Math. No. 60, Springer, 31-36.

ON CERTAIN SOLUTIONS OF AN INTEGRODIFFERENTIAL EQUATION

George Seifert

Department of Mathematics
Iowa State University
Ames, Iowa

Consider the equation

$$x'(t) = Ax(t) + \int_{-\infty}^{t} B(t-s)g(x(s))ds; \tag{1}$$

here $x(t)$ is an R^n-valued function of the time t, $A = (a_{ij})$ and $B(t) = (b_{ij}(t))$ are real $n \times n$ matrices, and $g(x)$: $R^n \to R^n$. We assume $B(t)$ continuous and integrable on $[0,\infty)$, g continuous and such that $g_i(x) = 0$ if $x_i = 0$, and $g_i(x) > 0$ if $x_i > 0$; here $g = (g_1,\ldots,g_n)$, $x = (x_1,\ldots,x_n)$.

For fixed $C > 0$, define $M_C = \{x \in R^n\colon\ x_i \geq 0,\ i = 1,\ldots,n,\ \sum_i x_i \leq C\}$. We give necessary and sufficient conditions that each solution $x(t)$ of (1) such that $x(t) \in M_C$ for $t \leq t_0$ satisfies $x(t) \in M_C$ for $t \geq t_0$ as long as it is defined. Since t_0 is arbitrary, this is clearly equivalent to having $x(t) \in M_C$ for t in some interval $(t_0,t_0 + b)$ whenever $x(t) \in M_C$ for $t \leq t_0$. By a solution $x(t)$ of (1) we mean a function on some interval $(-\infty,t_0+b)$, continuous and bounded there, whose derivative exists on $[t_0,t_0+b)$ and satisfies (1) there. The function $\phi(t) = x(t+t_0)$, $t \leq 0$, is called the initial value or past

ISBN 0-12-434180-2

history of the solution at t_0. A problem of this sort could arise if (1) models the population growth of a system of interacting species where x_i denotes the population of the ith specie.

In [1] we obtained some results for so-called positive, or flow, invariance of closed subsets of a state space for a more general class of delay-differential equations, where the state space X is a real Hilbert space. We denote by CB the space of continuous and bounded functions on $(-\infty, 0]$ to X. For a function $x(t)$ on $(-\infty, b)$ to X, we denote by x_t the function $x(t+s)$, $s \leq 0$. The norm $|\phi| = \sup\{|\phi(s)|: s \leq 0\}$ for $\phi \in CB$ makes CB a Banach space. Let $f(t,\phi)$ be a function on $R \times CB$ to X; we say that the function $x(t)$: $(-\infty, t_0+b) \to X$ is a solution of

$$x'(t) = f(t, x_t), \quad t_0 \leq t < t_0 + b, \tag{2}$$

$$x_{t_0} = \phi \in CB$$

if it satisfies (2).

Definition 1. Let $M \subset X$; we say M is positively invariant for (2) if for each $(t_0, \phi) \in CB$ with $\phi(s) \in M$ for $s \leq 0$, each solution $x(t)$ of (2) satisfies $x(t) \in M$ on some interval $[t_0, t_0+\delta)$, $0 < \delta \leq b$.

In what follows, we always assume M is closed. Clearly if M is open, it is positively invariant for (2) if and only if (2) has a solution.

For X infinite dimensional, we also assume that M is a distance set; i.e., for each $x \notin M$, there exists a point $y(x)$ on ∂M, the boundary of M such that $\operatorname{dist}(x,M) = |x-y(x)|$; and that each sequence $x_k \to x \in \partial M$, as $k \to \infty$, $x_k \notin M$, has a subsequence $\{x_{kj}\}$ such that

$$|x_{kj} - y(x_{kj})|^{-1}(x_{hj} - y(x_{hj})) \to u \quad \text{as} \quad j \to \infty;$$

here $y(x)$ is some nearest point to $x \notin M$. We denote the set of all such limits u by $N(x,M)$.

Clearly, if $X = R^n$, any closed subset of X is a distance set and $N(x,M)$ is nonempty for each $x \in \partial M$. However, simple examples show that for infinite dimensional X, $N(x,M)$ may be empty.

In [1], conditions on f are derived under which a necessary and sufficient condition that a convex closed set M be invariant for (2) is that for each $(t,\phi) \in R \times CB$ with $\phi(s) \in M$ for $s \leq 0$, $\phi(0) \in \partial M$, and $u \in N(\phi(0),M)$ we have

$$\langle f(t,\phi),u\rangle \leq 0; \tag{3}$$

here $\langle x,y\rangle$ denotes the inner product of x and y in X.

To simplify our treatment of the special case (1) of (2), we use a slight modification of this result, the following Theorem 1. First we need a definition.

Definition 2. Let $M_0 \subset M$, M_0 closed. M is said to be positively invariant for (2) relative to M_0 if for any solution $x(t)$ of (2) with $\phi(s) \in M_0$ for $s \leq 0$, we have $x(t) \in M$ for t in some interval $[t_0,t_0+\delta)$, $0 < \delta \leq b$.

It is clearly no loss of generality to assume in Definition 2 that $\partial M_0 \cap \partial M$ is nonempty.

We also introduce the following hypothesis on f, cf. [1];

(P) There exists a function $h(x)$: $X \to X$, such that $\langle h(x),u\rangle < 0$ for each $x \in \partial M$ and $u \in N(x,M)$, and such that for each ε $0 \leq \varepsilon \leq 1$, and $(t_0,\phi) \in R \times CB$ with $\phi(s) \in M$ for $s \leq 0$ and $\phi(0) \in \partial M$

$$x'(t) = f(t,x_t) + \varepsilon h(x(t)), \quad t \geq t_0, \tag{2ε}$$

$$x_{t_0} = \phi$$

has a solution $x^{\varepsilon}(t)$ on some interval $[t_0,t_0+\delta]$ such that $x^{\varepsilon}(t) \to x(t)$ as $\varepsilon \to 0$ for all t on that interval.

Theorem 1. Under hypothesis (P) on f and M convex, a necessary and sufficient condition that M be positively invariant for M is that (3) hold. Also if (3) holds for $(t,\phi) \in R \times CB$ with $\phi(s) \in M_0 \subset M$ for $s \leq 0$, M_0 a closed subset of M, then M is positively invariant for (2) relative to M_0.

This result is not proved in [1], but follows using the methods of [1].

We now state and prove a theorem about (1).

Theorem 2. In addition to the above-mentioned conditions on $B(t)$ and $g(x)$, suppose each initial value problem for (1) with $(t_0,\phi) \in R \times CB$, where $\phi(s) \in M$ for $s \leq 0$, has a unique solution. Also suppose without loss of generality that $A_1 \leq A_j$ where $A_j = -\sum_i a_{ij}$, $j = 1,\ldots,n$. Then M_C is positively invariant for (1) if

(i) $a_{ij} \geq 0$, $i \neq j$, $b_{ij}(t) \geq 0$ for $t \geq 0$, and all i,j, and

(ii) $L_C\int_0^\infty(\sum_i B_j^2(s))^{1/2}ds \leq A_1C$,

where $B_j(t) = \sum_i b_{ij}(t)$, $j = 1,\ldots,n$, and

$$L_C = \max\{(\sum_j g_j^2(x))^{1/2}: \ x \in M_C\}.$$

Conversely, if M_C is positively invariant for (1), then (i) holds, and also

(iii) $g_j(Ce_j)\int_0^\infty B_j(s)ds \leq A_jC$, $j = 1,\ldots,n$; here $e_j = (0\ldots0,1,0,\ldots,0)$, 1 in the jth place.

Proof. Our proof for the sufficiency of (i) and (ii) uses the fairly obvious fact that if M_1 and M_2 are positively invariant, then so is $M_1 \cap M_2$. We take $M_1 = \{x \in R^n: \ x_i \geq 0,$ $i = 1,\ldots,u\}$ and $M_2 = \{x \in R^n: \ \sum_i x_i \leq C\}$; then $M_C = M_1 \cap M_2$. We observe that for $x \in \partial M_2$, $N(x,M_2)$ consists of the single vector $n^{-1/2}(1,\ldots,1)$, while if $x \in \partial M_1$, $N(x,M_1)$ consists of unit vectors $u = (u_1,\ldots,u_n)$ with $u_i \le 0$ for $i = 1,\ldots,n$. Also (1) is of the form of (2) with

$$f(t,\phi) = A\phi(0) + \int_0^\infty B(s)g(\phi(-s))ds. \tag{4}$$

The assumption of the existence and uniqueness of solutions of the initial value problem for (1), and the above observations on the nature of $N(x,M_1)$ and $N(x,M_2)$ imply that (P) holds for (1); choose $h \equiv 1$ for M_1 and $h \equiv -1$ for M_2.

To show that (i) implies the positive invariance of M_1 is easy; we omit the details. We show that (ii) implies the positive invariance of M_2 relative to $M_1 \cap M_2 = M_C$ by using Theorem 1. Let $\phi \in CM$, $\phi(s) \in M_C$ for $s \le 0$, and $\phi(0) \in \partial M_2$. Recalling that $N(\phi(0), M_2) = \{n^{-1/2}(1,\ldots,1)\}$ we have, with f as in (4), and $u \in N(\phi(0), M_2)$,

$$\begin{aligned}\langle n^{1/2} f(t,\phi),u\rangle &= \sum_i \sum_j a_{ij}\phi_j(0) + \int_0^\infty \sum_i \sum_j b_{ij}(s)g_j(\phi(-s))ds \\ &= - \sum_j A_j\phi_j(0) + \int_0^\infty \sum_j B_j(s)g_j(\phi(-s))ds \\ &\le -A_1C + \int_0^\infty (\sum_j B^2{}_j(s))^{1/2}(\sum_j g^2{}_j(\phi(-s)))^{1/2}ds \\ &\le - A_1C + L_C\int_0^\infty (\sum_j B_j{}^2(s))^{1/2}ds.\end{aligned}$$

Using (ii) we obtain (3) and conclude from Theorem 1 that M_2 is positively invariant for (1) relative to M_C.

Now since M_1 is positively invariant for (1) and M_2 is positively invariant for (1) relative to $M_1 \cap M_2 = M_C$, we see easily that M_C is positively invariant for (1), for if the history of any solution $x(t)$ before t_0 is in M_C and it leaves M_C at $t_1 > t_0$, $t_1 - t_0$ arbitrarily small, then it leaves either M_1 or M_2 at t_1, which is impossible. Hence the sufficiency of (i) and (ii) is proved.

Now suppose M_C positively invariant for (1) and that (i) fails to hold. Then either $a_{i_1 j_1} < 0$ for some indices $i_1 \neq j_1$, or $b_{i_2 j_2}(t_0) < 0$ for some indices i_2, j_2 and $t_0 \geq 0$. Without loss of generality we assume $t_0 > 0$. In the first case we choose $\phi \in CB$ such that $\phi_{j_1}(0) = C/2$, $\phi_i(s) = 0$, $s \leq 0$ for $i \neq j_1$, and $\phi_{j_1}(s) = 0$ for $s \leq -\delta$ where $\delta > 0$ is so small that

$$|\int_{-\infty}^{0} b_{i_1 j_1}(-s) g_{j_1}(\phi(s))ds| = |\int_{-\delta}^{0} b_{i_1 j_1}(-s) g_j(\phi(s))ds| \tag{5}$$

$$< - a_{i_1 j_1} C/2.$$

Since $\phi(0) \in \partial M_C$ and $i_1 \neq j_1$, it follows that $-e_{i_1} \in N(\phi(0), M_C)$. But for such a ϕ and $t = 0$, the left side of (3) eventually reduces to

$$-a_{i_1 j_1} C/2 - \int_{-\infty}^{0} b_{i_1 j_1}(-s) g_{j_1}(\phi(s))ds,$$

which, by (5), is positive contradicting (3). So $a_{ij} \geq 0$ for $i \neq j$.

In the second case, there exists $\delta > 0$, $\delta < t_0$, so small that $b_{i_2 j_2}(t) < 0$ for $t \in [t_0-\delta, t_0+\delta]$. Choose $\phi \in CB$ such that $\phi(0) = 0$, $\phi_i(s) = 0$ for $s \leq 0$ $i \neq j_2$, $\phi_{j_2}(s) > 0$ for $s \in (-t_0-\delta, -t_0+\delta)$ and $\phi_{j_2}(s) = 0$ for $s \notin (-t_0-\delta, -t_0+\delta)$. Also choose $u = -e_{i_2} \in N(0, M_C)$. Then if $t = 0$ the left side of (3) reduces to

$$-\int_{-t_0-\delta}^{-t_0+\delta} b_{i_2 j_2}(-s) g_{j_2}(\phi(s))ds$$

which is positive, again contradicting (3).

It remains to show that (iii) holds. Suppose it does not for some $j = j_1$. Choose $\phi \in CB$ such that $\phi_{j_1}(s) = C$ for $s \leq 0$, and $\phi_i(s) = 0$ for $s \leq 0$ and $i \neq j_1$. Then $\phi(0) \in \partial M_C$, and since $\sum_i \phi_i(0) = C$, we may choose $u = n^{-1/2}(1,\ldots,1) \in N(\phi(0), M_C)$. Then if $t = 0$, the left side of (3) reduces to

$$-A_{j_1}C + g_{j_1}(Ce_{j_1})\int_0^\infty B_{j_1}(s)ds$$

which is positive, and again contradicts (3). This completes the proof.

Some concluding remarks are in order. First we note that the conditions on A and $B(t)$ in Theorem 2 imply that the diagonal elements a_{ii} of A must all be negative; from both (ii) and (iii) it follows that $A_1 > 0$ unless $A = 0$ and $B(t) = 0$ for all t, a case we obviously exclude, and since $A_1 \leq A_j$, $j \neq 1$, and $a_{ij} \geq 0$, $i \neq j$, this assertion follows. Loosely speaking, the diagonal elements of A must be sufficiently negative to assure our solutions do not escape M_C on its planar boundary $\sum_i x_i = C$, while the off diagonal elements of A and all of $B(t)$, being nonnegative keep the solutions from leaving M_C via the coordinate planes.

Also, more complicated conditions than (ii) and (iii) are clearly possible that would strengthen the theorem; our aim however is not in the direction of generality.

Finally, instead of inner product conditions, socalled subtangential conditions can be used for the positive invariance of closed sets in spaces more general than Hilbert spaces; i.e., in Banach spaces. For ordinary differential equations in Banach spaces, much work has been done on the positive invariance of closed sets; cf. [2]-[5]; for delay-differential equations, cf. [6] and [7].

REFERENCES

[1] Seifert, G. "Positive invariance of closed sets for generalized Volterra equations", (unpublished).

[2] Martin, R. H. (1973). "Approximation and existence of solutions to ordinary differential equations in Banach spaces", *Funk. Ekv. 16*, 195-211.

[3] Martin, R. H. (1975). "Invariant sets for perturbed semigroups of linear operators", *Ann. Mat. pura appl. 105*, 221-239.

[4] Volkmann, P. (1976). "Über die positive Invarianz einer abgeschlossenen Teilmenge eines Banachschen Raumes bezüglich der Differentialgleichung $u' = f(t,u)$", *J. für reine u. angew. Math. 282*, 59-65.

[5] Redheffer, R., and Walter, W. (1975). "Flow invariant sets and differential inequalities in normal spaces", *Applicable Analysis 5*, 149-161.

[6] Seifert, G. (1976). "Positively invariant closed sets for systems of delay differential equations", *J. Diff. Eqs. 22 (2)*, 292-304.

[7] Chang, Y. F. (1978). "Flow invariance for delay differential equations", Ph.D. Thesis, (Iowa State University, Ames, Iowa 50011).

A GREEN'S FORMULA FOR WEAK SOLUTIONS OF VARIATIONAL PROBLEMS

R. E. Showalter

Department of Mathematics
University of Texas at Austin
Austin, Texas

I. INTRODUCTION

Weak solutions of certain operator equations can be characterized by partial differential equations together with "natural" or "variational" boundary conditions. Let's recall how this arises in a classical treatment of the Neumann problem

$$\lambda u(x) - \Delta u(x) = F(x), \quad x \in G, \tag{1.1.a}$$

$$\frac{\partial u(s)}{\partial \nu} = g(s), \quad s \in \partial G. \tag{1.1.b}$$

For any strong solution $u \in H^2(G)$ of (1.1) it follows from the Green's formula

$$\int_G (\nabla u \cdot \nabla v + \Delta u \cdot v)dx = \int_{\partial G} \frac{\partial u}{\partial \nu} v \, ds \tag{1.2}$$

that

$$\int_G (\lambda uv + \nabla u \cdot \nabla v)dx = \int_G Fvdx + \int_{\partial G} gvds, \quad v \in H^1(G). \tag{1.3}$$

By the elementary techniques of functional analysis one finds there is a unique solution $u \in H^1(G)$ of (1.3). This solution

ISBN 0-12-434180-2

clearly satisfies (1.1.a) in $\mathcal{D}'(G)$ and

$$\int_G (\nabla u \cdot \nabla v + \Delta u \cdot v)dx = \int_{\partial G} gvds, \quad v \in H^1(G), \qquad (1.1.\text{b}')$$

so it is a *weak solution* of (1.1). The classical observation is that Green's formula (1.2) is meaningful not only for $u \in H^2(G)$ but also for those $u \in H^1(G)$ with $\Delta u \in L^2(G)$. Of course regularity theory for the Laplacian shows that these conditions are equivalent. However, we shall indicate how this provides an elementary extension of Green's formula which is useful in problems without regularity. Moreover the extension has little to do with linearity and applies as well to (possibly multi-valued) non-linear operator equations and inequalities.

II. GREEN'S FORMULA

Let the (trace) operator γ be a strict homomorphism of the topological vector space V onto B. Denote the kernel of γ by V_0; then $\gamma^*(g) \equiv g \circ \gamma$ defines an isomorphism of the dual B' onto the annihilator $V_0^\perp \subset V'$. Let $m_0\colon\ V \times V \to \mathbb{R}$ be continuous bilinear and non-negative, a semi-scalar-product, and $|\cdot|$ a continuous seminorm on V. Denote by F the space V with the seminorm $|v|_F \equiv m_0(v,v)^{1/2} + |v|$, $v \in V$. Then $F' \subset V'$. Assume V_0 is dense in F; then we identify $F' \subset V_0'$ by restriction. In our applications, V_0' is a space of distributions on G, and B is a space of functions on ∂G.

Let $M_1\colon\ B \to B'$ and define $\mathcal{M}u(v) = m_0(u,v) + M_1(\gamma u)(\gamma v)$, $u,v \in V$. For each $u \in V$, let Mu be the restriction of $\mathcal{M}u \in V'$ to V_0; then $Mu \in F'$ and we have

$$\mathcal{M}u(v) = Mu(v) + M_1(\gamma u)(\gamma v), \quad u,v \in V. \qquad (2.1)$$

Thus, in our applications the functional $\mathcal{M}u \in V'$ is given as a spatial part $Mu \in V_0'$ and a boundary part $M_1(\gamma u)$. This is a Green's formula for the rather special $\mathcal{M}\colon\ V \to V'$.

Let $L: V \to V'$ be an arbitrary function. Define Lu to be the restriction of $Lu \in V'$ to V_0, hence, $Lu \in V_0'$, and set $D \equiv \{u \in V: Lu \in F'\}$. For each $u \in D$, we have $Lu - Lu \in V_0^\perp$ so there is a unique $\partial u \in B'$ for which $Lu - Lu = \gamma^*(\partial u)$. This determines the unique function $\partial: D \to B'$ for which

$$Lu(v) = Lu(v) + \partial u(\gamma v), \quad u \in D, \quad v \in V. \tag{2.2}$$

This is a corresponding Green's formula which realizes $Lu \in V'$ as a spatial part $Lu \in V_0'$ and a boundary part $\partial u \in B'$.

III. APPLICATIONS

We illustrate with the simplest examples how (2.1) and (2.2) can be used to characterize solutions of functional equations. With the notation of Section 2, suppose we are given $F \in F'$ and $g \in B'$. Define $f = F + \gamma^* g \in V'$ and consider the functional equation

$$u \in V: \quad \lambda Mu + Lu = f \quad \text{in} \quad V'. \tag{3.1}$$

If u is a solution of (3.1), then applying (3.1) to those $v \in V_0$ yields

$$u \in V: \quad \lambda Mu + Lu = F \quad \text{in} \quad V_0'. \tag{3.2.a}$$

Since λMu and F belong to F', it follows $u \in D$ and we may apply (2.1) and (2.2) to the difference of (3.1) and (3.2.a) to obtain

$$\lambda M_1(\gamma u) + \partial(u) = g \quad \text{in} \quad B'. \tag{3.2.b}$$

This little exercise is easily reversed, so we find that (3.1) is equivalent to (3.2).

Our first example illustrates a situation with non-regular data. Let $V = \{v \in H^1(0,2): v(2) = 0\}$, the indicated Sobolev space over the interval $(0,2)$ with functions vanishing at the right end. Then $\gamma v = v(0)$ is trace at left end and $V_0 = H_0^1(0,2)$, $B = B' = \mathbb{R}$. If $\sum_{n=1}^{\infty} a_n$ is absolutely convergent, then

$$|v|_F \equiv \sum_{n=1}^{\infty} |a_n v(\tfrac{1}{n})|, \quad v \in V,$$

is a continuous seminorm on V; an exercise with truncation and regularization shows V is dense in $F = \{V, |\cdot|_F\}$. (We are setting $M \equiv 0$ so $|\cdot|_F = |\cdot|$.) Define $L\colon V \to V'$ by

$$Lu(v) \equiv \int_0^2 u'(x)v'(x)dx, \quad u,v \in V;$$

L is a bijection since it is V-coercive. The distribution $f \equiv \sum_{n=1}^{\infty} a_n \delta_{\frac{1}{n}}$ belongs to $F' \subset V'$; let's characterize the unique solution of

$$u \in V\colon \quad Lu = f \quad \text{in} \quad V'. \tag{3.3}$$

The formal part (3.2.a) shows

$$-u'' = \sum_{n=1}^{\infty} a_n \delta_{\frac{1}{n}}$$

so u'' is not integrable in any neighborhood of the origin; nevertheless, $u'(0^+)$ is defined and

$$u'(x) = u'(0^+) - \sum_{n=1}^{\infty} a_n H(x - \tfrac{1}{n}), \quad \text{a.e. } x \in (0,2).$$

($H(\cdot)$ is the Heaviside functional.) Since $u'' = 0$ on each interval $(1/n+1, 1/n)$, we compute $Lu(v) - Lu(v) = -u'(0)v(0)$, hence, $\partial u = -u'(0)$. Thus the boundary function ∂ takes the proper value even in this irregular situation.

A similar but somewhat more striking example for the degenerate elliptic equation

$$-(uu')' = f$$

occurs, wherein the boundary value of $-(uu')(0) = \partial u$ is meaningful but $u'(x)$ oscillates between $+1$ and -1 as $x \to 0^+$. One can, of course, contrive more exotic examples in $\mathbb{R}^n$ of the type that arise from the consideration of operators on weighted Sobolev spaces. These necessarily arise in certain degenerate or singular problems

IV. PSEUDOPARABOLIC EQUATION

Let $V = H^1(G)$, $V_0 = H_0^1(G)$ and $\gamma\colon V \to H^{1/2}(\partial G)$ be the usual trace operator. Define for $u, v \in V$

$$Mu(v) = Mu(v) = (u,v)_{L^2(G)} \qquad (M_1 \equiv 0)$$

$$Lu(v) = \int_G \nabla u \cdot \nabla v \, dx$$

so we have in the notation of Section 2

$$Lu = -\Delta u, \quad \partial u \cong \frac{\partial u}{\partial \nu}.$$

For each $\varepsilon > 0$, $L_\varepsilon \equiv (M+\varepsilon L)^{-1}L$ is a bounded linear operator on V so for each $u_0 \in V$, $u(t) \equiv \exp(-tL_\varepsilon)u_0$ is the unique solution in $C^1(\mathbb{R}, V)$ of

$$(M+\varepsilon L)u'(t) + Lu(t) = 0, \quad t \in \mathbb{R}, \quad u(0) = u_0.$$

Since $L(\varepsilon u'(t) + u(t)) = -Mu'(t) \in F'$ this solution is characterized by

$$\begin{cases} \dfrac{\partial u}{\partial t} - \varepsilon\Delta\dfrac{\partial u}{\partial t} - \Delta u = 0 & \text{in } \mathcal{D}'(G), \quad t \in \mathbb{R}, \\ \partial(\varepsilon\dfrac{\partial u}{\partial t} + u) = 0 & \text{in } B' = H^{-1/2}(\partial G), \\ u(x,0) = u_0(x) & \text{in } V. \end{cases}$$

Note that $\varepsilon u'(t) + u(t) \in D$ for all t even though $u(t) \in H^2(G)$ if and only if $u_0 \in H^2(G)$. The above holds as well for nonlinear operators and others for which regularity theory is inadequate.

V. NONLINEAR PROBLEMS

In keeping to the theme of the conference we shall explore some extensions to other types of nonlinear situations. Suppose the spaces V, V_0, B, the operators γ, M, L, and data F, g are given in Sections 2 and 3. Let C be a non-empty closed convex subset of B and define $K = \{v \in V:\ \gamma v \in C\}$. Let's characterize solutions of the variational equality

$$u \in K: \quad (\lambda M+L)u(v-u) \leq f(v-u), \quad v \in K. \tag{5.1}$$

Since K is invariant under addition of elements of V_0, we find that u is a solution of (5.1) if and only if

$$u \in V, \quad (\lambda M+L)u = F \text{ in } F', \quad \gamma u \in C, \tag{5.2.a}$$

$$(\lambda M_1+\partial)u(\psi-\gamma u) \leq g(\psi-\gamma u), \quad \psi \in C. \tag{5.2.b}$$

That is, (5.1) is equivalent to an equation in F' and a variational inequality in the boundary space B'.

The only necessarily linear operator in the preceding development has been the part M in F' of M; this occurred since M is obtained from a semi-scalar-product, m_0. However, we can let m_0 be a continuous semi-norm on V. If $\phi_1\colon B \to \mathbb{R} \cup \{\infty\}$ is proper, convex, and lower-semi-continuous, then so also is $\phi(v) \equiv m_0(v) + \phi_1(\gamma v)$, $v \in V$. The variational inequality

$$u \in \mathrm{dom}(\phi): \quad \langle f-Lu,\ v-u\rangle \leq \phi(v) - \phi(u), \quad v \in V \tag{5.3}$$

can be written in the multi-valued operator form

$$f - Lu \in \partial\phi(u) \quad \text{in } V', \tag{5.4}$$

the right side denoting the *subdifferential* of ϕ at u. Our Green's formulae can then be used to show that (5.4) is equivalent to the pair

$$F - Lu \in \partial m_0(u) \quad \text{in} \quad F', \tag{5.4.a}$$

$$g - \partial u \in \partial \phi_1(\gamma u) \quad \text{in} \quad B'. \tag{5.4.b}$$

Note that each of the "inclusions" is equivalent to a corresponding variational inequality and each involves a pair of nonlinear operators.

APPLICATION OF FIXED POINT THEOREMS IN APPROXIMATION THEORY

S. P. Singh

Department of Mathematics, Statistics
and Computer Science
Memorial University of Newfoundland
St. John's, Newfoundland, Canada

Recently, a few results in Approximation Theory using fixed point theorems have been given (see Cheney [2]). The aim of this paper is to extend and unify results due to Meinardus [5] and Brosowski [1]. Meinardus and Brosowski have taken mappings of Lipschitz's class, whereas in the present paper, mappings need not be continuous. In the end, a result for proximity maps in non-compact setting is proved.

The following definition will be needed.

Let X be a normed linear space and C a subset of X. C is said to be starshaped if there is at least one $p \in C$ such that, if $x \in C$ and $0 < \lambda < 1$, then $(1 - \lambda)p + \lambda x \in C$.

The main theorem is as follows:

<u>*Theorem 1*</u>. Let X be a normed linear space and $T: X \to X$ be a mapping. Let C be a subset of X such that C is a T-invariant and let $\bar{x}$ be a T-invariant point in X. If D, the set of best C-approximants to $\bar{x}$ is non-empty, compact and starshaped and T is:

(i) continuous on D;

ISBN 0-12-434180-2

(ii) $\|x - y\| \leq d(\bar{x}, C) \Rightarrow \|Tx - Ty\| \leq \|x - y\|$ for x, y in $D \cup \{\bar{x}\}$; where $d(\bar{x}, C)$ denotes the distance of $\bar{x}$ from C; then it contains a T-invariant point, which is a best approximation to $\bar{x}$ in C.

Proof. First, we show that $T: D \to D$. Let $y \in D$. Then

$$\|Ty - \bar{x}\| = \|Ty - T\bar{x}\| \leq \|y - \bar{x}\|$$

implying that $Ty \in D$.

Let p be the star centre of D. Then $\lambda p + (1 - \lambda)x \in D$ for all $x \in D$. Let $\{k_n\}$ be a sequence with $0 \leq k_n < 1$, such that $k_n \to 1$ as $n \to \infty$. Define $T_n: D \to D$ as follows:

$$T_n x = k_n Tx + (1 - k_n)p \quad \text{for all } x \in D.$$

Then

$$\|T_n x - T_n y\| = k_n \|Tx - Ty\|$$

$$\leq k_n \|x - y\|; \quad \text{for } \|x - y\| \leq d(\bar{x}, C).$$

Since D is compact and starshaped, and T_n is $(d(\bar{x}, C), k_n)$ uniform local contraction for each $n = 1, 2, \ldots$; each T_n has a unique fixed point, say x_n (see [4], [7]). Since D is compact, $\{x_n\}$ has a subsequence $x_{n_i} \to z$ say. We claim that $Tz = z$. Now,

$$x_{n_i} = T_{n_i} x_{n_i} = k_{n_i} Tx_{n_i} + (1 - k_{n_i})p.$$

Since T is continuous on D, taking limit as $i \to \infty$, we get

$$z = Tz.$$

We derive the following known results as corollaries.

Corollary 1. Let T be a non-expansive mapping on a normed linear space X. Let C be a T-invariant subset of X and $\bar{x}$ a T-invariant point. If D, the set of best C approximants to

$\bar{x}$ is non-empty, compact and convex, then it contains a T-invariant point [1].

A non-expansive mapping T satisfies the condition $\|x - y\| \leq \varepsilon$ then $\|Tx - Ty\| \leq \|x - y\|$, and is necessarily continuous on X and, hence, on D. A convex set is starshaped.

We give the following simple example to illustrate the fact that T need not be continuous on X.

Let $X = R$ and $C = [0, 1/2]$. Define $T\colon R \to R$ as follows:

$$\begin{aligned} Tx &= x - 1, \quad x < 0 \\ &= x, \quad 0 \leq x \leq \frac{1}{2} \\ &= \frac{x+1}{2}, \quad x > \frac{1}{2}. \end{aligned}$$

Clearly, $T(C) = C$ and $T(1) = 1 = \bar{x}$ say.

$$D = d(\bar{x}, C) = \{\tfrac{1}{2}\}.$$

T has a fixed point in D, a best approximation to $\bar{x}$ in C. However, T is not continuous on R.

Corollary 2. Let T be a non-expansive mapping on a normed linear space X. Let C be a T-invariant subset of X and $\bar{x}$ a T-invariant point in X. If D, the set of best C-approximants to $\bar{x}$ is non-empty, compact and starshaped, then it contains a T-invariant point [6].

Corollary 3. Let X be a normed linear space and $T\colon X \to X$ be a non-expansive mapping. Let T have a fixed point, say $\bar{x}$, and leaving a finite dimensional subspace C of X invariant. Then T has a fixed point which is a best approximation to $\bar{x}$ in C.

Clearly, D, the set of best C-approximants to $\bar{x}$, is non-empty. Also, D is closed, bounded and convex. Since C is finite dimensional, D is compact and the result follows from Corollary 2.

The following well-known result of Meinardus [5] follows from Corollary 3:

Let $T: B \to B$ be continuous where B is a compact metric space. If $C[B]$ is the space of all continuous real or complex functions on B with the sup. norm. Let $A: C[B] \to C[B]$ be of Lipschitz class with Lipschitz constant 1. Suppose further that $Af(T(x)) = f(x)$, $Ah(T(x)) \in V$, whenever $h(x) \in V$, where V is a finite dimensional subspace of $C[B]$.

Then there is a best approximation g of f with respect to V such that

$$Ag(T(x)) = g(x).$$

It is evident that the mapping

$$F: C[B] \to C[B],$$

defined by $F(g(x)) = A(g(T(x)))$ satisfies conditions of Corollary 3.

Lemma 1. Let C be a complete, convex subset of an inner product space X. Then the proximity map P is non-expansive, i.e.

$$\|Px - Py\| \leq \|x - y\|,$$

equality holds only if $\|x - Px\| = \|y - Py\|$ [3].

Now we prove a result in an inner product space where compactness condition is relaxed. We get a well-known theorem of Cheney and Goldstein [3] as a corollary.

Theorem 2. If A and B are complete, convex subsets of an inner product space X, P_1 and P_2 are proximity maps on A and B, respectively. Let T be the composition of P_1 and P_2. If the sequence of iterates $T^{n+1}x_0 = Tx_n$ has a convergent subsequence then $T^n x_0$ converges to a fixed point of T.

Proof. Since a proximity map, by Lemma 1, is non-expansive, so is $T = P_1P_2$. If $y = Tx \neq x$, $x \in A$, then $\|y - P_2y\| \leq \|y - P_2x\| < |x - P_2x|$. Now

$$\begin{aligned} ||Tx - Ty|| &= ||P_1P_2x - P_1P_2y|| \\ &\leq ||P_2x - P_2y|| \\ &\leq ||x - y|| \qquad \text{(by Lemma 1).} \end{aligned}$$

Since T is non-expansive and

$$||Tx - TTx|| < ||x - Tx||, \quad x \neq Tx;$$

and $\{T^n x\}$ has a convergent subsequence, hence T has a fixed point [3]. The sequence $\{T^n x_0\}$ itself coverges to a fixed point of T.

In case one of the sets is compact, then each sequence has a convergent subsequence and, therefore, the theorem due to Cheney and Goldstein [3] given below follows as a corollary.

Let A and B be two closed, convex subsets of a Hilbert space X and $T = P_1P_2$, a composition of their proximity maps, then convergence of $\{T^n x\}$ to a fixed point of T is assured when either: (i) one set is compact or (ii) one set is finite dimensional and the distance is attained.

REFERENCES

[1] Brosowski, B. (1969). "Fix punktsatze in der approximations theorie", Mathematica (Cluj) *11*, 195-220.

[2] Cheney, E. W. (1976). "Applications of fixed point theorems to approximation theory", Theory of Approximation with Applications, Academic Press, 1-8.

[3] Cheney, E. W., and Goldstein, A. A. (1959). "Proximity maps for convex sets", *Proc. Amer. Math. Soc. 10*, 448-450.

[4] Edelstein, M. (1961). "An extension of Banach's contraction principle", *Proc. Amer. Math. Soc. 12*, 7-10.

[5] Meinardus, G. (1963). "Invarianz bei linearen appriximationen", *Arch. Rat. Mech. Anal. 14*, 301-303.

[6] Singh, S. P. "An application of fixed point theorem to approximation theory", to appear, *J. Approx. Theory*.

[7] Subrahmanyam, P. V. (1977). "An application of a fixed point theorem to best approximation", *J. Approx. Theory 20*, 165-172.

EQUIVALENCE OF CONJUGATE GRADIENT METHODS AND QUASI-NEWTON METHODS

R. A. Tapia

Department of Mathematical Sciences
Rice University
Houston, Texas

ABSTRACT

The well-known equivalence between conjugate gradient methods and quasi-Newton methods for quadratic problems will be discussed. This will be followed by a discussion on the not so well-known equivalence between the preconditioned conjugate gradient methods and quasi-Newton methods for quadratic problems. Some thoughts on how this latter equivalence can be used to handle large scale optimizations will be presented.

ISBN 0-12-434180-2

APPROXIMATE SOLUTION OF ELLIPTIC BOUNDARY VALUE PROBLEMS BY SYSTEMS OF ORDINARY DIFFERENTIAL EQUATIONS

Russell C. Thompson

Department of Mathematics
Utah State University
Logan, Utah

and

Department of Mathematical Sciences
Northern Illinois University
DeKalb, Illinois

INTRODUCTION

In this paper, we obtain some results on monotone approximations of solutions of an elliptic boundary value problem on an unbounded domain in R^2 by applying some recent results on maximal and minimal solutions of infinite systems of ordinary differential equations. As a model problem, we consider the following boundary value problem for a nonlinear perturbation of the Laplace equation:

$$[P]\begin{cases} -\Delta u + g(x,y,u,u_x,u_y) = 0, & (x,y) \in \Omega \equiv (0,1) \times (0,\infty) \\ u(0,y) = \phi(y), \quad u(1,y) = \psi(y), & y \in (0,\infty) \\ u(x,0) = f(x), & x \in [0,1] \end{cases}$$

ISBN 0-12-434180-2

An approximate solution of $[P]$ can be obtained from the solution of the following related, infinite system of second order ordinary differential equations:

$$[P(h)]\quad \begin{cases} -\Delta_h u_i + g(x,y_i,u_i,u_i',\delta u_i) = 0, \quad x \in (0,1), \quad i \in Z+, \\ h > 0 \\ u_i(0) = \phi(y_i), \quad u_i(1) = \psi(y_i) \\ u_0(x) = f(x), \quad x \in [0,1], \end{cases}$$

where Z^+ denotes the set of positive integers. In $[P(h)]$, $\Delta_h u_i = u_i'' + h^{-2}\{u_{i+1}-2u_i+u_{i-1}\}$, $\delta u_i = h^{-1}\{u_{i+1}-u_i\}$, and $y_i = ih$, $i \in Z^+$. The components of a solution to this system approximates the solution to $[P]$ along the lines $y = y_i$ and consequently this approximation procedure has been named the method of lines. A discussion of the convergence of this method for nonlinear elliptic problems and a list of additional references on the method appears in reference [5]. In addition, in reference [4] this type of approximation is used to investigate existence, uniqueness and approximation of a problem similar to $[P]$. In the present paper, we will focus our attention on results about the monotone approximation of solutions to the infinite system $[P(h)]$.

I. MONOTONE APPROXIMATION OF SOLUTIONS TO $[P(h)]$

Since $[P(h)]$ is a boundary value problem for an infinite system of second order ordinary differential equations, any practical application of this system will require a reduction to a finite dimensional system of equations. Let $\underset{\sim}{w}(x)$ denote a sequence of continuous functions $\underset{\sim}{w}(x) = (w_1(x),w_2(x),\ldots)$ and consider the finite dimensional boundary value problem

$$[P(h,w,n)]\begin{cases} -\Delta_h u_i + g(x,y_i,u_i,u_i',\delta u_i) = 0, & x \in (0,1), \\ & i = 1,\ldots,n; \\ u_i(0) = \phi(y_i), \quad u_i(1) = \psi(y_i) & i = 1,\ldots,n \\ u_0(x) = f(x), \quad u_i(x) = w_i(x), & i \geq n+1, \\ & x \in [0,1]. \end{cases}$$

This system is obtained from $[P(h)]$ by truncating after n equations and closing the resulting finite dimensional system with elements from the sequence $\underset{\sim}{w}(x)$. By a solution of $[P(h,\underset{\sim}{w},n)]$ we will mean a sequence $\underset{\sim}{u}(x)$ of functions whose first n components satisfy the equations and boundary conditions in $[P(h,\underset{\sim}{w},n)]$ and whose components with index from $n+1$ on, are elements of the sequence $\underset{\sim}{w}(x)$.

Our main result is the following theorem on the convergence of solutions of $[P(h,\underset{\sim}{w},n)]$ to a solution of $[P(h)]$ as $n \to \infty$. We denote by F_γ the set of functions

$$F_\gamma = \{h(x,y): \; |h(x,y)| \leq Ce^{\gamma y}, \quad \gamma \geqq 0, \quad C > 0, \quad (x,y) \in \Omega\}.$$

<u>Theorem 1</u>. *In problem* $[P]$ *let* $\phi,\psi \in C(0,\infty) \in F_{\gamma_0}$ *for some* $\gamma_0 > 0$ $g(x,y,0,0,0) \in F_{\gamma_0}$ *and* $g,g_u,g_p,g_q \in C(\Omega \times R^3)$ *where* $g = g(x,y,u,p,q)$. *Let there exist* $L_1 > 0$ *and* $L_2 > 0$ *such that* $|g_p(x,y,u,p,q)| \leqq L_1$ *and* $|g_q(x,y,0,0,q)| \leqq L_2$ *and let there exist a* $\gamma_1 > 0$ *such that* $g(x,y,u(x,y),p(x,y),q(x,y)) \in F_{\gamma_1}$ *for all* $(x,y,u(x,y),p(x,y),q(x,y) \in \Omega \times F_{\gamma_0}^3$. *If there exists a* $\gamma > 0$ *such that* $g_u(x,y,u,p,q) > \gamma > \gamma_0^2 + \gamma_1^2 + L_2(\gamma_0+\gamma_1)$ *then* $[P_h]$ *has a unique solution* $\underset{\sim}{u}(x)$ *for* $h > 0$ *sufficiently small. Moreover, if we denote by* $\underset{\sim}{\beta}(x)$, *the sequence defined by*

$$\beta_i(x) = Ae^{\gamma y_i},$$

and by $\underset{\sim}{\alpha}(x)$, *the sequence defined by* $\underset{\sim}{\alpha}(x) = -\underset{\sim}{\beta}(x)$ *and if we let* $\{\underset{\sim}{\beta}^n(x)\}$ *and* $\{\underset{\sim}{\alpha}^n(x)\}$ *denote solutions to* $[P(h,\underset{\sim}{\beta},n)]$ *and* $[P(h,\underset{\sim}{\alpha},n)]$ *respectively, then* $\{\underset{\sim}{\beta}^n(x)\}$ *and* $\{\underset{\sim}{\alpha}^n(x)\}$ *converge monotonically to* $\underset{\sim}{u}(x)$ *from above and below respectively.*

In the following section we will state a result on the existence of maximal and minimal solutions to infinite dimensional systems of boundary value problems which forms a major ingredient in the proof of Theorem 1.

II. EXTREMAL SOLUTIONS FOR INFINITE SYSTEMS

The existence of maximal and minimal solutions and the monotone approximation of solutions to boundary value problems for infinite dimensional systems has been an area of active research in the past couple of years. References [1], [2] and [3] give some of the recent results in this area. The result which we use in the proof of Theorem 1, is found in reference [6]. Before stating the result we establish some notation.

Let $\underset{\sim}{p}$ be a positive sequence ($p_i > 0$ for all $i \in Z^+$) and let $E_{\underset{\sim}{p}}$ and $E^0_{\underset{\sim}{p}}$ denote the Banach spaces of sequences:

$$E_{\underset{\sim}{p}} = \{\underset{\sim}{w}: \ |\underset{\sim}{w}|_p < \infty\}; \quad |\underset{\sim}{w}|_p = \sup_{i\ Z^+}\{p_i|w_i|\},$$

$$E^0_{\underset{\sim}{p}} = \{w \in E_{\underset{\sim}{p}}: \ \lim_{i\to\infty} p_i|w_i| = 0\}.$$

A function $\underset{\sim}{h}: [0,1] \times E_{\underset{\sim}{p}} \times E_{\underset{\sim}{p}} \to E_{\underset{\sim}{p}}$ is said to $\Gamma_{\underset{\sim}{p}}$ continuous if it is weakly continuous in the following sense:

$$\left\{\begin{array}{l} \text{for every } i \in Z^+ \text{ the limits, } x \to x^0, \ y_j \to y_j^0 \\ z_j \to z_j^0 \text{ for each } j \in Z^+, \text{ imply the limit} \\ h_i(x,\underset{\sim}{y},\underset{\sim}{z}) \to h_i(x^0,\underset{\sim}{y}^0,\underset{\sim}{z}^0) \end{array}\right.$$

The space of $\Gamma_{\underset{\sim}{p}}$-continuous functions from $[0,1]$ into $E_{\underset{\sim}{p}}$ will be denoted by $C\Gamma_p([0,1],E_p)$ and higher order spaces denoted similarly by $C\Gamma_p^j([0,1],E_p)$. The space of strongly continuous function between these spaces will be denoted by $C([0,1],E_p)$.

We will say that $F\colon [0,1] \times E_p \times E_p \to E_p$ satisfies condition (H) on a subset $\Sigma \leq [0,1] \times E_p \times E_p$ if

$$(x,u,z),(x,\overline{u},\overline{z}) \in \Sigma,$$

$$p_i(u_i-\overline{u}_i) = \inf\{p_j(u_j-\overline{u}_j);\ j \in Z^+\} \leq 0, \quad z_i = \overline{z}_i \tag{1}$$

implies

$$F_i(x,u,z) \leq F_i(x,\overline{u},\overline{z}). \tag{2}$$

Further, we say that F_i satisfies (H') if strict inequality in (1) implies strict inequality in (2).

A function $f\colon [0,1] \times E_p \times E_p$ is said to satisfy a Nagumo condition with respect to z with Nagumo function $\Psi(s)$ if there exists a continuous function $\Psi(s)$ which is positive, nondecreasing and satisfies $\lim_{s\to\infty} \frac{s^2}{\Psi(s)} = \infty$ and $|f(x,u,z)|_p \leq \Psi(|z|_p)$.

Weak first and second derivatives, Du and D^2u, of a function $u\colon [0,1] \to E_p$ are defined by the limits

$$\lim_{h\to 0}\{h^{-1}[u(x+h)-u(x)] - Du(x)\} = 0$$

and

$$\lim_{h\to 0}\{h^{-2}[u(x+h)-2u(x)+u(x-h)] - D^2u(x)\} = 0.$$

If $w \in E_p$, $u \in R^n$, then we denote by $[w;u] \in E_p$ the sequence defined by

$$[w;u] = \begin{cases} u_i; & i = 1,\ldots,n \\ w_i; & i = n+1,n+2,\ldots \end{cases}$$

The element $\underset{\sim}{e}^i \in E_{\underset{\sim}{p}}$ is defined by the equation $e^i_j = \delta_{ij}$.

Consider the boundary value problem in $E_{\underset{\sim}{p}}$

$$\left.\begin{aligned} -D^2\underset{\sim}{u} = \underset{\sim}{f}(x,\underset{\sim}{u},D\underset{\sim}{u}) &= \underset{\sim}{0}, \quad x \in (0,1) \\ \underset{\sim}{G}(\underset{\sim}{u}(0),D\underset{\sim}{u}(0)) &= \underset{\sim}{0} = \underset{\sim}{H}(\underset{\sim}{u}(1),D\underset{\sim}{u}(1)) \end{aligned}\right\} \tag{3}$$

where $f_i(x,\underset{\sim}{u},\underset{\sim}{v}) = f_i(x,\underset{\sim}{u},v_i)$, $G_i(\underset{\sim}{u},\underset{\sim}{v}) = G_i(\underset{\sim}{u},v_i)$ and $H_i(\underset{\sim}{u},\underset{\sim}{v}) = H_i(\underset{\sim}{u},v_i)$ and consider also the related sequence of finite dimensional problems;

$$\left.\begin{aligned} -u_i'' + f_i(x,[\underset{\sim}{w}(x);\underset{\sim}{u}],u_i') = 0, \quad x \in (0,1), \\ G_i([\underset{\sim}{w}(0);\underset{\sim}{u}(0)],u_i'(0)) = 0 = H_i([\underset{\sim}{w}(1);\underset{\sim}{u}(1)],u_i(1)). \end{aligned}\right\} \tag{4}$$

(The inequality in $E_{\underset{\sim}{p}}$ is taken to be the usual component-wise inequality.)

Lemma 1 (Theorem 6.1 of reference [6]). Let there exist $\underset{\sim}{\alpha},\underset{\sim}{\beta} \in C([0,1],E^0_{\underset{\sim}{p}}) \cap C\Gamma_{\underset{\sim}{p}}{}^2([0,1],E_{\underset{\sim}{p}})$ satisfying $\underset{\sim}{\alpha}(x) \leq \underset{\sim}{\beta}(x)$ and the inequalities

$$\begin{cases} -D^2\underset{\sim}{\alpha}(x) + \underset{\sim}{f}(x,\underset{\sim}{\alpha}(x),D\underset{\sim}{\alpha}(x)) \leq \underset{\sim}{0}, \quad x \in (0,1) \\ \underset{\sim}{G}(\underset{\sim}{\alpha}(0),D\underset{\sim}{\alpha}(0)) \leq \underset{\sim}{0} \leq \underset{\sim}{H}(\underset{\sim}{\alpha}(1),D\underset{\sim}{\alpha}(1)) \end{cases}$$

and

$$\begin{cases} -D^2\underset{\sim}{\beta}(x) + \underset{\sim}{f}(x,\underset{\sim}{\beta}(x),D\underset{\sim}{\beta}(x)) \geq \underset{\sim}{0}, \quad x \in (0,1) \\ \underset{\sim}{G}(\underset{\sim}{\beta}(0),D\underset{\sim}{\beta}(0)) \geq \underset{\sim}{0} \geq \underset{\sim}{H}(\underset{\sim}{\beta}(1),D\underset{\sim}{\beta}(1)). \end{cases}$$

Let $\underset{\sim}{f}(x,\underset{\sim}{u},\underset{\sim}{z})$ satisfy a Nagumo condition with respect to $\underset{\sim}{z}$ with Nagumo function $\Psi(s)$ on the set $\Sigma = \{(x,\underset{\sim}{u},\underset{\sim}{z}) \in [0,1] \times E_{\underset{\sim}{p}} \times E_{\underset{\sim}{p}}\colon \ \underset{\sim}{\alpha}(x) \leq \underset{\sim}{u} \leq \underset{\sim}{\beta}(x)\}$ and satisfy a Lipschitz condition on closed bounded subsets of Σ with respect to z. Let $\underset{\sim}{f}\colon \Sigma \to E_{\underset{\sim}{p}}$ and $\underset{\sim}{G},\underset{\sim}{H}\colon \Sigma \to E_{\underset{\sim}{p}}$ be $\Gamma_{\underset{\sim}{p}}$-continuous on Σ. Further, let $\underset{\sim}{f}(x,\underset{\sim}{u},\underset{\sim}{z})$ satisfy (*H*) and let $-\underset{\sim}{G}(\underset{\sim}{u},\underset{\sim}{z})$ and $\underset{\sim}{H}(\underset{\sim}{u},\underset{\sim}{z})$ satisfy (*H'*) on the set Σ and let $G_i(\underset{\sim}{u},z_i)$ and $H_i(\underset{\sim}{u},z_i)$ be nondecreasing in z_i for each $\underset{\sim}{u} \in E_{\underset{\sim}{p}}$, $i \in Z^+$. Finally let

$\underset{\sim}{f}(x,\underset{\sim}{u},\underset{\sim}{z})$, $\underset{\sim}{G}(\underset{\sim}{u},\underset{\sim}{z})$ and $\underset{\sim}{H}(\underset{\sim}{u},\underset{\sim}{z})$ be semi-continuous with respect to $\underset{\sim}{u}$ in the following sense: there exists a continuous function $d\colon R \to [0,\infty)$ such that $d(0) = 0$, $d(s)$ is increasing for $s > 0$, $d(-s) = d(s)$ and such that

$$f_i(x,\underset{\sim}{u}-\sigma\underset{\sim}{e}^i,z_i) - f_i(x,\underset{\sim}{u},z_i) \geq -d(\sigma)$$

$$G_i(\underset{\sim}{u},z_i) - G_i(\underset{\sim}{u}-\sigma\underset{\sim}{e}^i,z_i) \geq -d(\sigma)$$

$$H_i(\underset{\sim}{u},-\sigma\underset{\sim}{e}^i,z_i) - H_i(\underset{\sim}{u},z_i) \geq -d(\sigma)$$

Then the sequences $\{\underset{\sim}{\beta}^n(x)\}$ and $\{\underset{\sim}{\alpha}^n(x)\}$, defined by the equations

$$\underset{\sim}{\beta}^0(x) = \underset{\sim}{\beta}(x), \quad \underset{\sim}{\beta}^n(x) = [\underset{\sim}{\beta}(x);\underset{\sim}{u}^n(x)]$$

$$\underset{\sim}{\alpha}^0(x) = \underset{\sim}{\alpha}(x), \quad \underset{\sim}{\alpha}^n(x) - [\underset{\sim}{\alpha}(x);\underset{\sim}{\overline{u}}^n(x)]$$

(where $\underset{\sim}{u}^n(x)$ is the solution of (4) with $\underset{\sim}{w}(x) = \underset{\sim}{\beta}(x)$ and $\underset{\sim}{\overline{u}}^n(x)$ is the solution of (4) with $\underset{\sim}{w}(x) = \underset{\sim}{\alpha}(x)$) converge monotonically from above and below, respectively, to the unique solution $\underset{\sim}{u}(x)$ of (3) in $C([0,1],E^0_{\underset{\sim}{p}}) \cap C\Gamma_{\underset{\sim}{p}}([0,1],E_{\underset{\sim}{p}})$.

III. PROOF OF THEOREM 1

Theorem 1 is a direct application of Lemma 1 to the boundary value problem $P(h)$. Therefore, it suffices to show that the hypotheses of Lemma 1 are satisfied.

Define $\underset{\sim}{G}(\underset{\sim}{u},\underset{\sim}{z})$ and $\underset{\sim}{H}(\underset{\sim}{u},\underset{\sim}{z})$ by the equations $G_i(\underset{\sim}{u},z_i) = \phi(y_i) - u_i$ and $H_i(\underset{\sim}{u},z_1) = u_i - \psi(y_i)$. Then it is not difficult to show that under the hypotheses of Theorem 1 for $h > 0$ sufficiently small and $A > 0$ sufficiently large, the inequalities, (5) and (6) are satisfied by $\underset{\sim}{\beta}(x)$ and $\underset{\sim}{\alpha}(x)$ respectively. Let $\lambda = \max\{\lambda_1,\lambda_0\}$ and let $\underset{\sim}{p}$ be the sequence $p_i = e^{-\lambda ih}$. Then $\underset{\sim}{\alpha},\underset{\sim}{\beta} \in C([0,1],E^0_{\underset{\sim}{p}})$. Moreover, the $\Gamma_{\underset{\sim}{p}}$ continuity and semi-continuity of $\underset{\sim}{f}$, $\underset{\sim}{G}$, and $\underset{\sim}{H}$ follow from the continuity of $\underset{\sim}{g}$ and the row finiteness of the problem. The Nagumo condition and the Lipschitz condition follow from the bound on g_p.

It remains, therefore, to show that the function $f_i(x,u,z_i) = -h^{-2}(u_{i+1}-2u_i+u_{i-1}) + g(x,y_i,u_i,z_i,\delta u_i)$ satisfies (H) (that $\underset{\sim}{G}$, $\underset{\sim}{H}$ satisfy (H') is immediate). Let $(x,\underset{\sim}{\overline{u}},\underset{\sim}{\overline{z}}),(x,\underset{\sim}{u},\underset{\sim}{z}) \in [0,1] \times E_{\underset{\sim}{p}} \times E_{\underset{\sim}{p}}$ be such that $p_i(u_i-\overline{u}_i) = \inf p_j(u_j-\overline{u}_j)\colon\ j \in Z^+ \le 0$, $z_i = \overline{z}_i$. Then

$$\begin{aligned} p_i[-h^{-2}\{(\overline{u}_{i+1}-u_{i+1}) - 2(\overline{u}_i-u_i) + (\overline{u}_{i-1}-u_{i-1})\} \\ + g(x,y_i,\overline{u}_i,\overline{z}_i,\delta\overline{u}_i) - g(x,y_i,u_i,z_i,\delta u_i)] \\ \ge p_i(u_i-\overline{u}_i)[h^{-2}\left(\frac{p_i}{p_{i-1}} + \frac{p_i}{p_{i+1}} - 2\right) + h^{-1}\left(\frac{p_i}{p_{i+1}} - 1\right)L_2 - \gamma] \\ \ge 0 \end{aligned}$$

for $h > 0$ sufficiently small. □

We note in conclusion that similar results hold for more general model problems than that considered here. In particular, the bound on γ is quite conservative and can be improved by making a different choice of $\underset{\sim}{\beta}(x)$ and $\underset{\sim}{\alpha}(x)$. (See for example [4]). Also, the bounds on the derivatives g_p and g_q can be replaced by weaker conditions. Because of the additional complexity we have not included details of these generalizations here. Finally, we note that Lemma 1 allows the boundary conditions to include certain types of nonlinear dependence on u and u_x.

REFERENCES

[1] Bernfeld, S. R., Lakshmikantham, V., and Leela, S. "A generalized comparison principle and monotone method for second order boundary value problems in Banach spaces", Proceedings U.S. Army Conference, to appear.

[2] Chandra, J., Lakshmikantham, V., and Leela, S. "A monotone method for infinite systems of nonlinear boundary value problems", *Arch. Rat. Mech. Anal.*, to appear.

[3] Chandra, J., Lakshmikantham, V., and Mitchell, A. R. "Existence of solutions to boundary value problems for nonlinear second order systems in a Banach space", Nonlinear Analysis, to appear.

[4] Schmitt, K., Thompson, R. C., and Walter, W. "Existence of solutions of a nonlinear boundary value problem via the method of lines", Nonlinear Analysis, to appear.

[5] Thompson, R. (1976). "Convergence and error estimates for certain nonlinear elliptic and elliptic parabolic equations", *SIAM J. Numer. Anal. 13*, 27-43.

[6] Thompson, R. "On extremal solutions to infinite dimensional nonlinear second order systems", submitted.

ASYMPTOTIC BEHAVIOR OF A CLASS OF DISCRETE-TIME MODELS IN POPULATION GENETICS

*H. F. Weinberger**

School of Mathematics
University of Minnesota
Minneapolis, Minnesota

I. INTRODUCTION

It has been known since the time of R. A. Fisher [3] that various one-dimensional models for the spread of an advantageous gene in a migrating population have associated with them certain wave speeds. These speeds have two characteristic properties. One is that for each speed at least as great as the wave speed there is a travelling wave with this speed, a result due to Fisher himself [3]. The second result, which was conjectured by Fisher, is that an initially confined population spreads out with a speed which is, in an asymptotic sense, the wave speed. When the initial distribution is a step function, this result goes back to the classical paper of Kolmogoroff, Piscounoff, and Petrowsky [4]. It was extended to monotone initial data by Kanel' [5,6]. Aronson and the author [1] showed that similar results are valid for arbitrary initial distributions which vanish outside a bounded interval. These results were also extended to

**This work was supported by the National Science Foundation through Grant MCS 76-06128 A01.*

ISBN 0-12-434180-2

the analogous model in two or more dimensions [2]. The latter observation is important because populations usually live either in a two-dimensional habitat (the surface of the earth) or possibly in a three dimensional one (the water in a lake).

The above results all concern a model of the form

$$\frac{\partial u}{\partial t} = D\nabla^2 u + f(u). \tag{1.1}$$

This model is intended to describe the development of a diploid species in which one gene occurs in two variant forms, called alleles, which we denote by a and A. There are then three genotypes, AA, aA, and aa, with respect to this gene, and it is assumed that the fitness of an individual depends only upon its genotype with respect to the gene under consideration. The variable $u(x,t)$ which occurs in (1.1) is the so-called gene fraction. If the densities of the three genotypes at x at time t are $\rho_{AA}(x,t)$, $\rho_{aA}(x,t)$, and $\rho_{aa}(x,t)$, respectively, then

$$u = \frac{2\rho_{AA}+\rho_{aA}}{2\rho_{AA}+2\rho_{aA}+2\rho_{aa}}.$$

That is, u is the fraction of all the alleles near x which are of the type A. The function $f(u)$ in (1.1) describes the relative fitnesses of the three genotypes, and D gives a rate of migration of the population.

Some difficulties with the Fisher model were pointed out in [1]. Because it is populations, not fractions, which migrate, one really needs a system of three equations for the three genotype densities, and one obtains the equation (1.1) only by making hypotheses which are inconsistent with the selection that results from the different fitnesses of the three genotypes.

A more serious objection to all models of the form (1.1) was raised in [10]. Namely, the model, which is deterministic, is intended to model a series of processes such as random mating and random death, which are by nature stochastic. Because of the law of large numbers one can hope to obtain deterministic information from a stochastic process if one waits for a time that is long compared to the mean time between the events of the process. However, the partial derivative in the model (1.1) requires one to assume that the process is deterministic over arbitrarily short time intervals, which does not seem to be justified. It is true that one can sometimes show that certain expected values associated with a stochastic process satisfy a partial differential equation like (1.1), but in our deterministic model u is not an expectation but the gene fraction, which is a deterministic quantity.

As a first step in overcoming this logical impasse, we have proposed in [10] a somewhat different model for the time evolution of the gene frequency. The basic idea is to describe the properties of the population not at all times but only at multiples of a certain unit of time. Thus we look not at $u(x,t)$ but only of $u(x,n\tau)$, which we write as $u_n(x)$. We assume that the generations are synchronized and that the time τ is a multiple of the duration of one generation. The quantity u_n is always to be measured during the same stage of the life cycle.

Our basic hypothesis is that $u_{n+1}(x)$ is uniquely determined by the values of $u_n(y)$ at all points of the habitat. We state this hypothesis mathematically by writing the equation

$$u_{n+1} = Q[u_n], \tag{1.2}$$

where Q is an operator which takes each admissible function $u(x)$ into some other admissible function. The particular model is defined by specifying a particular operator Q.

An objection similar to that which we have raised against using a continuous time variable can also be raised against using continuous space variables. A very small neighborhood will contain few or no individuals, so that the law of large numbers can not be expected to render the variance negligible. Moreover, one can count the individuals of various genotypes in a fixed region, but not "at a point". Therefore one can associate a gene fraction u with a region which is not too small, but not with a point. For these reasons we shall consider a class of operators Q which permit space as well as time to be discrete. Our main point is that such models can be treated in a mathematically satisfactory manner, so that there is no need to approximate them with models which involve a partial differential equation.

The following stepping stone model (see [7,8,9]) is an example of the class of models we have in mind.

The Euclidean plane R^2 is divided into squares of side 1 which are centered at the points (i,j) with integer coordinates. The unit of measurement (that is, the size of the squares) is chosen so large that during all of the life cycle except for a brief period of migration the interaction between individuals in different squares is negligible. Each square is named after its center $x \sim (i,j)$.

We assume that all individuals of one generation are born at essentially the same time. Let u be the gene fraction of the individuals born in a particular square. That is, u is the ratio of the number of A alleles in these individuals to the total number of alleles of types A and a. (The latter is, of course, twice the number of individuals.) We suppose that the number of new-born individuals is so large relative to the carrying capacity of the square that the number of individuals of each genotype which survive to maturity does not depend upon the number of births. The numbers of survivors N_{AA}, N_{aA}, and N_{aa} of the various genotypes do, however, depend upon the ratios of the genotypes at birth. We assume that mating is random so that by the Hardy-Weinberg law these ratios are u^2: $2u(1-u)$: $(1-u)^2$. Because the different genotypes have different fitnesses, the ratios N_{AA}: N_{aA}: N_{aa} may well be different, but the three numbers N_{AA}, N_{aA}, and N_{aa} are fixed functions of u. The fitnessesses of the three genotypes depend, of course, upon the environment, so that the number of survivors will, in general, also depend upon the particular square x. However, in this work we shall only consider the case of a homogeneous environment, so that these numbers do not depend explicitly on x. Thus, we have functions $N_{AA}(u)$, $N_{aA}(u)$, and $N_{aa}(u)$, which can be measured experimentally.

We next consider the migration scheme. For the sake of simplicity, we suppose that a fraction $\alpha(x,y)$ of the mature individuals of each genotype in the square centered at y migrates to the square centered at x. The homogeneity of the habitat requires that α depends only on the relative location $x - y$ of the two squares. Let $u_n(y)$ be the gene fraction of the individuals born in the square centered at y in the nth generation. We see from the preceding discussion that at the next migration $\alpha(x-y)N_{AA}(u_n(y))$ individuals of genotype AA arrive at square x from square y, with similar expressions for the other two genotypes. Therefore the gene fraction at x after the migration is given by

$$\frac{\sum_y \alpha(x-y)[2N_{AA}(u_n(y))+N_{aA}(u_n(y))]}{\sum_y \alpha(x-y)[2N_{AA}(u_n(y))+2N_{aA}(u_n(y))+2N_{aa}(u_n(y))]}.$$

In any realistic model, $\alpha(x-y) = 0$ when $|x-y|$ is too large, so that for each x the sums are over finitely many squares y. (Of course, $\alpha(0)$ represents the fraction of individuals who do not migrate.)

We now suppose that just after the migration the individuals who are at x mate at random and die. Then the gene fraction $u_{n+1}(x)$ of the next generation born in square x is equal to the above gene fraction. Thus, we have a recursion of the form $u_{n+1} = Q[u_n]$, where

$$Q[u](x) \equiv \frac{\sum_y \alpha(x-y)[2N_{AA}(u(y))+N_{aA}(u(y))]}{\sum_y \alpha(x-y)[2N_{AA}(u(y))+2N_{aA}(u(y))+2N_{aa}(u(y))]}. \quad (1.3)$$

We see that Q is an operator which takes any function $u(y)$ which is defined at the grid points y and has values on the interval $[0,1]$ into another function of this kind.

This model can clearly be generalized to permit the migration patterns to depend upon genotype and upon the total population, and to allow for additional attrition and selection between migration and mating.

The operator Q looks simpler if one assumes that the total surviving population $N_{AA}(u) + N_{aA}(u) + N_{aa}(u)$ is a constant independent of u. Q is then linear in the single function $g(u) = (2N_{AA}+N_{aA})/(2N_{AA}+2N_{aA}+2N_{aa})$. If one takes a suitable limit as the side of the squares approaches zero, one arrives at the recursion $u_{n+1} = Q[u_n]$ where

$$Q[u](x) \equiv \int \alpha(x-y)g(u(y))dy \tag{1.4}$$

with $\alpha(x)$ a probability distribution and u a function of the continuous variable x. A special class of such operators was discussed in [10].

One can place the partial differential equation model (1.1) in the same framework by defining Q to be the solution operator for a finite time interval τ. That is, if w is the solution of the initial value problem

$$\frac{\partial w}{\partial t} = \nabla^2 w + f(w),$$

$$w(x,0) = \phi(x),$$

we define

$$Q[\phi](x) = w(x,\tau). \tag{1.5}$$

Then if $u_n(x) = u(x,n\tau)$ and $u(x,t)$ satisfies the differential equation (1.1) we have the recursion $u_{n+1} = Q[u_n]$.

In fact, this idea can still be applied if D and f depend explicitly upon t, as long as the dependence is periodic of period τ. This allows one to treat seasonal variation of the rates of growth and migration in the continuous-variable case.

In this paper we shall summarize some of the salient features of the large-time behavior of a class of models of the form $u_{n+1} = Q[u_n]$, and sketch the ideas involved in the derivation of these results. Details will be published elsewhere.

II. DEFINITION OF THE WAVE SPEED

We begin by formulating some hypotheses about the recursion operator Q.

We consider the habitat H to be a subset of the Euclidean space R^N ($N = 1, 2$, or 3) with the property that if x and y are in H, then so are $x + y$ and $x - y$. That is, H is an additive subgroup of R^N. In particular, the origin 0 is in H. We shall suppose that H has at least one other point x. Since H contains all integral multiples of x, it is unbounded. H may, for example, be all of R^N, or the set of points with integral coordinates.

Let B be the set of continuous functions on H with values on the interval $[0,1]$. (Of course, if H is discrete, all functions defined on H are continuous.) For y in H we define the translation operator

$$T_y[u](x) \equiv u(x-y).$$

Let Q be an operator which takes B into B and which has the following properties.

$$\begin{aligned}
&\text{(i)} \quad Q[0] = 0, \quad Q[1] = 1.\\
&\text{(ii)} \quad QT_y = T_yQ \quad \forall\, y \in H.\\
&\text{(iii)} \quad u \le v \Longrightarrow Q[u] \le Q[v].\\
&\text{(iv)} \quad \text{For constant } \alpha \in (0,1), \quad Q[\alpha] > \alpha.\\
&\text{(v)} \quad u_k \to u \text{ uniformly on bounded subsets of } H \Longrightarrow\\
&Q[u_k] \to Q[u] \text{ at each point of } H.
\end{aligned} \tag{2.1}$$

Property (i) states that our model does not consider mutation into either of the allelic types A or a. That is, if no A (or a) is present in one generation, none will be present in the next generation.

Property (ii) states that the habitat is homogeneous, so that only relative positions matter.

Property (iii), which states that the operator Q is order-preserving, is a consequence of the maximum principle for parabolic equations when Q is the solution operator (1.5). When Q is of the form (1.4) with α a probability distribution, this property is equivalent to the hypothesis that $g(u)$ is nondecreasing. For Q of the form (1.3) we obtain (iii) by assuming that the function $2N_{AA}(u) + N_{aA}(u)$ is nondecreasing and that $N_{aA}(u) + 2N_{aa}(u)$ is nonincreasing. That is, when the gene fraction in a fixed square is increased, the number of alleles A that survive to maturity is increased while the number of surviving alleles a is decreased. This is very likely if the total number $N_{AA} + N_{aA} + N_{aa}$ is constant, and plausible but not necessarily true otherwise. That is, (iii) is a genuine biological assumption.

Hypothesis (iv) is also a biological assumption. We deal with constant initial data so that the effects of migration cancel, and we are only concerned with what happens in the absence of migration. For the solution operator (1.5) this hypothesis is valid if $f(u) > 0$ for $u \in (0,1)$. For the operator (1.4) it follows from the hypothesis $g(u) > u$ for $u \in (0,1)$. For the operator (1.3) one assumes that the gene fraction

$$(2N_{AA}(u)+N_{aA}(u))/(2N_{AA}(u)+2N_{aA}(u)+2N_{aa}(u))$$

is greater than u for $u \in (0,1)$. That is, the gene fraction always increases during the growth and selection process. This will be the case, provided the heterozyote genotype aA is more fit than the homozygote aa, while the homozyote AA is still more fit. For this reason we speak of the situation where the hypotheses (iv) is valid as the heterozygote intermediate case.

Hypothesis (v) is, of course, a mathematical hypothesis which states that Q is continuous. We note that when H is discrete, uniform convergence is just pointwise convergence.

We shall now show how a wave speed in the direction of any unit vector ξ may be defined.

Choose any function $\phi(s)$ of one variable with the properties

$$\begin{aligned} &\phi(s) \text{ is continuous and nonincreasing} \\ &\phi(s) \equiv 0 \quad \text{for} \quad s \geq 0 \\ &\phi(s) \equiv \alpha_0 \in (0,1) \quad \text{for} \quad s \leq -1. \end{aligned} \tag{2.2}$$

Also choose a unit vector ξ and a constant c, and define the operator

$$\hat{R}_{c,\xi}[a](s) \equiv \max\{\phi(s), Q[a(x\cdot\xi+s+c)](0)\} \tag{2.3}$$

on continuous functions of one variable with values in $[0,1]$. The operator Q acts on $a(x\cdot\xi+s+c)$, considered as a function of x in H. The maximum just means to take the larger of the two numbers for any particular s.

It follows from (2.1, iii) that $\hat{R}_{c,\xi}$ is also order-preserving:

$$a \geq b \Longrightarrow \hat{R}_{c,\xi}[a] \geq \hat{R}_{c,\xi}[b].$$

For fixed ϕ, c, and ξ, we now define a sequence of functions $a_n(s) = a_n(c,\xi;s)$ by means of the recursion

$$a_{n+1} = \hat{R}_{c,\xi}[a_n],$$

$$a_0(s) = \phi(s).$$

Our principal tool is the following simple comparison theorem.

Proposition 1. Let R be an operator on some function space, and suppose that R is order-preserving in the sense that

$$v \geq w \Longrightarrow R[v] \geq R[w]. \tag{2.4}$$

Let v_n and w_n be two sequences which satisfy the inequalities

$$v_{n+1} \geq R[v_n], \tag{2.5}$$

$$w_{n+1} \leq R[w_n], \tag{2.6}$$

and

$$v_0 \geq w_0.$$

Then

$$v_n \geq w_n$$

for all n.

The proof is by induction. Suppose $v_n \geq w_n$. Then by (2.5), (2.4), and (2.6),

$$v_{n+1} \geq R[v_n] \geq R[w_n] \geq w_{n+1}.$$

We see from the definition (2.3) of $\hat{R}_{c,\xi}$ that $a_1 \geq \phi = a_0$. We apply Proposition 1 with $R = \hat{R}_{c,\xi}$, $v_n = a_{n+1}$, and $w_n = a_n$ to see that $a_{n+1} \geq a_n$ for all n. That is, a_n is nondecreasing in n.

A simple induction proof shows that $a_n(s)$ is nonincreasing in s. By hypothesis this is true of $a_0 = \phi$. If a_n is nonincreasing, then for positive t

$$\begin{aligned} a_{n+1}(s+t) &= \max\{\phi(s+t), Q[a_n(x\cdot\xi+s+t+c)](0)\} \\ &\leq \max\{\phi(s), Q[a_n(x\cdot\xi+s+c)](0)\} \\ &= a_{n+1}(s) \end{aligned}$$

because ϕ is nonincreasing and Q is order-preserving. Thus each a_n is nonincreasing in s.

It follows that for $c' > c$

$$\hat{R}_{c',\xi}[a_n] \leq \hat{R}_{c,\xi}[a_n] = a_{n+1}.$$

Proposition 1 then shows that $a_n(c,\xi;s)$ is also nonincreasing in c.

Since $a_n(c,\xi;s)$ is nondecreasing in n and bounded above by 1, it has a limit

$$a(c,\xi;s) = \lim_{n\to\infty} a_n(c,\xi;s).$$

$a(c,\xi;s)$ is again nonincreasing in c and s. (It need not be continuous).

We define the sequence of constants α_n by

$$\alpha_{n+1} = Q[\alpha_n],$$

with $\alpha_0 = \phi(-\infty)$. Because of the property (2.1, iv) this sequence increases to 1.

It is easily seen from the property (2.1, v) that $a_n(c,\xi;-\infty) = \alpha_n$. Therefore

$$a(c,\xi;-\infty) = 1.$$

The function $a(c,\xi;s)$ may or may not be identically 1. Because it is nonincreasing in c, the set of c where $a(c,\xi;s) \equiv 1$ is connected. We define

$$c^*(\xi) = \sup\{c \mid a(c,\xi;s) \equiv 1\}. \tag{2.7}$$

The following lemma is useful in computing $c^*(\xi)$ and in finding its properties.

Lemma 1. $a(c,\xi;s) \equiv 1$ if and only if there is some n_0 such that

$$a_{n_0}(c,\xi;0) > \alpha_0 \tag{2.8}$$

Proof. It is obvious that if $a_n(c,\xi;s)$ approaches 1 for all s, there must be such an n_0.

Suppose, conversely, that there is an n_0 such that (2.8) is satisfied. Because a_{n_0} is continuous in s (this follows from the property (2.1, v)) and because $\phi < \alpha_0$ and $\phi = 0$ for $s > 0$, there is a positive δ such that

$$a_{n_0}(c,\xi;s+\delta) \geq \phi(s) = a_0(c,\xi;s).$$

Since a_n is nondecreasing in n and nonincreasing in s, we see that for $m \geq 0$

$$a_{n_0+m+1}(c,\xi;s+\delta) = Q[a_{n_0+m}(c,\xi;x\cdot\xi+s+\delta+c)](0)$$

$$= \hat{R}_{c,\xi}[a_{n_0+m}(c,\xi;s+\delta)].$$

Since $a_{n_0}(c,\xi;s+\delta) \geq a_0(c,\xi;s)$, Proposition 1 shows that

$$a_{n_0+m}(c,\xi;s+\delta) \geq a_m(c,\xi;s)$$

for all positive m. We let m approach infinity to see that

$$a(c,\xi;s+\delta) \geq a(c,\xi;s).$$

Because a is nonincreasing in s and δ is positive, this implies that a is constant. Since $a(c,\xi;-\infty) = 1$, the constant must be 1 and the Lemma is proved.

Because $a_1(0,\xi;-\infty) = \alpha_1 > \alpha_0$, one can choose c_0 so small that $a_1(0,\xi;c_0) > \alpha_0$. We then see from the definition of $\hat{R}_{c,\xi}$ that

$$a_1(c_0,\xi;0) = a_1(0,\xi;c_0) > \alpha_0,$$

so that $a(c_0,\xi;s) \equiv 1$. Thus $c^*(\xi)$ is defined for all ξ. It is possible that $c^*(\xi) = +\infty$ for some or all ξ. However, if the distance of migration in one season is bounded, $c^*(\xi)$ lies below this bound for all ξ.

It is easy to see from Proposition 1 and the fact that $a_n(c,\xi;-\infty) = \alpha_n$ that the function $c^*(\xi)$ does not depend upon the choice of α_0 or of the function ϕ.

III. THE PROPAGATION OF GENETIC CHANGE

In this section we shall indicate without proof the properties of the wave speed $c^*(\xi)$ which make it useful.

We define the closed convex set

$$S = \{x \in R^N \mid x \cdot \xi \leq c^*(\xi) \ \forall \ \xi \text{ with } |\xi| = 1\}.$$

We shall assume that this set is bounded. (This is certainly true if $c^*(\xi)$ is bounded. The case of unbounded S is discussed in [11].) For any positive n we define the dilatation nS by

$$nS = \{x \in R^N \mid \frac{1}{n} x \in S\}$$

$$= \{x \subset R^N \mid x \cdot \xi \leq nc^*(\xi)\}.$$

Our results state roughly that if $u_0(x)$ vanishes outside a bounded set, u_n is very small far from the set nS and that, if u_0 is not too small, u_n is large far inside of nS.

Theorem 1. Suppose $u_0 = 0$ outside a bounded set, and $u_0 < 1$. Let u_n be defined by the recursion $u_{n+1} = Q[u_n]$. Let S be bounded, and let S' be any open set which contains S. Then

$$\lim_{n\to\infty} \max_{x \notin nS'} u_n(x) = 0.$$

Theorem 2. Let S'' be any closed subset of the interior of S. For any positive γ there exists a radius r_γ such that if $u_0 \geq \gamma$ on a ball of radius r_γ,

$$\lim_{n\to\infty} \min_{x \in nS''} u_n(x) = 1. \tag{3.1}$$

Theorem 1 is obtained by noting that if $u_0 = 0$ for $|x| \geq \rho$, then Proposition 1 implies that u_n is bounded by all the functions $a(c,\xi;x\cdot\xi-nc-\rho-1)$ and by showing that $a(c,\xi;+\infty) = 0$ for $c \geq c^*(\xi)$.

The proof of Theorem 2 is more difficult. One sees from Lemma 1 that if $c < c^*(\xi)$, then there is an n_0 such that $a_{n_0}(c,\xi;s) > \phi(s)$. By an approximation process, one can assume without loss of generality that $Q[u](x)$ is independent of the bahavior of u outside a (large) bounded neighborhood of x. It then follows that $a_{n_0} \equiv \alpha_{n_0}$ for all sufficiently small s and $a_{n_0} \equiv 0$ for sufficiently large s. Thus when the constant A is very large, the function

$$b_n(x) = a_{n_0}(c,\xi;x\cdot\xi-nc-A)$$

depends only on the variable $x \cdot \xi$, is equal to α_{n_0} in an arbitrarily large neighborhood of the origin, and satisfies the inequality

$$b_{n+1} \leq Q[b_n].$$

One patches together these functions corresponding to different directions ξ to obtain a sequence $w_n(x)$ with the following properties:

(i) each w_n vanishes outside a bounded set.

(ii) $w_n = \alpha_{n_0}$ on the set $n(1+\varepsilon)S''$, where ε is positive.

(iii) There is an integer ℓ_0 such that $w_{n+k_0} \leq Q^{\ell_0}[w_n]$

where Q^{ℓ_0} means the ℓ_0th iterate of Q.

A continuity argument which uses the hypothesis (2.1, v) shows that if $u_0 > \gamma$ on a sufficiently large ball, then there is a k_0 such that $u_{k_0+n} \geq w_n$ for $0 \leq n \leq \ell_0$. Proposition 1 then shows that $u_{k_0+n} > w_n$, so that u_{k_0+k} is uniformly positive on $n(1+\varepsilon)S''$. Another continuity argument then shows that, for any positive δ, there are a j_0 and an m_0 such that $u_{k_0+j_0+n} > 1-\delta$ on $n(1+\frac{1}{2}\varepsilon)S''$ when $n \geq m_0$, which implies Theorem 2.

The extra condition $u_0 < 1$ in Theorem 1 can be removed by means of an additional reasonable hypothesis on Q.

Under stronger conditions on Q one can show that (3.1) holds whenever $u_0 \not\equiv 0$. A system with such behavior is said to display a hairtrigger effect. For the Fisher model (1.1) this property is known to hold when

$$\liminf_{u \searrow 0} f(u)u^{-\gamma} > 0$$

for $\gamma > (n+2)/n$. (See [2]).

If the condition (2.1, v) is supplemented by the condition that the range of Q is an equicontinuous family (which is trivial when H is discrete), one can also show that for each $c \geq c^*(\xi)$ there is a travelling wave solution $u_n = W(x \cdot \xi - nc)$ of the recursion $u_{n+1} = Q[u_n]$ with W nonincreasing, $W(-\infty) = 1$, and $W(\infty) = 0$.

REFERENCES

[1] Aronson, D. G., and Weinberger, H. F. (1975). "Nonlinear diffusion in population genetics, combustion, and nerve propagation", Partial Differential Equations and Related Topics, ed. J. Goldstein, Lecture Notes in Mathematics Vol. 446, Springer, 5-49.

[2] Aronson, D. G., and Weinberger, H. F. "Multidimensional nonlinear diffusion arising in population genetics", to appear in *Adv. in Math.*

[3] Fisher, R. A. (1937). "The advance of advantageous genes", *Ann. of Eugenics 7*, 355-369.

[4] Kolmogoroff, A., Petrovsky, I., and Piscounoff, N. (1937). "Étude de l'équations de la diffusion avec croissance de la quantité de matière et son application a un problème biologique", *Bull. Univ. Moscow, Ser. Internat., Sec. A, 1, #6*, 1-25.

[5] Kanel', Ja. I. (1961). "The behavior of solutions of the Cauchy problem when time tends to infinity, in the case of quasilinear equations arising in the theory of combustion", *Akad. Nauk S.S.S.R. Dokl. 132*, 268-271; *Soviet Math. Dokl. 1*, 535-536.

[6] Kanel', Ja. I. (1961). "Certain problems on equations in the theory of burning", *Akad. Nauk S.S.S.R., Dokl. 136*, 277-280; *Soviet Math. Dokl. 2*, 48-51.

[7] Karlin, S. (1976). "Population subdivision and selection migration interaction", Proceedings of the International Conference on Population Genetics and Ecology, Academic Press, New York.

[8] Malécot, G. (1969). "The Mathematics of Heredity", W. H. Freeman, San Francisco.

[9] Nagylaki, T. (1977). "Selection in One- and Two-Locus Systems", Lecture Notes in Biomathematics 15, Springer, Berlin.

[10] Weinberger, H. F. (1978). "Asymptotic behavior of a model in population genetics", Nonlinear Partial Differential Equations and Applications, ed. J.M. Chadam, Lecture Notes in Mathematics Vol. 648, Springer, 47-96.

[11] Weinberger, H. F. "Genetic wave propagation, convex sets, and semiinfinite programming", to appear in Constructive Approaches to Mathematical Models, a Symposium in Honor of R. J. Duffin, Academic Press, New York.

CONTRIBUTED PAPERS

THE VOLUME OF DISTRIBUTION IN SINGLE-EXIT COMPARTMENTAL SYSTEMS

*David H. Anderson**

Department of Mathematics
Southern Methodist University
Dallas, Texas

I. INTRODUCTION

In recent years, the use of mathematical compartmental models to describe biological phenomena has become more prevalent, particularly with respect to tracer kinetics. In a living organism the introduced tracer creates an observable transient that will provide information about certain aspects of the system. This paper shall be concerned with the identification of certain parameters of the system, in particular the volumes of distribution of the tracer, on the basis of these experimental observations.

The class of compartmental systems that will be treated in this article is concerned with a problem that is frequently encountered in physiology - the injection of an isotope by the intravenous route into one compartment at time zero and the sequential sampling of that same compartment to estimate the tracer

**Assistant Professor at the Department of Medical Computer Science, The University of Texas Health Science Center, Dallas, Texas 75235.*

ISBN 0-12-434180-2

concentration as a function of time and then utilize this concentration function to provide information about the characteristics of the system.

II. DEVELOPMENT OF THE MATHEMATICAL MODEL

A compartmental model is now set up by partitioning a portion of the human body into n compartments, each considered as a kinetically homogeneous and well-mixed amount of a material relative to the tracer being studied. Linear time-invariant compartmental systems are considered representable by a system of linear differential equations [1, p. 48],

$$\dot{q}_i(t) = \sum_{j=1}^{n} a_{ij} q_j(t) + b_i(t), \quad t \geq 0, \quad i = 1,2,\ldots,n$$

in which the ith equation gives the time rate of change of the amount $q_i(t)$ of drug in compartment i at time t in terms of the present amounts in each of the connecting compartments, input $b_i(t) \geq 0$ from the external environment, and excretion $a_{0i} q_i(t)$ of the chemical to the outside. Here the elements of the $n \times n$ *compartmental matrix* A satisfy

(a) $a_{ij} \geq 0, \quad i \neq j;$

(b) $a_{0j} \geq 0, \quad j = 1,2,\ldots,n;$

(c) $a_{jj} = -a_{0j} - \sum_{i \neq j} a_{ij} < 0, \quad j = 1,2,\ldots,n.$

Thus

$$|a_{jj}| \geq \sum_{i \neq j} a_{ij}, \quad j = 1,2,\ldots,n,$$

(A is diagonally dominant with respect to columns) and equality holds in the jth case only if there is no excretion $(a_{0j} = 0)$ from the jth compartment.

Let us now suppose that an amount D of tracer is in the body and has had time to reach distribution equilibrium throughout all of the compartments which it can enter. Also assume that the concentration of the tracer in one of the compartments can be measured. If the equilibrium concentration in this reference is denoted c_{eq}, then the total volume of distribution of the tracer is defined as D/c_{eq}.

If the compartmental system is *closed* ($a_{0j} = 0$ for all j), the problem is simple. A known dose D of tracer can be administered at $t = 0$. This tracer will not be lost from the system and an equilibrium concentration c_{eq} will be approached asymptotically. Then the total volume of distribution is D/c_{eq}. Realistically, however, the system is *open* ($a_{0j} > 0$ for at least one j), for some tracer is lost by metabolism, excretion, or traps, so that a steady-state of equilibrium is not approached in time after a single dose. In practice this problem is often circumnavigated by approximating a steady-state of equilibrium by infusing the tracer intraveneously at a constant rate over a long time. For this method to be valid, all *exit* of tracer must be from a *single* compartment which is the same as the one into which the drug is infused. When the interchange between the various compartments is reasonably rapid, equilibrium will be approximated in a short time. Upon abruptly discontinuing the infusion the quantity D_{eq} of tracer distributed throughout the entire system at equilibrium can be estimated by sampling the exit compartment sequentially in time. Then the total volume of distribution is D_{eq} / c_{eq}.

There has been some discussion in the literature [11] that results such as the upcoming Theorem 4 suggest that existing experimental techniques can occasionally be substantially simplified in that the steady-state whole-body mass of certain substances can be estimated from more realistic measurements, such as recording the plasma concentration *following a single tracer dose at time zero*, instead of the more cumbersome recordings of whole-body retention curves.

The above discussion serves to motivate the following restrictions to be placed on the model. At $t = 0$ there is administered an injection of infinitesimal duration of amount D of tracer into a compartment designated as number one; thus $b_i(t) = 0$ for all i and $q_i(0) = 0$, $i \neq 1$. The amount of chemical introduced is assumed so small as to leave unaltered the steady-state behavior of the system. Homogeneity of the mixture is assumed, and when the labeled material enters a compartment, it is assumed to be mixed with the unlabeled material instantaneously. Hence $q_1(0) = D$. Also loss of the tracer will occur only from the first compartment ($a_{01} > 0$, $a_{0j} = 0$ if $j \neq 1$). Let V_i be the volume (mass) of the ith compartment; it is assumed to remain constant over time. Thus if $c_i(t)$ is the tracer concentration in compartment i at time t, then for all i,

$$q_i(t) = V_i c_i(t).$$

The concentration of the drug in the first compartment is measured at various times and therefore $c_1(t)$ is assumed known for $t \geq 0$. Therefore $V_1 = D/c_1(0)$ is an observable parameter.

The system shall now be referred to as a *single-exit compartmental* (SEC) *system*. The associated matrix A is called a *single-exit compartmental* (SEC) *matrix* [2]. Thus the final form of the mathematical model of the system is

$$\dot{q} = Aq, \quad a(0)^T = (D,0,\ldots,0)^T \tag{1}$$

where A is a SEC matrix in which the sum of the first column is $-a_{01}$, and all other column sums are zero.

III. BASIC FACTS ABOUT THE MODEL

It is assumed that all irreversible loss of tracer in the system occurs from the first compartment (in particular, the system contains no traps [1, p. 53]). Then $q_i(t) \to 0$ as $t \to \infty$ for each $i = 1,2,\ldots,n$. This is a necessary and sufficient condition that A is a stable matrix and that the real part of any eigenvalue λ of A is negative [3, Chap. 8]. Hence λ cannot be zero and so A is nonsingular. It can be shown that every entry of $-A^{-1}$ is nonnegative [4, p. 49], [2]. Moreover, in [2] it is shown that each entry in the first row of $-A^{-1}$ is $1/a_{01}$. Hearon [5, p. 72] has also shown that the mean residence time of tracer,

$$\mu_1 = \int_0^\infty tc_1(t)dt \Big/ \int_0^\infty c_1(t)dt,$$

in a system with elimination from only the first compartment and where initially only compartment one is loaded, is given by the first column sum of $-A^{-1}$.

IV. IDENTIFICATION OF THE TRANSFER COEFFICIENTS

This section deals with the identification problem for the transfer constants a_{ij}. From (1), ther obtains

$$\sum_{i=1}^{n} \dot{q}_i = \sum_{j=1}^{n} q_j \sum_{i=1}^{n} a_{ij} = -a_{01}q_1.$$

Upon integrating both sides of this equation over the time interval $[0,\infty)$, and applying the appropriate boundary conditions on q_i, it is seen that

$$q_1(0) = a_{01}V_1\int_0^\infty c_1(t)dt,$$

from which a_{01} can be estimated. Also a_{11} can be identified from the given measurements, for

$$V_1\dot{c}_1(0) = \dot{q}_1(0) = \sum_{j=1}^{n} a_{1j}q_j(0) = a_{11}V_1c_1(0)$$

yields $a_{11} = \dot{c}_1(0)/c_1(0)$.

For the remainder of this section assume that system (1) is further restricted to catenary structure. A *catenary system* is a compartmental system in which the compartments are arranged in a series with the understanding that each compartment exchanges only with the immediately adjacent compartments [1, p. 53]. In this case the associated catenary matrix A is tridiagonal.

Theorem 1. If in (1) the matrix A is catenary, then

$$a_{i+1,i} = q_{i+1}^{(i)}(0)/q_i^{(i-1)}(0), \quad i = 1,\dots,n-1.$$

Proof: The differential equation in (1) implies that $q^{(i)} = Aq^{(i-1)}$, where $q^{(i)}$ is the ith time derivative of q. Then for all $i = 1,2,\dots,n-1$,

$$q_{i+1}^{(i)}(0) = \sum_{j=1}^{n} a_{i+1,j}V_j c_j^{(i-1)}(0).$$

Since (1) is a catenary system, the derivative $c_\gamma^{(k)}(0) = 0$ provided $k < \gamma - 1$ [6, p. 78], [7, p. 127]. Thus the sum in the last equation terminates at $j = i$. Moreover, because A is catenary and so $a_{i+1,j} = 0$ whenever $j < i$, then

$$q_{i+1}^{(i)}(0) = a_{i+1,i}V_i c_i^{(i-1)}(0)$$

which completes the proof.

Theorem 2. If system (1) has catenary structure, then the matrix A is identifiable.

Proof: Since the system is catenary, the element a_{21} is given by the known combination $-a_{11} - a_{01}$. Thus all elements in the first column of A are estimatable.

Using the fact that each element in the top row of A^{-1} is $-1/a_{01}$, the first equation in $A^{-1}\dot{q} = q$ implies

$$\sum_{j=1}^{n} \dot{q}_j = -a_{01}q_1. \tag{2}$$

Since $\dot{c}_j(0) = 0$ if $j > 2$, then (2) can be rewritten as

$$\dot{q}_2(0) = -a_{01}q_1(0) - \dot{q}_1(0),$$

from which $\dot{q}_2(0)$ can be determined. Then from

$$\ddot{q}_1(0) = a_{11}\dot{q}_1(0) + a_{12}\dot{q}_2(0)$$

the parameter a_{12} is estimated. Hence

$$\dot{q}_1 = a_{11}q_1 + a_{12}q_2$$

yields $q_2^{(i)}(0)$ for all $i \geq 2$. In

$$\ddot{q}_2(0) = a_{21}\dot{q}_1(0) + a_{22}\dot{q}_2(0) + a_{23}\dot{q}_3(0) \quad (3)$$

the quantity $\dot{q}_3(0) = V_3\dot{c}_3(0) = 0$ and a_{22} can be replaced by $-a_{12} - a_{32}$. Thus a_{32} can be computed from (3). Hence the second column of A has now been identified.

To get estimates on the elements in the third column of A, start by considering

$$q_2^{(3)}(0) = a_{21}\ddot{q}_1(0) + a_{22}\ddot{q}_2(0) + a_{23}\ddot{q}_3(0) \quad (4)$$

in which $\ddot{q}_3(0)$ is replaced by $a_{32}\dot{q}_2(0)$ via Theorem 1. Equation (4) can then be used to find a_{23}. From

$$\dot{q}_2 = a_{21}q_1 + a_{22}q_2 + a_{23}q_3,$$

the derivative $q_3^{(i)}(0)$ is known for $i \geq 2$. The last term of the equation

$$q_3^{(3)}(0) = a_{32}\ddot{q}_2(0) + a_{33}\ddot{q}_3(0) + a_{34}\ddot{q}_4(0) \quad (5)$$

is zero. If $-a_{23} - a_{43}$ is substituted for a_{33}, then (5) can be used to calculate a_{43}. Hence the third column of A is identified.

The proof is completed by mimicking the procedures of the above paragraph, for then all remaining columns of A can be successively identified.

A related result appears in [8, p. 337] where it is shown that a catenary n-compartment system is identifiable provided the tracer is injected into compartment n, the concentration of the nth compartment is the output, and excretion occurs from the first compartment. Also Theorem 2 is an extension of Bright's work [6, p. 76] since to identify A he assumes that the ratios $a_{ij}V_j/a_{ji}V_i$ are all known.

V. ESTIMATION OF VOLUME FROM EXPERIMENTAL OBSERVATIONS

It is now possible to estimate the total volume of distribution of the tracer in system (1) provided the internal couplings of the system are further constrained.

At equilibrium, the rate of transfer from compartment i to compartment j must be exactly equal to the rate of transfer from j to i. Hence

$$a_{ij}V_j = c_{i,eq}a_{ji}V_i/c_{j,eq} . \tag{6}$$

For all $i, j,$ let

$$\gamma_{ij} \equiv c_{i,eq}/c_{j,eq}. \tag{7}$$

Now introduce $(V\text{ dist})_j$ as the *volume of distribution of the tracer in compartment* j with reference to the concentration of the tracer in compartment 1 as measured by sampling, i.e.,

$$(V\text{ dist})_j \equiv q_{j,eq}/c_{1,eq}, \quad j = 2,3,\ldots,n,$$

and define

$$(V\text{ dist}) \equiv V_1 + \sum_{j=2}^{n} (V\text{ dist})_j$$

as the *total volume of distribution* of the tracer in system (1). Thus from (7) and (6),

$$(V\ \mathrm{dist})_j = \gamma_{j1}V_j = a_{j1}V_1/a_{1j} \tag{8}$$

whenever $a_{1j} \neq 0$, $j = 2,3,\ldots,n$. The next theorem is an immediate consequence of (8) since a_{21} and a_{12} can be estimated (Theorem 2).

Theorem 3. If system (1) has catenary structure, then

$$(V\ \mathrm{dist})_2 = a_{21}V_1/a_{12}$$

is always identifiable.

Let B be the $n \times n$ matrix with (i,j) - entry $b_{ij} \equiv \gamma_{ji}a_{ij}$. From (6) and (7), it is now seen that the volumes V_j satisfy the system

$$\sum_{j=1}^{n} b_{ij}V_j = \begin{cases} -a_{01}V_1, & i = 1 \\ 0, & i = 2,\ldots,n \end{cases}$$

of linear algebraic equations, or in vector form,

$$Bv = -a_{01}V_1 e \tag{9}$$

where $v^T \equiv (V_1, V_2, \ldots, V_n)^T$ and $e^T \equiv (1,0,\ldots,0)^T$.

A known result [6, equation (17)] precipitates from equation (9) if it is assumed (just for this paragraph) that system (1) is such that if all irreverisble outflow is stopped (set $a_{01} = 0$), then the concentration of the tracer would become uniform. Under this assumption, the ratio $\gamma_{ij} = 1$ for all i, j, the matrix $B = A$, and $(V\ \mathrm{dist})_j = V_j$ from (8). Denoting the elements of A^{-1} by $a_{ij}^{(-1)}$ and using the ℓ^1-vector norm, it follows from (9) that

$$(V\ \mathrm{dist}) = \|v\|_1 = a_{01}V_1\|A^{-1}e\|_1 = a_{01}V_1 \sum_{i=1}^{n} |a_{i1}^{(-1)}| = a_{01}V_1\mu_1,$$

so that $(V\ \mathrm{dist})$ is given in terms of identifiable parameters.

As a second case, assume that the configuration of compartments in system (1) is of the mammillary type [1, p. 55]. The matrix A is thus assumed to have positive entries on its first row, first column, and along the diagonal, and zeros elsewhere.

Theorem 4. If (1) is a mammillary system, then the total volume of distribution is given by $a_{01}V_1\mu_1$.

Proof: The $(i,1)$ - entry of $AA^{-1} = I$ is

$$0 = a_{i1}a_{11}^{(-1)} + a_{ii}a_{i1}^{(-1)} + \sum_{k\neq 1,i} a_{ik}a_{k1}^{(-1)}$$

for any $i \geq 2$. The summation term is zero since A is a mammillary matrix. Thus

$$0 = a_{i1}a_{11}^{(-1)} + a_{ii}a_{i1}^{(-1)} = (a_{i1} / -a_{01}) - a_{1i}a_{i1}^{(-1)} \tag{10}$$

for $i \geq 2$. Hence from (8) and (10),

$$\begin{aligned}
(V \text{ dist}) &= V_1 + V_1 \sum_{i=2}^{n} a_{i1} / a_{1i} \\
&= V_1(1 + a_{01} \sum_{i=2}^{n} |a_{i1}^{(-1)}|) \\
&= V_1(1 + a_{01}(\mu_1 - |a_{11}^{(-1)}|)) \\
&= V_1(1 + a_{01}(\mu_1 - 1/a_{01})) = a_{01}V_1\mu_1.
\end{aligned}$$

VI. AN APPLICATION

The application of concern here deals with the therapy of shock [9]. Shock of all forms appears to be related to inadequate tissue perfusion. There exists a low-flow state in vital organs and a loss in the volumes of fluids. All the responses to the reduction of fluid volumes eventually result in a decrease in flow to tissues and the initiation of compensatory mechanisms directed at correction of the low-flow state. One such compensation is the movement of extracellular fluid (ECF) into the circulation. Recent evidence that ECF plays an important role in hemorrhagic shock has invoked new attention to the study of the interstitial space of the body. In treating this problem, it is essential to get some measurement of the ECF volume. The space is somewhat inaccessible to analysis and so the development of methodology has taken some years. As it has turned out, sulfur - 35 - labeled sodium sulfate appears to be a satisfactory and convenient tracer for obtaining such information as the estimation of the ECF volume in man as well as certain animals such as dogs [10] and baboons. The human system has been modeled in the form of (1) with a 3-compartment catenary structure in which the first compartment includes blood plasma, the second is the nonplasma ECF, and the third includes intracellular fluid. With this model, a measurement of the ECF volume $(V\ \mathrm{dist})_2$ can be made via Theorem 3.

The S^{35} curves tend to show a rapid exponential slope followed by a second slower exponential disappearance curve. Visually a change in the slope of the plotted curve revealing the presence of an interacting third compartment (intracellular fluid) cannot be seen, if at all, until the elapsed time is much greater than *180* minutes. For clinical reasons there is strong interest in reducing the time period of observations. For these reasons, but especially because of the apparent lack of cellular penetration during the specified time period, the system has also been modelled as a two-compartment system. In this 2-compartment model both $(V\ \mathrm{dist})_2$ and $(V\ \mathrm{dist})$ can be calculated using the results obtained in Theorems 3 and 4.

VII. FINAL REMARKS

It is worth observing that certain of the parameters affiliated with system (1), notably a_{01}, μ_1, and - when it can be calculated - $(V\ \mathrm{dist})$ are given in terms of

$$\int_0^\infty t^j c_1(t)dt$$

and thus can be computed numerically. This avoids the touchy problem of curve-fitting (usually sums of exponentials) to get a closed form for $c_1(t)$. Another problem is present: truncation error. This is a consequence of the fact that data for $c_1(t)$ is collected only over a finite time-interval $[0,T]$ and not $[0,\infty)$. The question of how to handle this error is an interesting and important problem that has not yet been fully resolved.

ACKNOWLEDGMENTS

I would like to thank Drs. J. W. Drane and G. T. Shires for their guidance and helpful remarks during the course of this research.

REFERENCES

[1] Jacquez, J. A. (1972). "Compartmental Analysis in Biology and Medicine", Elsevier Publishing Co., Amsterdam.

[2] Anderson, D. H. "Iterative inversion of single-exit compartmental matrices", to appear, *Comput. Biol. Medicine*.

[3] Lancaster, P. (1969). "Theory of Matrices", Academic Press, New York.

[4] Hearon, J. Z. (1963). "Theorems on linear systems", *Ann. of the N.Y. Acad. of Sci.*, *108*, 36-68.

[5] Hearon, J. Z. (1972). "Residence times in compartmental systems and moments of a certain distribution", *Math. Biosci.* *15*, 69-78.

[6] Bright, P. B. (1973). "The volumes of some compartment systems with sampling and loss from one compartment", *Bull. Math. Biol.* *35*, 69-79.

[7] Hearon, J. Z., and London, W. (1972). "Path lengths and initial derivatives in arbitrary and Hessenberg compartmental systems", *Math. Biosci. 14,* 121-134.

[8] Bellman, R., and Åström, K. J. (1970). "On structural identifiability", *Math. Biosci. 7,* 329-339.

[9] Shires, G. T., and Carrico, C. J. (1966). "Current status of the shock problem", Current Problems in Surgery, M. M. Ravitch (ed.), Year Book Medical Publishers, Chicago, Ill.

[10] Walser, M., et al. (1963). "An evaluation of radiosulfate for determination of the volume of extracellular fluid in man and dogs", *J. Clin. Inves. 32,* 299.

[11] Bergner, P. E. (1977). "Application of a theorem by Bright to the generalized tracer system", *Bull. Math. Biol. 39,* 167-178.

ON IDENTIFICATION OF COMPARTMENTAL SYSTEMS

S. Sandberg
D. H. Anderson[1]
J. Eisenfeld[2]

Department of Medical Computer Science
University of Texas Health Science Center
Dallas, Texas

1. INTRODUCTION

A linear time-invariant compartmental model, which is frequently used to describe a physiological process [7], can be represented by the system of differential equations ($t \geq 0$)

$$\dot{x}(t) = Ax(t) + Bu(t) \tag{1a}$$

$$x(0) = 0 \tag{1b}$$

$$y(t) = Cx(t) \tag{1c}$$

where $A = [a_{ij}]$ is an $n \times n$ matrix and $B = [b_{ij}]$ and $C = [c_{ij}]$ are matrices of appropriate dimensions. In biomedical compartmental problems we often have the following interpretations: $x(t) = [x_i(t)]$ is the $n \times 1$ vector whose i^{th} component is the amount of tracer in compartment i at time t; $y(t)$ is

[1]*Also affiliated with the Department of Mathematics, Southern Methodist University, Dallas, Texas 75275.*

[2]*Also affiliated with the Department of Mathematics, The University of Texas at Arlington, Arlington, Texas 76019.*

ISBN 0-12-434180-2

the vector of observations; $[Bu(t)]_i$ is the amount of tracer input to compartment i at time t; a_{ij} is the fractional transfer coefficient of tracer from compartment j to compartment i. (For specific examples of compartmental physiological models see [1,3,7,9].)

The purpose of this paper is to point out some properties of certain types of compartmental systems. In addition to the general case for the matrix A, we will be concerned with two subclassifications of A: the mammillary system and the catenary system [7,6]. The paper will address two basic questions. The first concerns the restriction of the eigenvalues of the compartmental matrix A to specified subsets of the complex plane. In some cases it may be deduced that the eigenvalues are negative and distinct. A second concern is to obtain some properties of the impulse-response function, $Ce^{tA}B$, in special cases.

II. GENERAL COMPARTMENTAL MATRICES

A *compartmental matrix* $A = [a_{ij}]$ is an $n \times n$ matrix having the following properties [7, p. 48]:

(1) $a_{ij} \geq 0$, $i \neq j$;

(2) $a_{ii} < 0$ for all $i = 1,2,\ldots,n$;

(3) $\sum_{i=1}^{n} a_{ij} = -a_{0j} \leq 0$, $1 \leq j \leq n$ (a_{0j} = excretion from component j).

The matrix A is diagonally dominant with respect to the columns,

$$|a_{ii}| \geq \sum_{j \neq i} a_{ji}, \quad i = 1,2,\ldots,n,$$

where the summation is over all $j = 1,2,\ldots,n$, except $j = i$. Equality holds in the ith inequality only if $a_{0i} = 0$. If the eigenvalues of A are denoted λ_i and ordered as

$$|\lambda_1| \geq |\lambda_2| \geq \ldots \geq |\lambda_n|,$$

then the diagonal dominance coupled with the Gerschgorin Circle Theorem imply that $\mathrm{Re}(\lambda_i) \leq 0$ and that no eigenvalue is purely imaginary [7, p. 50]. Moreover, if $\lambda_n \neq 0$, then A is an invertible matrix and $-A^{-1} \geq 0$, i.e., each entry of the matrix $-A^{-1}$ is non-negative [6, p. 49]. If A is similar to a symmetric matrix or, equivalently, symmetrizable, then all of its eigenvalues must be real and nonpositive. In applications the eigenvalues are usually taken to be real; this raises the question of whether there are compartmental matrices with nonreal eigenvalues. The answer is affirmative, for consider

$$A = \begin{bmatrix} -2.0 & 0.5 & 0.0 & 0.75 \\ 1.0 & -1.0 & 0.25 & 0.10 \\ 0.0 & 0.25 & -1.0 & 0.15 \\ 0.0 & 0.25 & 0.75 & -1.0 \end{bmatrix}$$

Here $\lambda_1 = -2.34$, $\lambda_2, \lambda_3 = -1.25 \pm 0.24\sqrt{-1}$, and $\lambda_4 = -0.21$.

For any given compartmental matrix A, define

$$c = \max_{1 \leq i \leq n} |a_{ii}|;$$

also I will denote the identity matrix.

Theorem 1. The spectral radius $\rho(A+cI)$ of $A + cI$ is $c + \lambda_n$.

Assuming irreducibility of $A + cI$, this result has been proved by Hearon [6, p. 44]. However, it can be shown that the irreducibility condition can be eliminated from the hypothesis.

Theorem 2. Let A be any compartmental matrix.

(1) If S and s are the largest and smallest row sums, respectively, of A, then $s \leq \lambda_n \leq S$.

(2) The eigenvalue λ_n satisfies

$$|\lambda_n| \leq \min_{1 \leq i \leq n} |a_{ii}|$$

and

$$\min_{1 \leq i \leq n} a_{0i} \leq |\lambda_n| \leq \max_{1 \leq i \leq n} a_{0i}.$$

Proof. The matrix $A + cI \geq 0$. By the standard continuity argument we can assume $A + cI > 0$. Therefore there exists a positive eigenvector $w = [w_i]$ corresponding to the eigenvalue $c + \lambda_n = \rho(A+cI)$. Let $w_\ell = \max_{1\leq i\leq n} w_i$ and $w_t = \min_{1\leq i\leq n} w_i > 0$. The equation $(A+cI)w = (c+\lambda_n)w$ implies

$$\sum_{j\neq i} a_{ij}w_j = (\lambda_n - a_{ii})w_i, \quad i = 1,2,\ldots,n. \tag{2}$$

If $i = \ell$, then (2) yields

$$(\sum_{j\neq \ell} a_{\ell j})w_\ell \geq (\lambda_n - a_{\ell\ell})w_\ell$$

or

$$S \geq \sum_{j=1}^{n} a_{\ell j} \geq \lambda_n.$$

Similarly,

$$\lambda_n \geq \sum_{j=1}^{n} a_{tj} \geq s$$

which completes the proof of the first result.

Since $a_{ij} \geq 0$, $i \neq j$, and $w_j > 0$, for all $j = 1,2,\ldots,n$, then we must conclude that $0 \leq \lambda_n - a_{ii}$, $i = 1,2,\ldots,n$, from (2). Thus

$$|\lambda_n| \leq \min_{1\leq i\leq n} |a_{ii}|$$

Applying the argument of the previous paragraph to the transpose of $A + cI$, we get the second result stated in part 2 of the theorem.

III. SINGLE INPUT-OBSERVATION COMPARTMENTAL (SIOC) SYSTEMS

In this section we report on new results concerning compartmental systems in which both observation and input occur in the same compartment, say compartment *1*, and only in that compartment.

Thus in Equation (1), we take $B^T = (\beta, 0, \ldots, 0)$ and $C = (\delta, 0, \ldots, 0)$ where $\beta > 0$ and $\delta > 0$. For convenience we may assume, without loss of generality, that

$$B^T = C = (1, 0, \ldots, 0). \tag{3}$$

We also assume that A is symmetrizable by a positive definite diagonal matrix $Q = \mathrm{diag}(q_1, q_2, \ldots, q_n)$. This class includes the catenary system and the mammillary system, both of which will be defined later. For such systems we have the following result.

Theorem 3. The impulse response function $I(t) = Ce^{tA}B$ has the form

$$I(t) = \sum_{i=1}^{n} \alpha_i e^{\lambda_i t} \tag{4}$$

where $\lambda_1 \leq \lambda_2 \leq \ldots \leq \lambda_n \leq 0$ are the eigenvalues of A and $\alpha_i \geq 0$, for $i = 1, 2, \ldots, n$.

Proof. Since A is symmetrizable, the λ_i are real and so $\lambda_1 \leq \lambda_2 \leq \ldots \leq \lambda_n \leq 0$. Let Φ^i, $i = 1, 2, \ldots, n$, be eigenvectors of A corresponding to the eigenvalues λ_i. Also due to the symmetrizability of A, we may assume that the Φ^i are orthonormal with respect ot the Q-inner product:

$$\langle \Phi^i, \Phi^j \rangle \equiv (\Phi^i)^T Q \Phi^j = \delta_{ij}.$$

The vector B can be expanded in terms of the Φ^i,

$$B = \sum_{i=1}^{n} b_i \Phi^i,$$

where each b_i is given by

$$\langle \Phi^i, B \rangle = \Phi_1{}^i q_1$$

in which $\Phi_1{}^i$ is the first component of Φ^i. Now

$$Ce^{tA}B = B^T e^{tA} \Sigma b_i \Phi^i = \Sigma b_i B^T e^{tA} \Phi^i$$

$$= \Sigma b_i B^T e^{\lambda_i t} \Phi^i = \Sigma q_1 (\Phi_1{}^i)^2 e^{\lambda_i t},$$

which completes the proof. (Observe that the coefficients $\alpha_i \equiv q_1(\Phi_1{}^i)^2$ are *strictly* positive if and only if $\Phi_1{}^i \neq 0$.)

IV. SPECIAL SYSTEMS

A. *Mammillary Systems*

The *mammillary compartmental system* consists of a main, or mother, compartment with all the other compartments connected to the mother compartment. None of the other compartments are connected to each other. The corresponding mammillary compartmental $n \times n$ matrix A is similar to a symmetric matrix and thus has real, nonpositive eigenvalues [6]. Sheppard and Householder [8, p. 514] have shown that the eigenvalues of a mammillary matrix are distinct provided that

$$a_{1i} \neq a_{1j} \quad \text{for all} \quad (i,j), \quad i \neq j \neq 1,$$

$$a_{1i} a_{i1} > 0 \quad \text{for} \quad i = 1,2,\ldots,n. \tag{5}$$

Theorem 4. Let (A,B,C) be a SIOC mammillary system of n compartments with excretion only from the main compartment. If condition (5) is not satisfied, then there is a realization $(\tilde{A},\tilde{B},\tilde{C})$ of smaller dimension such that $\tilde{A}$ is a SIOC mammillary matrix with the same main compartment.

Proof. *Case 1:* $a_{1i} = a_{1j}$ for some (i,j), $i \neq j \neq 1$. The system of differential equations for a mammillary system with excretion only from the first compartment $(a_{01} > 0)$ is

$$\begin{aligned}
\dot{x}_1 &= a_{11}x_1 + a_{12}x_2 + \ldots + a_{1n}x_n + d_1 \\
\dot{x}_2 &= a_{21}x_1 - a_{12}x_2 + \ldots \qquad + d_2 \\
&\vdots \\
\dot{x}_n &= a_{n1}x_1 + \ldots \qquad - a_{1n}x_n + d_n
\end{aligned}$$

where

$$d_i = \sum_{j=1}^{n} b_{ij}u_j, \quad a_{11} = -\sum_{i\neq 1} a_{i1} - a_{01}.$$

Let $a_1 = a_{1j} = k$ and $z(t) = x_i(t) + x_j(t)$. The equations now become a $(n-1)$ system of differential equations corresponding to a $(n-1)$ compartmental system.

$$\begin{aligned}
\dot{x}_1 &= a_{11}x_1 + a_{12}x_2 + \ldots + kz + \ldots + a_{1n}x_n + d_1 \\
\dot{x}_2 &= a_{21}x_1 - a_{12}x_2 + \ldots \qquad + d_2 \\
&\vdots \\
\dot{z} &= (a_{i1}+a_{j1})x_1 + \ldots \quad - kz + \ldots \qquad + d_i + d_j \\
&\vdots \\
\dot{x}_n &= a_{n1}x_1 + \ldots \qquad - a_{1n}x_n + d_n
\end{aligned}$$

Case 2: $a_{1i}a_{i1} = 0$ for some $i, i \neq 1$. If either $a_{1i} = 0$ or $a_{i1} = 0$ it can be shown that x_i can be deleted from the system and then one obtains a $(n-1)$ - consistent system of equations.

Corollary 1. If condition (5) is not satisfied, then there exists a realization $(\tilde{A},\tilde{B},\tilde{C})$ such that condition (5) is satisfied and $\tilde{A}$ is a SIOC mammillary matrix of order $< n$.

Proof. It follows from repeated application of Theorem 4.

Theorem 5. The representation (4) of $I(t)$ in the case of a SIOC mammillary matrix with excretion only from compartment 1 has n linearly independent terms if and only if condition (5) is satisfied.

Proof. First the necessary condition. Suppose condition (5) is violated. Then by applying corollary 1 it can be seen that there is a realization with less than n compartments and thus $I(t)$ has less than n linearly independent terms.

Now for the sufficient condition. Suppose condition (5) holds. Then it is known that all the eigenvalues are distinct [8]. It is now necessary to show $\alpha_i > 0$, $i = 1,2,\ldots,n$. From the remark following the proof of Theorem 3, it suffices to show that for each eigenvector ϕ of A, the first component ϕ_1 is not equal to zero. Let $A\phi = \lambda\phi$ and suppose to the contrary that $\phi_1 = 0$. Then $-a_{12}\phi_2 = \lambda\phi_2$. Thus either $\lambda = -a_{12}$ or $\phi_2 = 0$. But $\lambda \neq -a_{1j}$ for any j [8, p. 514]. Hence $\phi_2 \equiv 0$. The same argument applies to all ϕ_i, $i = 3,4,\ldots,n$ so that $\phi \equiv 0$. But this contradicts the fact that ϕ is nonzero since it is an eigenvector of A.

Since there is a realization where the number of components is precisely equal to the number of linearly independent terms in the representation (4) of $I(t)$, this implies that we can always model the system in such a fashion that *all* the eigenvalues are observable.

This comment brings up the problem of determining the number of terms in the representation (4) of $I(t)$. To this end we point out that for SIOC systems we have the identification problem which consists of two equations, the representation equation (4) and the convolution equation

$$y(t) = \int_0^t I(t-s)u(s)ds$$

where $y(t) = x_1(t)$ is observed and $u(t)$ is the input. These two equations are precisely the same equations which arise in the analysis of fluorescence decay data [5], where special techniques have been developed for parameter identification including the number of linearly independent terms in (4).

B. Catenary Systems

The *catenary* or *cascading compartmental system* is a group of compartments in a row connected in such a manner that each compartment interacts only with its adjacent neighbors.

The associated catenary compartmental matrix, which is a tridiagonal matrix, is similar to a symmetric matrix and thus its eigenvalues are real and nonpositive [6]. These eigenvalues are distinct [4, p. 30] provided that

$$\begin{matrix} a_{i,i+1} \neq 0 \\ a_{i+1,i} \neq 0 \end{matrix} \qquad 1 \leq i \leq n. \tag{6}$$

Theorems 4 and 5 and Corollary 1 will also be true if the word catenary is substituted for mammillary and condition (6) for condition (5). The proofs are obtained in a similar way.

REFERENCES

[1] Anderson, D. H., et al (1976). "The Mathematical Analysis of a Four-Compartment Stochastic Model of Rose Bengal Transport Through the Hepatic System", Proceedings of the International Conference on Non Linear Systems, Academic Press.

[2] Bellman, R. (1970). "On Structural Identifiability", *Math. Biosciences 7*, 329-339.

[3] Dominquez, R. (1950). "Kinetics of Elimination, Absorption and Volume of Distribution in the Organism", *Med. Physics 2*, Ed. by O. Glosser, Year Book Publishers, Inc., Chicago, 476-489.

[4] Faddeev, D. K., and Faddeeva, V. N. (1963). "Computational Methods of Linear Algebra", W. H. Freeman and Company, San Francisco.

[5] Ford, C. C. "Some Unresolved Questions Pertaining to the Mathematical Analysis of Fluorescence Decay Data", Applied

Nonlinear Analysis, V. Lakshmikantham, Ed., Academic Press, New York.

[6] Hearon, J. Z. (1963). "Theorems on Linear Systems", *Annals of the New York Academy of Sciences 108*, 36-68.

[7] Jacquez, J. A. (1972). "Compartmental Analysis in Biology and Medicine", Elsevier Publishing Co., Amsterdam, The Netherlands.

[8] Sheppard, C. W., and Householder, A. S. (1951). The Mathematical Basis of the Interpretation of Tracer Experiments in a Closed Steady-State System", *J. of Applied Physics 22, 11*, 510-520.

[9] Sheppard, C. W. (1962). "Basic Principles of the Tracer Method", John Wiley and Sons, New York.

PRECONDITIONING FOR CONSTRAINED OPTIMIZATION PROBLEMS WITH APPLICATIONS ON BOUNDARY VALUE PROBLEMS

*Owe Axelsson**

Center for Numerical Analysis
The University of Texas at Austin
Austin, Texas

ABSTRACT

Constrained boundary value problems appear in problems where an incompressibility constraint is valid as for rubber-like materials, where det $J(\vec{u}) = 1$, J the local Jacobi transformation matrix, $dV = FdV_0$. However, a constraint may also be introduced due to a mixed variable formulation, as in the stream function (ψ) - vorticity (ω) formulation, where $\omega = \Delta\psi$.

The constrained problem may be formulated as a Lagrangian functional problem. The Lagrangian multiplier corresponds then to the hydrostatic pressure p.

To prove boundedness of p may be difficult for nonlinearly constrained problems. A constructive approach for this is presented, which at the same time provides a method for the numerical solution of the problem.

**On leave from Chalmers University of Technology, Gothenburg, Sweden.*

ISBN 0-12-434180-2

The initial approximation is calculated from the penalized functional corresponding to a perturbed Lagrangian with perturbation parameter ε. A modified minimal residual conjugate gradient method, preconditioned by a linearized perturbed Lagrangian operator, is proved to converge if ε is small enough. A slightly weakened form of the so called Brezzi-Babuška condition is used in the proof.

The discretized perturbed operator is factorized approximately by an incomplete factorization. This is achieved by the use of a preconditioning by incomplete factorization for the corresponding primal variable (or for the dual variable) formulation.

Preliminary numerical results for this latter approach applied on a stream function, vorticity formulation (i.e., a first biharmonic problem) is also presented.

EVALUATION OF QUASI-LINEAR TECHNIQUES FOR NONLINEAR PROCESSES WITH RANDOM INPUTS

M. Balakrishna

Rockwool Industries, Inc.
Belton, Texas

David A. Hullender

Mechanical Engineering
The University of Texas at Arlington
Arlington, Texas

I. INTRODUCTION

Several approaches are mentioned in literature [1-5] for modeling of nonlinear processes with either deterministic or stochastic inputs. Quasilinear modeling techniques have been popular especially for systems with random inputs because of their simplicity in obtaining closed form solutions. These techniques require that the probability density functions for the system variables either be known or assumed. In this paper, a model is formulated for a nonlinear process that is representative of a guideway profile with constrained irregularities. Relative merits of two quasilinear techniques in approximating the nonlinearity is presented. The results of this study indicate that for this application, the quasilinear approximation based on the probability density function approach is more

ISBN 0-12-434180-2

accurate compared to that based on the traditional random input describing function approach.

II. MODEL FORMULATION

As mentioned earlier, the nonlinear process under study relates to a guideway profile with constrained irregularities. The proposed method for constraining the guideway irregularities is illustrated in Figure 1. The elevation of the guideway profile at each check point is constrained such that the difference between its elevation and the elevation of a number, m, of previous check points does not exceed a specified value, δ. Thus, for $m = 0$, the guideway remains unconstrained and for $m = \infty$, the entire guideway profile falls within the specified band. The model that is being formulated relates to the later case and is useful in determining the power spectral density (PSD) of the guideway profile with constrained irregularities.

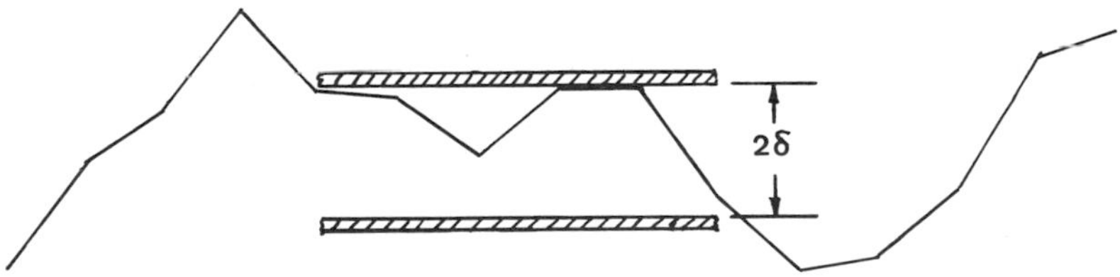

FIGURE 1a. Finite length constraint used to bound guideway irregularities within a band with width of 2δ.

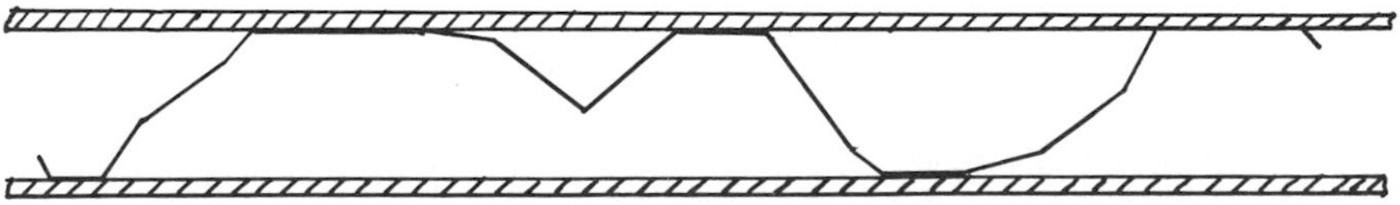

FIGURE 1b. Infinitely long constraint used to bound the entire guideway profile within a band with width of 2δ.

The block diagram used to formulate the analytical model for the PSD of a guideway profile with constrained irregularities is shown in Figure 2. The modeling process begins by sampling a white noise random process at intervals of h. A saturation nonlinearity is utilized to constrain the signal within bounds of $\pm\,\delta$. A constrained random walk process is generated by summing the input sequences and the output signal of the nonlinearity delayed by a sampling period, h. The output of the nonlinearity is processed through a polygonal hold which generates a guideway profile with strainght sections.

Making a quasilinear approximation for the nonlinearity and replacing it by an equivalent constant gain, C, the block diagram reduces to the following transfer function for $X(s)$:

$$X(s) = \frac{e^{sh}(1-e^{-sh})^2}{hs^2}\,\frac{C}{(1-Ce^{-sh})}\,A^*(s). \tag{1}$$

Thus the power spectral density for x can be written as follows:

$$S_x(s) = \frac{e^{sh}(1-e^{-sh})^2C}{hs^2(1-Ce^{-sh})}\,\frac{e^{-sh}(1-e^{sh})^2C}{h(-s)^2(1-Ce^{sh})}\,S_a(s). \tag{2}$$

Using trignometric identities equation (2) simplifies to

$$S_x(s) = \frac{4C^2}{h^2s^4}\,\frac{(1-\cosh sh)^2}{(1+C^2-2C\cosh sh)}\,S_a(s). \tag{3}$$

From reference [7], the spectrum for an uncorrelated sequence of random numbers sampled at an interval of h with a standard deviation of σ_a is given by

$$S_a(s) = \sigma_a^2/h. \tag{4}$$

Substituting equation (4) into (3) and replacing s by $j\Omega$ and then multiplying by two in order to achieve a one sided spectrum, the expression for the guideway PSD becomes

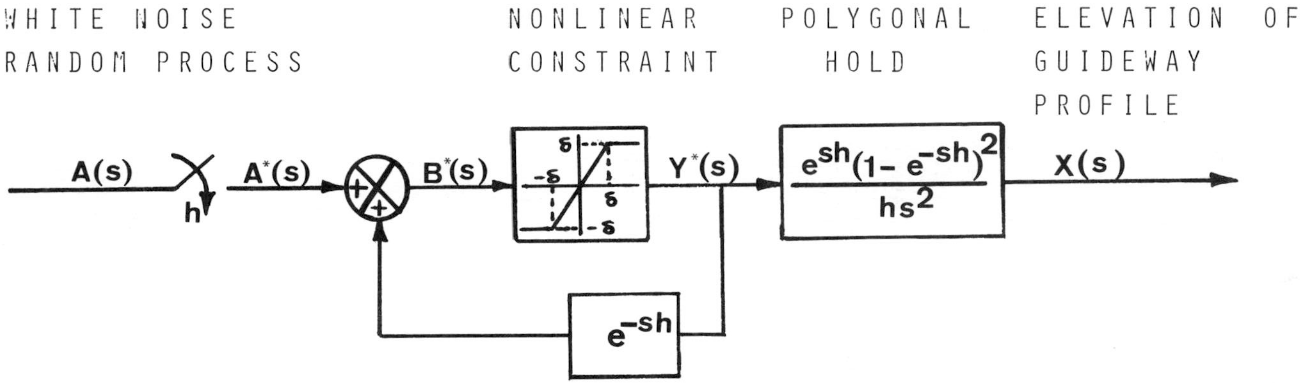

FIGURE 2. Block diagram of a constrained random walk process representative of guideway profile irregularities.

$$S_x(\Omega) = \frac{8\sigma_a^2 C^2 (1 - \cos \Omega h)^2}{h^3 \Omega^4 (1 + C^2 - 2C \cos \Omega h)} . \tag{5}$$

In order to make use of equation (5), the magnitude of C is needed. However, the magnitude of C depends on the statistical characteristics of the input, σ_b, to the nonlinearity. In addition, the statistics of σ_b are a function of the statistical characteristics of the output, σ_y, of the nonlinearity. In the next section, two methods are presented to evaluate the equivalent gain C.

III. EVALUATION OF EQUIVALENT GAIN, C

In this section, two approaches are presented for approximating the saturation nonlinearity shown in Figure 2. They are:

(1) Random Input Describing Function (RIDF) approach,

(2) Probability Density Function (PDF) approach.

The Random Input Describing Function (RIDF) approach, also called the minimum mean squared error describing function approach, was presented by Booton [2] in 1953. He obtained an equivalent gain function of a nonlinear device by comparing the output of the nonlinearity with the output of an equivalent gain function. The equivalent gain was chosen as the one which would yield the minimum RMS difference between the two outputs. Figure 3 summarizes the RIDF approach.

The Probability Density Function (PDF) approach, also called the output variance equivalent describing function approach, was presented by Barrett and Coales [3] in 1955. The same approach was used by Axelby [4] in the treatment of biased random inputs. The method utilizes the probability density function for the nonlinearity to compute the RMS value of the output of the nonlinearity. Knowing the RMS value of the input, the equivalent gain is evaluated by taking the ratio of the output and input standard deviations.

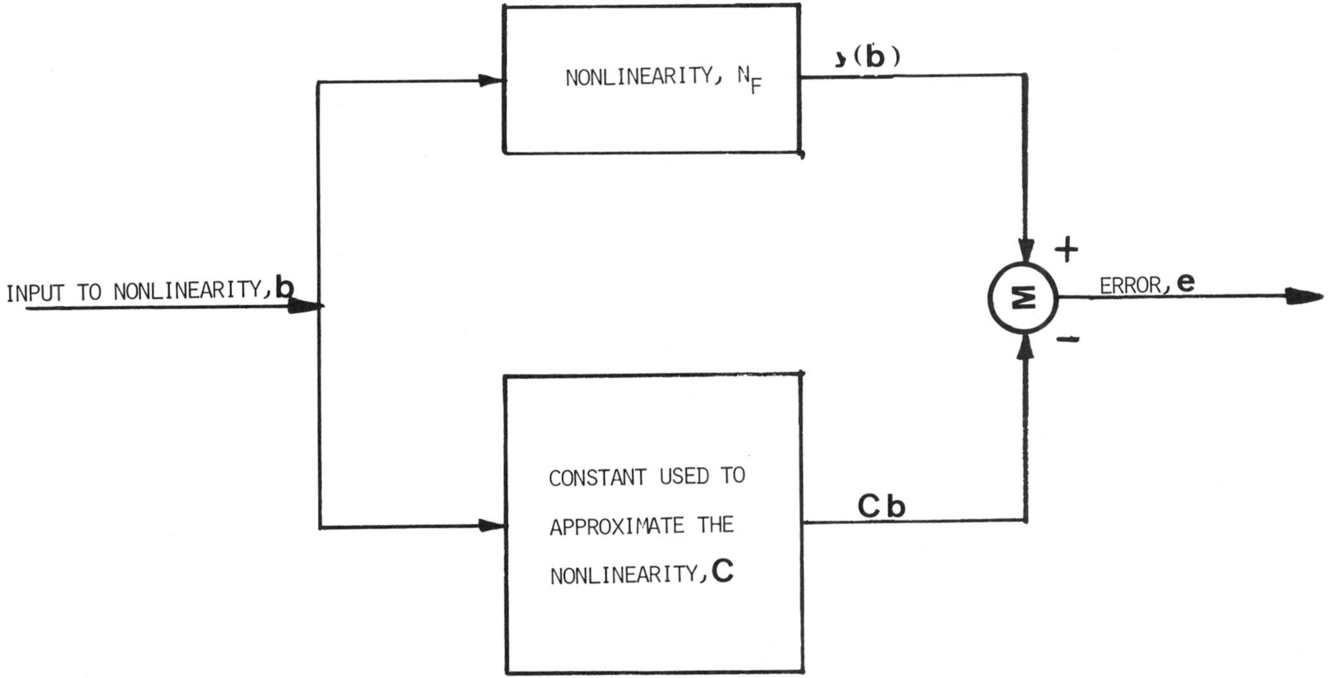

FIGURE 3. Block diagram demonstrating that the RIDF error is equal to the true output of the nonlinearity minus the approximation for the output.

A. *RIDF Approach*

Denoting the output of the nonlinearity by y and the quasi-linear approximation to the output of the nonlinearity by $C\,b$, then the mean square error, $\overline{e^2}$, associated with the approximation is given by

$$\overline{e^2} = \overline{(y-Cb)^2} = \overline{y^2} - 2C\overline{yb} + C^2\overline{b^2}\,. \tag{6}$$

As demonstrated in reference [3], $\overline{e^2}$ is minimum if C is defined in terms of the probability density function, $f(b)$, as follows

$$C = \frac{1}{\overline{b^2}} \int_{-\infty}^{\infty} by f(b)\,db = \frac{\overline{by}}{\overline{b^2}}\,. \tag{7}$$

Assuming the input to the nonlinearity, b, to be Gaussian, the equation for C simplifies to

$$C = erf\left(\frac{\delta}{\sqrt{2}\sigma_b}\right) \tag{8}$$

where $erf(\cdot)$ denotes the error function defined by

$$erf(\gamma) = \frac{2}{\sqrt{\pi}} \int_0^{\gamma} e^{-\beta^2}\,d\beta\,. \tag{9}$$

In order to evaluate C using equation (8), the standard deviation σ_b associated with the input to the nonlinearity is needed. But σ_b is in turn a function of C. Thus an additional equation relating C and σ_b is needed and this is derived in Appendix A. The result is

$$\sigma_b^2 = \frac{1}{1-C^2}\,\sigma_a^2\,. \tag{10}$$

Substituting σ_b from (10) into (8) gives

$$C = erf(\delta\sqrt{1-C^2}/\sqrt{2}) \tag{11}$$

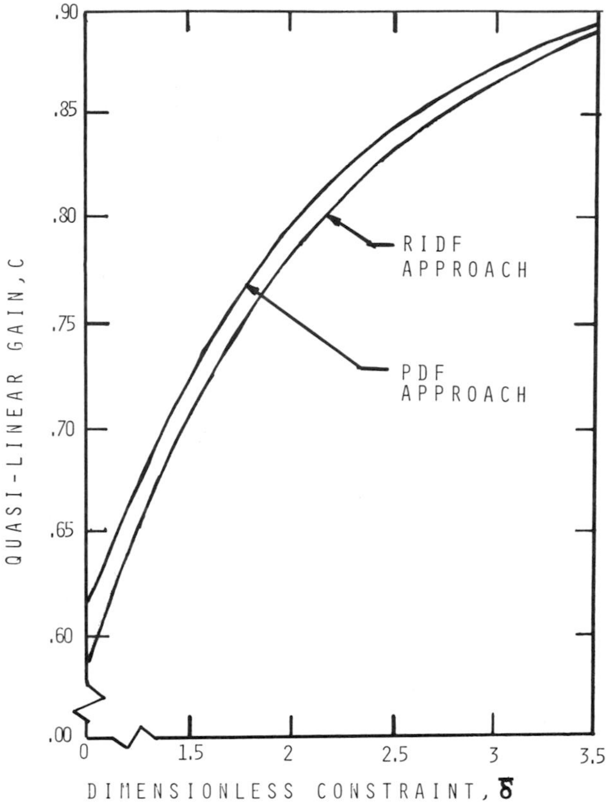

FIGURE 4. Graphical representation of equivalent gain, C, as a function of the dimensionless constraint parameter, $\overline{\delta}$.

where $\overline{\delta} = \delta/\sigma_a$.

It can be observed that equation (11) is transcendental in nature thus requiring an iterative type solution. The result is shown in Figure 4 as a function of the dimensionless parameter $\overline{\delta}$.

B. PDF Approach

Assuming that the input b to the nonlinearity to be Gaussian, the probability density function, $f(y)$ for the output of the nonlinearity can be shown to be

$$f(y) = \begin{cases} 0 & y < -\delta \\ \frac{1}{2}\left[1 - erf\left(\frac{\delta}{\sqrt{2}\sigma_b}\right)\right]\delta_D(y+\delta) & y = -\delta \\ \frac{1}{\sqrt{2\pi}\sigma_b}\exp(-y^2/2\sigma_b^2) & -\delta < y < \delta \\ \frac{1}{2}\left[1 - erf\left(\frac{\delta}{\sqrt{2}\sigma_b}\right)\right]\delta_D(y-\delta) & y = \delta \\ 0 & y > \delta \end{cases} \tag{12}$$

where $\delta_D(\cdot)$ denotes the impulse function. This nonlinear effect on the probability density function is shown in Figure 5.

The output variance, σ_y^2, of the nonlinearity is defined by

$$\sigma_y^2 = \int_{-\infty}^{\infty} y^2 f(y)dy. \tag{13}$$

Substituting (12) into (13) and integrating gives

$$\sigma_y^2 = \delta^2 + (\sigma_b^2 - \delta^2)erf\left(\frac{\delta}{\sqrt{2}\sigma_b}\right) - \frac{2\delta\sigma_b}{\sqrt{2\pi}}\exp\left(-\frac{\delta^2}{2\sigma_b^2}\right). \tag{14}$$

As derived in Appendix A, the input variance, σ_b^2, and the output variance, σ_y^2, of the nonlinearity are related by the expression

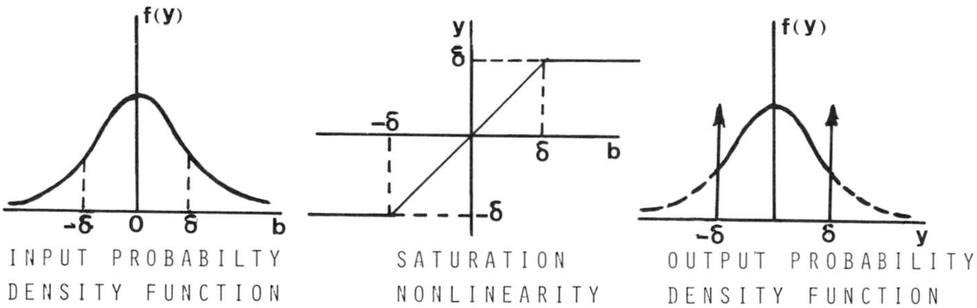

FIGURE 5. Input and output probability density functions for the saturation nonlinearity.

$$\sigma_b^2 = \sigma_a^2 + \sigma_y^2 \; . \tag{15}$$

The quasilinear approximation, C, for the PDF approach is equal to the ratio of the output standard deviation to the input standard deviation, i.e.,

$$C = \sigma_y / \sigma_b \; . \tag{16}$$

Combining equations (14), (15) and (16) and nondimensionalizing gives

$$C^2 = \overline{\delta}^2(1-C^2) + (1-\overline{\delta}^2(1-C^2))\,erf\left(\frac{\overline{\delta}\sqrt{1-C^2}}{\sqrt{2}}\right) - \frac{2\overline{\delta}\sqrt{1-C^2}}{\sqrt{2\pi}}\exp\left(-\frac{\overline{\delta}^2(1-C^2)}{2}\right) . \tag{17}$$

Equation (17) is plotted in Figure 4 for typical values of $\overline{\delta}$.

IV. COMPARISON OF THE TWO APPROACHES

The purpose of this section is to compare the two quasilinear approximations for the nonlinearity and to determine the relative accuracy of one over the other. This has been accomplished by resorting to two methods.

(1) Comparison based on guideway PSD; and

(2) Comparison based on nonlinearity output variance.

A. Accuracy Comparison Based on Guideway PSD

The PSD is computed using equation (5) with the gain C evaluated using the two approaches. The results are shown in Figure 6 for $\overline{\delta} = 1$. Also shown is the PSD obtained using numerical simulation. The numerical simulation results were obtained by following the block diagram shown in Figure 2. The exact characteristics of the saturation nonlinearity were included in order to determine the accuracy of the quasilinear analytical models. A total of *16,384* guideway profile elevation samples were

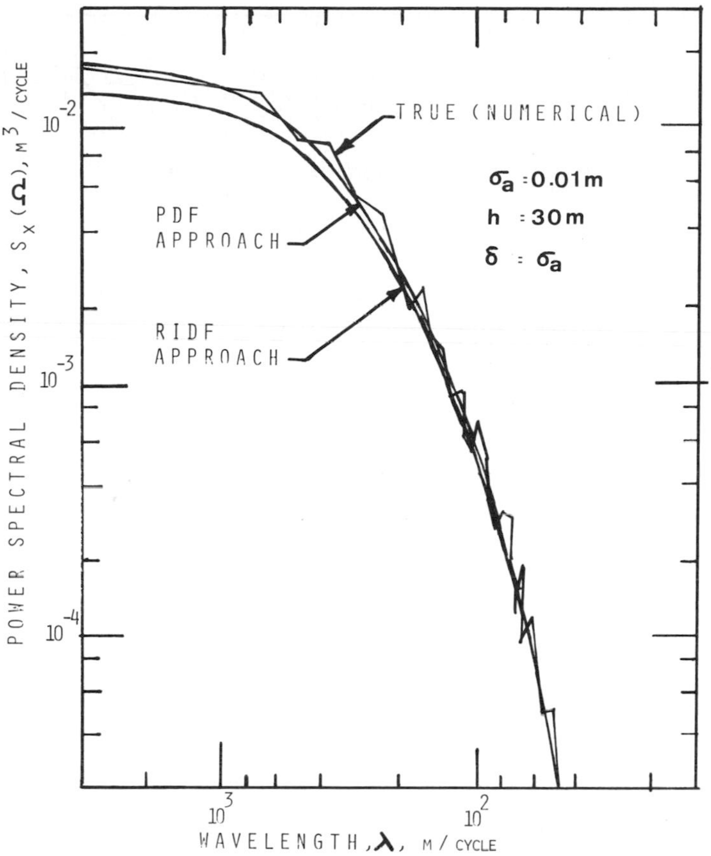

FIGURE 6. Accuracy comparison of the RIDF approach and the PDF approach based on the guideway PSD.

obtained and processed using a Fast Fourier Transform algorithm to compute the numerical PSD. The normalized standard error associated with the numerical results is *12.5%*.

Comparison of the curves shown in Figure 6 demonstrates the superiority of the PDF approach over the RIDF approach. The accuracy is more pronounced at the longer wavelengths where the roll over characteristic occurs.

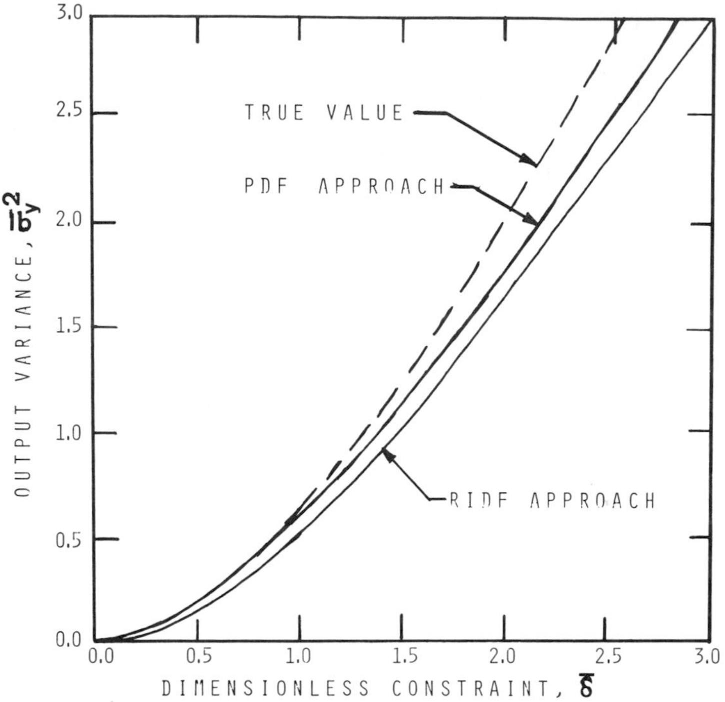

FIGURE 7. Accuracy comparison of the RIDF approach and the PDF approach with the true value in evaluating the nonlinearity output variance.

B. *Accuracy Comparison Based on Nonlinearity Output Variance*

This section presents an accuracy comparison of the RIDF approach and the PDF approach in computing the output variance of the saturation nonlinearity.

The nonlinearity output variance, σ_y^2 is obtained by combining equations (14) and (15). The result is

$$\sigma_y^2 = \frac{C^2}{1-C^2}\,\sigma_a^2 \quad . \tag{18}$$

where C is defined by equation (11) in the case of RIDF approach and by equation (17) in the case of PDF approach. The results for each case are shown in Figure 7. Also shown in the figure is the true value of σ_y^2 determined numerically. A total of *16,384* samples of the random variable "a" were used to generate the same number of values for y. These random variables were then used to compute σ_y^2.

The results have demonstrated the relative accuracy of the PDF approach over the RIDF approach. For instance, for $\overline{\delta} = 1$ the RIDF error is *6.1%* whereas the PDF error is only *0.6%*. For $\overline{\delta} = 2$, the corresponding errors are *7.1%* and *4.0%* respectively. Thus, for the process under consideration, the quasilinear approximation based on the PDF approach is significantly more accurate than that based on the RIDF approach.

V. SUMMARY AND CONCLUSIONS

The purpose of this paper is to evaluate the relative accuracies of two quasilinear techniques for nonlinear processes with stochastic inputs. Specifically, the nonlinear process under study related to a guideway profile with constrained irregularities. A model has been formulated and an equation has been derived to determine the PSD of guideway irregularities that have been constrained to fall within a band of width 2δ. The modeling process required the use of a saturation nonlinearity. A quasilinear approximation was used for the nonlinearity and two methods were presented to evaluate the quasilinear approximation. They were:

(1) the random input describing function (RIDF) approach;

(2) the probability density function (PDF) approach.

The accuracy of the two techniques was determined by evaluating the guideway PSD using the analytical model and then comparing the results to those obtained using the numerical simulation. Another comparison was by determining the output variance of the nonlinearity using the two approaches and comparing the results with the true value obtained numerically.

Based on these results, it is concluded that for the process under consideration, the quasilinear approximation based in the PDF approach is more accurate than that based on the RIDF approach.

VI. APPENDIX A

The purpose of this appendix is to derive Equation (10) presented in the text. Referring to Figure (2), the following relationship can be written

$$B^*(s) = A^*(s) + e^{-sh}Y^*(s). \tag{19}$$

Equation (19) can also be written as a difference equation

$$b(i) = a(i) + y(i-1). \tag{20}$$

Squaring both sides of (20) gives

$$b^2(i) = a^2(i) + 2a(i)y(i-1) + y^2(i-1). \tag{21}$$

Taking the expected value of both sides of (21) yields

$$E\{b^2(i)\} = E\{a^2(i)\} + 2E\{a(i)y(i-1)\} + E\{y^2(i-1)\}. \tag{22}$$

Since a is zero mean and stationary, then

$$E\{b^2(i)\} = \sigma_b^2 \tag{23}$$

$$E\{a^2(i)\} = \sigma_a^2 \tag{24}$$

$$E\{y^2(i-1)\} = \sigma_y^2 . \tag{25}$$

Also a is a white noise process and, thus, successive samples of a are independent and uncorrelated; that is

$$E\{a(i)y(i-1)\} = 0. \tag{26}$$

Substituting Equations (23)-(26) into (22) yields

$$\sigma_b^2 = \sigma_a^2 + \sigma_y^2 . \tag{27}$$

But $y = Cb$, and thus, $\sigma_y = C^2\sigma_b^2$. Substituting this result in (27) results in the desired equation for σ_b^2, that is,

$$\sigma_b^2 = \frac{1}{1-C^2}\,\sigma_a^2\,. \tag{28}$$

REFERENCES

[1] Gelb, A., and Vander Velde, W. E. (1968). "Multiple Input Describing Functions and Nonlinear System Design", McGraw-Hill Book Company.

[2] Booton, R. C., Jr., Mathews, M. V., and Seifert, W. W. (August, 1953). "Nonlinear Servomechanisms with Random Inputs", *M.I.T. DCAL Report 70.*

[3] Barrett, J. F., and Coals, B. F. (November, 1955). "An Introduction to the Analysis of Nonlinear Control Systems with Random Inputs", *Proc. of the IEEE, Vol. C-103, No. 3,* 190-199.

[4] Axelby, G. S. (November, 1959). "Random Noise with Bias Signals in Nonlinear Devices", *IRE Transactions on Automatic Control, Vol. AC-4, No. 2,* 167-172.

[5] Crandal, S. H. (August 1, 1976). "On Statistical Linearization for Nonlinear Oscillators", Department of Mechanical Engineering, M.I.T., Cambridge, Massachusetts.

[6] Barber, M. N., and Ninham, B. W. (1970). "Random Walks", Gordon and Breach, New York.

[7] Hullender, D. A., and Bartley, T. M. (September, 1974). "Guideway Roughness as Related to Design Tolerances and Profile Constraints", Final Report to U.S. Department of Transportation, *PB 244-073.*

[8] Balakrishna, M., and Hullender, D. A. (Dec., 1976). "Analytical Model for Guideway Surface Roughness", Journal of Dynamic Systems, Measurements and Control, *Vol. 98, Series G, No. 4,* 425-431.

[9] Balakrishna, M. (May, 1978). "Analytical Modeling of Guideway Roughness", Ph.D. Dissertation, Department of Mechanical Engineering, The University of Texas at Arlington, Arlington, Texas.

TWO PROBLEMS IN NONLINEAR FINITE ELEMENT ANALYSIS[1]

G. F. Carey
T. T. Pan
R. Renka

Texas Institute for Computational Mechanics
The University of Texas at Austin
Austin, Texas

I. INTRODUCTION

[2]"There is more in a common soap bubble than those who have only played with them commonly imagine." We use this classical minimal surface problem as a prototype nonlinear example that is of fundamental importance in the analysis and approximate solution of nonlinear elliptic problems. Two specific applications are considered: firstly, the deflection of thin membranes, as this is described directly by the minimal surface equation; and secondly, a compressible flow problem in which the minimal surface equation is a particular form that corresponds to the choice of a quite fictitious gas for the flow problem.

[1]*This research has been supported in part by AFOSR Grant F-49620-78-C-0083.*

[2]*"Soap Bubbles," Sir Charles Vernon Boys, New and Enlarged Edition, London, 1931.*

ISBN 0-12-434180-2

The study of minimal surfaces has long been an active area of research, the celebrated example of the soap film being associated with the Belgian physicist Plateau who conducted experiments concerning this phenomenon in the 1800's. The corresponding mathematical problem concerns the solution of a nonlinear elliptic partial differential equation in a weak or generalized sense - that is, minimization of an associated functional. In the case of the soap film the functional to be minimized is the area; for large deflection of a membrane it is the potential energy; and for compressible flow it is the pressure integral. A good description of the mathematical problem is presented by Courant [1].

In this article we are concerned with the construction of an appropriate weak variational statement of each nonlinear problem, finite element formulation for computing an approximate solution, and determination of *a priori* error estimates. For brevity we will indicate only the main formulative steps and results. The reader may consult the referenced reports for further details.

II. ANALYSIS AND METHOD

A. *Membrane Problem* [2]

Consider a portion of a membrane of negligible thickness and which is unable to sustain shear. Let $u(x,y)$ denote the deflected shape of the membrane. From considerations of equilibrium, we find that the internal membrane force T is constant and obtain the minimal surface equation

$$(1+u_y^2)u_{xx} - 2u_x u_y u_{xy} + (1+u_x^2)u_{yy} = 0 \tag{1}$$

In the following treatment we consider approximations to a generalized solution of (1) in a strictly convex domain Ω of the (x,y) plane having smooth (twice-differentiable) boundary $\partial\Omega$ and with prescribed surface height as Dirichlet data on the boundary.

A classical solution to (1) also satisfies the weaker requirement that the generalized solution minimize the functional

$$I = \int_{\Omega} (1+u_x^2+u_y^2)^{1/2} d\Omega \tag{2}$$

Let the data u_0 be the restriction to $\partial\Omega$ of a function in the Sobolev space $W_q^2(\Omega)$ for some $q > 2$. If u_0 is of bounded slope, then there is a unique function $u \in W_q^2(\Omega)$ which minimizes I over all Lipschitz functions u in Ω such that $u = u_0$ on $\partial\Omega$. Equivalently, we can express the minimization problem by the alternative variational statement that

$$\int_{\Omega} \frac{u_x v_x + u_y v_y}{(1+u_x^2+u_y^2)^{1/2}} d\Omega = 0 \tag{3}$$

for admissible functions $u(x,y)$ and test functions $v(x,y)$.

1. *Finite Element Analysis.* Let Ω^* be a triangulation of Ω consisting of the usual conforming finite elements [3]. The general form of the finite element approximation on Ω^* may be written as

$$\tilde{u}(x,y) = \sum_{j=1}^{n} q_j p_j(x,y) \tag{4}$$

where q_j are generalized degrees of freedom, usually nodal variables, and the patch functions $p_j(x,y)$ have compact support in the interior of Ω^*. In particular, if we utilize a simple Lagrange basis on a rectilinear triangulation Ω^*, then $q_j = u_j$ and p_j are piecewise polynomials on the patch of elements adjacent to node j such that $p_j(x_i,y_i) = \delta_{ij}$.

Substituting the approximation $\tilde{u}(x,y)$ into I the variational condition $\delta I = 0$ implies $\partial I/\partial u_k = 0$, or

$$\int_{\Omega_k^*} [1+\{u_j(p_j)_x\}^2 + \{u_j(p_j)_y\}^2]^{-1/2}[u_i\{(p_i)_x(p_k)_x + (p_i)_y(p_k)_y\}]d\Omega = 0 \quad (5)$$

where Ω_k^* denotes the patch of elements adjacent to node k.

If $p_j(x,y)$ are piecewise linear, the element contribution to the vertex nodes is

$$\underset{\sim}{q}_e = A_e[1+\underset{\sim}{u}_e^T \underset{\sim}{M}_e \underset{\sim}{u}_e]^{-1/2}[\underset{\sim}{M}_e]\underset{\sim}{u}_e \quad (6)$$

where A_e is element area, $u_e(x,y) = \underset{\sim}{L}^T(x,y)\underset{\sim}{u}_e$ and $\underset{\sim}{M}_e = \underset{\sim}{L}_x\underset{\sim}{L}_x^T + \underset{\sim}{L}_y\underset{\sim}{L}_y^T$ is a 3×3 matrix.

Combining these element contributions at the nodes yields the nonlinear finite element system $\underset{\sim}{g}(\underset{\sim}{u}) = \underset{\sim}{0}$ in equation (5). To show that there exists a unique minimizing function $\tilde{u} \in S_n$, we need merely note that the function $f(\underset{\sim}{x}) = (1+|\underset{\sim}{x}|^2)^{1/2}$, $\underset{\sim}{x} = (x_1,x_2) \in \mathbb{R}^2$, $|\underset{\sim}{x}|^2 = x_1^2 + x_2^2$ is strictly convex which implies that the mapping on $I: \tilde{u} \to \int f(\underset{\sim}{\nabla}\tilde{u})d\Omega$ is also strictly convex. Since f is continuous and $\tilde{u} \in S_n$, a finite dimensional subspace, then there exists a unique minimizer $\tilde{u}$.

2. *Error Estimates* [4]. We state the following theorems concerning error estimates:

<u>*Theorem 1*</u>. Let $u \in W_2^2(\Omega) \cap W_\infty^1(\Omega)$. Then there is a constant C_1 such that for $0 < h < 1$

$$||\underset{\sim}{\nabla}u - \underset{\sim}{\nabla}\tilde{u}||_{0,2,\Omega^*} \leq C_1 h \quad (7)$$

<u>*Theorem 2*</u>. Let $u \in W_q^2(\Omega)$ for some $q > 2$ and $u_0 \in W_1^2(\partial\Omega)$. Then for any p with $1 \leq p < 2$, there is a constant C_2 such that, for $0 < h < 1$,

$$||u - \tilde{u}||_{0,p,\Omega^*} \leq C_2 h^2 \quad (8)$$

Observe that for the linear membrane problem $u_x, u_y << 1$ and with $p = 2$, inequality (8) reduces to the standard error estimate of linear theory.

3. *Numerical Solution.* We solve the nonlinear finite element system $\underset{\sim}{g}(\underset{\sim}{u}) = \underset{\sim}{0}$ in equation (5) by Newton-Raphson iteration:

$$\underset{\sim}{J}(\underset{\sim}{u}^{i+1} - \underset{\sim}{u}^{i}) = -\underset{\sim}{g}(\underset{\sim}{u}^{i}), \tag{9}$$

where $\underset{\sim}{J} = (\partial g_i / \partial u_j)$ is the Jacobian matrix of the system $\underset{\sim}{g}(\underset{\sim}{u})$. The contributions to the Jacobian matrix $\underset{\sim}{J}$ of (9) can be developed directly on each element from (6) and then mapped into $\underset{\sim}{J}$ in the usual way.

From (6) we obtain

$$(J_{ij})^e = \frac{\partial q_i^{(e)}}{\partial u_j} = A_e(1+u_r M_{rs} u_s)^{-1/2} M_{ij}$$
$$- A_e(1+u_r M_{rs} u_s)^{-3/2} (M_{ik} u_k)(M_{j\ell} u_\ell)$$

or

$$\underset{\sim}{J}_e = A_e(1+\underset{\sim}{u}_e^T \underset{\sim}{M}_e \underset{\sim}{u}_e)^{-1/2} \underset{\sim}{M}_e - A_e(1+\underset{\sim}{u}_e^T \underset{\sim}{M}_e \underset{\sim}{u}_e)^{-3/2} (\underset{\sim}{M}_e \underset{\sim}{u}_e)(\underset{\sim}{M}_e \underset{\sim}{u}_e)^T . \tag{10}$$

The scheme is readily implemented in a finite element program and numerical examples are described in the concluding section of this article.

B. Compressible Flow Problem [5]

1. *Formulation.* There is considerable current interest in computing compressible flows using finite elements. The motivating physical problems are shocked transonic flows and arise in transonic airfoil design. The full transonic flow is difficult to analyze or compute being of mixed type due to the presence of supersonic regions. As a first step towards analysis and computations related to this problem we consider compressible subsonic flows. The class of flows is nonlinear but remains elliptic.

The full potential equation describing compressible, potential flow is

$$\{1 + M_\infty^2(\frac{\gamma-1}{2})(1-\Phi_{,j}^{\,2})\}\Phi_{,ii} - M_\infty^2\Phi_{,i}\Phi_{,j}\Phi_{,ij} = 0 \tag{11}$$

where Φ is a velocity potential, $q_i = U_\infty\Phi_{,i}$, U_∞ is the free stream flow and M_∞ is the Mach number in the far field.

An alternative form to (11), that emphasizes the qualitative similarity to the minimal surface problem, is obtained by defining the potential by $q_i = \phi_{,i}$ and expanding (11) to

$$(a^2-\phi_x^2)\phi_{xx} - 2\phi_x\phi_y\phi_{xy} + (a^2-\phi_y^2)\phi_{yy} = 0 \tag{12}$$

where $a^2 = a_\infty^2 + \frac{(\gamma-1)}{2}(U_\infty^2-\phi_x^2-\phi_y^2)$ defines the local speed of sound, a_∞ is the sound speed in the undisturbed flow, and γ is the ratio of specific heats of the gas.

It is not easy to construct an appropriate variational problem from (11) or (12), but using a parametric form for the functional that is based on the minimal surface integral, we obtain

$$I = \int_\Omega [1 - M_\infty^2(\frac{\gamma-1}{2})(1-\Phi_{,i}^{\,2}/U_\infty^2)]^{\gamma/(\gamma-1)}d\Omega \tag{13}$$

2. *Finite Element Method.* We follow the approach used in the previous example to develop an approximate analysis for the compressible flow problem. Employing the same type of triangulation and piecewise linear approximation, the element contribution to the nonlinear system $\partial I/\partial\Phi_i = g_i = 0$ is

$$\frac{\partial I_e}{\partial \underset{\sim}{\Phi}_e} = \underset{\sim}{g}_e = \frac{2\gamma}{(\gamma-1)} A_e\{1 - M_\infty^2(\frac{\gamma-1}{2})(1-\underset{\sim}{\Phi}_e^T\underset{\sim}{M}_e\underset{\sim}{\Phi}_e)\}^{\frac{1}{\gamma-1}}\underset{\sim}{M}_e\underset{\sim}{\Phi}_e \tag{14}$$

where $\underset{\sim}{M}_e$, A_e are defined following equation (6).

The contribution to the element Jacobian $\underset{\sim}{J}_e = (\partial g_i/\partial\Phi_j)_e$ is

$$J_e = \frac{\partial q_e}{\partial \Phi_e} = \frac{2\gamma}{(\gamma-1)} A_e\{1 - M_\infty^2(\frac{\gamma-1}{2})(1-\Phi_e^T M_e \Phi_e)\}^{\frac{1}{\gamma-1}} M_e$$

$$+ \frac{2\gamma}{(\gamma-1)^2} A_e\{1 - M_\infty^2(\frac{\gamma-1}{2})(1-\Phi_e^T M_e \Phi_e)\}^{\frac{2-\gamma}{\gamma-1}} (M\Phi)_e (M\Phi)_e^T \quad (15)$$

Combine element contributions to the Jacobian to complete the description of the Newton iteration.

III. NUMERICAL RESULTS

Sample results are presented below for each of the membrane and compressible flow examples described above.

As a test problem, we solve for the catenoid extending between two concentric rings and compare this with the exact solution. The finite element solution is computed on the discretization of a single "wedge" of elements defined by two radial lines with natural boundary conditions on these radial sides. The solution and error are graphed in Figure 1.

Next the finite element solution to

$$(1+u_y^2)u_{xx} - 2u_x u_y u_{xy} + (1+u_x^2)u_{yy} = -8[1+16x(1-x)(3x^2-3x+1)]$$

on the unit square for data $u = 0$ on sides $x = 0,1$, $y = 0$ and $u = x(1-x)/4$ on $y = 1$ is presented in Figure 2. Approximately six Newton iterations are required for solution to an iterate error of order 10^{-9} with a mesh of size 8×8. The exact solution is $xy(1-x)/4$ and the 1, 2 and ∞ norms of the error are $.304$, $.0585$ and $.0209$.

In the compressible flow computations a sequence of numerical experiments are conducted at low and moderate Mach numbers. At low Mach numbers and for a slender obstacle the compressibility effects are small and the Newton method converges within a couple of iterations. At progressively higher Mach numbers, even when

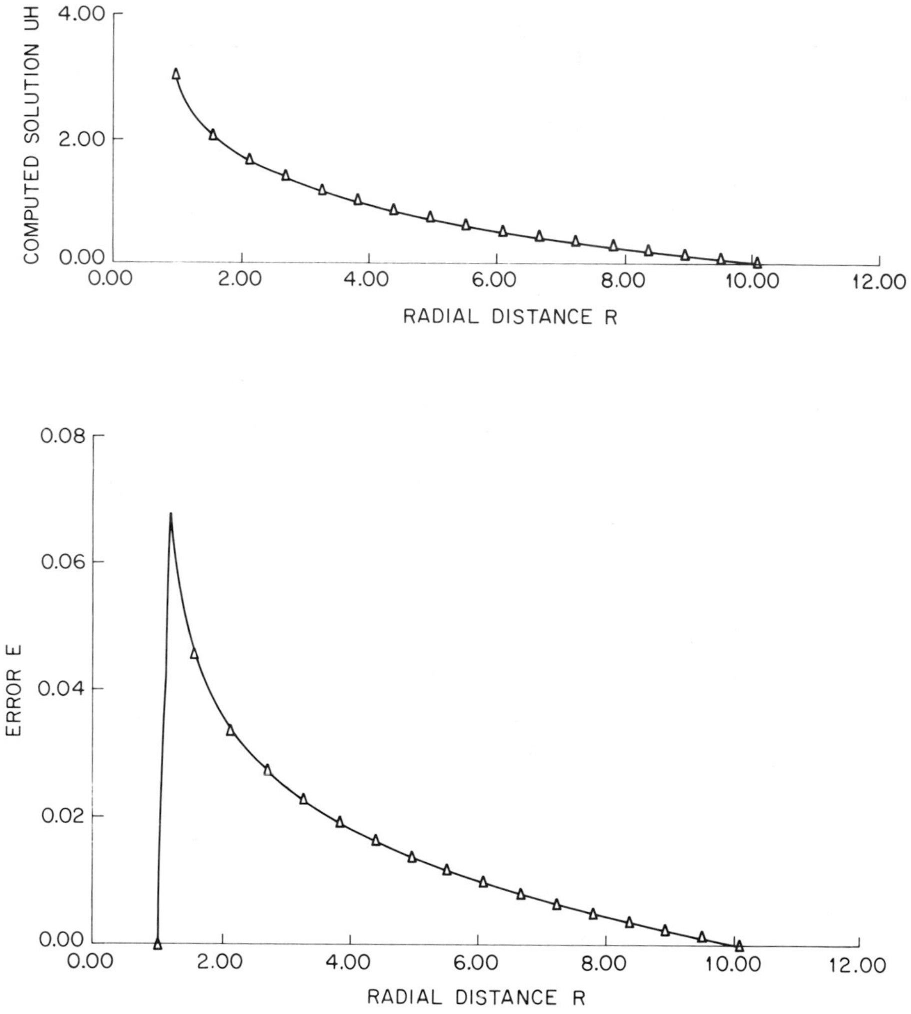

FIGURE 1. Catenary solution, and error curves for wedge angle $\theta = 0.35$, and $N = 48$ equal radial segments.

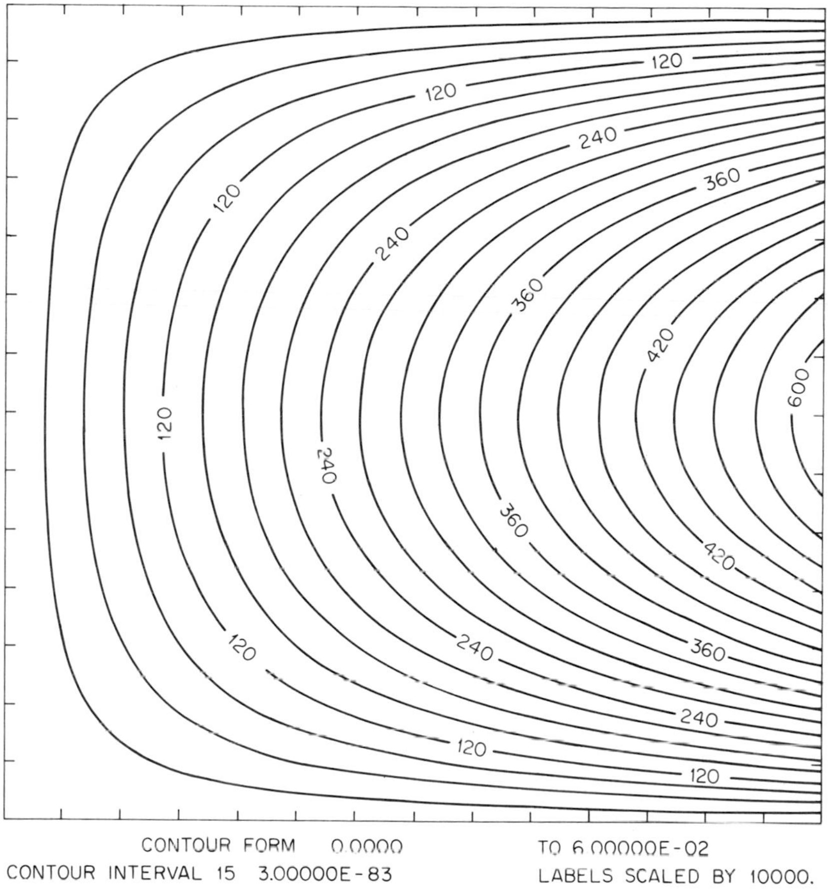

FIGURE 2. Elevation contours for finite element solution. Exact solution is $xy(1-x)/4$.

flow remains subsonic, the iteration is quite sensitive to the starting guess. As we approach the critical Mach number (one supersonic point), it is necessary to develop the solution incrementally in incident Mach number M_∞. Mach line contours for flow about a symmetrical *10%* biconvex airfoil are presented in Figure 3 at incident Mach number of *.6*.

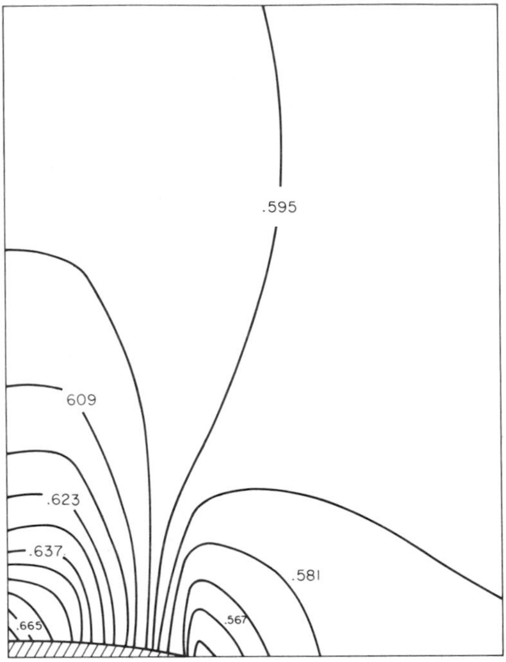

FIGURE 3. Isomach contours in quarter-plane for flow over a 10% biconvex airfoil at incident Mach number $M_\infty = 0.6$.

REFERENCES

[1] Courant, R. (1950). "Dirichlet's Principle", Wiley Interscience, New York.

[2] Carey, G. F., and Renka, R. (1978). "Nonlinear Membrane Analysis", TICOM Report, University of Texas, Austin, (in press).

[3] Martin, H. C., and Carey, G. F. (1973). "Introduction to Finite Element Analysis", McGraw-Hill.

[4] Johnson, Cl., and Thomée, V. (April, 1975). "Error Estimates for a Finite Element Approximation of a Minimal Surface", *Math. Comp. 29, 130,* 343-349.

[5] Carey, G. F. (1978). "Transonic Aerodynamics", Chapter 9 of Finite Elements and Fluids, Vol. 3, R. H. Gallagher et al (eds.), John Wiley & Sons.

FIXED POINT THEORY AND INWARDNESS CONDITIONS

James Caristi

Department of Mathematics
Texas Lutheran College
Seguin, Texas

Integral operators resulting from differential equations can not be expected to have images which are always contained in their domains. Hence the hypotheses of many fixed point theorems include assumptions which guarantee that the image and domain of the operators have enough in common. An inwardness condition is one which asserts that, in some sense, points from the domain are mapped "toward" the domain. For example, for a mapping T which maps K into a larger space X it is sometimes assumed that T takes the boundary of K back into K. Possibly the weakest of the inwardness conditions, the Leray-Schauder boundary condition, is the assumption that T map points x of ∂K anywhere except to the outward part of the ray originating at some interior point of K and passing through x.

Let X be a topological linear space and K a subset of X. If x belongs to K we define the inward set, $I_K(x)$, as follows: $I_K(x) = \{x + c(u-x): u \in K \text{ and } c \geq 1\}$. A mapping $T: K \to X$ is said to be inward if Tx belongs to $I_K(x)$ for each x in K. We say that T is weakly inward in case Tx belongs to the closure of $I_K(x)$ for each x in K. For K a convex subset of a normed space, we have that T is weakly inward if and only if

ISBN 0-12-434180-2

$\lim_{h\to 0^+} h^{-1}d((1-h)x+hTx,K) = 0$ for each x in K [4, Theorem 1.2].

In order for some kind of iteration to take place, it is necessary to make assumptions on T which guarantee that for each x in K there exists y in K such that $d(y,Ty) < d(x,Tx)$ (see [7] for further discussion). The following theorem involves this type of assumption and has been found to be somewhat useful in proving other fixed point theorems for mappings satisfying inwardness conditions.

Theorem 1 [7]. *Let* M *be a complete metric space,* $f\colon M \to M$ *an arbitrary function, and* $\phi\colon M \to R^+$ *a lower semi-continuous function.*

(*) If $d(x,f(x)) \le \phi(x) - \phi(f(x))$ *for all* x *in* M *then* f *has a fixed point in* M.

Since the appearance of this result in [7], C. S. Wong [16] has simplified the original transfinite induction argument, Browder [3] has presented a proof which avoids the use of transfinite induction, Siegel [15] has presented a simple constructive proof, Brézis and Browder [1] have given a general theorem on ordered sets which includes Theorem 1, above, and other results in nonlinear functional analysis, and Kasahara [9] has also proved an analogue of Theorem 1 in the setting of partially ordered sets. Theorem 1 is actually implicit in Brϕndsted's work [2] and is essentially equivalent to a theorem of Ekeland [8] (which is not formulated as a fixed point theorem). A generalization is given by Downing and Kirk in [5]. Kirk has used Theorem 1 to obtain results concerning normal solvability [11] and metric convexity [10] (in the latter, Kirk shows that Theorem 1 characterized completeness). A recent application is the following important extension of a theorem of Lim [12].

Theorem 2 (Downing and Kirk [6], Reich [13]). *Let* X *be a uniformly convex Banach space,* K *a nonempty bounded closed convex subset of* X, *and* Γ *a nonexpansive set-valued mapping defined on* K *and taking values in the family of nonempty compact subsets of* X. *If* $\Gamma x \subset I_K(x)$ *for all* x *in* K, *then there exists* x_0 *in* K *such that* x_0 *belongs to* Γx_0.

It is not known whether Theorem 2 remains true upon replacing $\Gamma x \subset I_K(x)$ with the assumption that $\Gamma x \cap I_K(x)$ is nonempty.

Another unsolved problem involves condensing mappings. Let K be a bounded closed convex subset of a Banach space X and let $T\colon K \to X$ be a condensing mapping. Reich [14] has proved that if either T is inward on K or T satisfies the Leray-Schauder boundary condition, then T has a fixed point. It is not known whether the result remains true if T is only assumed to be weakly inward. However, given any $\varepsilon > 0$ there exists a subset H of K such that T is weakly inward on H and the measure of noncompactness of H is less than ε.

Weakly inward mappings seem to be significantly more difficult to manage than inward mappings. For metric spaces, inward and weakly inward mappings can be defined in a natural way, and we have that an inward contraction mapping defined on a closed subset of a metric space always has a fixed point [4, Theorem 2.2]. It is not known whether this result is true for weakly inward contraction mappings, although it is true for weakly inward contraction mappings defined on a closed convex subset of a Banach space [4, Theorem 2.3].

In attempting to improve Theorem 1, Kirk has raised the question of whether f continues to have a fixed point if we replace $d(x,f(x))$ in (*) by $(d(x,f(x))^p$ where $p > 1$. It is known that $\inf\{d(x,f(x))\colon\ x \in M\} = 0$.

REFERENCES

[1] Brézis, H., and Browder, F. E. (1976). "A general principle on ordered sets in nonlinear functional analysis", Advances in Mathematics 21, 666-676.

[2] Brφndsted, A. (1974). "On a lemma of Bishop and Phelps", *Pacific J. Math.*, *55*, 335-341.

[3] Browder, F. E. (1976). "On a theorem of Caristi and Kirk", Proc. Seminar on Fixed Point Theory and Its Applications, Dalhousie University, June 1975, Academic Press, 23-27.

[4] Caristi, J. (1976). "Fixed point theorems for mappings satisfying inwardness conditions", *Trans. Amer. Math. Soc.*, *215*, 241-251.

[5] Downing, D., and Kirk, W. A. (1977). " A generalization of Caristi's theorem with applications to nonlinear mapping theory", *Pacific J. Math.*, *69*, -39-346.

[6] Downing, D., and Kirk, W. A. (1977). "Fixed point theorems for set-valued mappings in metric and Banach spaces", *Math. Japonica*, *22*, 99-112.

[7] Eisenfeld, J., and Lakshmikantham, V. (1977). "Fixed point theorems on closed sets through abstract cones", *Applied Math. and Computation*, *3*, 155-167.

[8] Ekeland, I. (1972). "Sur les problemes variationnels", *Comptes Rendus Acad. Sci. Paris*, *275*, 1057-1059.

[9] Kasahara, S. (1975). "On fixed points in partially ordered sets and Kirk-Caristi theorem", Math. Seminar Notes XXXV, Kobe University, 2.

[10] Kirk, W. A. (1976). "Caristi's fixed point theorem and metric convexity", *Colloquium Math.*, *36*, 81-86.

[11] Kirk, W. A. (1976). "Caristi's fixed point theorem and the theory of normal solvability, Seminar on Fixed Point Theory and Its Applications, Dalhousie University, June 1975, Academic Press.

[12] Lim, T. C. (1974). "A fixed point theorem for multivalued nonexpansive mappings in a uniformly convex Banach space", *Bull. Amer. Math. Soc.*, *80*, 1123-1126.

[13] Reich, S. (1978). "Approximate selections, best approximations, fixed points, and invariant sets", *J. Math. Anal. Appl.*, *62*, 104-113.

[14] Reich, S. (1973). "Fixed points of condensing functions", *J. Math. Anal. Appl.*, *41*, 460-467.

[15] Siegel, J. (1977). "A new proof of Caristi's fixed point theorem", *Proc. Amer. Math. Soc.*, *66*, 54-56.

[16] Wong, C. S. (1976). "On a fixed point theorem of contractive type", *Proc. Amer. Math. Soc.*, *57*, 283-284.

Applied Nonlinear Analysis

A DIRECT COMPUTATIONAL METHOD FOR ESTIMATING THE PARAMETERS OF A NONLINEAR MODEL

Stephen W. Cheng[1]
Jerome Eisenfeld[2]

Department of Mathematics
University of Texas at Arlington
Arlington, Texas

I. INTRODUCTION

We consider the system identification problem which deals with determining information about the unknown constant matrices A, B, and C for the linear dynamical system:

$$\dot{z}(t) = Az(t) + Be(t), \quad 0 < t < T, \quad z(0) = 0,$$

$$f(t) = Cz(t), \quad 0 < t < T,$$

where $e(t)$ and $f(t)$ are given discretely on the interval $[0,T]$. This leads to a nonlinear estimation problem, or a so-called deconvolution problem, which deals with estimating the number of components N, the decay rates λ_i, $i = 1, 2, \ldots, N$, and the corresponding amplitudes $\alpha_i > 0$, $i = 1, 2, \ldots, N$ in the model:

[1]*Partially supported by NIH Grant HL 10078.*

[2]*Also affiliated with the Department of Medical Computer Science, University of Texas Health Science Center at Dallas, Dallas, Texas 75235.*

ISBN 0-12-434180-2

$$x(t) = \sum_{i=1}^{N} \alpha_i \exp(\lambda_i t), \quad 0 \le t < \infty, \tag{1.1}$$

$$f(t) = \int_0^t e(s)x(t-s)ds, \quad 0 \le t < \infty, \tag{1.2}$$

from the given discrete data $f(t_i)$, $e(t_i)$, $i = 1, 2, \ldots, q$, [1]-[3]. Deconvolution problems are important in a number of fields. One of these is the analysis of fluorescence decay experimental data [3].

Many numerical methods are available for approaching this problem. The Isenberg method of moments (or IMOM for short) was found to be the best among these methods in most cases [2]-[7]. A direct computational method which is essentially based on the general moment theory has been developed [7]. This method which was called the direct computational method (or DCM for short) alleviates some long standing difficulties arising from IMOM. Two of these are the cut-off error incurred when integration is required to infinite time and lack of flexibility in weighting the data.

In this paper, the theory is presented with the specification of the admissible weighting functions and the estimation of the amplitudes α_i using the linear least squares method as an option. The filtering scheme, which can be used to filter out one known decay term and increase the accuracy of the estimation, is presented. Numerical examples are provided.

II. DCM (DIRECT COMPUTATIONAL METHOD)

We consider a model of the form

$$x(t) = \sum_{i=1}^{N} \alpha_i \exp(\lambda_i t), \quad 0 \le t \le T, \tag{2.1}$$

$$f(t) = \int_0^t e(s)x(t-s)ds, \quad 0 \le t \le T, \tag{2.2}$$

instead of equations (1.1)-(1.2). The $\{\lambda_i\}$ are distinct and the $\{\alpha_i\}$ are greater than zero. It has been shown in [7] that by DCM the given discrete data $f(t_i)$, $e(t_i)$, $i = 1, 2, \ldots, q$ are mapped onto a moment sequence $\{s_k\}$, $k = 0, 1, 2, \ldots, m$, and s_k can be calculated by

$$s_k = (-1)^k \int_0^T f(t)h^{(k)}(t)dt, \quad k = 0, 1, \ldots, m \tag{2.3}$$

where $h(t)$ is an admissible weighting function and satisfying:
(i) h is m times continuously differentiable where m is sufficiently large so that $m \geq 2N + 1$ (m can be chosen so that $m \geq 2N - 1$ in the cases of estimating λ_i and α_i),
(ii) $h^{(k)}(T) = 0$ and

$$\int_0^T e(t)h^{(k)}(t)dt = 0, \quad k = 0, 1, \ldots, m - 1. \tag{2.4}$$

It also has been shown in [7] that the moment sequence $\{s_k\}$ can be expressed by

$$s_k = \sum_{i=1}^{N} a_i\lambda_i^k, \quad k = 0, 1, \ldots, m \tag{2.5}$$

where

$$a_i = \alpha_i \int_0^T \exp(\lambda_i t) \left(\int_0^{T-t} e(s)h(s+t)ds \right) dt. \tag{2.6}$$

Once the moment sequence $\{s_k\}$ is obtained, the model parameters N, λ_i, and α_i can be estimated by applying the general moment theory, see [2], [7]-[10].

It is natural to choose the admissible weighting function $h(t)$ in the form

$$h(t) = (T-t)^m(t^m + d_{m-1}t^{m-1} + \ldots + d_0). \tag{2.7}$$

Substitution of Equation (2.7) into Equation (2.4) gives rise to an $m \times m$ system of equations for $d_0, d_1, \ldots, d_{m-1}$, any solution of which yields an admissible weighting function $h(t)$ [7].

III. FILTERING SCHEME

Consider the case with one known λ_i in Equation (2.1). Without loss of generality, we assume that λ_N is known. Let $\{s_k\}$ be a moment sequence defined by Equation (2.5). Then

$$s_{k+1} - \lambda_N s_k = \sum_{i=1}^{N} a_i \lambda_i^{k+1} - \lambda_N \sum_{i=1}^{N} a_i \lambda_i^k$$
$$= \sum_{i=1}^{N-1} a_i(\lambda_i - \lambda_N)\lambda_i^k = \sum_{i=1}^{N-1} b_i \lambda_i^k$$

where $b_i = a_i(\lambda_i - \lambda_N)$, $i = 1, 2, \ldots, N-1$. Let $s_{k+1} - \lambda_N s_k$ be denoted by $\hat{s}_k$ then $\{\hat{s}_k\}$ forms a moment sequence. Again, by applying the general moment theory, N, the remaining λ_i, and all α_i can be estimated. Thus we have the following result:

Theorem 1. Let $\{s_k\}$ be a moment sequence defined by (2.5), and let $\hat{s}_k = s_{k+1} - \lambda_N s_k$ then $\{\hat{s}_k\}$ forms a moment sequence. Moreover, let $\{\hat{G}_k\}$ denote the Hankel matrices

$$\hat{G}_k = \begin{pmatrix} \hat{s}_0 & \hat{s}_1 & \cdots & \hat{s}_{k-1} \\ \hat{s}_1 & \hat{s}_2 & \cdots & \hat{s}_k \\ \vdots & \vdots & & \\ \hat{s}_{k-1} & \hat{s}_k & \cdots & \hat{s}_{2k-2} \end{pmatrix}$$

and let $|\hat{G}_k|$ denote the determinant of $\hat{G}_k$, then the following statements are true:

(1) $|\hat{G}_{N-1}| \neq 0$

$|\hat{G}_k| = 0$ if $k > N - 1$.

In the event that $\{\hat{s}_k\}$ is positive we have further:

$|\hat{G}_k| > 0$ if $K \leq N - 1$.

(2) The numbers λ_k, $k = 1, 2, \ldots, N-1$, are the roots of a polynomial

$$P_{N-1}(\lambda) = \lambda^{N-1} + c_{N-2}\lambda^{N-2} + \cdots + c_0$$

where the coefficients c_i are given by

$$\begin{pmatrix} c_0 \\ c_1 \\ \vdots \\ c_{N-2} \end{pmatrix} = -\hat{G}_{N-1}^{-1} \begin{pmatrix} \hat{s}_{N-1} \\ \hat{s}_N \\ \vdots \\ \hat{s}_{2N-3} \end{pmatrix}$$

(3) The numbers b_i are given by

$$\begin{bmatrix} b_1 \\ b_2 \\ \vdots \\ b_{N-1} \end{bmatrix} = V_{N-1}^{-1} \begin{bmatrix} \hat{s}_0 \\ \hat{s}_1 \\ \vdots \\ \hat{s}_{N-2} \end{bmatrix}$$

where V_{N-1} is the Vandermonde matrix

$$V_{N-1} = \begin{bmatrix} 1 & 1 & \cdots & 1 \\ \lambda_1 & \lambda_2 & & \lambda_{N-1} \\ \vdots & \vdots & & \vdots \\ \lambda_1^{N-2} & \lambda_2^{N-2} & \cdots & \lambda_{N-1}^{N-2} \end{bmatrix}.$$

(4) The numbers a_i are given by

$$a_i = \frac{b_i}{\lambda_i - \lambda_N}, \quad i = 1, 2, \ldots, N-1,$$

and

$$a_N = \frac{s_k - \sum_{i=1}^{N-1} a_i \lambda_i^k}{\lambda_N^k}.$$

Theorem 1 is essentially an algorithm for the purposed filtering scheme. If we know one of λ_i, we can filter out λ_i from the calculated moment sequence, and form a new moment sequence. Then by applying the moment theory, we can estimate N, the remaining λ_i (i.e., $\lambda_1, \lambda_2, \ldots, \lambda_{N-1}$), and all α_i (i.e., $\alpha_1, \alpha_2, \ldots, \alpha_N$). Practically, this leads to increase accuracy of the estimation. Numerical examples are presented in Section 5.

IV. LINEAR LEAST SQUARES METHOD FOR ESTIMATING α_i

Rewriting Equations (2.1)-(2.2), we have

$$f(t) = \int_0^t e(t-s)\left[\sum_{i=1}^{N} \alpha_i \exp(\lambda_i s)\right] ds$$

$$= \sum_{i=1}^{N} \alpha_i \left[\int_0^t e(t-s)\exp(\lambda_i s)ds\right],$$

$$f(t) = \sum_{i=1}^{N} \alpha_i \phi_i(t) \qquad (4.1)$$

where $\phi_i(t) = \int_0^t e(t-s)\exp(\lambda_i s)ds$. Once λ_i are known, Equation (4.1) becomes a linear model, then the linear least squares method (or LLSM for short) can be used to estimate α_i as an option [11]. Numerical examples using LLSM for estimating α_i as an option are presented in the next section.

V. NUMERICAL EXAMPLES

In this section we present a number of examples using computer simulated data as illustrations of the performance by DCM and of the various points raised in this paper. For comparison, we also present the results obtained by analysis of the data by IMOM. All computer programs used in this study were written in FORTRAN-10 and have been performed on the DEC System-10 and System-20 in time-sharing conversational fashion. All computations were performed in double precision.

The model is of the form as in Equations (2.1)-(2.2) with $-1/\tau_i$ instead of λ_i, and let $e(t)$ be chosen so that

$$e(t) = \beta \sin \pi t/\hat{T}, \quad 0 \leq t \leq \hat{T},$$

$$= 0, \qquad \hat{T} < t.$$

The data sets for $f(t)$ with noise distributed in a Gaussian fashion and $e(t)$ were generated by choosing $\beta = 10^5$ and $\hat{T} = 30$ as described in [2]. Many different analyses were carried out for the purpose of comparing the results of various combinations of parameters $(N, \alpha_i, \tau_i, \text{and } T)$. The admissible weighting functions are chosen as Equation (2.7) for all analyses in this section. Table I gives a comparison of the expected parameters and the calculated results obtained by analysis of the data by DCM for different sampling intervals and a variety of cases. Table II gives a comparison of the accuracy of the estimated parameters with the expected values; the relative errors (in percent) of the estimated parameters are presented along with the results obtained by IMOM. Observe that DCM not only provides nice results for the sampling interval $[0, 500]$, but also provides acceptable results even in the cases of smaller sampling intervals. In most cases, especially in the cases of larger N, the larger the sampling interval, the greater the accuracy. In the cases of larger N we need more samples to get acceptable results. Table III shows the performance of the filtering scheme; the expected values and the relative errors (in percent) of the calculated results, are presented. Study of Table III reveals that the filtering scheme increases the accuracy of the estimation in most cases. Table IV gives a comparison of the accuracy of the estimated parameters α_i with the expected values using DCM and LLSM. Evidently, the error committed by DCM is far less than LLSM option.

Comments and conclusion about DCM and the comparison between DCM and IMOM can be found in [7]. Although DCM as described in [7] has already shown itself to be an effective and efficient method for solving nonlinear estimation problems as well as system identification problems, the new scheme and option described in this paper enhance its applicability.

TABLE I. Comparison of the Estimated Parameters with the Expected Values by DCM

		Expected Values	[0,100]	[0,200]	[0,300]	[0,400]	[0,500]
1	α_1	1.00	1.046	1.013	1.010	1.009	1.008
	τ_1	7.00	7.256	7.051	7.022	7.011	7.005
	α_2	1.00	0.952	0.990	0.994	0.995	0.996
	τ_2	18.50	18.835	18.547	18.525	18.518	18.515
2	α_1	1.00	1.016	1.000	0.999	0.999	0.999
	τ_1	15.00	15.252	15.008	14.998	14.995	14.996
	α_2	0.50	0.481	0.499	0.499	0.499	0.499
	τ_2	75.00	78.133	75.094	75.019	75.013	75.013
3	α_1	0.50		0.512	0.506	0.505	0.505
	τ_1	5.00		5.204	5.116	5.088	5.074
	α_2	0.25		0.250	0.245	0.245	0.246
	τ_2	20.00		21.420	20.458	20.305	20.242
	α_3	0.10		0.084	0.097	0.098	0.099
	τ_3	45.00		47.232	45.337	45.173	45.125

		Expected Values	[0,500]	[0,600]	[0,700]	[0,800]	[0,900]
4	α_1	0.10	0.105	0.104	0.103	0.103	0.103
	τ_1	5.00	5.211	5.164	5.137	5.118	5.105
	α_2	0.034	0.034	0.033	0.033	0.033	0.033
	τ_2	15.00	17.117	16.380	16.033	15.834	15.706
	α_3	0.017	0.013	0.014	0.015	0.016	0.016
	τ_3	30.00	33.266	31.399	30.824	30.564	30.422
	α_4	0.0084	0.0078	0.0082	0.0083	0.0083	0.0084
	τ_4	60.00	60.677	60.198	60.090	60.050	60.031

TABLE II. Comparison of the Accuracy of the Parameters as Estimated by Different Methods (Relative error in percentage, expected values given in Table I)

		DCM					IMOM
		[0,100]	[0,200]	[0,300]	[0,400]	[0,500]	[0,500]
1	$\Delta\alpha_1$	4.64	1.27	0.98	0.88	0.84	7.14
	$\Delta\tau_1$	3.66	0.73	0.32	0.16	0.08	1.13
	$\Delta\alpha_2$	4.77	0.98	0.61	0.48	0.42	2.22
	$\Delta\tau_2$	1.81	0.26	0.13	0.10	0.08	0.03
2	$\Delta\alpha_1$	1.65	0.00	0.06	0.05	0.06	0.74
	$\Delta\tau_1$	1.68	0.05	0.02	0.03	0.02	0.37
	$\Delta\alpha_2$	3.02	0.23	0.11	0.10	0.10	0.12
	$\Delta\tau_2$	4.18	0.13	0.03	0.02	0.02	0.03
3	$\Delta\alpha_1$		2.37	1.20	0.98	0.97	11.31
	$\Delta\tau_1$		4.07	2.33	1.77	1.47	3.90
	$\Delta\alpha_2$		0.04	2.03	1.84	1.66	16.94
	$\Delta\tau_2$		7.10	2.29	1.52	1.21	1.66
	$\Delta\alpha_3$		15.68	3.15	1.75	1.29	11.21
	$\Delta\tau_3$		4.96	0.75	0.39	0.28	0.42

		DCM				
		[0,500]	[0,600]	[0,700]	[0,800]	[0,900]
4*	$\Delta\alpha_1$	4.77	3.81	3.28	2.93	2.69
	$\Delta\tau_1$	4.21	3.28	2.73	2.37	2.10
	$\Delta\alpha_2$	0.08	3.12	3.93	4.10	4.07
	$\Delta\tau_2$	14.12	9.20	6.88	5.56	4.71
	$\Delta\alpha_3$	25.57	14.90	10.12	7.57	6.04
	$\Delta\tau_3$	10.89	4.66	2.75	1.88	1.41
	$\Delta\alpha_4$	6.81	2.22	1.11	0.68	0.48
	$\Delta\tau_4$	1.13	0.33	0.15	0.08	0.05

*Four-component case for IMOM is not available.

*TABLE III. Comparison of the Accuracy of the Parameters as Estimated Without/With Filtering Scheme (Relative error in percentage, * indicates the filtering out decay term)*

		Expected Values		Without Filtering	With Filtering 1	2	3	4
1	α_1	1.00	$\Delta\alpha_1$	0.32	0.06	0.04		
	τ_1	15.00	$\Delta\tau_1$	0.13	*	0.09		
	α_2	1.00	$\Delta\alpha_2$	0.38	0.18	0.07		
	τ_2	30.00	$\Delta\tau_2$	0.07	0.03	*		
2	α_1	1.00	$\Delta\alpha_1$	2.82	0.21	0.78	1.67	
	τ_1	7.50	$\Delta\tau_1$	2.09	*	0.46	1.17	
	α_2	0.50	$\Delta\alpha_2$	4.68	0.82	1.71	3.06	
	τ_2	18.50	$\Delta\tau_2$	2.79	0.69	*	1.10	
	α_3	0.24	$\Delta\alpha_3$	2.76	1.73	0.90	0.44	
	τ_3	34.80	$\Delta\tau_3$	0.38	0.34	0.21	*	
3	α_1	0.10	$\Delta\alpha_1$	4.77	0.17	1.25	2.20	2.77
	τ_1	5.00	$\Delta\tau_1$	4.21	*	1.00	1.89	2.45
	α_2	0.034	$\Delta\alpha_2$	0.08	5.45	5.52	5.04	4.48
	τ_2	15.00	$\Delta\tau_2$	14.12	3.50	*	3.37	5.62
	α_3	0.017	$\Delta\alpha_3$	25.57	10.37	3.77	2.88	7.43
	τ_3	30.00	$\Delta\tau_3$	10.89	4.02	2.12	*	1.60
	α_4	0.0084	$\Delta\alpha_4$	6.81	3.09	1.96	6.86	0.28
	τ_4	60.00	$\Delta\tau_4$	1.13	0.57	0.38	0.16	*

TABLE IV. Comparison of the Accuracy of the Estimated Parameters α_i With the Expected Values by Different Methods (Relative Error in Percentage)

		Expected Values	DCM Estimated Values	DCM Relative Errors	LLSM Estimated Values	LLSM Relative Errors
1	α_1	1.00	1.002	0.24	1.026	2.60
	τ_1	15.00	14.993			
2	α_1	1.00	1.000	0.00	1.059	5.89
	τ_1	30.00	30.027			
	α_2	1.00	0.998	0.21	0.982	1.78
	τ_2	60.00	60.014			
3	α_1	1.00	1.000	0.01	1.061	0.61
	τ_1	15.00	14.997			
	α_2	0.45	0.449	0.16	0.442	1.71
	τ_2	45.00	45.017			
4	α_1	0.10	0.101	0.62	0.074	25.59
	τ_1	5.00	4.974			
	α_2	0.017	0.0170	0.08	0.036	114.51
	τ_2	30.00	30.063			
	α_3	0.0084	0.0084	0.49	0.0022	73.36
	τ_3	60.00	60.049			

REFERENCES

[1] Sage, A. P., and Melsa, J. L. (1971). "System Identification", Academic Press, New York.

[2] Eisenfeld, J., Bernfeld, S. R., and Cheng, S. W. (1977). "System identification problems and the method of moments", *Math. Biosciences, 36,* 199-211.

[3] Ford, C., and Eisenfeld, J. "Some unresolved questions pertaining to the mathematical analysis of fluorescence decay data", Applied Nonlinear Analysis, V. Lakshmikantham, Ed., Academic Press, New York.

[4] Isenberg, I., and Dyson, R. D. (1969). "The analysis of fluorescence decay by a method of moments", *Biophys. J., 9,* 1337-1350.

[5] Isenberg, I., Dyson, R. D., and Hanson, R. (1973). "Studies on the analysis of fluorescence decay data by the method of moments", *Biophys. J., 13,* 1090-1115.

[6] Small, E. W., and Isenberg, I. (1976). "The use of moment index displacement in analyzing fluorescence time-decay data", *Biopolymers, 15,* 1093-1100.

[7] Eisenfeld, J., and Cheng, S. W. "General moment methods for a class of nonlinear models", submitted for publication.

[8] Gantmacher, F. R. (1959). "The Theory of Matrices, Vol. I & II", Chelsea, New York.

[9] Eisenfeld, J., and Soni, B. (1977). "Linear algebraic computational procedure for system identification problems", Proceedings of the First International Conference on Math. Modeling (Vol. I), X. J. R. Avula, Ed., Univ. of Missouri Press, 551-561.

[10] Hallmark, J., and Eisenfeld, J. "Separation and monotonicity results for the roots of the moment problem", Applied Nonlinear analysis, V. Lakshmikantham, Ed., Academic Press, New York.

[11] Draper, N. R., Smith, H., Jr. (1967). "Applied Regression Analysis", Wiley, New York.

A NOTE ON THE ASYMPTOTIC BEHAVIOR OF NONLINEAR SYSTEMS

*Kuo-Liang Chiou**

Department of Mathematics
Wayne State University
Detroit, Michigan 48202

In this paper we shall be concerned with asymptotic relationships between the solutions of the system

$$\frac{dy(t)}{dt} = Ay(t), \quad t \geq 0 \tag{1}$$

and those of the nonlinear system

$$\frac{dx(t)}{dt} = Ax(t) + f(t,x(t)), \quad t \geq 0 \tag{2}$$

where x, y, and f are n-vectors in R^n, A is a constant matrix in $R^{n\times n}$ for $t \geq 0$, and $f(t,x)$ is a continuous function of t and x for $t \geq 0$ and $\|x\| < \infty$. Here $\|\cdot\|$ denotes any appropriate vector (or matrix) norm.

Throughout this paper we shall always call the following conditions "Assumption A":

(i) $\alpha(t)$ is a positive continuous functions on $J = [0,\infty)$; and

**This work was supported by the U.S. Army Research Office Grant DAAG29-78-G-0042 at Research Triangle Part, N.C.*

ISBN 0-12-434180-2

(ii) $\omega(t,s)$ is nonnegative, continuous on $J \times J$, and is nondecreasing in s for $s > 0$ and fixed $t \in J$.

The purpose of this paper is that: suppose a solution $y(t)$ of (1) is given, we are interested in knowing if there exists a solution $x(t)$ of (2) such that $\|x(t) - y(t)\| = o(\alpha(t))$ as $t \to \infty$.

Now we shall prove the following theorem via the Schander-Tychonoff theorem.

Theorem. Let $y(t)$ be an arbitrary nontrivial solution of (1). Suppose that there exists $\alpha(t)$, and $\omega(t,s)$ satisfying Assumption A;

for an arbitrary positive constant $\varepsilon < 1$, there exists

$$t_0 > 0 \text{ such that } \|y(t)\| \leq (1-\varepsilon)\alpha(t), \quad t \geq t_0; \tag{3}$$

$$\|f(s,x)\|\exp(-\|A\|s) \leq \omega(s, \|x\|\alpha^{-1}(s)), \quad \text{for } t_0 \leq t \leq s; \tag{4}$$

and

$$\lim_{t\to\infty} \int_t^\infty \omega(s,1)ds = 0. \tag{5}$$

Then there exists a solution $x(t)$ of (2) such that

$$\|x(t) - y(t)\| = o(\alpha(t)) \quad \text{as} \quad t \to \infty \tag{6}$$

and

$$\|x(t)\| \leq \alpha(t) \quad \text{as} \quad t \to \infty. \tag{7}$$

Proof. For a given positive constant ε in (3), (5) implies that there exists a large $T_0 (> t_0)$ such that

$$\int_t^\infty \omega(s,1)ds < \varepsilon \quad \text{for} \quad t \geq T_0. \tag{8}$$

Via the Schander-Tychonoff theorem (see [3, p. 9]) we will establish the existence of a solution of the integral equation

$$x(t) = \Phi(t)c - \Phi(t)\int_t^\infty \Phi^{-1}(s)f(s,x(s))ds, \quad t \geq T_0$$

where $\Phi(t) = \exp(At)$ and $\Phi(t)c = y(t)$. Consider the set

$$F = \{u:\ u(t) = \alpha^{-1}(t)x(t) \text{ where } x(t) \text{ is continuous on } J_0 = [T_0,\infty) \text{ and } \|u(t)\| \le 1 \text{ for } t \ge T_0\}$$

and define the operator T by

$$Tu(t) = \frac{\Phi(t)c}{\alpha(t)} - \frac{\Phi(t)}{\alpha(t)} \int_t^\infty \Phi^{-1}(s)f(s,\alpha(s)u(s))ds. \tag{9}$$

Since $\Phi(t)\Phi^{-1}(s) = \exp(A(t-s)$, using (4) we obtain for $t \le s$

$$\begin{aligned} \|\Phi(t)\Phi^{-1}(s)f(s,x)\| &\le \|\Phi(t)\Phi^{-1}(s)\|\ \|f(s,x)\| \\ &\le \exp(\|A\|(s-t) \cdot \|f(s,x)\| \\ &\le \exp(-\|A\|t) \cdot \exp(\|A\|s) \cdot \|f(s,x)\|. \end{aligned} \tag{10}$$

First we will establish that $TF \subset F$. Taking the norm to both sides of (9) and using (3), (4) and (10), we obtain

$$\begin{aligned} \|Tu(t)\| &\le \frac{\|\Phi(t)c\|}{\alpha(t)} + \frac{1}{\alpha(t)} \int_t^\infty \|\Phi(t)\Phi^{-1}(s)f(s,\alpha(s)u(s))\|ds \\ &< \frac{\|y(t)\|}{\alpha(t)} + \frac{\exp(-\|A\|t)}{\alpha(t)} \int_t^\infty \omega(s,\|u(s)\|)ds \\ &\le \frac{\|y(t)\|}{\alpha(t)} + \varepsilon \le 1 \end{aligned}$$

It is clear that $\alpha(t)Tu(t)$ is continuous on $J_0 = [T_0,\infty)$. This proves $TF \subset F$.

Second we will show that T is continuous. Suppose that the sequence $\{u_n\}$ in F converges uniformly to u in F on every compact subinterval of J_0. We claim that Tu_n converges uniformly to Tu on every compact subinterval J_1 of J_0. Let ε_1 be a small positive number satisfying $\varepsilon_1 < 1$. (5) implies that there exists $T_1 > T_0$ so that for $t \ge T_1$

$$\int_t^\infty \omega(s,1)ds < \frac{\varepsilon_1}{4}. \tag{11}$$

Then using (9) we obtain the following inequalities, for $t \in J_0$.

$$\|Tu_n(t) - Tu(t)\| \leq \frac{1}{\alpha(t)} \left\| \int_t^\infty \Phi(t)\Phi^{-1}(s)f(s,\alpha(s)u_n(s))ds - \int_t^\infty \Phi(t)\Phi^{-1}(s)f(s,\alpha(s)u(s)ds \right\|$$

$$\leq \frac{\|\Phi(t)\|}{\alpha(t)} \cdot \int_t^{T_1} [\|\Phi^{-1}(s)\| \cdot \|f(s,\alpha(s)u_n(s)) - f(s,\alpha(s)u(s))\|]ds$$

$$+ \frac{1}{\alpha(t)} \int_{T_1}^\infty [\|\Phi(t)\Phi^{-1}(s)f(s,u_n(s)\alpha(s))\| + \|\Phi(t)\Phi^{-1}(s)f(s,\alpha(s)u(s))\|]ds. \qquad (12)$$

Now using (4) and (11), the second integral on the right side of (12) satisfies

$$\frac{1}{\alpha(t)} \int_{T_1}^\infty [\|\Phi(t)\Phi^{-1}(s)f(s,\alpha(s)u_n(s))\| + \|\Phi(t)\Phi^{-1}(s)f(s,\alpha(s)u(s))\|]ds \qquad (13)$$

$$\leq \frac{\exp(-\|A\|t)}{\alpha(t)} \int_{T_1}^\infty [\omega(s,\|u_n(s)\|) + \omega(s,\|u(s)\|]ds$$

$$\leq 2 \int_{T_1}^\infty \omega(s,1)ds < \frac{\varepsilon_1}{2}.$$

By the uniform convergence there is an $N = N(\varepsilon,T_1)$ such that if $n \geq N$ then

$$\|f(t,\alpha(t)u_n(t)) - f(t,\alpha(t)u(t))\| < \frac{\varepsilon_1}{2M_1M_2(T_1-T_0)} \qquad (14)$$

where $M_1 = \sup_{T_0 \leq t \leq T_1} \|\Phi^{-1}(t)\|$ and $M_2 = \sup_{t \in J_1} \frac{\alpha(t)}{\|\Phi(t)\|}$.
Combining (12), (13), and (14) yields for $t \in J_1$

$$\|Tu(t) - Tu_n(t)\| < \varepsilon_1 \quad \text{for} \quad n \geq N.$$

This shows that Tu_n converges uniformly to Tu on compact subintervals J_1 of J_0. Hence T is continuous.

Third we claim that the functions in the image set TF are equicontinuous and bounded at every point of J_0. Since $TF \subset F$, it is clear that the functions in TF are uniformly bounded. Now we show that they are equicontinuous at each point of J_0. For each $u \in F$, the function $z(t) = \alpha(t)Tu(t)$ is a solution of the linear system below

$$\frac{dv}{dt} = Av + f(t,\alpha(t)u(t)).$$

Since $\|z(t)\| \leq \alpha(t)\|Tu(t)\| \leq \alpha(t)$ and $\|f(t,\alpha(t)u(t))\|$ is uniformly bounded for $u \in F$ on any finite t interval, we see that $\frac{dv}{dt}$ is uniformly bounded on any finite interval. Therefore, the set of all such z is equicontinuous on any finite interval. To see that the functions in TF are equicontinuous at every point in J_0, consider

$$\|Tu(t_1) - Tu(t_2)\| = \|\alpha^{-1}(t_1)z(t_1) - \alpha^{-1}(t_2)z(t_2)\| \tag{15}$$

$$\leq \|\alpha^{-1}(t_1)\| \; \|z(t_1) - z(t_2)\| + \|\alpha^{-1}(t_1) - \alpha^{-1}(t_2)\| \cdot \|z(t_2)\|$$

where t_1, t_2 are in some finite interval. The right side of (15) can be made small by virtue of the equicontinuity of the family $\{z(t)\}$ and the continuity of $\alpha^{-1}(t)$. Thus the functions in TF are equicontinuous at each point of J_0.

All of the hypotheses of the Schauder-Tychonoff theorem are satisfied. Thus there exists a $u \in F$ such that $u(t) = Tu(t)$; that is, there exists a solution $x(t)$ of

$$x(t) = y(t) - \Phi(t) \int_t^\infty \Phi^{-1}(s)f(s,x(s))ds.$$

Therefore $x(t)$ is a solution of (2) and possesses the asymptotic behavior of (6) and (7). This proves the Theorem.

Remark 1. The above theorem is an improvement of Theorem 1 in [4] with $\Delta(t) = I$ and $A(t)$ constant matrix.

Remark 2. We now apply the above theorem to the following equation

$$\left.\begin{aligned} v_1\theta_1''(t) + c\theta_1'(t) + \alpha_1\theta_1(t)(1-\theta_2(t)) &= 0 \\ v_2\theta_2''(t) + c\theta_2'(t) + \alpha_2\theta_2(t)(\theta_1(t)-1) &= 0 \end{aligned}\right. \quad t > 0 \tag{16}$$

where $v_i > 0$, $\alpha_i > 0$ for $i = 1,2$ and $c > 0$. Rewrite (16) as

$$P'(t) = Ap(t) + f(t,p), \quad t > 0 \tag{17}$$

where

$$A = \begin{pmatrix} 0 & 0 & 1 & 0 \\ 0 & 0 & 0 & 0 \\ -v_1^{-1}\alpha_1 & 0 & -v_1^{-1}c & 0 \\ 0 & v_2^{-1}\alpha_2 & 0 & -v_2^{-1}c \end{pmatrix} \text{ and } f(t,p) = \begin{pmatrix} 0 \\ 0 \\ \alpha_1 v_1^{-1}\theta_1\theta_2 \\ -\alpha_2 v_2^{-1}\theta_1\theta_2 \end{pmatrix}$$

Thus

$$\|f(t,p)\| \le c\|p\|^2$$

The eigenvalues of A are $\dfrac{-c \pm \sqrt{c^2-4v_1\alpha_1}}{2v_1}$ and $\dfrac{-c \pm \sqrt{c^2+4v_2\alpha_2}}{2v_2}$.

Therefore from the above theorem we obtain that there exists a solution $P(t)$ of (17) such that $\|P(t)\| \le \exp(\mu t)$ as $t \to \infty$ where $\lambda_m \le \mu < \frac{\lambda_M}{2}$ and λ_M and λ_m are the largest and smallest eigenvalues of A respectively. Note that $\frac{1}{2}\lambda_M > \lambda_m$.

REFERENCES

[1] Brauer, F., and Wong, J. S. W. (1969). "On the asymptotic relationships between solutions of two systems of ordinary differential equations", *J. Differential Equations 6*, 142-153.

[2] Cooper, W. G., and Hallam, G. T. (1975). "Nonlinear perturbation of a linear system of differential equations with a regular singular point", *Rocky Mountain J. Math., Vol. 5*, 247-254.

[3] Coppel, W. A. (1965). "Stability and asymptotic behavior of differential equations", Heath and Co., Boston.

[4] Hallam, T. G., and Heidel, J. W. (1970). "The asymptotic manifolds of a perturbed linear system of differential equations", *Translations Amer. Math. Soc. 149*, 233-241.

A SECOND-STAGE EDDY-VISCOSITY CALCULATION FOR THE FLAT-PLATE TURBULENT BOUNDARY LAYER*

Sue-Li Chuang
Fred R. Payne

Department of Aerospace Engineering
University of Texas at Arlington
Arlington, Texas

I. INTRODUCTION

The concept of the "Dual Structure" of Turbulence, i.e., "Large Eddies" and "Small Eddies," was developed extensively by Townsend[1] in his monograph. Lumley[2] gave the first objective mathematical definition of the large eddies; this definition also provides a scheme whereby these structures can be removed from experimental, two-point, velocity co-variance data usually obtained via hot-wire or laser-Doppler anemometry. Payne[3] and Lemmerman[4] applied Lumley's definition to two flow prototypes: The 2-D Wake and Flat-Plate Boundary Layer.

Based upon the large-eddy velocity data calculated by Lemmerman[4-5], its two-point correlation was constructed.[7] This provided an overwhelming mass of data for full 6-dimensional analysis. In this paper, emphasis is upon analyzing one-point correlations of the large-eddy component, i.e., a 3-dimensional

**The current effort was partly supported (1975-78) by NASA Ames Grant NSG-2077, Dr. M. W. Rubesin, technical monitor.*

ISBN 0-12-434180-2

analysis. The subsequent calculation of small-eddy viscosity[6] yields an approximately linear relation[7] between "small-eddy viscosity" and the corrdinate, y, normal to the wall in the logarithmic region of 2-D turbulent boundary layers with zero pressure gradient. More extensive 3-D analysis of Chuang's results is underway and a 6-D analysis is planned. Implications for improved modelling of turbulent flows are significant.

See previous paper by Payne[8] for a survey of the mathematical bases of PODT-SAS (Proper Orthogonal Decomposition Theorem-Structural Analysis System) as well as physical interpretation of the eigen-values and eigen-functions extracted from empiric data by PODT-SAS. Herein is presented a brief recapitulation of Lumley's eigen-value problem (PODT), the Lumley-Payne small eddy-viscosity, ν_{se} and Chuang's work[7] in construction of the large eddy co-variance tensor, B_{ij} and subsequent calculations of ν_{se}.

II. PODT-SAS AND THE SMALL-EDDY VISCOSITY

The kernel R_{ij} in the integral equation of Lumley's (1965)[2] Proper Orthogonal Decomposition Theorem (PODT) is the total measured two-point Reynolds stress tensor, which is defined as:

$$R_{ij} = \overline{u_i(\underline{x})u_j(\underline{x}')}; \quad i,j = 1,2,3 \tag{1}$$

where $u_i(\underline{x})$ and $u_j(\underline{x}')$ are the fluctuating, turbulent velocity components at the space-points $\underline{x}$ and $\underline{x}'$ respectively. The over-bar in equation (1) represents taking the time average over a period of time such that the turbulence stochastics can be considered stationary, the "Ergodic Hypothesis."

In Lemmerman's paper (1976)[4], the large-eddy velocity components have been extracted from the total measured two-point Reynolds stress tensor by solving numerically Lumley's linear, integral, eigen-value equation (PODT):

$$\int R_{ij}\phi_j dv = \lambda\phi_i \tag{2}$$

The "large-eddy velocity correlation tensor", B_{ij}, is then defined analogously to R_{ij}, from the calculated large-eddy velocity components, $u_{LE_i}(\underline{x})$; $i = 1,2,3$.

$$B_{ij} = u_{LE_i}(\underline{x})u_{LE_j}(\underline{x}'); \quad i,j = 1,2,3 \tag{3}$$

where B_{ij} is the quantity contributed by the large-eddy motion.

The two comparable quantities, B_{ij} and R_{ij}, have the same physical dimensions. If B_{ij} is subtracted from R_{ij}, intuitively what is left must be the contribution due to small-eddy motion:

$$\text{Small-eddy contribution} = R_{ij} - B_{ij} \tag{4}$$

Analogous to the usual Boussinesq[9] (1877) eddy viscosity, ν_T,

$$-\overline{u_1(\underline{x})u_2(\underline{x})} = \nu_T \frac{dU}{dy} \tag{5}$$

the small-eddy viscosity, ν_{se}, is defined and calculated by the equation:

$$R_{ij} = B_{ij} + \nu_{se}\frac{dU}{dy} \quad (i,j = 1,2,3) \tag{6}$$

where U is the time-averaged velocity field and the y-direction is normal to the flat plate of the fully-developed 2-D turbulent boundary layer. Note that for more general shear flows in which statistical homogeneity prevails in two of the three spatial directions can also use the simple phenomenological model, equation (6).

III. CALCULATIONS

The schematic in Figure 1 shows that four sources of data are needed for the small-eddy viscosity calculation. The experimental turbulent intensities throughout the boundary layer are necessary in order to denormalize the measured two-point Reynolds stress tensor from both Grant's (1958)[10] and Tritton's (1967)[11] experiments. The large-eddy velocity field is taken directly from

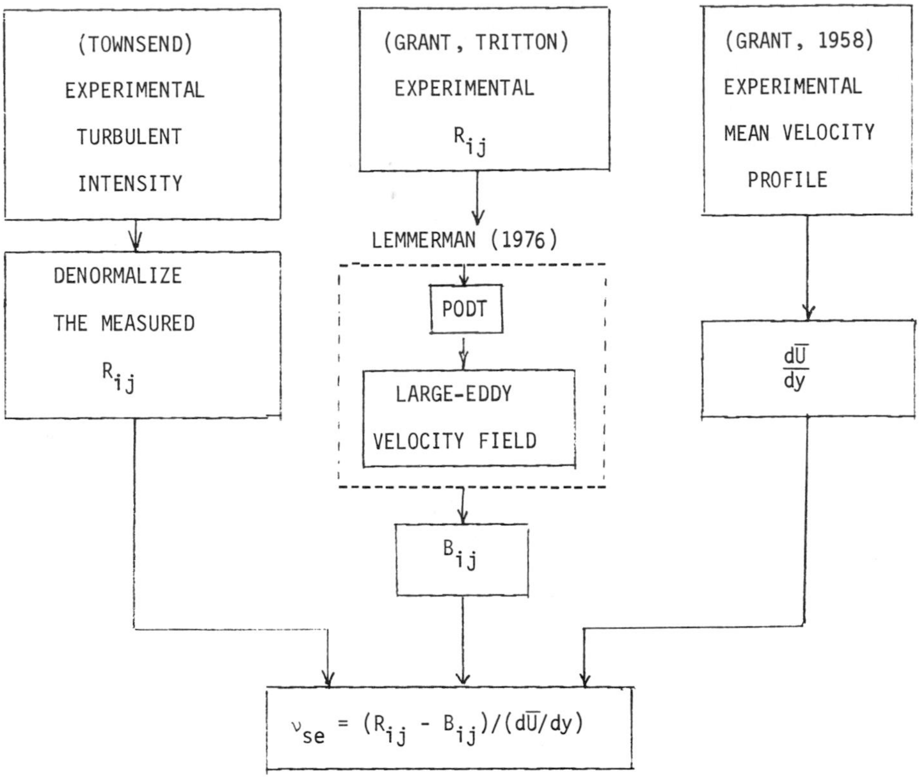

FIGURE 1. Block diagram of the small-eddy calculation (Chuang, 1978a, 1978b).

Lemmerman (1976)[4] and Lemmerman and Payne (1977)[5] without any modification or interpolation. The mean velocity profile $U(y)$, where y is the distance from the wall, is taken from Grant's experimental results (1958)[10].

Due to a lack of regularity in the calculated B_{ij} and ν_{se}, an average is taken over points in each plane parallel to the wall and the final calculations are restricted to one-point correlations, i.e., a 3-D rather than 6-D analysis.

IV. RESULTS

Single-point large-eddy correlation (B_{ij}) calculations, for the indices (i, j) being equal ($1,1$), ($2,2$) and ($3,3$), lead to physical quantity identical to twice the kinetic energy per unit mass. The sum of the three components, B_{11}, B_{22} and B_{33}, at a point is twice the total kinetic energy per unit mass contained in the large-scale motions at that space point.

From Figure 2, the spectrum with two peaks suggests the probability of two discrete families of large eddies. The one with the smaller scale (first peak) is energetically more intensive than the one of large scale (second peak); this is probably the major contributor to the near-wall Reynolds' stress which drives and sustains the turbulent fluctuations.

Figure 3 shows an approximately linear relation between ν_{se}, the "small-eddy viscosity," and the coordinate y normal to the

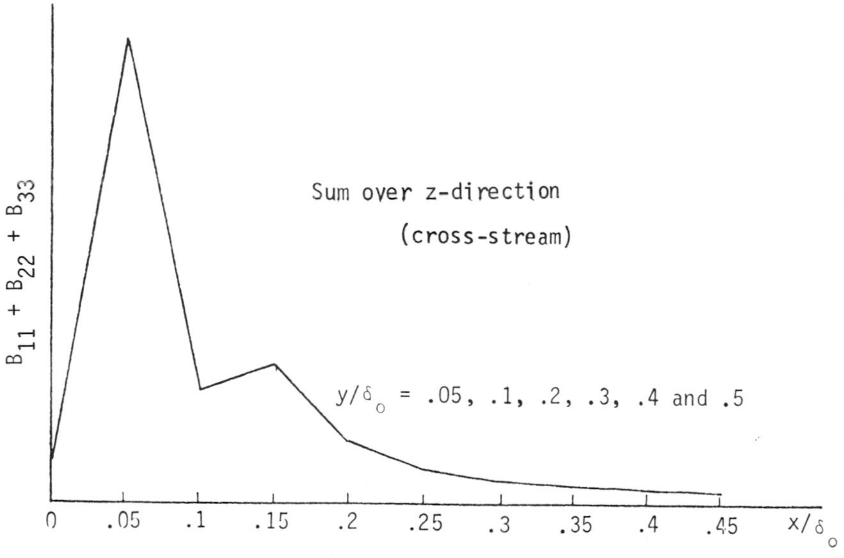

FIGURE 2. Generic shape of big eddy kinetic energy ($B_{11} + B_{22} + B_{33}$) averaged over the lateral direction, z.

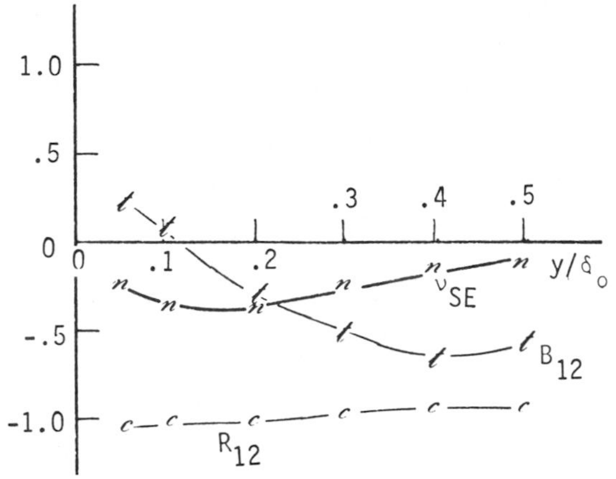

FIGURE 3. Plots of Reynolds stress R_{12}, *big eddy component* B_{12} *and small-eddy viscosity,* ν_{se}.

wall for the closed interval $y/\delta_0 = [0.05, 0.65]$, which encompasses about 60% of the entire turbulent boundary layer, since the "edge" of the boundary-layer is approximately $y/\delta_0 = 1.2$ and $y/\delta_0 = 0.0$ is the flat wall.

V. CONCLUSIONS

One must exercise caution in physical interpretation of the results herein presented due to the following:

(1) Since the data base (Grant,[10] 1958; Tritton,[11] 1967) derives from experiments performed *prior* to Lumley's mathematically rational definition [2] (1966) or "Large Eddy", considerable ingenuity was required (Lemmerman,[4] 1976; Lemmerman and Payne,[5] 1977) to both interpolate and extrapolate data to sufficient density in $(y, \underline{r})$ space of four dimensions that Lumley's eigenvalue problem might be solved (Lemmerman,[4] 1976) with a reasonable degree of precision.

(2) The only present source of "Large Eddy" components for the boundary-layer (Lemmerman,[4] 1976, Lemmerman and Payne,[5] 1977) provides these components on a uniform grid spacing unlike the empirical R_{ij} data[10-11] which were taken on a non-uniform space grid in 4-space, $(\underline{r}, y)$. Hence, Chuang[7] (1978a,b) required further interpolation of R_{ij} data to use and compare with Lemmerman's results.

(3) Throughout Lemmerman's calculations[4-5] and hence the present work[7], an unmeasured but constant numeric factor proportional to τ_0, the wall shear stress, is implicit. Therefore, a significant amount of work remains to recover the absolute levels of dimensional quantities. This effort is in progress at the University of Texas at Arlington by the authors.

(4) In view of Item 3, the penultimate comparison of present work with recent "conditional sampling" studies[12] cannot yet be made on a quantitative basis, e.g. to verify or not that the frequencies and wave lengths of the Lumley-Payne-Lemmerman-Chuang large eddies[2-7] correlate with either the "bursting" phenomena[12] or the subsequent "sweep" concept.[12] Again, this is in progress at UTA concurrently with Item 3 above.

(5) Recall that the data base is limited to the inner 60% of the boundary layer; i.e., not all of the large scale structure has been measured and extracted for the entire boundary layer.

However, one can state unequivocally that:

(1) Results do not contradict other recent results and models.[12-15]

(2) Results are consistent with the dual role of "large" structures in the boundary layer; i.e., as major contributor to the Reynolds' stress (probably the smaller scale eddy in Figure 2) and the sweep-entrainment processes,[12-15] i.e., boundary layer "Blow-off" due to some stress release mechanism(s), and subsequent sweeping-in of "new" laminar or quasi-laminar/turbulent fluid from outside the nominal boundary layer interface with the external flow.

(3) The linear portion of ν_{se} (Fig. 3) is identical in *shape* with the usual boundary layer ν_T profiles;[16] however, the slope appears much smaller than the usual value of Von Karman's constant, $k = 0.41$.[17]

VI. PROGNOSIS

One of us (FRP) has wrestled with PODT and, later, SAS implications for the past decade and a half and for rougly this period of time has been a single "voice in the wilderness"[3,6,8,18-30] crying out for:

(1) Application of PODT-SAS to other data sets in turbulent experiments; this cry has been heeded by Reed (1977)[31] for the flow in a round-jet.

(2) Investigation of the fundamentals of PODT and R_{ij} measurements by applied mathematicians.[3,6,8,20,24,25,30,32]

(3) Estension of application of PODT-SAS to *any* stochastic process.[3,6,8,32]

The implications for, specifically, modeling of turbulent flows are readily apparent to initiates some of whom are now attempting to "calculate" turbulent flows to higher orders of accuracy[33-36] than via "engineering approximations" such as Boussinesq eddy-viscosity, "*k-e*" models,[33] etc.

Here is a developed methodology[2,3,6,8] which will extract *any* pre-dominate "structure" from *any* set of data; generation of data bases via nature, experiment, or digital/analog simulation is immaterial to PODT-SAS.

ACKNOWLEDGMENTS

(a) John Lumley for PODT formulation and his encouragement 1964-66 as dissertation supervisor of one of us (FRP) and co-worker/consultant since that later time.

(b) USAFIT, USN/ONR, USAF Aero Propulsion Lab, NSF, General Dynamics/ Ft. Worth, NASA/Ames and UTA Sundry financial supports for the time period 1961 to the present.

REFERENCES

[1] Townsend, A. A. (1956). "The Structure of Turbulent Shear Flows", University Press, Cambridge; 2nd Edition 1976.

[2] Lumley, J. L. (1966). "The Structure of Inhomogeneous Turbulent Flow", Proc. Inter. Colloq. on Fine-Scale Processes in the Atmos..., Doklady Akademia Nauk SSR, Moscow.

[3] Payne, F. R. (1966). "Large Eddy Structure of a Turbulent Wake", Ph.D. Thesis, Penn State Univ. and Contractor Report to USN/ONR and partly summarized in HOF 1967.

[4] Lemmerman, L. A. (1976). "Extraction of the Large Eddy Structure of a Turbulent Boundary Layer", Ph.D. Thesis, University of Texas at Arlington.

[5] Lemmerman, L. A. and Payne, F. R. (1977). *AIAA Paper No. 77-717*, 10th AIAA Fluids and Plasma Conference, Albuquerque.

[6] Payne, F. R. (1977). UTA/RCAS Conference on Industrial Mathematics, Arlington, Texas, May 1977, and SIAM Fall 1977, Albuquerque.

[7] Chuang, S. L. (1978a). MSAE Thesis, UTA; partly summarized by Chuang (1978b) at S. W. Region AIAA Graduate Division Student Paper Competition, April, 1978, Arlington, Texas.

[8] Payne, F. R. (1978). "A Second Stage Eddy-Viscosity Model for Turbulent Fluid Flows: Or, a Universal Statistical Tool?", Proc. Applied Non-Linear Analysis International Conference sponsored by RCAS/UTA, April 1978, (Published in this volume as the immediately preceding paper.)

[9] Boussinesq, J. (1877). *Mem. Pres. Acad. Sci. Paris 23*, 46, as cited in J. O. Hinze, "Turbulence" (McGraw-Hill, New York, 1975), 2nd ed., 23.

[10] Grant, H. L. (1958). *J. Fluid Mech. 4,* 149.

[11] Tritton, D. J. (1967). *J. Fluid Mech. 28,* 439.

[12] Willmarth, W. W. (1975). "Structure of Turbulence in Boundary Layers", Advances in Applied Mechanics, *15*, 159-254.

[13] Offen, G. R., and Kline, S. J. (1975). *J. Fluid Mech., 70,* 209.

[14] Kline, S. J., et al, (1967). *J. Fluid Mech., 30,* 741.

[15] Blackwelder, R. F., and Kaplan, R. E. (1972). *Int. Union Theor. Appl. Mech., 12th.*

[16] Hinze, J. O. (1975). "Turbulence", 2nd ec., McGraw-Hill, 644-646.

[17] White, F. M. (1974). "Viscous Fluid Flow", McGraw-Hill, (A good modern overview of the general subject of viscous flow.), 472.

[18] Payne, F. R., and Lumley, J. L. (1965). *Bull. Am. Phy. Soc.*

[19] Payne, F. R., and Lumley, J. L. (1967). "Physics of Fluids", *Sup., S-189, Sept.*

[20] Payne, F. R. (1968a). "Predicted Large-Eddy Structure of a Turbulent Wake", contractor report to USN/ONR Fluid Mechanics Branch, Contract NONR 656(33).

[21] Payne, F. R. (1968b). "Large Eddy Structure of Turbulences", General Dynamics/Ft. Worth Internal Report MRP-271.

[22] Payne, F. R. (1968c). "Turbulence", ibid., Report MRP-272.

[23] Payne, F. R. (1968d). "Comparison of Total Pressure to Velocity RMS Values", ibid., MRP-258.

[24] Payne, F. R. (1969). "Generalized Gram-Charlier Method for Curve Fitting Statistical Data", *AIAAJ, October.*

[25] Payne, F. R. (1975a). "In Search of Big Eddies", Research in Progress Report, 14th Mid-West. Mech. Conf., Norman.

[26] Payne, F. R. (1975b), Rep. No. AE/TRL-75-1 to NASA/Ames on Grant NSG-2077, December.

[27] Payne, F. R. (1976). Progress Report to NASA/Ames (NSG-2077), August.

[28] Payne, F. R., and Lemmerman, L. A. (1977). "Large Eddy Structure of a Flat-Plate Boundary-Layer", Proc. 15th Mid-West. Mech. Conf., Chicago.

[29] Payne, F. R. (1977a). Proc. 1st Int. Symp. on Turbulent Shear Flows, Vol. I, Penn State Univ., 16.5.

[30] Payne, F. R., and Lemmerman, L. A. (1977). Poster Session, SIAM Fall/77, Albuquerque.

[31] Reed, X. B., et al, (1977). Proc. 1st Int. Symp. on Turb. Shear Flows, Vol. I, 2.33.

[32] Payne, F. R. (1977b). Proc. Non-Linear Equations in Abstract Spaces, RCAS/UTA, June, Academic Press.

[33] Launder B. E., and Spalding, D. D. (1972). "Lectures in Mathematical Models of Turbulence", Academic Press.

[34] Ceheci, T., and Bradshaw, B. (1977). "Momentum Transfer in Boundary Layers", Hemisphere/McGraw-Hill.

[35] Ceheci, T., and Bradshaw, B. (1977). Proc. 1st Int. Symp. on Turb. Shear Flows, Penn State. (Many papers of pertinence to subject area.) (In press.).

[36] Ceheci, T., and Bradshaw, B. (1979). Proc. 2nd Int. Symp. on Turb. Shear Flows, London (to be held July 2-4 at Imperial College). (Many papers of pertinence to subject area.)

NONLINEAR OPTIMIZATION AND EQUILIBRIA IN POLICY FORMATION GAMES WITH RANDOM VOTING

Peter Coughlin

Department of Economics
Harvard University
Cambridge, Massachusetts

I. INTRODUCTION

Random voting has been analyzed in spatial models of electoral competition by Hinich, Ledyard and Ordeshook (1972, 1973), Slutsky (1975), Katz and Denzau (1977), Katz and Hinich (1977) and Hinich (1977). The existence of a pure condidate equilibrium has been shown when strong assumptions of concavity and convexity are made on individual probability functions. The general conclusion, however, is that the introduction of indeterminancy in voter behavior has rendered the median voter results of deterministic models to the status of artifacts. This paper demonstrates that when individual probability functions satisfy a weak symmetry condition there exists a median voter result under the same conditions which yielded this result in the earlier models.

Sections II and III follow the notation of McKelvey (1975) and extend the Davis-Hinich models presented therein to include probabilistic voting. The structure for voting behavior follows the assumptions explored in Hinich (1977). The results are derived for elections with abstentions due to pure indifference and hence include the full participation model of Hinich as a special

ISBN 0-12-434180-2

case. Lemma 1 shows that when randomness in voting is due to non-systematic non-policy factors the resulting expected plurality voting games are symmetric and zero sum. Theorem 1 proves that under conditions analogous to those which resulted in deterministic median voter results there are corresponding median random voter results. This theorem shows that median voter results are inappropriate when the electorate deviates from symmetry properties, but that these results are compatible with the presence of indeterminancy in election outcomes.

II. ELECTIONS WITH PROBABILISTIC VOTING

An electorate, $\xi(\Omega, S, \{U_\alpha\})$ or ξ, is a set of alternative policies S and a set of voters Ω such that for each $\alpha \in \Omega$ there is a cardinal utility function $U_\alpha\colon\ S \to R$.

Following Hinich (1977) and McKelvey-Wendell (1976), let $\Omega = \{1, 2, \ldots, n\}$.

A two candidate contest in ξ with probabilistic voting, $C_2(\xi, \{V_\alpha\})$ or C_2, is an electorate together with an index set $C = \{1, 2\}$ for the candidates such that for each $\alpha \in \Omega$ there is a random vector $V_\alpha\colon\ S^2 \to \{(1,0,0), (0,1,0), (0,0,1)\}$ with the associated probabilities $P^i_\alpha(s,t) = Pr\{V^i_\alpha(s,t) = 1\}$ for $i \in C$ and $i = 0$ (abstention) for each $(s,t) \in S^2$.

C has probabilistic policy related voting if for all $\alpha \in \Omega$ and all $\bar{s} \in S^2$ and all $i, j \in C$, $i \neq j$:

(i) $P^i_\alpha(\bar{s}) > P^j_\alpha(\bar{s}) \Rightarrow U_\alpha(s_i) > U_\alpha(s_j)$ and

(ii) $P^i_\alpha(s,t) = P^j_\alpha(t,s)$.

The condition of probabilistic policy related voting states that the individual voter probabilities of voting for the particular candidates are conditional upon the candidates' positions. Candidate positions do not, however, fully determine voting behavior and citizens may cast ballots on heuristic factors, social influences or the nature of the communication process. Since the non-policy factors are non-systemmatic, if candidates had taken each other's positions initially the corresponding likelihoods of voting for them would also be switched.

Probabilistic policy related voting can be seen to imply McKelvey's policy related voting assumption when voting behavior is deterministic. Note in particular that $P^0_\alpha(s,t) = 1 - P^1_\alpha(s,t) - P^2_\alpha(s,t) = 1 - P^2_\alpha(t,s) - P^1_\alpha(t,s) = P^0_\alpha(t,s)$. We therefore have $V^0_\alpha(s,t) = V^0_\alpha(t,s)$ for every $(s,t) \in S^2$ when each $P^i_\alpha(s,t)$ is 0 or 1.

If C_2 has probabilistic voting, then for $i \in C$ the expected vote is $V_i(s,t) = \sum_{\alpha=1}^n P^i_\alpha(s,t)$ for each $(s,t) \in S^2$. The expected plurality for $i \in C$ is therefore $Pl_i(s,t) = V_i(s,t) - V_j(s,t)$ for each $(s,t) \in S^2$ when $j \in C$ and $j \neq i$.

If C_2 has probabilistic voting the associated expected plurality voting game is therefore $\Gamma(S,S,M)$ where $M(\bar{s}) = (Pl_1(\bar{s}),Pl_2(\bar{s}))$ for each $\bar{s} \in S^2$.

The following lemma establishes that every election game with probabilistic policy related voting is symmetric. This demonstrates that the structure is analogous to the deterministic models through McKelvey (1975) and provides the initial result in deriving the median random voter results.

Lemma 1. Let C_2 have probabilistic policy related voting. Then $\Gamma(S,S,M)$ is a symmetric zero-sum game.

Proof: M is zero sum since $Pl_i(\bar{s}) = -Pl_j(\bar{s})$ for each $\bar{s} \in S^2$. For any $(s,t) \in S^2$, $V_i(s,t) = \sum_{\alpha=1}^n P^i_\alpha(s,t) = \sum_{\alpha=1}^n P^j_\alpha(t,s) = V_j(t,s)$. Thus $Pl_i(s,t) = V_i(s,t) - V_j(s,t) = V_j(t,s) - V_i(t,s) = Pl_j(t,s)$. But $Pl_j(t,s) = -Pl_i(t,s)$, so $Pl_i(s,t) = -Pl_i(t,s)$ and hence M is symmetric

QED

Lemma 1 implies that the value of any expected plurality voting game with probabilistic policy related voting is zero. This means that the expected outcome from equilibrium pure or mixed strategies is a tie. Additionally we have the property that if there exists a pure equilibrium then there exists an equilibrium in pure strategies with identical condidate positions. Theorem 1 will show that if the electorate, ξ, satisfies symmetry properties analogous to those which implied deterministic median voter results then there is an equilibrium in pure strategies when each candidate selects the median of the distribution of voter ideal points.

III. MEDIAN RANDOM VOTER RESULTS

C_2 is a regular spatial model iff $S = R^m$ and there is a function $X\colon \Omega \to R^m$ which associates an ideal point $x_\alpha \in R^m$ with each $\alpha \in \Omega$. $p_X(x)$ will denote the proportion of voters with the ideal point $X(\alpha) = x$, and p_X will denote the discrete probability function on S defined by $p_X(x)$.

A two condidate contest has quadratic based loss iff there is a positive definite $n \times n$ matrix A and a monotone decreasing function $\phi_\alpha\colon R \to R$ for each $\alpha \in \Omega$ such that $U_\alpha(x) = \phi_\alpha(\|x - x_\alpha\|A)$. C_2 has concave quadratic utility iff $\phi_\alpha(t)$ is a concave function for each $\alpha \in \Omega$.

An electorate has symmetric preferences iff p_X is radially symmetric and $\phi_\alpha = \phi_\beta$ when $(x_\alpha + x_\beta)/2 = \mu$, where μ is the median of p_X.

C_2 has utility based probabilistic voting iff there is a monotonically increasing function $\psi_\alpha\colon R \to [0,1]$ for each $\alpha \in \Omega$ such that $P^i_\alpha(s_1,s_2) = \psi_\alpha(U_\alpha(s_i) - U_\alpha(s_j))$ for each $i, j \in C$, $i \neq j$, and $(s_1,s_2) \in S^2$.

A two condidate contest has policy related abstention iff there exists a function $W\colon R^2 \to R$ such that $P^0_\alpha(s_1,s_2) = W(U_\alpha(s_1),U_\alpha(s_2))$ for each $(s_1,s_2) \in S^2$. C_2 has abstention due to pure indifference if $W(u_1,u_2)$ is a monotone decreasing function of $|u_1 - u_2|$.

An electorate has symmetric probabilistic voting iff $\psi_\alpha(-y) = 1 - W(|y|) - \psi_\alpha(y)$ for every $y \in R$ and $\psi_\alpha = \psi_\beta$ when $(x_\alpha + x_\beta)/2 = \mu$, where μ is the median of P_X.

This structure follows the usual assumptions for a Davis-Hinich electorate in which deterministic median voter results hold, and introduces analogous symmetry properties on the individual probability functions. The assumption of symmetric preferences is also weaker than the usual assumption of homogeneous utility functions. The following theorem proves that the median voter result is not an artifact under these conditions. This result emphasizes that the absence of median voter results in models with probabilistic voting is due to the absence of symmetry properties, rather than due to the incorporation of indeterminancy in election outcomes.

Theorem 1. Let C_2 be a regular spatial model with symmetric concave quadratic utility based probabilistic voting and abstention due to pure indifference. If ξ has symmetric preferences then (μ,μ) is a pure strategy equilibrium for Γ.

Proof: We need to show $Pl_1(\theta_1,\mu) \leq Pl_1(\mu,\mu) \leq Pl_1(\mu,\theta_2)$. Utility based probabilistic voting implies probabilistic policy related voting so Γ is symmetric by Lemma 1. So $Pl_1(\mu,\mu) = -Pl_1(\mu,\mu) = 0$. We can therefore equivalently show $V_1(\theta_1,\mu) \leq V_2(\theta_1,\mu)$ and $V_2(\mu,\theta_2) \leq V_1(\mu,\theta_2)$ for all θ_1, $\theta_2 \in S$ where $\theta_1 \neq \mu$, and $\theta_2 \neq \mu$.

Suppose $\theta \neq \mu$.

For any voter with $x_\alpha = \mu$, $y_\alpha(\theta_1,\mu) = (U_\alpha(\theta_1) - U_\alpha(\mu)) \leq 0$ since ϕ is monotone decreasing. Since ψ_α is monotone increasing we thus have $P^1_\alpha(\theta_1,\mu) = \psi_\alpha(U_\alpha(\theta_1) - U_\alpha(\mu)) \leq \psi_\alpha(U_\alpha(\mu) - U_\alpha(\theta_1)) = P^2_\alpha(\theta_1,\mu)$. Letting $M = \{\alpha \in \Omega\colon x_\alpha = \mu\}$, we have

$$\Sigma_M P^1_\alpha(\theta_1,\mu) \leq \Sigma_M P^2_\alpha(\theta_1,\mu). \tag{1}$$

Since the electorate has symmetric preferences, for every $x_\alpha \neq \mu$ there is an $x_\beta \neq \mu$ so that $(x_\alpha + x_\beta)/2 = \mu$ and

$p_X(x_\alpha) = p_X(x_\beta)$. We can thus consider voters in symmetric pairs. For any such pair let α denote the voter closer to θ_1, i.e., $\|\theta_1 - x_\alpha\|_A \leq \|\theta_1 - x_\beta\|_A$.

$$\|\theta_1 - x_\beta\|_A = \|x_\beta - \theta_1\|_A = \|x_\alpha + x_\beta - \theta_1 - x_\alpha\|_A$$
$$= \|2\mu - \theta_1 - x_\alpha\|_A,$$

so $U_\beta(\theta_1) = \phi_\beta(\|\theta_1 - x_\beta\|_A) = \phi_\alpha(\|2\mu - \theta_1 - x_\alpha\|_A) = U_\alpha(2\mu - \theta_1)$. Additionally, $\mu = \frac{1}{2}(\theta_1) + \frac{1}{2}(\mu + (\mu - \theta_1))$ so $U_\alpha(\mu) \geq \frac{1}{2}U_\alpha(\theta_1) + \frac{1}{2}U_\alpha(\mu + (\mu - \theta_1))$ since ϕ is concave. Thus $U_\alpha(\mu) \geq \frac{1}{2}U_\alpha(\theta_1) + \frac{1}{2}U_\beta(\theta_1)$, and so $2U_\alpha(\mu) \geq U_\alpha(\theta_1) - U_\beta(\theta_1)$. Since $U_\alpha(\mu) = U_\beta(\mu)$ by symmetric preferences and quadratic based loss,

$$U_\alpha(\mu) - U_\alpha(\theta_1) \geq U_\beta(\theta_1) - U_\beta(\mu). \tag{2}$$

By (2), if $U_\alpha(\mu) - U_\alpha(\theta_1) \leq 0$, then $|U_\beta(\theta_1) - U_\beta(\mu)| \geq |U_\alpha(\theta_1) - U_\alpha(\mu)|$. Suppose $U_\alpha(\mu) - U_\alpha(\theta_1) \geq 0$. $\|x_\alpha - x_\beta\|_A = \|x_\beta - \theta_1 + \theta_1 - x_\alpha\|_A$, so $\|x_\alpha - x_\beta\|_A \leq \|x_\beta - \theta_1\|_A + \|\theta_1 - x_\alpha\|_A$. But $x_\alpha = 2\mu - x_\beta$, so $x_\alpha - x_\beta = 2\mu - x_\beta$. Hence $\|x_\alpha - x_\beta\|_A = 2\|\mu - x_\beta\|_A$. Therefore $2\|\mu - x_\beta\|_A \leq \|x_\beta - \theta_1\|_A + \|\theta_1 - x_\alpha\|_A$, so $\|\theta_1 - x_\alpha\|_A \leq \|\theta_1 - x_\beta\|_A$ implies $\|x_\beta - \theta_1\|_A \geq \|\mu - x_\beta\|_A$. Therefore $U_\beta(\mu) \geq U_\beta(\theta_1)$, and so $U_\beta(\theta_1) - U_\beta(\mu) \leq 0$. Thus $|U_\beta(\theta_1) - U_\beta(\mu)| = U_\beta(\mu) - U_\beta(\theta_1) \geq U_\alpha(\theta_1) - U_\alpha(\mu) = |U_\alpha(\theta_1) - U_\alpha(\mu)|$, by (2).

Now $W(|U_\alpha(\theta_1) - U_\alpha(\mu)|) \geq W(|U_\beta(\theta_1) - U_\beta(\mu)|)$ since W is monotone decreasing. Additionally, (2) implies $\psi_\alpha(U_\alpha(\mu) - U_\alpha(\theta_1)) \geq \psi_\beta(U_\beta(\theta_1) - U_\beta(\mu))$ since $\psi_\alpha = \psi_\beta$ is monotone increasing. Therefore

$$P_\alpha^1(\theta_1,\mu) = 1 - W(|U_\alpha(\theta_1) - U_\alpha(\mu)|) - \psi_\alpha(U_\alpha(\mu) - U_\alpha(\theta_1))$$
$$\leq 1 - W(|U_\beta(\theta_1) - U_\beta(\mu)|) - \psi_\beta(U_\beta(\theta_1) - U_\beta(\mu)).$$

But $W(|U_\beta(\theta_1) - U_\beta(\mu)|) = 1 - \psi_\beta(U_\beta(\theta_1) - U_\beta(\mu)) - \psi_\beta(U_\beta(\mu) - U_\beta(\theta_1))$, so $P_\alpha^1(\theta_1,\mu) \leq \psi_\beta(U_\beta(\mu) - U_\beta(\theta_1)) = P_\beta^2(\theta_1,\mu)$. By (2) and symmetric probabilistic voting, $P_\beta^1(\theta_1,\mu) = \psi_\beta(U_\beta(\theta_1) - U_\beta(\mu)) \leq \psi_\alpha(U_\alpha(\mu) - U_\alpha(\theta_1)) = P_\alpha^2(\theta_1,\mu)$ since $\psi_\alpha = \psi_\beta$ is monotone increasing. Thus

$$P_\alpha^1(\theta_1,\mu) + P_\beta^1(\theta_1,\mu) \leq P_\beta^2(\theta_1,\mu) + P_\alpha^2(\theta_1,\mu). \tag{3}$$

By (1) and (3) we have $V_1(\theta_1,\mu) = \sum_{\alpha=1}^{n} P_\alpha^1(\theta_1,\mu)$ $\leq \sum_{\alpha=1}^{n} P_\alpha^2(\theta_1,\mu) = V_2(\theta_1,\mu)$.

By a similar argument with θ_2 and (μ,θ_2) we have $V_1(\mu,\theta_2) \geq V_2(\mu,\theta_2)$.

QED

We thus have median random voter results in spatial models of electoral competition with probabilistic voting. If we let $W(u_1,u_2) = 0$ be the monotone decreasing (not strictly) abstention function then we additionally have median random voter results in the models explored by Hinich (1977).

This result has not used the strong assumptions of concavity and convexity for the individual probability functions which led to the existence of a pure equilibrium in Hinich, Ledyard and Ordeshook (1972), 1973), but which were subsequently criticized for lack of empirical meaning in Slutsky (1975).

IV. CONCLUSION

The literature on the spatial model of electoral competition has focused on the incorporation of alternative assumptions into existing models of elections and the development of sufficient conditions for the existence of pure candidate equilibria. This paper incorporates probabilistic voting into the models due to McKelvey of Hinich-Davis electorates and demonstrates that under conditions analogous to those which implied earlier median voter results there are median random voter results. This additionally extends the recent work on random non-policy factor in voting by Hinich and others.

REFERENCES

[1] Coleman, J. (1972). "The Positions of Political Parties in Elections" in Probability Models of Collective Decision Making, (Niemi and Weisberg, eds.), Columbus, Ohio: Merrill.

[2] Coughlin, P. (1977). "Mathematical Models of Elections With Incomplete Information and Sequences of Decisions" in Proceedings of the First International Conference on Mathematical Modeling, (Avula, ed.), University of Missouri at Rolla.

[3] Denzau, A., and Kats, A. (1977). "Expected Plurality Voting Equilibrium and Social Choice Functions", Review of Economic Studies 44.

[4] Fishburn, P., and Gehrlein, W. (1976). "Win Probabilities and Simple Majorities in Probabilistic Voting Situations", Mathematical Programming 11.

[5] ________. (1977). "Toward a Theory of Elections With Probabilistic Preferences", Econometrica 45.

[6] Hinich, M., and Davis, O. (1966). "A Mathematical Model of Policy Formation in a Democratic Society" in Mathematical Applications in Political Science, II, (Bernd, ed.), Dallas: Southern Methodist University Press.

[7] ________. (1967). "Some Results Related to a Mathematical Model of Policy Formation in a Democratic Society" in Mathematical Applications in Political Science, III, (Bernd, ed.), Charlottesville: University Press of Virginia.

[8] Hinich, M., Ledyard, J., and Ordeshook, P. (1972). "Nonvoting and the Existence of Equilibrium Under Majority Rule", Journal of Economic Theory 4.

[9] Hinich, M., Ledyard, J., and Ordeshook, P. (1973). "A Theory of Electoral Equilibrium: A Spatial Analysis Based on the Theory of Games", Journal of Politics 35.

[10] Hinich, M., and Kats, A. (Summer 1977). "Uncertainty in Candidate Objective Functions", Econometric Society Meeting.

[11] Hinich, M. (1977). "Equilibrium in Spatial Voting: The Median Voter Result Is an Artifact", Journal of Economic Theory.

[12] McKelvey, R. (1975). "Policy Related Voting and Electoral Equilibria", Econometrica 43.

[13] McKelvey, R., and Wendell, R. (1976). "Voting Equlibria in Multidimensional Choice Spaces", Mathematics of Operations Research 1.

[14] Slutsky, S. (1975). "Abstentions and Majority Equilibrium", Journal of Economic Theory 11.

ON THE BOUNDED SOLUTIONS OF A NONLINEAR CONVOLUTION EQUATION[1]

Odo Diekmann[2]

Mathematisch Centrum
Amsterdam, The Netherlands

Hans G. Kaper[3]

Applied Mathematics Division
Argonne National Laboratory
Argonne, Illinois

ABSTRACT

This investigation is concerned with the nonlinear convolution equation $u(x) - (g \circ u) * k(x) = 0$ on the real line $\mathbb{R}$. The kernel k is nonnegative and integrable on $\mathbb{R}$, with $\int_{\mathbb{R}} k(x)dx = 1$; the function g is real-valued and continuous on $\mathbb{R}$, $g(0) = 0$, and there exists a $p > 0$ such that $g(x) > x$ for $x \in (0,p)$ and $g(p) = p$. Sufficient conditions are given for the non-existence of bounded nontrivial solutions.

[1]*This paper will be published in Nonlinear Analysis, Theory, Methods and Applications, Vol. 2 (1978).*

[2]*Work performed under the auspices of the Netherlands Organization for the Advancement of Pure Research (Z.W.O.).*

[3]*Work performed under the auspices of the U.S. Department of Energy.*

ISBN 0-12-434180-2

Implications for the solution of the inhomogeneous equation $u(x) - (g \circ u) * k(x) = f(x)$, $x \in \mathbb{R}$, are discussed. Finally, uniqueness (modulo translation) is shown to hold. The results are applied to a problem of mathematical epidemiology.

SOME UNRESOLVED QUESTIONS PERTAINING TO THE MATHEMATICAL ANALYSIS OF FLUORESCENCE DECAY DATA

Corey C. Ford

Department of Physiology
University of Texas Health Science Center
Dallas, Texas

*Jerome Eisenfeld**

Department of Mathematics
The University of Texas at Arlington
Arlington, Texas

I. INTRODUCTION

Several papers in these conference proceedings deal with compartmental analysis identification problems [1], [2], [3], [4], [5], [6]. In this paper we discuss applications of these techniques to the analysis of multi-exponential fluorescence decay data. The problem is to determine the amplitudes and time constants for the N exponential components present in the data. Treatment of non-random errors such as scattered light, zero time

**Also affiliated with the Department of Medical Computer Science, University of Texas Health Science Center at Dallas, Dallas, Texas 75235.*

ISBN 0-12-434180-2

shifts, and cutoff errors are discussed. The method of analysis used is based on moment techniques [7], [4], [5], [6].

II. FLUORESCENCE DECAY ANALYSIS; A DESCRIPTION OF THE PROBLEM

In our experiments a sample containing fluorescent molecules is excited with a short pulse of light, $E(t)$. Many of the excited molecules emit photons fluorescently and a single photon is detected randomly by the instrument. The time between excitation and the detected emission is measured and the event is recorded. This process is repeated until a histogram is developed representing the fluorescence decay profile, $F(t)$. Each fluorescent component is assumed to decay exponentially. The impulse response, $I(t)$, is given as

$$I(t) = \sum_{i=1}^{N} \alpha_i \exp(-t/\tau_i) \tag{1}$$

where the α_i are pre-exponential constants or amplitudes, τ_i are the fluorescence decay times, and N is the number of components.

We have used the phenomenon of fluorescence energy transfer to measure the distance separating two fluorescent molecules attached to specific sites on muscle proteins. When excitation energy from the donor fluorophore (D) is transferred to an acceptor (A), the decay times of the donor decrease. The efficiency of transfer is defined as

$$\varepsilon = 1 - \frac{\tau}{\tau_0} \tag{2}$$

where τ_0 is the decay time of D in the absence of A and τ the decay time of D in the presence of A. If a molecule gives multiple decay times any or all may be shortened by energy transfer. We generally look at the longest component present since it is usually easiest to measure. The efficiency ε is related to the separation distance (R) as

$$R = (\varepsilon^{-1} - 1)^{1/6} R_0 \tag{3}$$

where R_0 is the distance at which 50% of excited donors transfer energy to acceptors [11]. An energy transfer pair can be thought of as a "spectroscopic ruler" [13].

III. METHOD OF ANALYSIS

In order to quantify a distance we need to measure the values of fluorescence decay time of D in the presence and absence of A. The impulse response function $I(t)$ would be directly obtained if the exciting impulse $E(t)$ was a delta function. Since $E(t)$ is of finite width, we observe $F(t)$ defined by the convolution integral

$$F(t) = \int_0^t I(u)\ E(t-u)du \tag{4}$$

Observe that equations (1) and (4) can be put in the standard form of the system identification problem [1], [4], [5].

$$x = A\ x(t) + B\ E(t), \tag{5}$$

$$F(t) = C\ x(t), \tag{6}$$

where

$$B = (\alpha_1, \alpha_2, \ldots, \alpha_N)^T, \tag{7}$$

$$C = (1, 1, \ldots 1), \tag{8}$$

$$A = \mathrm{diag}(-\tau_1^{-1}, -\tau_2^{-1}, \ldots -\tau_N^{-1}), \tag{9}$$

and

$$I(t) = C\ \exp(tA)\ B. \tag{10}$$

Based on discrete data for the impulse $E(t)$ and the output $F(t)$, we wish to identify the matrices A and B which contain the amplitudes and decay times of $I(t)$.

We use a modified Isenberg method of moments to compute the components of $I(t)$ [7], [8]. Using the moments of $F(t)$ and $E(t)$ defined respectively as

$$\mu_k = \int_0^\infty t^k F(t)dt \qquad k = 1,2,\ldots , \tag{11}$$

$$m_k = \int_0^\infty t^k E(t)dt \qquad k = 1,2,\ldots , \tag{12}$$

we can determine the moments G_k of $I(t)$ [7] defined as

$$G_k = \int_0^\infty t^k I(t)dt \qquad k = 1,2,\ldots . \tag{13}$$

The relationship between μ_k, m_k and G_k is [7]

$$\frac{\mu_k}{k!} = \sum_{s=1}^{k+1} \frac{G_s m_{k+1-s}}{(k+1-s)!} \qquad k = 1,2,\ldots . \tag{14}$$

The G_k are related to the set α_i, τ_i by [7]

$$G_k = \sum_{i=1}^{N} \alpha_i \tau_i^k \qquad k = 1,2,\ldots . \tag{15}$$

The problem is now reduced to the classical moment problem [9]. For numerical methods of estimating the α_i, τ_i from the sequence of G_k's see [7], [4].

Efficiency and accuracy of the calculation are improved by careful choice of weighting functions available as options in the method of moments. These weighting functions include the exponential depression parameter (λ) and the moment displacement index (M_D) [7], [10]. A description of these options appears later in the paper.

IV. DESCRIPTION OF THE DATA

The moments program, appropriately weighted, is now applied to real data collected in an energy transfer experiment. Figure 1 shows the impulse *E(t)* and the convoluted response *F(t)* obtained for *D* in the presence of *A*. Now we consider some of the complicating aspects of this fluorescence decay data.

First, the observed decay curve *F(t)* is only known over a finite interval $[0,T]$. Furthermore, the curvature in this semi-log plot is suggestive of multiexponential decay. Each channel in the histogram of *E(t)* and *F(t)* is known to a finite number of counts. Counting error is assumed to be Poisson distributed and the standard deviation is approximated by the square root of the number of counts. Also, the first *50* channels have been used to collect random background counts resulting from noise in the detector. *F(t)* is thus sitting on a flat baseline which is removed by averaging the counts in channels *1* to *50* and subtracting this value from each subsequent data point.

In this example some scattered light from *E(t)* is able to leak through the system along with *F(t)*. Scatter occurs

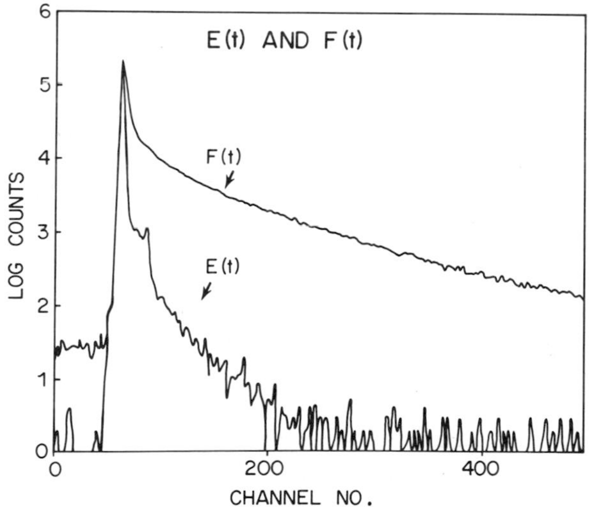

FIGURE 1

instantaneously and $F(t)$ therefore follows $E(t)$ closely in the initial part of the decay [10]. A zero time shift error may also be present and will be discussed later [12].

V. ANALYSIS OF THE DATA

Analysis by the method of moments begins with the determination of $\hat{N}$, the number of components in the impulse response $I(t)$. We form Hankel matrices of the set G_k as follows

$$\left\{G_{i+j-1+M_D}\right\} \quad (i,j = 1,2,\ldots,N; \quad M_D = 1,2,\ldots) .$$

As shown elsewhere in the proceedings [5], there will be N successive, positive determinants, D, of these matrices for N components. Values of these determinants obtained for the data of Figure 1 are given in Table 1 for moment displacements of 0, 1 and 2. Calculated values for the set α_i, τ_i are summarized in Table 2A. For an M_D of 0 it can be seen that the analysis for 2 components results in sub-eigenvalues [4] bounded by the values obtained for a 3 component analysis. Likewise, a one component analysis is a sub-eigenvalue of a 2 component analysis. The determinants used to predict N are estimates and, although an M_D of 2 is predicted to give 2 components, the program will only converge for 1. There are thus some errors in this method of determining N which are probably due to the cutoff error discussed later.

In Table 2B we present the result of successive moment displacement applied to a one component analysis of our date in Figure 1. As predicted [4], the single component as well as its amplitude converges to the maximum eigenvalue in $I(t)$. In this case the maximum eigenvalue is the same as that calculated in the 3 component analysis given in Table 2A. The amplitude, was determined to be 0.0074 in both cases. This procedure is

TABLE 1. Determining the Number of Components

M_D	$\hat{D}_1$	$\hat{D}_2$	$\hat{D}_3$	$\hat{D}_4$	$\hat{D}_5$	$\hat{N}$
0	1.000	1.545	0.480	-9.284	-36.4556	3
1	1.000	0.798	-1.566	-5.259	51.317	2
2	2.545	1.203	-18.152	-74.244	1637.243	2

TABLE 2A. Analysis of Data in Figure 1. Lifetimes in Nanoseconds (λ = 0.008)

Analyzed Components	$M_D = 0$			$M_D = 1$			$M_D = 2$		
	τ_1	τ_2	τ_3	τ_1	τ_2	τ_3	τ_1	τ_2	τ_3
1	16.9			58.2			72.8		
2	3.0	74.3		12.7	77.4		*	*	
3	0.5	25.5	96.0	*	*	*	*	*	*

* *Incomplete Convergence.*

TABLE 2B. Convergence of 1 component analysis to the maximum eigenvalue with successive moment index displacement (λ = 0.015).

M_D	AMPLITUDE	DECAY TIME (nsec)
0	0.192	11.48
1	0.026	48.19
2	0.018	62.59
3	0.013	73.89
4	0.011	81.09
5	0.0096	86.38
6	0.0087	90.20
7	0.0080	93.00
8	0.0076	94.94
9	0.0073	95.96
10	0.0074	95.68

very useful since the maximum eigenvalue can then be filtered from the analysis [6] and calculations for the remaining components will be improved. We conclude that there are at least three components in our data and analysis for more is not possible with our program and the data of Figure 1.

The data analyzed in Tables 2A and 2B corresponds to the measurement of D lifetime in the presence of acceptor. Considering only the long component, we use $\tau = 96$ nsec and $\tau_0 = 98.6$ nsec (determined in a separate experiment). The energy transfer efficiency ε is 2.6% from equation (2). Using the equations of Forster [11] we calculated R_0 to be 23 Å, and therefore, from equation (3) the donor and acceptor molecules are separated by 42 Å. This value is in agreement with our model and has provided the starting point for additional studies.

VI. NON-RANDOM ERRORS

As mentioned earlier, there are non-random errors in fluorescence decay data. The most serious of these are scatter and the zero time shift [10]. Cutoff error is peculiar to the Isenberg method of moments [7]. Scatter and the zero time shift enter the description of $F(t)$ as

$$F(t) = \int_0^t I(u)\ E(t-u+s)du + \gamma E(t+s) \tag{17}$$

where γ is the scatter coefficient and s the zero time shift [10]. Scatter is due to imperfect isolation of light from $E(t)$ in the collection of $F(t)$. A zero time shift can occur when $E(t)$ and $F(t)$ are collected at very different wavelengths, since the instrument response time is wavelength dependent [12]. As a result, times zero for $E(t)$ and $F(t)$ are not coincident leading to errors in the deconvolution of $I(t)$.

These two errors are largely corrected by an M_D of 1 or 2 [10]. Using an M_D replaces the set G_k with the set G_{k+M_D}. The drawback to using moment index displacement is that fast decay components may be lost to the analysis since M_D weights against data at early times. An example of this is seen in Table 2A where the shortest decay time is lost when M_D is increased from 0 to 1 and the middle decay time is lost when the M_D is incremented from 1 to 2. Isenberg [10] has discussed ways of determining γ and s for a given set of data.

The cutoff error arises from moment integrals in equation (11) and (12) which must be carried to infinity for data known only to a finite time T. The method of moments successively approximates a correction to the moments from T to infinity as

$$\Delta\mu_k = \int_T^\infty t^k \sum_{i=1}^{N} \alpha_i \exp(-t/\tau_i)dt \tag{18}$$

where the α_i, τ_i are estimates obtained using moments calculated from 0 to T. Corrections to the set m_k are not needed because $E(t)$ is assumed to reach 0 before time T. The approximations are iterated until internal self-consistency is reached between the moments μ_k and m_k and the set α_i, τ_i. More sophisticated weighting techniques [6] can eliminate the cutoff error by taking "moments" which need only be integrated from 0 to T [6], [8].

The cutoff error can be reduced by using exponential depression. This weighting function replaces $F(t)$ and $E(t)$ with $F(t) = \exp(-\lambda t)F(t)$ and $E(t) = \exp(-\lambda t)E(t)$ respectively [7]. The constant λ is referred to as the exponential depression parameter and the deconvolution then results in $I(t) = \exp(-\lambda t)I(t)$. Use of exponential depressions weights against data at the tail of a decay curve reducing the cutoff correction $\Delta\mu_k$ and improving the analysis.

VII. UNRESOLVED PROBLEMS

1. We would like to calculate all components of $I(t)$ as well as the zero time shift and scatter coefficient in a single experiment. This requires additional analytic capability and some knowledge of how much data is neccessary.

2. Two components close together cannot be resolved and are seen as *1* average decay time by the program. It would be useful to have a method of predicting this situation even if the analysis cannot be performed.

3. Components very different from each other lead to stiffness in the equations and create bandwidth problems experimentally. It is important to know the range over which $F(t)$ should be collected to maximize resolution of both components.

4. Weighting functions are needed which remove errors without eliminating information. At the very least, we would like to know how to choose these functions to optimize the tradeoff.

5. All instrumentation has a certain non-linearity associated with it. It is not known how much damage such non-linearity could have on an analysis.

   A great deal of progress has been made in the analysis of multi-exponential decay data. It remains, however, the responsibility of the experimentalist to provide data which is within the limitations of present analytic ability.

REFERENCES

[1] Jacquez, J. A. "Models of Biological Systems: Linear and Nonlinear", Applied Nonlinear Analysis, V. Lakshmikantham, Ed., New York: Academic Press, to appear.

[2] Anderson, D. "The Volume of Distribution in Single-Exit Compartmental Systems", Applied Nonlinear Analysis, V. Lakshmikantham, Ed., New York: Academic Press, to appear.

[3] Sandberg, S., Anderson, D. H., and Eisenfeld, J. "On Identification of Compartmental Systems", Applied Nonlinear Analysis, V. Lakshmikantham, Ed., New York: Academic Press, to appear.

[4] Hallmark, J., and Eisenfeld, J. "Separation and Monotonicity Results for the Roots of the Moment Problem", Applied Nonlinear Analysis, V. Lakshmikantham, Ed., New York: Academic Press, to appear.

[5] Soni, B., and Eisenfeld, J. "System Identification of Models Exhibiting Exponential, Harmonic and Resonant Modes", Applied Nonlinear Analysis, V. Lakshmikantham, Ed., New York: Academic Press, to appear.

[6] Cheng, S., and Eisenfeld, J. "A Direct Computational Method for Estimating the Parameters of a Nonlinear Model", Applied Nonlinear Analysis, V. Lakshmikantham, Ed., New York: Academic Press, to appear.

[7] Isenberg, I., Dyson, R. D., and Hanson, R. (1973). "Studies on the Analysis of Fluorescence Decay Data by the Method of Moments", Biophysical Journal, Vol. 13, 1090-1115.

[8] Eisenfeld, J., and Cheng, S. "General Moment Methods for a Class of Nonlinear Models", submitted for publication.

[9] Gantmacher, F. R. (1964). "The Theory of Matrices", Vol. II, New York: Chelsea.

[10] Small, E. W., and Isenberg, I. (1977). "On Moment Index Displacement", *J. Chem. Phys., Vol. 66,* 3347.

[11] Forster, Th. (1948). "Intermolecular Energy Transfer and Fluorescence", *Ann. Phys., Vol. 2*, 55.

[12] Wahl, Ph., Auclet, J. C., and Donzel, B. (1974). "The Wavelength Dependence of the Response of a Pulse Fluorometer using the Single Photoelectron Counting Method", *Rev. Sci. Inst., Vol. 45*, 28-32.

[13] Stryer, L., and Haugland, R. P. (1967). "Energy Transfer: A Spectroscopic Ruler", *Biochem., Vol. 58*, 719-726.

SEPARATION AND MONOTONICITY RESULTS FOR THE ROOTS OF THE MOMENT PROBLEM

James Hallmark
Jerome Eisenfeld[1,2]

Department of Mathematics
University of Texas at Arlington
Arlington, Texas

I. INTRODUCTION

Consider the system identification problem

$$\dot{x}(t) = Ax(t) + Bu(t)$$

$$y(t) = Cx(t) \quad 0 < t < T,$$

where $u(t)$ and $y(t)$ are given discretely on the interval $[0,T]$ and we wish to determine information about the unknown constant matrices A, B, and C.

Many techniques are available for approaching this problem. Several of these methods lead to what is known as the moment problem, [1]-[6].

[1]*Partially supported by University of Texas of Arlington Organized Research.*

[2]*Also affiliated with the Department of Medical Computer Science, University of Texas Health Science Center at Dallas, Dallas, Texas 75235.*

ISBN 0-12-434180-2

This paper pertains to the moment problem which will now be introduced.

II. MOMENT PROBLEM

Assume we are given a moment sequence $\{s_k\}$, $k = 0,1,2,\ldots$. Several previously mentioned methods used in solving the system identification problem will give us such a sequence. From the analysis involved in these methods, it can be shown that the terms in the moment sequence have the form

$$s_k = \sum_{i=1}^{N} \alpha_i \lambda_i^k \tag{2.1}$$

where $\alpha_i > 0$, λ_i are real and distinct, and $0 < \lambda_1 < \lambda_2 < \ldots < \lambda_N$. Thus the moment problem is: given a moment sequence $\{s_k\}$, where s_k has the form (2.1), determine N, α_i, λ_i, $i = 1,2,\ldots,N$. The values of $\lambda_1,\lambda_2,\ldots,\lambda_N$ will be called the eigenvalues of the moment problem. Some useful results will now be stated about the moment problem.

Methods exist for determining N, the number of components, and are discussed in [3]. Assuming now that N is known, then the eigenvalues of the moment problem are the N real distinct roots of the equation

$$\det\left(G_N^{k+} - \lambda G_N^k \right) = 0 \tag{2.2}$$

where G_N^k is the $N \times N$ matrix

$$G_N^k = \begin{pmatrix} s_k & s_{k+1} & \cdots & s_{k+N-1} \\ s_{k+1} & & & \cdot \\ \vdots & & & \vdots \\ s_{k+N-1} & & \cdots & s_{k+2N-2} \end{pmatrix}$$

It may be shown that $\lambda_1,\lambda_2,\ldots,\lambda_N$ are also the N roots of the N-th degree polynomial

$$\det \begin{bmatrix} 1 & \lambda & \lambda^2 & \cdots & \lambda^N \\ s_k & s_{k+1} & s_{k+2} & \cdots & s_{k+n} \\ \vdots & & & & \\ s_{k+N-1} & & & \cdots & s_{k+2N-1} \end{bmatrix} = 0. \qquad (2.3)$$

Changes in k are referred to as *moment displacements*. This terminology is due to Isenberg [4].

Roots of equations (2.2) and (2.3) are independent of k. For a discussion of these results see [2]-[4].

Several problems are encountered when solving the moment problem. Because of insufficient data or numerical errors, the value of N may be difficult to compute. This is certainly the case if N is 'large'. Even if N is known, solutions of equations (2.2) and (2.3) may contain large numerical errors if N is large, say $N \geq 5$. The following question now arises. If we assume only n terms $(n = 1,2,\ldots,N)$ in model (2.1) and then solve equation (2.2) or (2.3), how will these n roots (to be called subeigenvalues) relate to the actual eigenvalues $\lambda_1,\lambda_2,\ldots,\lambda_N$? The remainder of this paper will be directed at answering this question.

III. THE SUBEIGENVALUE PROBLEM

Consider the $n \times n$ matrices

$$G_n^k = \begin{bmatrix} s_k & s_{k+1} & \cdots & s_{k+n-1} \\ s_{k+1} & & & \vdots \\ \vdots & & & \\ s_{k+n-1} & & \cdots & s_{k+2n-2} \end{bmatrix} \qquad (3.1)$$

where $n = 1,2,3,\ldots,N$ and $k = 0,1,2,\ldots$.

Let $\lambda^k_{n,i}$ $(i = 1,2,\ldots,n)$ be the n roots of the equation

$$\det\left(G^{k+1}_n - \lambda G^k_n \right) = 0. \tag{3.2}$$

with $\lambda^k_{n,1} \leq \lambda^k_{n,2} \leq \ldots \leq \lambda^k_{n,n}$. These values of $\lambda^k_{n,i}$ will be referred to as the subeigenvalues of the moment problem. As before, an equivalent method to determine these subeigenvalues is to determine the n roots of the polynomial

$$\det \begin{pmatrix} 1 & \lambda & \lambda^2 & \cdots & \lambda^n \\ s_k & s_{k+1} & s_{k+2} & \cdots & s_{k+n} \\ \vdots & & & & \vdots \\ s_{k+n-1} & & & \cdots & s_{k+2n-1} \end{pmatrix} = 0. \tag{3.3}$$

These are several ways to represent $\lambda^k_{n,i}$. Two such representations, which will be of help later, are:

$$\lambda^k_{n,i} = \max_{y_j \in R^n} \min_{\substack{(x,y_j)=0 \\ x\neq 0}} R^k_n(x) \quad j = 1,2,\ldots,i-1 \tag{3.4}$$

$$= \min_{y_j \in R^n} \max_{\substack{(x,y_j)=0 \\ x\neq 0}} R^k_n(x) \quad j = 1,2,\ldots,n-i \tag{3.5}$$

where (x,y_j) is the inner product in R^n of x with y_j and

$$R^k_n(x) = \frac{\left(G^{k+1}_n x, x\right)}{\left(G^k_n x, x\right)} = \frac{x^T G^{k+1}_n x}{x^T G^k_n x}.$$

Henceforth, when $i = 1$ equation (3.4) will become

$$\lambda^k_{n,1} = \min_{\substack{x\in R^n \\ x\neq 0}} R^k_n(x)$$

and when $i = n$, equation (3.5) will be replaced by

$$\lambda^k_{n,n} = \max_{\substack{x\in R^n \\ x\neq 0}} R^k_n(x).$$

For verification of the above results, see [5].

Using (3.1) we have

$$\left(G_n^k x, x\right) = \sum_{i,j=1}^{n} s_{k+i+j-2} x_i x_j \tag{3.6}$$

where $x = (x_1, x_2, \ldots, x_n)^T$. Since s_k has the form (2.1), we have

$$\left(G_n^k x, x\right) = \sum_{i=1}^{N} \alpha_i \lambda_i^k \gamma_i(x) \tag{3.7}$$

where $\gamma_i(x) = \left(\sum_{j=1}^{n} \lambda_i^{j-1} x_j \right)^2$.

We now have

$$R_n^k(x) = \frac{\left(G_n^{k+} x, x\right)}{\left(G_n^k x, x\right)} = \frac{\sum_{i=1}^{N} \alpha_i \lambda_i^k \gamma_i(x) \lambda_i}{\sum_{i=1}^{N} \alpha_i \lambda_i^k \gamma_i(x)} \tag{3.8}$$

Note that $\alpha_i \lambda_i^k \gamma_i(x) \geq 0$, so that (3.8) is a weighted center of mass of $\lambda_1, \lambda_2, \ldots, \lambda_N$ with λ_i having weight $\alpha_i \lambda_i^k \gamma_i(x)$. Since $\lambda_1 < \lambda_2 < \ldots < \lambda_N$, it follows from (3.8) that

$$\lambda_1 \leq R_n^k(x) \leq \lambda_N \tag{3.9}$$

for $n = 1, 2, \ldots, N$. Combining (3.4) and (3.9), we have the following result:

Theorem 1. Let $\lambda_{n,i}^k$ be the n roots of (3.3) with $\lambda_{n,1}^k \leq \lambda_{n,2}^k \leq \ldots \leq \lambda_{n,n}^k$. Then

$$\lambda_1 \leq \lambda_{n,i}^k \leq \lambda_N$$

for $n = 1, 2, \ldots, N$ and $k = 0, 1, 2, \ldots$.

We will now prove a separation result about the subeigenvalues.

Theorem 2. As n increases, $n = 2, 3, \ldots, N$, the subeigenvalues of the moemtn problem interlace for a fixed moment displacement k. That is

$$\lambda_{n,i}^k \leq \lambda_{n-1,i}^k \leq \lambda_{n,i+1}^k \quad \text{for } i = 1, 2, \ldots, n-1$$

Proof: First it will be shown that

$$\lambda^k_{n,i} \leq \lambda^k_{n-1,i} \quad \text{for fixed } i > 1. \tag{3.10}$$

Let s^n = any set of $i - 1$ vectors from R^n

S^n = collection of all sets s^n

$T^n = \{x \in R^n \mid (x,y) = 0$ for all $y \in s^n$ and $x \neq 0\}$ for a given s^n

$\hat{T}^n = \{x \in T^n \mid nth$ component $x = 0\}$

Using this notation and (3.4) we have the following:

$$\lambda^k_{n,i} = \max_{s^n \in S^n} \min_{x \in T^n} R^k_n(x) \tag{3.11}$$

Since $\hat{T}^n \subset T^n$, we have

$$\max_{s^n \in S^n} \min_{x \in T^n} R^k_n(x) \leq \max_{s^n \in S^n} \min_{x \in \hat{T}^n} R^k_n(x).$$

Thus to prove (3.10), it suffices to show that

$$\max_{s^n \in S^n} \min_{x \in \hat{T}^n} R^k_n(x) \leq \max_{s^{n-1} \in S^{n-1}} \min_{x \in T^{n-1}} R^k_{n-1}(x). \tag{3.12}$$

Now (3.12) will be true if it can be shown that for each $s^n \in S^n$ there exists $s^{n-1} \in S^{n-1}$ such that $\{R^k_n(z) \mid z \in \hat{T}^n\} = \{R^k_{n-1}(w) \mid w \in T^{n-1}\}$ where $\hat{T}^n$ is determined by s^n and T^{n-1} is determined by s^{n-1}. Let $s^n \in S^n$ be fixed. Define the map $\phi\colon \hat{R}^n \to R^{n-1}$ by $\phi(x) = x^*$ where $x = (x_1, x_2, \ldots, x_{n-1}, 0)$ and $x^* = (x_1, x_2, \ldots, x_{n-1})$ and $\hat{R}^n = \{x \in R^n \mid nth$ component of $x = 0\}$. Let $s^{n-1} = \phi(s^n) \in S^{n-1}$. Then for all $z \in \hat{T}^n$, $\phi(z) = w \in T^{n-1}$ and furthermore $R^k_n(z) = R^k_{n-1}(w)$. Also for all $w \in T^{n-1}$, $\phi^{-1}(w) = z \in T^n$ and again $R^k_n(z) = R^k_{n-1}(w)$. Thus

$$\{R^k_n(z) \mid z \in \hat{T}^n\} = \{R^k_{n-1}(w) \mid w \in T^{n-1}\} \text{ and (3.12) is true.}$$

Thus it has been shown that $\lambda^k_{n,i} \leq {}^k_{n-1,i}$. If we use representation (3.5) for $\lambda^k_{n,i}$, an argument similar to the one above yields

$$\lambda_{n-1,i}^{k} \leq \lambda_{n,i+1}^{k}. \tag{3.13}$$

Combining (3.10) and (3.13), the proof is complete.

The remaining results have to do with moment displacements.

Theorem 3. As the moment displacements increase, so do the subeigenvalues. That is; for fixed n $(n = 1,2,\ldots,N)$ and for fixed i $(i = 1,2,\ldots,n)$

$$\lambda_{n,i}^{k} \leq \lambda_{n,i}^{k+1} \quad \text{for} \quad k = 0,1,2,\ldots .$$

Proof: Using (3.7) and letting $\beta_i(x) = \alpha_i\gamma_i(x)$, we have

$$(G_n^k x,x) = \sum_{i=1}^{N} \beta_i(x)\lambda_i^k. \tag{3.14}$$

That is $\{(G_n^k x,x)\}$ is a moment sequence for fixed n. Observe that the positive definiteness of G_n^k implies that $(G_n^k x,x) > 0$ for $x \neq 0$. Thus the moment sequence (3.14) is positive for $x \neq 0$. It follows that the 2×2 Hankel matrices formed by $\sigma_k = (G_n^k x,x)$ are positive definite [5] and hence

$$\begin{vmatrix} \sigma_k & \sigma_{k+1} \\ \sigma_{k+1} & \sigma_{k+2} \end{vmatrix} > 0. \tag{3.15}$$

This immediately yields

$$\frac{\left(G_n^{k+1}x,x\right)}{\left(G_n^k x,x\right)} \leq \frac{\left(G_n^{k+2}x,x\right)}{\left(G_n^{k+1}x,x\right)} \tag{3.16}$$

for $x \in R^n$, $x \neq 0$ and for $n = 1,2,\ldots,N$, $k = 0,1,2,\ldots$. Using (3.4), we now have the required result.

Theorems 1 and 3 imply that for fixed n and i, $\lambda_{n,i}^{k}$ converges as k increases. The next two theorems will be concerned with the convergence of the subeigenvalues.

From (3.1), $G_1^k = s_k$ and therefore solving equation (3.2), with $n = 1$, we have $\lambda_{1,1}^{k} = s_{k+1}/s_k$. Applying (2.1) we have

$$\lambda_{1,1}^k = \frac{\sum_{i=1}^{N} \alpha_i \lambda_i^{k+1}}{\sum_{i=1}^{N} \alpha_i \lambda_i^k} = \frac{\sum_{i=1}^{N} \alpha_i \left(\frac{\lambda_i}{\lambda_N}\right)^k \lambda_i}{\sum_{i=1}^{N} \alpha_i \left(\frac{\lambda_i}{\lambda_N}\right)^k} .$$

Letting k increase and recalling that $\lambda_1 < \lambda_2 < \ldots < \lambda_N$, we have

$$\lambda_{1,1}^k \to \lambda_N \quad \text{as} \quad k \quad \text{increases.} \tag{3.17}$$

Theorem 4. Assuming n $(n = 1,2,\ldots,N)$ terms in model (2.1), then the largest of the n subeigenvalues derived from (3.3) converges to the largest eigenvalue λ_N as the moment displacement k increases. That is,

$$\lambda_{n,n}^k \to \lambda_N \quad \text{as} \quad k \to +\infty \quad \text{for} \quad n = 1,2,\ldots,N.$$

Proof: Using theorem 2 we have $\lambda_{1,1}^k \leq \lambda_{n,n}^k \leq \lambda_N$. Combining this with (3.17), the proof is complete.

Theorem 5. Assuming n $(n = 1,2,\ldots,N)$ terms in model (2.1), then the next to largest subeigenvalue derived from (3.3) converges to the next to largest eigenvalue λ_{N-1} as the moment displacement k increases. That is,

$$\lambda_{n,n-1}^k \to \lambda_{N-1} \quad \text{as} \quad k \to +\infty \quad \text{for} \quad n = 2,3,\ldots,N$$

Proof: We first show that $\lambda_{2,1}^k \to \lambda_{N-1}$. Consider the two roots of $\det\left(G_2^{k+1} - \lambda G_2^k\right) = 0$, denoted by $\lambda_{2,1}^k$ and $\lambda_{2,2}^k$. Using (3.3) we have $\lambda_{2,1}^k$ and $\lambda_{2,2}^k$ are the two roots of

$$\lambda^2\left(s_k s_{k+2} - s_{k+1}^2\right) - \lambda\left(s_k s_{k+3} - s_{k+1} s_{k+2}\right) + \left(s_{k+1} s_{k+3} - s_{k+2}^2\right) = 0.$$

The sum of the two roots of this equation must equal $\frac{\left(s_k s_{k+3} - s_{k+1} s_{k+2}\right)}{\left(s_k s_{k+2} - s_{k+1}^2\right)}$. Using (2.1), we have

$$\lambda_{2,1}^k + \lambda_{2,2}^k = \frac{\sum_{i,j=1}^{N} \alpha_i \alpha_j (\lambda_j - \lambda_i) \lambda_i^k \lambda_j^{k+2}}{\sum_{i,j=1}^{N} \alpha_i \alpha_j (\lambda_i - \lambda_j) \lambda_i^k \lambda_j^{k+1}}$$

Dividing each term by λ_{N-1} and letting k increase we have

$$\lim_{k\to+\infty}\left(\lambda_{2,1}^{k} + \lambda_{2,2}^{k}\right) = \lambda_N + \lambda_{N-1}$$

Using Theorem 4, we then have

$$\lim_{k\to+\infty}\lambda_{2,1}^{k} = \lambda_{N-1}. \tag{3.18}$$

Now applying the separation result in Theorem 2 yields

$$\lambda_{2,1}^{k} \le \lambda_{n,n-1}^{k} \le \lambda_{N-1}. \tag{3.19}$$

(3.18) and (3.19) yield the required result.

In conclusion, we have shown that assuming n terms in model (2.1) and generating n subeigenvalues $\lambda_{n,1}^{k}, \lambda_{n,2}^{k}, \ldots, \lambda_{n,n}^{k}$, the two largest subeigenvalues converge to the two largest eigenvalues λ_N and λ_{N-1} as the moment displacement increase. We hope that future work will generalize these results to show that these n subeigenvalues converge to the n largest eigenvalues of (2.1). The results obtained here have been applied in the analysis of fluorescence decay data, as can be seen in [6]. Numerical examples that illustrate these results will now be presented.

IV. NUMERICAL EXAMPLE

Consider the model (2.1) with three components.

Let the eigenvalues, λ_i, be $.5$, 2.0, and 3.0 with respective amplitudes, α_i, of 2.0, 1.0, and 2.0. In this example we will assume the above model to have only 1 and 2 terms and generate subeigenvalues under these assumptions.

Assume $n = 1$

moment displacement	$\lambda^k_{1,1}$
$k = 0$	1.8000
$k = 1$	2.5000
$k = 2$	2.7667
$k = 3$	2.8614
$k = 4$	2.9084
$k = 5$	2.9379
$k = 6$	2.9579
$k = 7$	2.9716
$k = 8$	2.9809
$k = 9$	2.9872
$k = 10$	2.9914
$k = 11$	2.9943
$k = 12$	2.9962

Assume $n = 2$

moment displacement	$\lambda^k_{2,1}$	$\lambda^k_{2,2}$
$k = 0$	.6013	2.8511
$k = 1$	.8468	2.9032
$k = 2$	1.3304	2.9492
$k = 3$	1.7527	2.9827
$k = 4$	1.9299	2.9962
$k = 5$	1.9817	2.9996
$k = 6$	1.9954	3.0000

For additional examples, see [6].

REFERENCES

[1] Hildebrand, F. B. (1974). "Introduction to Numerical Analysis", New York: McGraw Hill.

[2] Eisenfeld, J., and Soni, B. (1977). "Linear algebraic computational procedures for system identification problems", Proceedings of the First International Conference on Math. Modeling (Vol. 1), Edited by X. J. R. Avula, Univ. Missouri Press, 551-561.

[3] Eisenfeld, J., and Cheng, S. "General Moment Methods for a Class of Nonlinear Models", submitted for publication.

[4] Isenberg, J., Dyson, R. D., and Hanson, R. (1973). "Studies on the analysis of fluorescence decay data by the method of moments", *Biophys. J.*, *13*, 1090-1115.

[5] Gantmacher, F. R. (1964). "The Theory of Matrices, Vol. I and II", New York: Chelsea.

[6] Ford, C., and Eisenfeld, J. "Some Unresolved Questions Pertaining to the Mathematical Analysis of Fluorescence Decay Data", Applied Nonlinear Analysis, V. Lakshmikantham, Ed., New York: Academic Press.

SYSTEM IDENTIFICATION OF MODELS EXHIBITING EXPONENTIAL, HARMONIC AND RESONANT MODES

B. Soni[1]
J. Eisenfeld[2]

Department of Mathematics
The University of Texas at Arlington
Arlington, Texas

I. INTRODUCTION

A classical problem arising in compartmental analysis is the so called identification problem or the inverse problem. One is presented with the linear time invariant compartment model

$$\dot{z}(t) = Az(t) + Bu(t), \quad t > 0, \quad z(0) = 0, \tag{1.1}$$

$$y(t) = Cz(t), \tag{1.2}$$

where A is a square matrix and the dimensions of B and C are consistent with that of A. The problem is to estimate a certain subset of the matrix elements $\{a_{ij}\}$, $\{b_{ij}\}$ and $\{c_{ij}\}$ from discrete observations of the input vector $u(t)$ and the

[1]*Partially supported by University of Texas at Arlington Organized Research Grant.*

[2]*Also affiliated with the Department of Medical Computer Science, University of Texas Health Science Center at Dallas, Dallas, Texas 75235.*

ISBN 0-12-434180-2

output vector $y(t)$. When instantaneous mixing is assumed the analogous equations (1.1) are replaced by

$$z(t) = Az(t), \quad t > 0, \quad z(0) = z_0. \tag{1.1'}$$

An intermeidate problem is that of identifying the respective impulse response matrix [3]-[6], i.e. the matrix

$$D(t) = C \exp(tA)B. \tag{1.3}$$

In the case of the model (1.1)' the corresponding problem is of identifying the matrix

$$\hat{D}(t) = C \exp(tA). \tag{1.3'}$$

In either case each entry $x(t) = d_{ij}(t)$ or $x(t) = \hat{d}_{ij}(t)$ of the matrix satisfies a differential equation,

$$x^{(N)}(t) + c_{N-1}x^{(N-1)}(t) + \dots + c_0 x(t) = 0, \tag{1.4}$$

with constant coefficients c_i, $i = 0,1,2,\dots,N-1$ where $N \leq$ the degree of the minimal polynomial of A. Thus $x(t)$ has the form

$$x(t) = \sum_{j=1}^{n} \sum_{i=1}^{n_j} \alpha_{ij} t^{i-1} e^{\lambda_j t} + \sum_{p=1}^{m} \sum_{u=1}^{m_p} t^{u-1} e^{\gamma_p t} \Big[(\eta_{up} \sin(\beta_p t) + \Psi_{up} \cos(\beta_p t) \Big], \tag{1.5}$$

where λ_j, $j = 1,2,\dots,n$, are distinct real numbers, $z_p = \gamma_p + i\beta_p$, $p = 1,2,\dots,m$, are distinct complex numbers such that $\beta_p \neq 0$, $p = 1,2,\dots,m$, and

$$N = \sum_{j=1}^{n} n_j + 2 \sum_{p=1}^{m} m_p. \tag{1.6}$$

This paper deals with the identification of all parameters and integers occurring in equation (1.5). One of the most challenging of the computational problems is to identify the number of components N as defined in (1.6).

The identification procedure to be presented is based on the method of moments. We call this method LAM (Linear algebraic method). The method requires the estimation of "generalized moments" (which we define in Section 2) denoted by s_k, $k = 1,2,\ldots,2L$. They have the form

$$s_k = \int_a^b x(t)w_k(t)dt, \quad k = 1,2,\ldots,2L, \tag{1.7}$$

where $w_k(t)$ represents respective weighting functions, the interval $[a,b]$ represents the interval on which observations are available and $L \geq N + 1$ is known. If $x(t)$ is not known directly, however data is collected for $y(t)$ and $u(t)$, then one may estimate the integrals s_k by a deconvolution method proposed in [2]. Hence we restate our problem as follows. One is presented with the discrete (noisy) observations $x(t_i)$, $i = 0,1,2,\ldots,q$, and the general model (1.5). The problem is to identify all parameters and integers occurring in (1.5).

The identification process may be separated into two stages. The first deals with mapping discrete data into a "generalized moment sequence". The second stage, discussed in Section 4, shows how such a mapping can be accomplished by means of a suitably chosen "weighting function." The identification algorithm is presented in Section 5. Sections 2 and 3 deal with the theory of moment sequences. Numerical examples are presented in Section 6.

This work is an extension of authors' previous paper [1] where $x(t)$ has the more special form

$$x(t) = \sum_{j=1}^{n} \alpha_j \exp(\lambda_j t), \tag{1.8}$$

where λ_j's are distinct and the a_j's are positive. A more extensive presentation of the algorithm, including the proofs of the theorems, will be given in [8].

II. GENERALIZED DISCRETE MOMENT SEQUENCE (g.d.m.s.)

We consider first the classical discrete moment problem [7]. Given a sequence $s_1, s_2, \ldots$ of real numbers, it is required to determine real numbers α_j and λ_j, $j = 1,2,\ldots,N$, such that the following equations hold:

$$s_k = \sum_{j=1}^{N} \alpha_j \lambda_j^{k-1}, \quad k = 1,2,\ldots, \tag{2.1}$$

$$\alpha_j > 0, \quad j = 1,2,\ldots, \tag{2.2}$$

$$0 < \lambda_1 < \lambda_2 < \ldots < \lambda_N. \tag{2.3}$$

The sequence of real numbers $\{s_k\}$, $k = 1,2,\ldots,$ satisfying (2.1)-(2.3) is called a discrete moment sequence [7]. We define a generalized discrete moment sequence as follows. Given a sequence, $s_1, s_2, \ldots, s_{2L}$, of real numbers, it is required to determine integers N, n, n_j, $j = 1,2,\ldots,n$; m, m_j, $j = 1,2,\ldots,m$; real numbers λ_j, $j = 1,2,\ldots,n$; α_{ij}, $i = 1,2,\ldots,n_j$, $j = 1,2,\ldots,n$; and complex numbers w_{ij}, $i = 1,2,\ldots,m_j$, $j = 1,2,\ldots,m$; and z_j, $j = 1,2,\ldots,m$, such that the following relations hold:

$$s_k = \left(\sum_{i=1}^{k} \sum_{j=1}^{n} \binom{k-1}{i-1} \alpha_{ij} \lambda_j^{k-i}\right) + \left(\sum_{i=1}^{k} \sum_{j=1}^{m} \binom{k-1}{i-1} \left(w_{ij} z_j^{k-i} + \bar{w}_{ij} \bar{z}_j^{k-i}\right)\right),$$

$$k = 1,2,\ldots,2L. \tag{2.4}$$

Here:

$$\binom{k-1}{i-1} = \frac{(k-1)!}{(i-1)!(k-i)!}, \quad i = 1,2,\ldots,k,$$

z_j, $j = 1,2,\ldots,m$, are distinct complex numbers such that

$$z_j = \gamma_j + i\beta_j, \quad \beta_j \neq 0, \quad j = 1,2,\ldots,m, \tag{2.5}$$

λ_j, $j = 1,2,\ldots,n$, are distinct real numbers, (2.6)

$$\alpha_{n_j j} > 0, \quad j = 1,2,\ldots,n, \tag{2.7}$$

$$\alpha_{ij} = 0 \quad \text{for all} \quad i > n_j, \quad j = 1,2,\ldots,n, \tag{2.8}$$

$$w_{ij} = 0 \quad \text{for all} \quad i > m_j, \quad j = 1,2,\ldots,m, \tag{2.9}$$

$$N = \sum_{j=1}^{n} n_j + 2\sum_{j=1}^{m} m_j, \tag{2.10}$$

$$L \geq N + 1. \tag{2.11}$$

The sequence of real numbers s_k, $k = 1,2,\ldots,2L$, satisfying (2.4)-(2.11) is called a generalized discrete moment sequence (g.d.m.s.). Let us define operations:

$$D_j s_k = s_{k+1} - \lambda_j s_k, \quad k = 1,2,3,\ldots,2L-1, \quad j = 1,2,\ldots,n, \tag{2.12}$$

$$E_j s_k = (s_{k+2} - z_j s_{k+1}) - \bar{z}_j (s_{k+1} - z_j s_k), \tag{2.13}$$

that is,

$$E_j s_k = s_{k+2} - 2\gamma_j s_{k+1} + (\gamma_j^2 + \beta_j^2) s_k, \quad k = 1,2,3,\ldots,2L-2,$$
$$j = 1,2,\ldots,m.$$

<u>*Remark.*</u> Let $\{s_k\}_{k=1}^{2L}$ be a g.d.m.s., then

$$\prod_{i=1}^{n} D_i^{n_i} \prod_{j=1}^{m} E_j^{m_j} s_k = 0, \quad k = 1,2,\ldots,n,\ldots,2L-N. \tag{2.14}$$

III. HANKEL MATRICES OF GENERALIZED DISCRETE MOMENT SEQUENCES

Let $\{s_k\}_{k=1}^{2L}$ be a g.d.m.s. Define the Hankel matrices

$$G_j^{(k)} = \begin{pmatrix} s_k & s_{k+1} & \cdots & s_{k+j-1} \\ \\ s_{k+j-1} & s_{k+j} & \cdots & s_{2j+k-2} \end{pmatrix}$$

$$j = 1,2,3,\ldots, \quad k = 1,2,\ldots,2(L-j+1). \tag{3.1}$$

One may obtain the following results:

Theorem 1. Let $\{s_k\}_{k=1}^{2L}$ be a g.d.m.s. and let the integer ρ be defined by

$$\rho = (1/2)\left(\sum_{j=1}^{n} n_j^2 - N\right) \tag{3.2}$$

then

$$\det G_j^{(k)} = 0 \quad \text{for} \quad j = N+1,\ N+2,\ \ldots,\ L,$$

$$k = 1,2,\ldots,2(L-j+1), \tag{3.3}$$

$\det G_N^{(1)} \neq 0$, and

$$\det G_N^{(1)} > 0 \quad \text{if} \quad \rho \quad \text{is even}, \tag{3.4}$$

$$\det G_N^{(1)} < 0 \quad \text{if} \quad \rho \quad \text{is odd}. \tag{3.5}$$

Theorem 2. Let $\{s_k\}_{k=1}^{2L}$ be a g.d.m.s., $\Omega_N(\mu) = \dfrac{q_N(\mu)}{(-1)^N \det G_N^{(1)}}$

and

$$q_N(\mu) = \det\begin{pmatrix} 1 & \mu & \mu^2 & \cdots & \mu^N \\ s_1 & s_2 & s_3 & \cdots & s_{N+1} \\ s_N & s_{N+1} & s_{N+2} & \cdots & s_{2N} \end{pmatrix}, \tag{3.6}$$

then:

(i) n = number of distinct real roots of $\Omega_N(\mu)$,

(ii) λ_j, $j = 1,2,\ldots,n$ are the distinct real roots of $\Omega_N(\mu)$ with respective multiplicities n_j, $j = 1,2,\ldots,n$,

(iii) m = number of distinct complex roots of $\Omega_N(\mu)$,

(iv) z_j, $j = 1,2,\ldots,m$, are the distinct complex roots of $\Omega_N(\mu)$ with respective multiplicities m_j, $j = 1,2,\ldots,m$.

Remarks. The Hankel matrices $G_j^{(k)}$ (3.1) can be decomposed as

$$G_j^{(k)} = P_j Q_j^{(k)} P_j^T, \quad j = 1,2,\ldots,2L, \quad k = 1,2,\ldots,2(L-j+1) \tag{3.7}$$

where P_j is of dimension $j \times N$, $Q_N^{(k)}$ is of dimension $N \times N$. In particular, for $j = N$, $G_N^{(1)}$ is nonsingular as in Theorem 1 and hence P_N and $Q_N^{(1)}$ both are nonsingular.

Theorem 3. Let $\{s_k\}_{k=1}^{2L}$ be a g.d.m.s. then

$$\underline{\alpha} = P_N^{-1} \underline{s} \tag{3.8}$$

where $\underline{\alpha} = [\alpha_{11},\alpha_{21},\ldots,\alpha_{nn}, w_{11},w_{21},\ldots,w_{mm}, \bar{w}_{11},\bar{w}_{21},\ldots,\bar{w}_{mm}]^T$. $\underline{s} = [s_1,s_2,\ldots,s_N]^T$ and P_N is the nonsingular square matrix defined in (3.7).

IV. GENERATION OF g.d.m.s.

Define set X as follows:

$$X = \left\{ \begin{array}{l} x(t) \mid a \leq t \leq b, \ x(t) \text{ is of the form (1.5)}, \\ \lambda_j, \ j = 1,2,\ldots,n, \ \text{are distinct real numbers}, \\ z_j = \gamma_j + i\beta_j, \ \beta_j \neq 0, \ j = 1,2,\ldots,m, \\ \text{are distinct complex numbers.} \end{array} \right\} \tag{4.1}$$

Notice that X is a finite dimensional vector space and the dimension of X is N as defined in (2.10). Let

$$W_L = \left\{ w(t) \;\middle|\; \begin{array}{l} w(t) > 0, \ t \in (a,b), \ w(t) \in C^{2L}[a,b], \\ w^{(i-1)}(a) = w^{(i-1)}(b) = 0, \ i = 1,2,\ldots,2L-1 \end{array} \right\}. \tag{4.2}$$

We call W_L a set of weighting functions.

A fixed $w(t) \in W_L$ generates a sequence

$$\langle x(t), w_k(t) \rangle = (-1)^{k-1} \int_a^b x(t) w^{(k-1)}(t)\,dt, \quad k = 1,2,\ldots,2L. \tag{4.3}$$

Denote

$$s_k = \langle x(t), w_k(t) \rangle, \quad k = 1,2,\ldots,2L. \tag{4.4}$$

Using above definitions one may prove the following.

Theorem 4. A sequence of real numbers generated in (4.4) is a g.d.m.s. if $\alpha_{n_j j} > 0$, $j = 1,2,\ldots,n$, for $x(t) \in X$.

V. IDENTIFICATION PROCEDURE

We are given noisy data $x(t_i)$, $i = 0,1,2,\ldots,q$. We assume that $L \geq N + 1$ is known and $\alpha_{n_j j} > 0$ for $j = 1,2,\ldots,n$. Our identification procedure is discussed in the following 3 steps.

Step 1. Choose $w(t) \in W_L$ as defined in (4.2). (5.1)

Estimate s_k, $k = 1,2,\ldots,2L$, using definitions (4.3)-(4.4) by an appropriate quadrature formula for integration. Compute det $G_j^{(1)}$, $j = 1,2,\ldots,L$. Using Theorem 1 and Theorem 4, we obtain

$$N = \max_{1 \leq j \leq L} \left\{ j \;\middle|\; \det G_j^{(1)} \neq 0 \right\}. \tag{5.2}$$

In pratical situations, because of experimental and computational errors, one does not get det $G_j^{(1)} = 0$ for all $j = N+1, N+2,\ldots,$ but one finds a sharp drop in the value of det $G_j^{(1)}$ to a number ε relatively small in absolute value. Hence instead of (5.2) we estimate

$$\hat{N} = \max_{1 \leq j \leq L} \left\{ j \;\middle|\; |\det G_j^{(1)}| > \varepsilon, \; 1 \leq j \leq L \right\}. \tag{5.3}$$

Step 2. Once $\hat{N}$ is estimated, using Theorem 2 we can theoretically identify n, n_j, $j = 1,2,\ldots,n$, m, m_j, $j = 1,2,\ldots,m$, λ_j, $j = 1,2,\ldots,n$, and z_j, $j = 1,2,\ldots,m$. Since the roots of the polynomial may be extremely sensitive to small changes in coefficients, the errors involved in computing coefficients of the monic polynomial $\Omega_N(\mu)$, defined in Theorem 2, may lead us to completely different roots and hence we may end up with completely different structure of $x(t)$. To handle this situation, along with Theorem 2 we use (3.4), (3.5) and (2.14), to identify the aforesaid parameters. Special cases are described below.

Let $\Omega_N(\mu) = \mu^N + c_{N-1}\mu^{N-1} + \dots + c_0$ be a monic polynomial defined in Theorem 2. For $N = 1$ we have only one possible structure.

Let $N = 2$. Now $\det G_2^{(1)} > 0$ implies that $\Omega_2(\mu)$ has two distinct real roots. If $\det G_2^{(1)} < 0$, we have two possible structures. We compute

$$\eta = \frac{s_2 - \sqrt{-\det G_2^{(1)}}}{s_1} \quad \text{and} \quad \gamma = \eta^2 s_2 - 2\eta s_3 + s_4. \tag{5.4}$$

Using (2.14), if $\gamma = 0$, we have one real root with multiplicity two, otherwise we have complex roots.

Let $N = 3$. Then $\det G_3^{(1)} > 0$ implies that $\Omega_3(\mu)$ has three distinct real roots. If $\det G_3^{(1)} < 0$ we compute the constants c_0, c_1, c_2 and the parameters

$$R = \frac{(AA)^3}{27} + \frac{(BB)^2}{4} \quad \text{where} \quad AA = \frac{1}{3}(3c_1 - c_2^2) \quad \text{and}$$

$$BB = \frac{1}{27}(2c_2^3 - 9c_1c_2 + 27c_0). \tag{5.5}$$

If $R > 0$ we have one real and two complex roots. If $R = 0$, we compute

$$\eta = \frac{\sqrt[3]{c_0} - \dfrac{c_2}{3} - \operatorname{sgn}\left(\dfrac{c_2}{3}\right)\sqrt{\dfrac{c_1}{3}}}{3} \quad \text{and} \quad \gamma = s_1\eta^3 - 3s_2\eta^2 + 3s_3\eta - s_4. \tag{5.6}$$

If $\gamma \neq 0$ we have one simple real root and one real root with multiplicity 2, otherwise we have one real root with multiplicity 3.

Step 3. Once Step 1 and Step 2 are finished, the problem of identifying α_{ij} and w_{ij} is of lower order of difficulty. We use Theorem 3 to identify these parameters. One may also use linear least squares to identify these parameters, as the respective model (1.5) is linear in these parameters.

VI. NUMERICAL EXAMPLES

In this section we present numerical examples using computer simulated data. All computations have been performed on the IBM 370-155 computer in FORTRAN, using single precision arithmetic. The noise which has a uniform distribution is also added to $x(t)$. For these examples we assume $L = 4$ and $N \leq 3$ and we choose the weighting function $w(t) \in W_4$ as

$$w(t) = \sin^7\left(\frac{\pi(t-a)}{(b-a)}\right). \tag{6.1}$$

For convenience we choose equally spaced data points $x(j)$, $j = 0,1,2,\ldots,70$. The g.d.m.s., $\{s_k\}$, $k = 1,2,\ldots,8$, is computed using the composit Simpson's rule of integration.

In these examples we consider all possible structures.

A. $N = 1$, we have only one possible case $x(t) = \alpha_{11}e^{\lambda_1 t}$. Examples for this case have been presented in [1].

B. $N = 2$, we have three possible structures of $x(t)$:

(i) $x(t) = \alpha_{11}e^{\lambda_1 t} + \alpha_{12}e^{\lambda_2 t}$, (examples for this case have been presented in [1]).

(ii) $x(t) = \alpha_{11}e^{\lambda_1 t} + \alpha_{21}te^{\lambda_1 t}$,

(iii) $x(t) = e^{\gamma_1 t}\left(\eta_{11}\sin\beta_1 t + \Psi_{11}\cos\beta_1 t\right)$.

For cases (ii) and (iii) we consider the following two examples.

Example 1. $x(t) = 5.0e^{-.01t} + 0.1te^{-.01t}$

Expected parameters: $N = 2$, $n = 1$, $n_1 = 2$, $m = 0$, $\lambda_1 = -0.01$, $\alpha_{11} = 5.0$, $\alpha_{21} = 0.1$.

Step 1. From the computed determinants, $\det G_i^{(1)}$, $i = 1, \ldots, 4$, which are respectively $0.12103900E{+}03$, $-0.20730250E{+}01$, $-0.62276790E{-}07$, $0.47945110E{-}14$, we estimate $\hat{N} = 2$.

Step 2. Since $\det G_2^{(1)} < 0$, $n \neq 2$. We compute $\eta = -0.10005300E{-}01$ and $\gamma = 0.28690043E{-}05$. Hence we estimate $\hat{n} = 1$, $\hat{n}_1 = 2$, $m = 0$ and $\hat{\lambda}_1 = \eta$.

Step 3. Solving (3.8) we get $\hat{\alpha}_{11} = 0.499778003{-}01$ and $\hat{\alpha}_{21} = 0.10019210E{+}00$.

Example 2. $x(t) = e^{-.05t}\big(\sin(.05t) + 0.5 \cos(.05t)\big)$.

Determinants: $0.34962540E{+}01$, $-0.39188754E{-}01$, $0.25200293E{-}08$, $0.90490722E{-}17$

$\eta = -0.66600716E{-}01$, $\gamma = -0.71629974E{-}03$.

	N	n	m	m_1	γ_1	β_1
Expected	2	0	1	1	-0.50000000E-01	0.50000000E-01
Estimated	2	0	1	1	-0.49998611E-01	0.49999814E-01

	n_{11}	ψ_{11}
Expected	0.10000000E+01	0.50000000E+00
Estimated	0.99956610E+00	0.50000972E+00

C. $N = 3$, we have four possible structures of $x(t)$.

Example 3. $x(t) = (1.0 + 10.0t + 2.0t^2)e^{-.05t}$.

Determinants: 0.94841328E+04, -0.13220200E+06, -0.35761895E+04, 0.40516555E-02,

R = 0.64110284E-13, η = -0.50000627E-01,

γ = -0.38993312E-06.

	N	n	m_1	m	λ_1	α_{11}
Expected	3	1	3	0	-0.50000000E-01	0.10000000E+01
Estimated	3	1	3	0	-0.50000627E-01	0.10328026E+01

	α_{21}	α_{31}
Expected	0.10000000E+02	0.20000000E+01
Estimated	0.99949007E+01	0.19999180E+01

Example 4. $x(t) = (2.0 + 0.01t)e^{.03t} + 0.50e^{-.05t}$.

Determinants: 0.14366412E+03, 0.15601616E+01, -0.28224720E-04, 0.90954327E-12

R = -0.38989700E-10, n = -0.17500531E-02,

γ = -0.42688232E-04.

	N	n	n_1	n_2	m	λ_1	λ_2
Expected	3	2	2	1	0	0.30000000E-01	-0.50000000E-01
Estimated	3	2	2	1	0	0.31871039E-01	-0.50268579E-01

	α_{11}	α_{21}	α_{12}
Expected	0.20000000E+01	0.50000000E+00	0.50000000E+00
Estimated	0.20348787E+01	0.88596896E-02	0.46314700E+00

Example 5. $x(t)=(2.0e^{-.05t})+e^{-.03t}(3.0\sin(0.02t)+0.01\cos(0.02t))$.

Determinants: 0.21071804E+02, -0.84263563E-01, -0.92183501E-06, -0.26345270E-14.

R = 0.13365693E-10

	N	n	n_1	m	m_1	λ_1	γ_1	β_1
Expected	3	1	1	1	1	-.50000000E-01	-.30000000E-01	.20000000E-01
Estimated	3	1	1	1	1	-.50671406E-01	-.29773440E-01	.19862771E-01

	α_{11}	η_{11}	ψ_{11}
Expected	0.20000000E+01	0.30000000E+01	0.10000000E-01
Estimated	0.34185597E-01	0.29853060E+01	0.10763079E-01

Example 6. $x(t) = 2.0e^{-.05t} + 3.0e^{-.03t} + .01e^{.02t}$.

Determinants: 0.30050079E+02, 0.10454530E+00, 0.33736160E-06, 0.23740237E-14

	N	n	m_1	m_2	m_3	λ_1	λ_2	λ_3
Expected	3	3	1	1	1	-.50000000E-01	-.30000000E-01	0.20000000E-01
Estimated	3	3	1	1	1	-.50520600E-01	-.30386200E-01	0.17941800E-01

REFERENCES

[1] Eisenfeld, J., and Soni, B. (1977). "Linear algebraic computational procedures for system identification problems", Proc. First International Conference on Math. Modeling (Vol. 1), Edited by X. J. R. Avula, Univ. Missouri Press, 551-561.

[2] Eisenfeld, J., and Cheng, S. W. "General moment methods for a class of nonlinear models", to appear.

[3] Khatwani, K. J., and Bajwa, J. S. (1976). "Identification of lineartime-invariant systems using exponential signals", *IEEE Trans. Automat. Contr., Vol. 20,* 146-148.

[4] Eisenfeld, J. "Identification of linear time invariant systems", UTA Math. Dept. Report 80.

[5] Wang, Chang-Yi (1977). "Resonance effect in compartmental analysis and its detection", *Math. Biosciences 36,* 109-117.

[6] Holt, J. and Antill, R. (1977). "Determining the numbers of terms in a prony algorithm exponential fit", *Math. Biosciences 36,* 319-332.

[7] Gantmacher, F. R. (1964). "The Theory of Matrices, Vol. II", New York - Chelsea.

[8] Soni, B. (1978). "A new method of system identification", Ph.D. Dissertation, Department of Mathematics, Univ. of Texas at Arlington, Texas.

DIFFERENTIAL EQUATION ALGORITHMS FOR MINIMIZING A FUNCTION SUBJECT TO NONNEGATIVE CONSTRAINTS

B. S. Goh

Mathematics Department
University of Western Australia
Nedlands, Australia*

I. INTRODUCTION

Most of the existing algorithms (variable metric and conjugate gradient algorithms) for an unconstrained minimization problem are based on a quadratic model of the objective function. Thus outside a small neighbourhood of the optimal solution, these algorithms cannot be expected to converge to the optimal solution if the objective function is not convex. Ideally, algorithms should converge globally to the optimal solution whenever the objective function has a unique optimal solution.

Generally, differential equation algorithms for solving optimization problems converge more slowly than variable metric algorithms. But there are several reasons for studying differential equation algorithms for solving optimization problems. Branin [1] reported that differential equation algorithms have good convergence behaviour for initial points far away from the optimal solution. The freedom to vary the step size makes a sequence of

**Address during 1978 is Department of Mathematics, University of British Columbia, Vancouver, B. C., Canada V6T 1W5.*

ISBN 0-12-434180-2

points generated by an optimization algorithm like that in the numerical solution of a related differential equation on a digital computer. At present, the theory and applications of stability (convergence) concepts for differential equations are better known than those for difference equations. Finally there are many well developed numerical methods for solving differential equations (see Lapidus and Seinfeld [2]). Each of these methods generate an optimization algorithm from a differential equation algorithm for computing the optimal solution.

A differential equation algorithm for solving a minimization problem is said to be globally convergent if the trajectories from every initial admissible point remain in the feasible region and converge on the optimal solution as the independent variable t tends to infinity. In this paper some general conditions for global convergence of some differential equation algorithms for solving an optimization problem subject to positive or nonnegative constraints are established. An interesting feature of these algorithms is that the feasibility of the trajectories is maintained without using penalty functions or discontinuous (switching) functions in the differential equations.

II. OPTIMIZATION WITH POSITIVE CONSTRAINTS

We shall briefly examine some differential equation algorithms for finding the minimum of a function subject to positive constraints. The more general problem of solving a system of algebraic equations subject to positive constraints is discussed in another paper, Goh [3].

The problem is to find x^* which minimizes the function,

$$f(x_1, x_2, \ldots, x_n) \tag{1}$$

subject to

$$x_i > 0, \quad i = 1, 2, \ldots, n. \tag{2}$$

It is assumed that $f(x)$ has continuous derivatives in the positive orthant, $R_+^n = \{x | x_i \geq 0, \quad i = 1, 2, \ldots, n\}$.

A differential equation algorithm for computing x^* is an initial value problem of the form,

$$\dot{x}_i = -x_i H_i(x), \quad x_i(0) = x_{i0}, \quad i = 1, 2, \ldots, n \tag{3}$$

where $\dot{x}_i = dx_i/dt$ and the initial vector $x_0 \in R_+^n$. The initial vector is chosen in an arbitrary manner.

Let $\nabla f(x) = (\partial f/\partial x_i)$. If $H(x) = \nabla f(x)$, the algorithm in (3) is none other than a modified version of the steepest descent algorithm. The modification makes the trajectories remain in the positive orthant for all finite values of t. Generally the standard steepest descent algorithm will generate trajectories which leave the positive orthant.

In general, $H = G\nabla f$ where $G(x)$ is a nonsingular matrix. If $G = (\partial^2 f/\partial x_i \partial x_j)^{-1}$, algorithm (3) is a modification of the Newton algorithm. If $H(x)$ is a continuous vector function, the factors $x_1, x_2, \ldots, x_n$ in front of $H_1, H_2, \ldots, H_n$ respectively, keep the trajectories of (3) in the positive orthant for all finite values of t.

Theorem 1. The algorithm (3) is globally convergent to the optimal solution x^* of (1) and (2) if there exists a positive and constant diagonal matrix D such that

$$W(x) = -(x - x^*)^T DH(x) \leq 0, \tag{4}$$

for all $x \in R_+^n$ and $W(x)$ does not vanish identically along a nontrivial solution of (3).

Proof: Let $D = \text{diag } (d_1, d_2, \ldots, d_m)$. A Liapunov function for (3) is

$$V(x) = \sum_{i=1}^{n} d_i[x_i - x_i^* - x_i^* ln(x_i/x_i^*)]. \tag{5}$$

Along solutions of (3),

$$\dot{V}(x) = -\sum_{i=1}^{n} d_i(x_i - x_i^*)H_i(x) = W(x). \tag{6}$$

By LaSalle's invariance principle, the initial value problem is globally stable (convergent) if there exists positive constants $d_1, d_2, \ldots, d_n$ such that condition (4) is satisfied (see LaSalle [4]).

Corollary 1.1. The modified steepest descent algorithm ((3) with $H = \nabla f$) is globally convergent if there exists a positive diagonal matrix D such that

$$-(x - x^*)^T D\nabla f(x) < 0 \tag{7}$$

for all $x \in R_+^n$ and $x \neq x^*$.

Example 1. Condition (7) is much weaker than the condition that $f(x)$ is convex. Let

$$f = (x_1 - 10)^2/2 + \int_0^{x_2} (y - 5)\exp\,[\sin(20)y]\, dy. \tag{8}$$

The equations of the modified steepest descent algorithm for this function are

$$\dot{x}_1 = -x_1(x_1 - 10) \tag{9a}$$

$$\dot{x}_2 = -x_2(x_2 - 5)\exp\,[\sin(20x_2)]. \tag{9b}$$

These equations satisfy corollary 1.1. with $d_1 = 1$, $d_2 = 1$. Hence the algorithm is globally convergent.

The derivative $\partial f/\partial x_2$ fluctuates rapidly as x_2 increases. Clearly $f(x)$ is not a convex function. A variable metric algorithm could have oscillatory behaviour in this problem.

III. OPTIMIZATION WITH NONNEGATIVE CONSTRAINTS

The problem is to compute x^* which minimizes $f(x_1, x_2, \ldots, x_n)$ subject to

$$x_i \geq 0, \quad i = 1, 2, \ldots, n. \tag{10}$$

It is assumed that $f(x)$ has continuous second order derivatives in the positive orthant.

Let P be a subset of $\{1, 2, \ldots, n\}$ and $Q = \{1, 2, \ldots, n\} - P$. Let $x_i^* > 0$ for all $i \in P$ and $x_i^* = 0$ for all $i \in Q$. Using the Kuhn-Tucker theorem (see Fiacco and McCormick [5]) it can be shown that necessary conditions for x^* to be optimal are

$$\partial f/\partial x_i = 0 \quad \text{for all} \quad i \in P, \tag{11a}$$

$$\partial f/\partial x_i \geq 0 \quad \text{for all} \quad i \in Q \tag{11b}$$

at the point x^*.

Under suitable conditions, the initial value problem,

$$\dot{x}_i = - x_i \partial f/\partial x_i, \quad x_i(0) = x_{i0} \tag{12a}$$

$$\text{and} \quad x_{i0} > 0, \quad i = 1, 2, \ldots, n \tag{12b}$$

generates trajectories which remain in the positive orthant and which converge to x^* as $t \to \infty$.

Theorem 2. Every trajectory of the initial value problem (12) remain in the positive orthant for all finite values of t and converges to x^* as $t \to \infty$ if there exists a positive diagonal matrix D such that condition (7) is satisfied.

Proof: A Lyapunov function which will establish this convergence result is

$$V(x) = \sum_{i \in P} [x_i - x_i^* - x_i^* ln(x_i/x_i^*)] + \sum_{i \in Q} d_i |x_i|. \tag{13}$$

In the positive orthant,

$$\dot{V}(x) = - (x - x^*)^T D \nabla f(x). \tag{14}$$

The standard local existence and uniqueness theorem implies that no trajectory of (12) which begins in the positive orthant will intersect a coordinate axis hyperplane for a finite value of t. This property and the condition that $\dot{V}(x)$ is negative definite, imply that every trajectory of (12) which begins in the positive orthant will remain in it for all finite value of t and converge to x^* as $t \to \infty$.

Corollary 2.1. If the hessian, $\nabla^2 f(x)$, is positive definite in the positive orthant and x^* is a local optimal solution, then every trajectory of the initial value problem in (12) remains in the positive orthant and converges to x^*.

Proof: Let D be the identity matrix. Using the mean value theorem, condition (14) implies that

$$\dot{V}(x) = -(x - x^*)^T \nabla^2 f(s)(x - x^*) - \sum_{i \in Q} d_i x_i \frac{\partial f(x^*)}{\partial x_i} \tag{15}$$

where s denotes a set of points between x and x^*.

Condition (11) and the assumption that $\nabla^2 f$ is positive definite imply that $\dot{V}(x)$ is negative definite in R^n_+. Hence Theorem 2 is satisfied.

The initial value problem for (12) is a modification of the standard steepest descent algorithm. Therefore, for points close to x^*, convergence in the component x_i where $i \in P$ would generally be slow. If the index set P is known, it is possible to replace (12) by another system which has second order convergence in x_i for all $i \in P$ and first order convergence in x_i for all $i \in Q$. The variable x_i for $i \in Q$ could converge rapidly to x_i^* even though it has first order rate of convergence. This is because this subset of variables satisfies the constraints in an inequality manner at x^*. But the index set P is usually not known beforehand. Therefore in the general case, another approach to construct an algorithm which converges rapidly must be used. This requires further research.

REFERENCES

[1] Branin, F. H. (1972). "Widely convergent method for finding multiple solutions of simultaneous nonlinear equations", *IBM J. Res. Develop.* *16*, 504-522.

[2] Lapidus, L., and Seinfeld, J. H. (1971). "Numerical Solution of Ordinary Differential Equations", Academic Press, New York.

[3] Goh, B. S. (1978). "Global convergence of some differential equation algorithms for solving equations involving positive variables", *BIT 18*, 84-90.

[4] LaSalle, J. P. (1976). "The Stability of Dynamical Systems", *SIAM*, Philadelphia.

[5] Fiacco, A. V., and McCormick, G. P. (1968). "Nonlinear Programming: Sequential Unconstrained Minimization Techniques", John Wiley, New York.

STABILITY OF A NONLINEAR DELAY DIFFERENCE EQUATION IN POPULATION DYNAMICS

B. S. Goh[*]

Mathematics Department
University of Western Australia
Nedlands, Australia

I. INTRODUCTION

Insect and fish (e.g. salmon) populations with nonoverlapping generations may be described by scalar first order nonlinear difference equations, (May [1], Ricker [2]). Recent mathematical studies by Li and York [3], and May [1] suggest that this class of population models can have a wide range of dynamical behavior. On the other hand, data for many insect populations, which have been assembled by Hassell, Lawton and May [4], suggest that most insect populations in the field have stable dynamics. Here very flexible and refined conditions for global stability in a scalar first order difference equation are described. These flexible conditions for stability lend support to the hypothesis that outside a small neighborhood of an equilibrium, most animal populations with nonoverlapping generations, have stable dynamics.

In general, the use of Liapunov functions for population models is limited to a system with one or two state variables.

[*]*Address during 1978 is Department of Mathematics, University of British Columbia, Vancouver, B. C., Canada, V6T 1W5.*

ISBN 0-12-434180-2

This is because it is usually impossible to verify that a function of three or more variables is negative definite. Here a special type of Liapunov function is used for a delay difference equation so that it is only necessary to establish that a function of two variables is negative definite, irrespective of the length of the delay. This special technique is used to show that the Antarctic fin whale population, *Balaenoptera physalus*, is globally stable.

II. POPULATION WITH NONOVERLAPPING GENERATIONS

Let $N(t)$ denote the number of reproductive females at time t. Assuming that the population has a constant sex ratio, a model of the population with nonoverlapping generations is

$$N(t+1) = F[N(t)], \quad t = 0, 1, 2, \ldots \tag{1}$$

Example 1. (Ricker [2]) A salmon population may be described by

$$N(t+1) = N(t) \exp [r(1-N(t)/K)] \tag{2}$$

where r and K are positive constants.

Equation (1) has an equilibrium at N^* if $N^* = F(N^*)$. The equilibrium N^* is a global attractor if every solution of (1) which begins in $R_+ = (0,\infty)$ remains in it and tends to N^* as $t \to \infty$. An equilibrium is globally stable if (i) it is locally stable and (ii) it is a global attractor.

Theorem 1. Let $V(N) : (0,\infty) \to (0,\infty)$ be a continuous function such that (i) $V(N)$ is strictly monotonic decreasing for all $N \in (0,N^*)$, (ii) $V(N)$ is strictly monotonic increasing for all $N \in (N^*,\infty)$, (iii) $V(N) \to \infty$ as $N \to 0+$ and as $N \to \infty$. If the function

$$\Delta V(N) = V[F(N)] - V(N) \tag{3}$$

is negative definite for all $N \in (0,\infty)$ then the equilibrium N^* of (1) is globally stable.

This theorem follows directly from corollary 1.2. in the paper by Kalman and Bertram [5]. In this theorem the condition $V(N) \to \infty$ as $N \to 0+$ replaces the condition $V(x) \to \infty$ as $x \to -\infty$ in the standard theorem because in population dynamics the state variable must be nonnegative.

Example 2. Using the Liapunov function

$$V(N) = [\ln(N/N^*)]^2 \tag{4}$$

it can be shown that N^* of (1) is globally stable if (i) $F(N^*) = N^*$, $N^* > 0$, (ii) $(N^*)^2/N > F(N) > N$ for all $N \in (0,N^*)$ and (iii) $N > F(N) > (N^*)^2/N$ for all $N \in (N^*,\infty)$.

Another candidate to act as a Liapunov function for (1) is

$$V(N) = (N^2 - N^{*2})/2 - N^{*2}\ln(N/N^*). \tag{5}$$

Theorem 2. Let $G(N) : [N^*,\infty) \to (0,N^*]$ be a strictly monotonic decreasing function such that $G(N^*) = N^*$ and $G(N) \to 0+$ as $N \to \infty$. Let $G^{-1}(N) : (0,N^*] \to [N^*,\infty)$ be the inverse function of $G(N)$. The equilibrium N^* of (1) is globally stable if (i) $F(N^*) = N^*$, $N^* > 0$, (ii) $G^{-1}(N) > F(N) > N$ for all $N \in (0,N^*)$, and (iii) $N > F(N) > G(N)$ for all $N \in (N^*,\infty)$.

Proof: A Liapunov function for (1) is

$$\begin{aligned} V(N) &= G^{-1}(N) - N \quad \text{for} \quad N \in (0,N^*) \\ &= N - G(N) \quad \text{for} \quad N \in [N^*,\infty). \end{aligned} \tag{6}$$

By considering separately each of the cases, (i) $G^{-1}(N) > F(N) > N^*$, $N^* > N > 0$, (ii) $N^* > F(N) > N$, $N^* > N > 0$, (iii) $N > F(N) > N^*$, $N > N^*$, (iv) $N^* > F(N) > G(N)$, $N > N^*$, and (v) $F(N) = N^*$, $N \neq N^*$, it can be shown that the function $\Delta V(N)$ is negative definite for all $N \in (0,\infty)$. Hence N is globally stable.

This theorem generalizes a result proved by Fisher, Goh and Vincent [6] in which $G(N)$ is a straight line.

Corollary 2.1. The equilibrium N^* is locally stable if (i) $F(N)$ is continuous at N^*, (ii) $1 > F'(N^* + 0) > -c$, and (iii) $1 > F'(N^* - 0) > -1/c$ where c is any positive constant.

Example 3. Equation (2) has a locally stable equilibrium at $N^* = K$ if $2 > r > 0$. This suggests that we let

$$G(N) = N \exp[2(1-N/K)] \quad \text{for} \quad N \in [K,\infty). \tag{7}$$

Unfortunately $G^{-1}(N)$ cannot be determined analytically. However the graph of $G^{-1}(N)$ is the image of the graph of $G(N)$ in the "mirror" $y = N$; this fact enables us to plot $G^{-1}(N)$ numerically. It can be shown graphically that

$$G^{-1}(N) > N \exp[r(1-N/K)] > N, \tag{8}$$

for all $N \in (0,K)$ if $2 > r > 0$. It follows that the equilibrium N^* of (2) is globally stable if $2 > r > 0$.

III. AGE-STRUCTURED POPULATION

Consider a population with a relatively long life span (e.g. whales). Assume that it has a constant sex ratio. Let $Y_m(t)$, where $m = 1, 2, \ldots, k$, denote the number of females which are m years of age at time t. Let $N(t)$ be the number of reproductive females at time t. Suppose there is a negligible number of old and nonreproductive females. A model of this type of population is

$$Y_1(t+1) = F_1[N(t)]$$

$$Y_2(t+1) = F_2[Y\ (t)]$$

$$\cdots\cdots\cdots\cdots$$

$$N(t+1) = S[N(t)] + F_{k+1}[Y_k(t)], \tag{9}$$

where $F_1, F_2, \ldots, F_{k+1}, S$ are linear or nonlinear functions.

Eliminating $Y_1, Y_2, \ldots, Y_k$ we get

$$N(t+1) = S[N(t)] + F[N(t-k)] \tag{10}$$

where $F(.) = F_{k+1}F_k\ldots F_1(.)$. This delay difference equation has been used by Clark [7] for formulating optimal management policies in the harvesting of whale populations.

Model (10) is equivalent to

$$N_1(t+1) = N_2(t)$$

$$N_2(t+1) = N_3(t)$$

$$\cdot\ \cdot\ \cdot\ \cdot\ \cdot\ \cdot\ \cdot\ \cdot$$

$$N_{k+1}(t+1) = S[N_{k+1}(t)] + F[N_1(t)]. \tag{11}$$

This system has an equilibrium $N^* = (n^*, n^*, \ldots, n^*)$ where $n^* = S(n^*) + F(n^*)$.

For convenience, let

$$U(N_i) = (N_i^2 - n^{*2})/2 - n^{*2}\ln(N_i/n^*), \tag{12}$$

$$Z(N_{k+1}) = N_{k+1}^2(N_{k+1} - n^*)^2, \tag{13}$$

$$R(N_1,N_{k+1}) = S(N_{k+1}) + F(N_1), \tag{14}$$

$$V(N) = \sum_{i=1}^{k} U(N_i) + bU(N_{k+1}) + cZ(N_{k+1}), \tag{15}$$

$$W(N_1,N_{k+1}) = -U(N_1) - (b-1)U(N_{k+1}) + bU[R(N_1,N_{k+1})] + c[Z(R(N_1,N_{k+1})) - Z(N_{k+1})]. \tag{16}$$

Theorem 3. If there exists positive constants b and c such that $W(N_1,N_{k+1})$ is negative definite in the positive quadrant of the (N_1,N_{k+1}) -space then the equilibrium N^* of (11) is globally stable

Proof: Along solutions of (10) we have

$$\Delta V(N) = \sum_{i=2}^{k+1} U(N_i) + bU[R(N_1,N_{k+1})] + cZ[R(N_1,N_{k+1})] - \sum_{i=1}^{k} U(N_i) - bU(N_{k+1}) - cZ(N_{k+1}) = W(N_1,N_{k+1}). \tag{17}$$

By assumption, $W(N_1, N_{k+1})$ is negative definite. Hence $\Delta V(N)$ is negative semidefinite in the positive orthant.

Using an extension of the direct method of Liapunov (corollary 1.3 in Kalman and Bertram [5]) the proof is completed by showing that $\Delta V(N)$ does not vanish identically along a nontrivial solution of (11). The assumption that $W(N_1, N_{k+1})$ is negative definite implies that $\Delta V(N)$ is equal to zero only at a point of the form

$$N = (n^*, N_2, \ldots, N_k, n^*). \tag{18}$$

If (11) has a solution of this form for $t = 0, 1, 2, \ldots, k-1,$ we have

$$\begin{aligned} N(0) &= (n^*, N_2(0), N_3(0), \ldots, N_k(0), n^*), \\ N(1) &= (n^*, N_2(1), N_3(1), \ldots, N_{k-1}(1), n^*, n^*), \\ &\cdots\cdots\cdots\cdots\cdots\cdots\cdots\cdots \\ N(k-1) &= (n^*, n^*, \ldots, n^*) = N^*. \end{aligned} \tag{19}$$

Therefore other than the equilibrium solution or a solution which reaches N^* in the manner shown in (19), the function $\Delta V(N)$ does not vanish identically along a solution of (11). It follows that N^* is globally stable.

Note that in practice, the constants b and c must be chosen by trial and error. Instead of the function in (12), $U(N_i)$ could be some other Liapunov function for a scalar first order difference equation, like those in section 2 of this paper.

IV. A FIN WHALE POPULATION

Two delay difference equations have been fitted to the data, provided by Allen [8], for the southern hemisphere stock of the fin whale, *Balaenoptera physalus*. The first equation is

$$N(t+1) = sN(t) + rN(t-8) \exp[-qN(t-8)] \tag{20}$$

where $s = 0.96$, $r = 0.12$ and $1/q = 3 \times 10^5$.

The second equation is

$$N(t+1) = sN(t) + rN(t-8)/[1 + pN(t-8)] \tag{21}$$

where $s = 0.96$, $r = 0.12$ and $p = 6 \times 10^{-6}$.

For convenience let $x = N/N^*$. Equation (20) implies

$$x_1(t+1) = x_2(t)$$

$$x_2(t+1) = x_3(t)$$

$$\cdot\ \cdot\ \cdot\ \cdot\ \cdot\ \cdot\ \cdot\ \cdot$$

$$x_9(t+1) = sx_9(t) + rx_1(t) \exp [- q^*x_1(t)] \tag{22}$$

where $q^* = \ln [r/(1-s)]$. This model has an equilibrium at $(1, 1, \ldots, 1)$.

A Liapunov function for this normalized model is $V(N)$ of (15) where $b = 15$ and $c = 1$. These constants are chosen by trial and error so that $W(x_1,x_9)$ of (16) is negative definite. We establish that $W(x_1,x_9)$ is negative definite by plotting a large number of its level sets and using the property that $W(x_1,x_9)$ is continuously differentiable.

We can also use other techniques for computing the global maximum of a function of two variables (see Dixon and Szego [9]). They lead to the conclusion that $W(x_1,x_9)$ has a unique global maximum at the point $(1,1)$ and that $W(1,1) = 0$. It follows that $W(x_1,x_9)$ is negative definite. By Theorem 3 model (20) is globally stable.

This exercise is repeated for model (21) with $b = 15$ and $c = 10$; the conclusion is that model (21) is globally stable. This is encouraging for it means that the conclusion that the fin whale population is globally stable is not critically dependent on the form of the function $F(N)$ which is fitted to the data.

For practical purposes the conclusion that the model is globally stable should not be accepted literally. This is because a deterministic model cannot be expected to be a good representation the dynamics of a population at low densities. Thus for practical purposes the global stability of a model implies only that the real population is stable relative to large perturbations, provided that the perturbed population is not too small.

ACKNOWLEDGMENTS

The author would like to thank Mr. T. T. Agnew for his assistance. This work was partially supported by the National Research Council of Canada (Grant number A-3990).

REFERENCES

[1] May, R. M. (1976). "Simple mathematical models with very complicated dynamics," *Nature 261*, 459-467.

[2] Ricker, W. E. (1954). "Stock and recruitment," *J. Fish. Res. Board Can.*, *11*, 559-623.

[3] Li, T. Y., and Yorke, J. A. (1975). "Period three implies chaos," *Amer. Math. Monthly 82*, 985-992.

[4] Hassell, M. P., Lawton, J. H., and May, R. M. (1976). "Patterns of dynamical behaviour in single-species populations," *J. Anim. Ecology*, *45*, 471-486.

[5] Kalman, R. E., and Bertram, J. E. (1960). "Control system analysis and design via the second method of Lyapunov. II Discrete time systems," *Trans. ASME Ser. D, J. Basic Engng.*, *82*, 394-400.

[6] Fisher, M. E., Goh, B. S., and Vincent, T. L. (1979). "Some stability conditions for discrete time single species models," *Bull. Math. Biol.*, in press.

[7] Clark, C. W. (1976). "A delayed-recruitment model of population dynamics, with an application to baleen whale populations," *J. Math. Biol.*, *3*, 381-391.

[8] Allen, K. R. (1973). "Analysis of stock-recruitment relations in Antarctic fin whales," *Rapp. P-V. Reun. Cons. int. Expl. Mer.*, *164*, 132-137.

[9] Dixon, L. C. W., and Szego, G. P. (1977). "Towards Global Optimization," North Holland Publ. Co., Amsterdam, 472 pp.

BILINEAR APPROXIMATION AND HARMONIC ANALYSIS OF ANALYTIC CONTROL/ANALYTIC STATE SYSTEMS

R. D. S. Grisell

University of Texas Medical Branch
Galveston, Texas

INTRODUCTION

Interest in bilinear approximation is motivated by many revealing applications, such as in modeling populations, economies, transmitter-mediated photoreceptive neurons, and membrane systems. Bilinear systems can be used to approximate general linear control/analytic state systems [1,2]. Some stimates of error in approximating analytic control/analytic state systems will be given, utilizing both time-domain and frequency-domain methods.

I. BILINEAR APPROXIMATION

Analogously to the approximation of differential equations by linearizations, there are conditions under which the approximation with linear and bilinear terms of an analytic control/analytic state system will be arbitrarily close in a sufficiently small neighborhood of a point in time, unless the system has certain kinds of critical behavior at that point. Control systems will be considered of the form

$$\dot{y} = f(u,y,t), \quad z = g(y) \tag{1.1}$$

ISBN 0-12-434180-2

with f and g analytic in state $y = (y_1,\ldots,y_n)$ and f analytic in control $u = (u_1,\ldots,u_N)$ but continuous in t. Bilinear systems will be used in matrix form

$$\dot{w} = L(t)w + \sum_{i=1}^{M} u_i B_i(t)w, \quad z = C(t)w \tag{1.2}$$

in a representation of linear, L, and quadratic forms, B_i, and with linear output function represented by a matrix, C.

The approximation of (1.1) with (1.2) can be broken into two steps [1]: firstly approximating (1.1) with a special, linear - analytic system

$$\dot{y} = f_0(y,t) + \sum_{j}^{P} u^j f_j(y,t), \quad v = g(y) \tag{1.3}$$

in which the monomials u^j are regarded as an enlarged set of controls; here $j \equiv (j_1,\ldots,j_N)$ indicates multinomial terms, and the summation is over all j such that $j_1 + j_2 + \ldots + j_N \leq P$. Secondly, (1.3) is approximated by (1.2). Most work has been done on the autonomous case, where quite procise bounds can be obtained. For example, Krener [2] has shown that for any integer $\varepsilon > 0$, there exists a bilinear realization of the form (1.2) with constant matrices, such that for $t \in [0,T]$

$$|v(t) - z(t)| < Mt^{\varepsilon+1}$$

for some constants M and $T > 0$. This is a surprisingly sharp result, but local. We will take a different approach to step 2, obtaining less sharp bounds, but ones which are global under certain conditions.

Some notation will be helpful in stating theorem 1: Let $L(t)$ be $N \times N$, and suppose $\det(L_0 - r\,\mathrm{Id}) = 0$ has k roots r of multiplicities $m_1,\ldots,m_k$, where Id is the identity matrix and det is the determinant operation. Let a be the minimum absolute value of the real parts of all the roots, and define

$$M(t) \equiv N \max_{1<i<k} t^{m_i - 1} e^{-(m_i - 1)},$$

$$L''(t) \equiv L(t) + \sum_{M}^{M} u'_j B_j(t),$$

$$K(t) \equiv M(t)\exp[M(t) \int_{t_0}^{t} L''(s)ds] \quad \text{and} \quad K \equiv K(\infty).$$

Let y satisfy (1.1) with initial condition $y(t_0) = y_0$ and let w satisfy (1.2). Finally, let $f'(u,y,t)$ be the terms of higher order than bilinear in the Taylor's expansion of $f(u,y,t)$ and let $z_0 \equiv w_0 - y_0$, with $w_0 \equiv w(t_0)$. Now the B_j will be taken as the coefficients of the bilinear terms.

Theorem 1. Given that (1) L_0 is stable, (2) there exist $\beta, \delta > 0$ such that $|y_0| \leq \delta/K$ and $|f'(u,y,t)| \leq \frac{a}{2K}|y|$ for $|u| \leq \mu$ and $t \geq t_0$, (3) $\int_{t_0}^{\infty} |L''(t)|dt < \infty$; it follows that

$$|y(t) - w(t)| \leq C\, K \exp[-\tfrac{1}{2}a(t-t_0)(1 - |y_0|/C)] \tag{1.7}$$

where $C \equiv \max\{|z_0|, \frac{1}{2}a(t - t_0)\}$.

Corollary 1. If in place of (3), $\int_{t_0}^{\infty}|L'(t)|dt < \infty$, and if controls are "impulsively small"

$$|\int_{t_0} e^{as} u_j^! B_j(s)ds| \leq D < \infty, \quad \text{for each } j,$$

then (1.7) follows with $D + |y_0|$ in place of $|y_0|$.

The proofs are fairly straightforward applications of classical techniques in stability. Bound (1.7) is quite sharp, and in some cases with impulse responses equality is approached. Another result along these lines is

Corollary 2. If in place of (3), $\int_{t_0}^{\infty}|\frac{dL''}{dt}|dt < \infty$, then (1.7) follows with $(a - a')$ in place of a, where $a' \equiv \max_{s<t}|L''(s)| + \int_{t_0}^{\infty}|\frac{dL''}{dt}(t')|dt'$.

Certainly, such results can be generalized in various ways. We note that hypothesis (2) which essentially involves the nonlinearity condition, can be varied; for instance by permitting an exponential growth factor

$$|f'(u,y,t)| \leq Me^{AT}|y|^m, \quad m \in I^+,$$

keeping the bound for u but relaxing the uniformity in t. However, (1.7) is particularly useful when, for proper z_0 and y_0, the bound at first increases and then decreases exponentially. This implies a bound on r.m.s. error for any time interval, which will be needed for frequency-domain considerations later.

Sharp bounds in theorem 1 are obtained at the price of complicated condition (2), and there is need for a method of obtaining optimal δ. In modeling applications we have found the following procedure convenient: The equality

$$|f'(u,y,t)| = a|y|/2k \tag{1.8}$$

can be approximated to any desired accuracy via a Taylor's expansion of f' in u and y. The triangle inequality and (1.8) give a polynomial equation in $|y|$ such that a minimal root comes out to the short side of an optimal δ. The root is, however, reasonably optimal, without extensive computations.

Periodic Coefficients and Controls

While theorem 1 provides bounds on the error in response of a bilinearized model to certain classes of aperiodic or transient stimuli, it is useful in harmonic analysis with periodic stimuli (controls) to have bounds and growth rates on the error of bilinearization. Assuming that the linear term $L(t)$ of the right-hand-side of (1.2), the control $u(t)$ and $B_j(t)$ $j = 1,\ldots,M$ are periodic in t (but not necessarily of the same period), deduction from classical results [6] leads to a matrix solution of (1.2) of the form

$$Y(t) = P(t)e^{At}, \quad w(t) = Y(t)w_0$$

where P is a periodic matrix of period τ and A is a constant matrix such that

$$e^{A\tau} = Y(t)^{-1}Y(t+\tau).$$

P solves

$$\dot{P} = L''P - Pa$$

simply as

$$\exp[\int_{t_0}^{t}(L + \sum_{i=1}^{M} u_i B_i)]P(t_0)\exp(-A(t-t_0)).$$

Theorem 2. If f' satisfies the nonlinearity condition

$$\lim_{y \to w_u(t)} \frac{|f'(u,y,t)|}{|y - w_u(t)|} = 0$$

uniformly for $|u| < \mu$ and $t > t_0$ with w_u satisfying (1.2) and for which u, L'' is stable and L_1, then (1.7) results but with $C = \max\{|z_0|, \frac{1}{2}N|y_0|\beta t\}$ where $M \equiv KN$, $K \equiv |P|$ and $N \equiv |P^{-1}|$, and with $[\frac{1}{2}N(1 + |z_0|/C) - 1]\beta$ in place of $(1 - |z_0|/C)a$, where $\beta \equiv -\max_i \mathrm{Re}\{\log(i\text{-th eigenvalue of } L)\}$.

The theorems can be used to estimate errors incurred in the frequency domain by means of Parseval's identity. If c_i is the i-th Fourier coefficient of $y - w$, then

$$\sum_{i=1}^{\infty} c_i^2 = \int |y - w|^2 dt. \tag{1.9}$$

Of course in practice, discrete transforms are used, and the sum must be truncated. Bessel's inequality then provides less sharp bounds for the least squares difference in transforms between the nonlinear response y and its bilinear approximation w, in terms of r.m.s. difference in the time domain on the right-hand-side of (1.9). For example, if y is the response to a sinusoidal u, and the first and all higher harmonics are small in comparison with the fundamental, then the least squares error in the fundamental peak can be estimated.

Conversely, given that the error is negligible in truncating the transform at the N-th term and given the least squares error (l.s.e.) in frequency-response functions in an approximation of y with w, it is possible to estimate r.m.s. error in multiple-impulse responses as functions of several lagged times, or response to several superimposed sinusoids. In addition to Parseval's equation which holds approximately in this case, George's association-reduction operation [7] must be applied to do this.

Remarks. A broad class of systems satisfying the hypotheses of theorem 2 is with coupled dynamical variables having linear system equations and linear coupling unless the variables come out of synchrony, when coupling becomes nonlinear. Models of this class can be found in such areas as planar-to-shear wave mode conversion in sound propagation, cooperative electrically conducting channels in nerve membrane, and cross-bridge attachment in the sliding filament hypothesis of muscle contraction. Computer tests in these modelling applications have shown theorem 2 to be quite sharp.

We have obtained some explicit and fairly sharp bounds on the rate of growth of error in a bilinear approximation when this is not stable under small perturbations of the control, as assumed for theorem 2. Generally, the error grows with the exponent of the exponent of βt, but the formulas are too complicated to be stated here and are probably limited in usefulness.

Finally, we note that Krener's result extends to the periodic case trivially if $\tau < T$, in which case $|y(t) - w(t)| < M\tau^{\xi+1}$.

II.   HARMONIC ANALYSIS OF ANALYTIC CONTROL/BILINEAR SYSTEMS

It is convenient when dealing with complex transfer functions to write a real sinusoidal control or stimulus of frequency w in terms of a complex amplitude a as:

$$A\cos wt - B\sin wt = \frac{(A + jB)e^{jwt}}{2} + \frac{(A + jB)e^{-jwt}}{2}$$

$$= \frac{ae^{jwt}}{2} + \frac{\overline{ae}^{-jwt}}{2} \tag{2.1}$$

If the system is linear, with transfer function H, then there is a well known formula for steady state response, assuming for cimplicity there is no DC term:

$$y(t) = \frac{a}{2} H(jw)e^{jw} + \frac{\overline{a}}{2} H(-jw)e^{-jw},$$

There is an analogous formula for a nonlinear system whose output can be expanded in a convergent Taylor's series in the output [7]:

$$\begin{aligned} y(t) = \text{Re}\ \{ & aH_1(j\omega)e^{j\omega t} + \frac{|a|^2}{2} H_2(j\omega,-j\omega) + \frac{a^2}{2} H_2(j\omega,j\omega)e^{j2\omega t} \\ & + \frac{3a^2\bar{a}}{4} H_3(j\omega,j\omega,-j\omega)e^{j\omega t} + \frac{a^3}{4} H_3(j\omega,j\omega,j\omega)e^{j3\omega t} \\ & + \frac{3}{8}|a|^4 H_4(j\omega,j\omega,-j\omega,-j\omega) + \frac{a^3\bar{a}}{2} H_4(j\omega,j\omega,j\omega,-j\omega)e^{j2\omega t} \\ & + \frac{5}{8} a^3\bar{a}^2 H_5(j\omega,j\omega,j\omega,-j\omega,-j\omega)e^{j\omega t} \\ & + \frac{5}{32} a^4\bar{a} H_5(j\omega,j\omega,j\omega,j\omega,-j\omega)e^{j3\omega t} + \dots \}. \end{aligned} \tag{2.3}$$

The formulas for a second degree nonlinearity are probably not so well known, and to illustrate the general method, the derivations will be given. The input signal will now be in the form:

$$x(t) = \frac{a_1}{2} e^{j\omega_1 t} + \frac{a_1}{2} e^{-j\omega_1 t} + \frac{a_2}{2} e^{j\omega_2 t} + \frac{\bar{a}_2}{2} e^{-j\omega_2 t}.$$

The Laplace transform is then

$$\begin{aligned} X(s) = & \frac{a_1}{2}\delta(s - j\omega_1) + \frac{\bar{a}_1}{2}\delta(s + j\omega_1) + \frac{\bar{a}_2}{2}\delta(s - j\omega_2) \\ & + \frac{a_2}{2}\delta(s + j\omega_2), \end{aligned}$$

where δ is the Dirac delta function. The response contribution from the first order transfer function is

$$\text{Re}\left\{ \sum_{i=1}^{2} \frac{a_i}{2} H(j\omega_i)e^{j\omega_i t} \right\} \tag{2.4}$$

as in the linear case. The contribution from the second order transfer function H_2 is obtained by double, inverse Laplace transformation:

$$\begin{aligned} \text{Re}\Bigg\{ \int_{\rho_1-j\infty}^{\rho_1+j\infty} \int_{\rho_2-j\infty}^{\rho_1+j\infty} & H_2(s_1,s_2)\ \tfrac{1}{4}a^2\delta(s_2 - w_1)\delta(s_2 - w_2) \\ & + \tfrac{1}{4}a\bar{a}\delta(s_1 - w_1)\cdot\delta(s_2 + w_2) \\ & + \tfrac{1}{4}\bar{a}a\delta(s_1 + w_1)\delta(s_2 - w_2) \\ & + \tfrac{1}{4}a^2\delta(s_1 + w_1)\delta(s_2 + w_2)\ e^{j(s_1+s_2)t}\, ds_1 ds_2 \Bigg\}. \end{aligned} \tag{2.5}$$

From the relation

$$H_2(s_1,s_2) = \frac{Y(s_1+s_2)X(s_1)X(s_2)}{2I^2}$$

with I as the power level of the input X; and from the property of Laplace transforms, $L(-s) = \overline{L}(s)$; the following relations hold:

$$H_2(s_1,-s_2) = \overline{H}(s_1,s_2), \quad H_2(-s_1,s_2) = \overline{H}_2(-s_1,-s_2).$$

These may be used to simplify (2.5):

$$\mathrm{Re}\left\{\tfrac{1}{4}\left[a^2H_2(jw_1,jw_2)e^{j(w_1+w_2)t} + a\overline{a}H_2(jw_1-jw_2)e^{j(w_1-w_2)t}\right.\right.$$
$$\left.\left.\overline{a}aH_2(-jw_1,jw_2)e^{j(w_1-w_2)t} + a^2H_2(-jw_1-jw_2)e^{j(-w_1-w_2)t}\right]\right\}$$
$$= \tfrac{1}{2}\mathrm{Re}\left\{a^2H_2(jw_1,jw_2)e^{j(w_1+w_2)t} + a\overline{a}H_2(jw_1,-jw_2)e^{j(w_1-w_2)}\right\}. \tag{2.6}$$

Example With Polynomial Output

The techniques used in this example will apply in a number of other commonly occurring biological systems, such as those involving a bimolecular or Michaelis-Menton kinetics with nonconstant substrate and those with nondynamic nonlinearities (e.g. rectification, satiration, etc.) [8]. The analysis will begin with the second equation of Table 1, for the n-system.

It is first necessary to expand n in terms of (2.3) and (2.6) and analogous expressions for higher order terms which will turn out to be unnecessary, fortunately, and dots will suffice:

$$n(t) = n_\infty + \mathrm{Re}\{a_1H_1(jw_1)e^{jw_1t} + a_2H_1(jw_2)e^{jw_2t}$$
$$+ \tfrac{1}{2}a_1a_2[H_2(jw_1,jw_2)e^{j(w_1+w_2)t} + H_2(jw_1,-jw_2)e^{j(w_1-w_2)t}]$$
$$+ \ldots\} \tag{2.7}$$

We will rewrite the system eqution for n in simpler form as:

$$\dot{n} = \alpha - \gamma n \tag{2.8}$$

where $\gamma \equiv (\alpha + \beta)$. It is next necessary to expand α and γ in Taylor's series:

$$\alpha = \alpha(V_H) + \alpha'(V_H)\ V + \frac{\alpha''}{2}(V_H)\Delta V^2 + \ldots \tag{2.9}$$

$$\gamma = \gamma(V_H) + \gamma'(V_H)\ V + \frac{\gamma''}{2}(V_H)\Delta V^2 + \ldots \tag{2.10}$$

where $\Delta V \equiv V - V_H$. The input or driving parameter is potential V, and is to be like (2.4). Then

$$\Delta V^2 = 2(|a_1|^2 + |a_2|^2) + 2\mathrm{Re}\{a_1^2 e^{j2w_1t} + a_2^2 e^{j2w_2t} + a_1a_2e^{j(w_1-w_2)t} + a_1a_2e^{j(w_1+w_2)t}\} \tag{2.11}$$

Again, higher order terms such as ΔV^3, ΔV^4, ... will turn out to be unnecessary. Substituting (2.7), (2.9), (2.10) and (2.11) in (2.8), and collecting similar sinusoids; with $R_i \equiv \mathrm{Re}\{H_1(jw_i)\}$, $I_i \equiv \mathrm{Im}\{H_1(jw_i)\}$, $\alpha_1 \equiv \alpha'$, $\alpha_2 = \alpha''$, $\gamma_1 = \gamma'$, etc.:

$$w_iR_i = \gamma_0 I_i \quad \text{for} \quad i = 1,2 \tag{2.12}$$

as the $\sin(w_it)$-terms;

$$w_1I_1 + \gamma_0R_1 = \gamma_1 n_\infty + \alpha_1$$
$$-w_2I_2 - \gamma_0R_2 = \gamma_1 n_\infty + \alpha_1 \tag{2.13}$$

as the $\cos(w_it)$-terms;

$$(w_1 + w_2)R_{12} - \gamma_0 I_{12} = \gamma_1(I_1 + I_2)$$
$$-(w_1 + w_2)I_{12} - \gamma_0R_{12} = 2(\gamma_2 n_\infty + \alpha_2) + \tfrac{1}{2}\gamma_1(R_1 + R_2) \tag{2.14}$$

sin and $\cos(w_1 + w_2)t$-terms; where $R_{12} \equiv \mathrm{Re}\{H_2(jw_1,jw_2)\}$ and $I_{12} \equiv \mathrm{Im}\{H_2(jw_1,jw_2)\}$;

$$(w_1 - w_2)R_{1-2} + \gamma_0 I_{1-2} = 0$$
$$-(w_1 - w_2)I_{1-2} - \gamma_0 R_{1-2} = 2(\gamma_2 n_\infty + \alpha_2) + \tfrac{1}{2}\gamma_1(R_1 + R_2) \tag{2.15}$$

sin and $\cos(w_1 - w_2)t$-terms; where $R_{1-2} = \mathrm{Re}\{H_2(jw_1,-jw_2)\}$, etc.

Since the determinants of the left-hand sides of the system (2.13) with substitutions as in (2.12) are $\gamma_0^2 + w_i^2$ (for $i = 1, 2$) these equations can be solved at all frequencies w for $H_1(jw) = R(w) + jI(w)$:

$$R = \frac{-(\gamma_1 n + \alpha_1)\gamma_0}{\gamma_0^2 + w^2}, \quad I = \frac{-w(\gamma_1 n_\infty + \alpha_1)}{\gamma_0^2 + w^2} \tag{2.16}$$

The determinant of the left-hand-sides of (2.14) is $(w_1 + w_2)^2 + \gamma_0^2$, which is nonzero, and

$$R_{12} = -\frac{[I(w_1)+I(w_2)]\gamma_1(w_1+w_2)-[2n_\infty\gamma_2+\alpha_2]+\frac{1}{2}[R(w_1)+R(w_2)]\gamma_1\gamma_0}{(w_1 + w_2)^2 + \gamma_0^2} \tag{2.17}$$

and

$$I_{12} = \frac{-\{2[n_\infty\gamma_2+\alpha_2]+\frac{1}{2}[R(w_1)+R(w_2)]\gamma_1\}(w_1+w_2)-[I(w_1)+I(w_2)]\gamma_0\gamma_1}{(w_1 + w_2)^2 + \gamma_0^2} \tag{2.18}$$

Similarly

$$R_{1-2} = -\frac{-\gamma_0[2(\gamma_2 n_\infty+\alpha_2)+\frac{1}{2}\gamma_1(R(w_1)+R(w_2))](w_1-w_2)}{(w_1 - w_2)^2 + \gamma_0^2} \tag{2.19}$$

and

$$I_{1-2} = \frac{\gamma_0[2(\gamma_2 n+\alpha_2)+\frac{1}{2}\gamma_1(R(W_1)+R(w_2))](w_1 - w_2)}{(w_1 - w_2)^2 + \gamma_0^2} \tag{2.20}$$

Substitution of $w = w_1 = w_2$ in (2.17) and (2.18) will provide explicit forms for $H_2(jw, jw)$. $H_2(jw, -jw)$ for use in (2.3) can be obtained from $H_2(jw_1, jw_2)$ and $H_2(jw_1, -jw_2)$ respectively by setting $w = w_1 = w_2$.

Having solved for the kernels of the dynamic part of the potassium n-system, the first in Table 1, the "ionic equation" presents relatively less of a problem because it is nondynamic in n. It is only necessary for harmonic analysis to insert a single sinusoidal response like (2.3) or a multiple sinusoidal response such as (2.5) into the n^m-term:

$$I = \text{Re } jwCae^{jwt} - g_L(\text{Re } ae^{jwt} - V_L)$$
$$- g_K(V_H - V_K)[n_\infty + \text{Re}\{aH_1(jw)e^{jwt} + \frac{|a|^2}{2} H_2(jw,-jw)$$
$$+ \frac{a^2}{2} H_2(jw,jw)e^{j2wt} + \ldots\}]^m \tag{2.21}$$

and then to use the multinomial expansion if closed form expressions are desired for the amplitudes and phases of various harmonics. Although (21) is adequate for computer evaluation of I, in some cases the computer may be avoided by considering the various parameter dependencies of the amplitudes of the first few harmonics.

For simplicity in the following, the mnemonic subscripting notation begun in the preceding frequency-domain calculations will be taken over now to the time-domain for terms in (2.2):

$$T_{p-q}[(p-q)wt] = \frac{(p+q)!}{p!\ q!} \frac{2^{\text{sign}|p-q|}}{2^{p+q}} H_{p+q}(\underbrace{jw,\ldots,jw}_{p}, \underbrace{-jw,\ldots,-jw}_{q})e^{(p-q)wt}$$

Collecting terms in (2.2) of like periodicity, and inserting (2.3) into (2.7),

$$\begin{aligned}
&[n_\infty + a^2T_{1-1} + a^4T_{2-2} + a^6T_{3-3} + \ldots \\
&\quad + aT_{1-0}(wt) + a^3T_{2-1}(wt) + a^5T_{3-1}(wt) + \ldots \\
&\quad + a^2T_{2-0}(2wt) + a^4T_{3-1}(2wt) + a^6T_{4-2}(2wt) + \ldots \\
&\quad + \ldots]^m \\
&= C_0^m + C_0^{m-1}C_1(wt) + C_0^{m-2}(C_1(wt)^2 + C_0C_2(2wt)) \\
&\quad + C_0^{m-3}(C_1(wt)^3 + C_1(wt)C_2(2wt)) \\
&\quad + C_0^{m-4}(C_1(wt)^4 + C_1(wt)^2C_2(2wt) + C_1(wt)C_3(3wt) + C_2(2wt)^2) \\
&\quad + \ldots + C_0^{m-k}\left(\sum_{\substack{\text{all } e_i, c_{j_i} \\ \text{such that } e_1 j_1 + \ldots + e_{m_k} j_{m_k} = k}} \prod_{i=1}^{m_k} C_{j_i}^{e_i}(j_i wt) \right) + \ldots
\end{aligned} \tag{2.22}$$

where

$$C_0 = n_\infty + \sum_{j=1}^{\infty} a^{2j} T_{j-j}, \quad \text{(first line of left-hand-side of (21))}$$

$$C_1 = a \operatorname{Re}\{H_1(w)e^{wt}\} + \sum_{j=2}^{\infty} a^{2j-1} T_{j-(j-1)},$$

$$\text{(since } T_{1-0} = \operatorname{Re}\{H_1(w)e^{wt}\})$$

$$C_2 = \sum_{j=2}^{\infty} a^{2(j-1)} T_{j-(j-2)}, \quad \text{(where } T_{2-0} = \operatorname{Re}\{H_2(jw,jw)e^{2wt}\})$$

$$C_3 = \sum_{j=3}^{\infty} a^{3(j-2)} T_{j-(j-3)}, \quad \text{etc.}$$

For experimental purposes, it is useful to consider the expressions for the amplitudes of the first few harmonics as functions of n_∞ and a (the first few terms on the right-hand-side of (2.22)). Since it is probably not feasible to accurately fit terms in a^p for p much greater than 5, considering the dynamic range under which membranes are likely to hold up for sufficient time, only the terms up to a will be written out explicitly:

Constant term:

$$\begin{aligned} C_0^m = {} & n_\infty^m + n_\infty^{m-1}(a^2 T_{1-1} + a^4 T_{2-2} + \sum_{j=3}^{\infty} a^{2j} T_{j-j}) \\ & + n_\infty^{m-2}(a^4 T_{1-1}^2 + \sum_{j=2}^{\infty} a^{4j} T_{j-j}^2 + \sum_{j \neq k}^{\infty} a^{2(j+k)} T_{j-j} T_{k-k}) \qquad (2.23) \\ & + n_\infty^{m-3}(\sum^{\infty} a^{2(j+k+p)} T_{j-j} T_{k-k} T_{p-p}) + \ldots + n_\infty^{m-q}(\sum_{j=1}^{\infty} a^{2j} T_{j-j})^q \end{aligned}$$

$(q = 4, 5, \ldots, m)$

First Harmonic:

$$C_0^{m-1}C_1(wt) = n_\infty^{m-1}\Big[a\mathrm{Re}\{H_1(w)e^{wt}\} + a^3T_{2-1} + a^5T_{3-2} + \sum_{j=4}^{\infty} a^{2j-1}T_{j-(j-1)}\Big]$$
$$+ n_\infty^{m-2}\Big[a^3T_{1-1}\mathrm{Re}\ H_1(w)e^{wt} + a^5(T_{1-1}T_{2-1} + T_{2-2}\mathrm{Re}\ H_1(w)e^{wt}) + \sum_{j=3} a^{2j+1}(\sum_{k=1}^{j} T_{k-k}T_{j-(j-k)})\Big]$$
$$+ n_\infty^{m-3}\Big[a^5T_{1-1}\mathrm{Re}\{H_1(w)e^{wt}\} + \sum_{j=2}^{\infty} a^{(2j)^2+1}\sum_{k,p}^{j} T_{p-p}T_{k-(k-p)}T_{j-(j-(k-p))}\Big]$$
$$+ \ldots + n_\infty^{m-q}\Big[\mathrm{Re}\{H_1(w)e^{wt}\}(\sum_{j=1}^{\infty} a^{2j+q}T_{j-j}) + (\sum_{j=2}^{\infty} a^{2j-1}T_{j-(j-1)})(\sum_{k=1}^{\infty} a^{2k}T_{k-k})^q\Big]$$
$$(q = 4,5,\ldots,m) \tag{2.23}$$

Second Harmonic:

$$C_0^{m-2}(C_1(w)^2 + C_0C_2(2wt)) = n_\infty^{m-1}\Big[\sum_{k=2}^{\infty} a^{2(k-1)}T_{k-(k-2)}\Big]$$
$$+ n_\infty^{m-2}\Big[\sum_{j=1}^{\infty} a^{2j}(T_{j-(j-1)}\mathrm{Re}\ H_1(w)e^{wt}) + \sum_{j=1,k=2}^{\infty} a^{2(j+k)-2}(T_{j-j}T_{k-(k-2)} + T_{j-(j-1)}T_{k-(k-1)})\Big] + \ldots + n_\infty^{m-q}\Big[\sum_{\substack{j_i=1\\k=2}}^{\infty} a^{2(j_1+\ldots+j_{q-1}+k)-2}$$
$$\cdot \left(\prod_{i=1}^{q-1} T_{j_i-j_i}\right)T_{k-(k-2)} + \sum_{\substack{j_i=1\\j=1}}^{\infty} a^{2(j_1+\ldots+j_q+j)-2}\left(\prod_{i=1}^{q} T_{j_i-j_i}\right)$$
$$\cdot\ T_{j-(j-1)}\mathrm{Re}\{H_1(w)e^{wt}\} + \sum_{\substack{j_i=1\\j=1\\k=2}}^{\infty} a^{2(j_1+\ldots+j_q+j+k)-2}\left(\prod_{i=1}^{q} T_{j_i-j_i}\right)$$
$$\cdot\ T_{j-j}T_{k-(k-2)} + T_{j-(j-1)}T_{k-(k-1)}\Big] \tag{2.25}$$

III. DIFFUSION-REACTION AND OUTPUT FUNCTIONALS

Diffusion-reaction problems often have outputs involving convolution with the Green's function for the diffusion [9], [10]. For simplicity of notation, we will write only equations for a planar, bulk-limited diffusive compartment, but the Green's function G may be taken as general.

$$\frac{\partial U}{\partial t} = -\alpha C + \beta(B - U) \tag{3.1}$$

$$D\frac{\partial^2 C}{\partial x^2} = \frac{\partial C}{\partial t} + \alpha CU - \beta(B - U) + S(x,t), \tag{3.2}$$

where: $C \equiv$ diffusant concentration; $D \equiv$ diffusion coefficient; $\alpha, \beta \equiv$ forward, resp. back rate constants; $U \equiv$ unbound site concentration; $B \equiv$ maximum available sites for binding diffusant; $S \equiv$ souce.

The bimolecular reaction is represented by a bilinear equation (3.1) and has a determined Volterra series [11], but we will only need the general expression

$$U(x,t) = U(x,0) + \int_0^t h_1(s)C(x,t-s)ds \tag{3.3}$$

$$+ \int_0^t \int_0^t h_2(s_1,s_2)C(x,t-s_1)C(x,t-s_2)ds_1 ds_2 + \ldots$$

which will be symbolized more briefly as $U = h[x](C)$. Let d be the diffusion functional, operating on the diffusant concentration profile, and let

$$A(x,t) \equiv \int_0^L G(x,y,t)C(y,t)dy,$$

integrating over the initial distribution in the compartment of length L. Then

$$k(S) = d(s + \frac{d}{dt}\, h(k(S)) + A \tag{3.4}$$

for the output diffusant concentration profile as a functional k of source. Transforming all functions and functionals to the frequency domain in (3.4) (frequency-response functions are capitalized; $t \to f$, $x \to s$), equating orders and solving for successive orders:

$$K_1(f,s) = \frac{D_1(f,s)}{1 - D_1(f,s)H_1[s](f)jf} \tag{3.7}$$

$$K_2(f_1,f_2,s_1,s_2)$$
$$= \frac{D_1(f_1+f_2,s_1+s_2)H_2(f_1,f_2,s_1,s_2)K_1(f_1,s_1)K_1(f_2,s_2) + D_2}{\{1 - D_1(f_1,s_1)H_1[s_1](f_1)\}\{1 - D_1(f_2,s_2)H_1[s_2](f_2)jf_2\}}$$
$$\cdot \frac{1}{1 - D_1(f_1+f_2,s_1+s_2)H_1[s_1+s_2](f_1+f_2)j(f_1+f_2)} \tag{3.8}$$

In (3.8) we have included a summand D_2 (or $D_2(f_1,f_2,s_1,s_2)$) in the numerator to show how nonlinear diffusion would come in, but of course (3.2) represents linear diffusion. The third order for nonlinear diffusion is considerably more complicated, but will be supplied on request. Only the linear case will be written out here:

$$K_3(f_1,f_2,f_3,s_1,s_2,s_3)$$
$$= \frac{D_1(f_1+f_2+f_3,s_1+s_2+s_3)}{1 - D_1(f_1+f_2+f_3,s_1+s_2+s_3)H_1[s_1+s_2+s_3](f_1+f_2+f_3)}$$
$$\cdot \Big[H_2(f_1,f_2+f_3,s_1,s_2+s_3)K_1(f_1,s_1)K_2(f_2,f_3,s_2,s_3)$$
$$+ H_2(f_1+f_2,f_3,s_1+s_2,s_3)K_2(f_1,f_2,s_1,s_2)K_1(f_3,s_3)$$
$$+ H_3(f_1,f_2,f_3,s_1,s_2,s_3)K_1(f_1+f_2+f_3,s_1+s_2+s_3)\Big].$$

The output function of position and time giving the concentration profile is obtained by inverse transforming the sum of the reduction-associations of the K_n. It can be seen from (3.7) as one would expect, the first order transfer characteristics K_1 are not sufficient to distinguish reaction nonlinearity. Whereas, (3.8) contains H_2 in the numerator, which is quite convenient for detection of bimolecularity once the poles due to the denominator have been removed. It is a fortunate circumstance, for this, that the denominators are such that a simple discriminant function for features peculiar to the reaction kinetics is the ratio

$$T(f_1, f_2, s_1, s_2) \quad \frac{K_2(f_1, f_2, s_1, s_2)}{K_1(f_1+f_2, s_1+s_2)K_1(f_1, s_1)K_1(f_2, s_2)} \; ;$$

assuming the diffusion is linear (however, it may be non-Fickian). In case the reaction is simply bimolecular, and D is small,

$$H_2 \doteq T = \frac{\alpha}{1 + j2\pi(f_1+f_2)\beta}$$

Unfortunately, a very similar function results for "quadratic coupling" of the reaction to diffusant, that is, a C^2 term. Depending on sign, this would reveal either autocatalysis or autoinhibition of the binding reaction. Although competition can be distinguished from simple Michaelis-Menton enzyme-substrate type kinetics by third and higher order response functions, these have generally (in our experience) been found both cumbersome to compute and difficult to obtain experimentally in slowly responding preparations, [12], [8].

CONCLUDING REMARKS

Error estimation can be obtained by making use of both time- and frequence-domain information, and this is particularly powerful when there are large higher order kernels which are typically difficult to estimate. Double sinusoidal stimulation at one pair of frequencies is sufficient, in principle, to pick off the four or five low order parameters of an analytic control/analytic state system, or all parameters of a simple bilinear system. Moreover, there would be an "intermediate small signal" range of amplitude over which contributions of higher order terms would be minimized but the second order would be still visible over the first order transfer characteristics.

REFERENCES

[1] Brockett, R. W. (1976). "Volterra series and geometric control theory", *Automatica 12*, 167-176.

[2] Krener, A. J. (1975). "Bilinear and nonlinear realizations of input-output maps", *SIAM J. Control 13(4)*, 827-834.

[3] Plant, R. E., and Kim, M. (1976). "Mathematical description of a bursting pacemaker neuron by a modification of the Hodgkin-Huxlec equations", *Biophys. J. 16*, 227-244.

[4] Connor, J. A., and Stevens, C. F. (1971). "Inward and delayed outward membrane currents in isolated neural somata under voltage clamp", *J. Physiol. 213*, 1-19.

[5] Adelman, W. J., Jr., and Palti, Y. (1969). "The effects of external potassium and long duration voltage conditioning on the amplitude of sodium currents in the giant axon of the squid, Loligo pealei", *J. Gen. Physiol. 53*, 589-606.

[6] Erugin, N. P. (1946). "Reducible systems (3,*)", *Trudy Mat. Inst. Steklov 13*, 1-95.

[7] George, D. A. (July 24, 1959). "Continuous nonlinear systems", Technical Report 355, M.I.T. Research Lab of Electronics.

[8] Grisell, R. D. (1977). "Optimal testing pulse sequences for estimation of characterizing features of higher order Wiener and Volterra kernels", Simulation and Identification in Biological Science, Pomona, Khalsa Publications, 102.

[9] Hallam, T. G., and Young, E. C. (1977). "Bounds for solutions of reaction-diffusion equations", Nonlinear Systems and Applications (Ed. V. Lakshmikantham), Academic Press, 163-171.

[10] Grisell, R. D., and Andresen, M. C. "Multicompartment diffusion and reaction modeling system applied to transmitter-mediated photoreception", *Comp. in Biol. and Med.*, (to appear).

[11] D'Alessandro, P., Isidori, A., and Ruberti, A. (1974). "Realizations and structure theory of bilinear dynamical systems", *SIAM J. Control 12(3)*, 517-535.

[12] Andresen, M. C., and Grisell, R. D. (5-9 November 1977). "Model for an extra-retinal photoreceptor", *30th ACEMB*, P3.32.

Applied Nonlinear Analysis

PERSISTENT SETS VIA LYAPUNOV FUNCTIONS

G. W. Harrison

Department of Mathematics
University of Georgia
Athens, Georgia

ABSTRACT

A modeler may understand the basic dynamics of a biological system, but there are always some forces, perhaps due to the effects of a random environment, that he does not understand or cannot predict. In this situation it may not be feasible to predict exact trajectories of the system, but it may be important to determine whether the system stays in an acceptable subset of the state space (e.g. that certain species do not go extinct). Let $x \in R^n$ be a vector of state variables of a system whose dynamics are given by

$$\dot{x} = f(x) + g(t,x) \tag{1}$$

where f is a known function but g is only known to be a member of a set of functions G, and represents the influence of an unpredictable environment.

Definition. A set $M \subset R^n$ is *persistent under* G if, for any function $g \in G$, every solution $x(t)$ of equation (1) with $x(t_0) \in M$ for some t_0, remains in M for all $t > t_0$.

Let $V \in C_1[R^n, R^+]$ and let $\dot{V}_f(x) = \frac{\partial V}{\partial x} \cdot f(x)$ and $\dot{V}_g(x) = \frac{\partial V}{\partial x} \cdot g(x)$.

ISBN 0-12-434180-2

Lemma. If for some $b > 0$, $L < V(x) < L + b$ implies that $\dot{V}_f(x) + \dot{V}_g(x) < 0$ for all $g \in G$, then $\{x: V(x) \leq L\}$ is persistent under G.

Several methods are given to show that the Lyapunov function V satisfies the hypothesis of this lemma. One method leads to the following result:

Theorem. If $\dot{V}_g(x) \leq U(x)$ for all $g \in G$ and

$$\dot{V}_f(x) + U(x) < 0 \quad \text{if} \quad r_1 < \|x\| < r_2, \tag{2}$$

if for any positive scalars α and β, $\alpha < \beta$ implies $V(\alpha x) < V(\beta x)$, and if $\max\limits_{\|x\|=r_1} V(x) \leq L < \min\limits_{\|x\|=r_2} V(x)$, then $M = \{x: V(x) \leq L\}$ is persistent under G.

This theorem can be interpreted as saying that the set M is persistent if, at its boundary, the known forces directed toward its interior, measured by the negativity of $\dot{V}_f(x)$, are stronger than the unpredictable forces, measured by $\dot{V}_g(x) \leq U(x)$. Nothing needs to be known about the behavior of the system in the interior of M.

The preceding results are applied to show the existence of persistent sets when f is linear and every $g \in G$ satisfies $\|g(t,x)\| \leq K\|x\|^{\alpha}$ for any $\alpha > 0$, and when f is a generalized Lotka-Volterra species interaction model and every $g \in G$ satisfies bounds that grow linearly with x.

SPATIAL HETEROGENEITY AND THE STABILITY OF PREDATOR-PREY SYSTEMS: POPULATION CYCLES

Alan Hastings

Department of Pure and Applied Mathematics
Washington State University
Pullman, Washington

I. INTRODUCTION

The interaction of space, dispersal, and local dynamics in the behavior of biological populations has been extensively studied by both theoreticians and experimentalists in recent years. These investigations stem from the seminal work of Skellam (1951) and Hutchinson (1951). The role of spatial heterogeneity in the stabilization of predator prey interactions was studied experimentally by Huffaker (1958) with mites and theoretically by Hutchinson (1959). More recent theoretical investigations include those of Vandermeer (1973), Levin (1974), Slatkin (1974), Maynard Smith (1974), Hilborn (1975), Hastings (1977, 1978), Ziegler (1977), Caswell (1978) using a 'patch-type' approach and by a number of other authors using a reaction-diffusion equation approach (see reviews by Levin, 1976a,b). For a discussion of more recent experimental work, see Hastings (1978).

This paper attempts to elaborate further the role of spatial heterogeneity by considering the existence and stability of periodic solutions to the equations describing the population dynamics of a one predator one prey system, in an environment consisting

ISBN 0-12-434180-2

of a large number of discrete, identical patches. In earlier papers, (Hastings 1977, 1978) I considered such systems, but concentrated instead on solutions in which the number of patches occupied was a locally stable equilibrium point for the equations; hence total population size was constant. However, both in Huffaker's (1958) system of mites on aranges and in more recent investigations by Luckinbill (1974) of a predator-prey system with paramecium that incorporated spatial factors, total population size fluctuated, and global extinctions occurred. In a recent paper, Gurney and Nisbet (1978) have shown how certain stochastic elements can lead to periodic behavior in models similar to the one considered here. The present paper complements the work of Gurney and Nisbet by considering the possibility of oscillations in a deterministic model.

The rationale behind using patch type models is discussed in Hastings (1977, 1978), Caswell (1978) and Levin (1976a,b). These models will be more useful than reaction diffusion systems in cases where dispersal is long range, as, for example, in intertidal organisms that have pelagic larvae. The models do contain a number of approximations that are made to create an analytically tractable model. The model could certainly be modified to include these, although the analysis would be complicated.

The organization of the rest of the paper is as follows. In Section 2, I derive the simplest model and summarize its dynamics. Section 3 considers one modification, 'age structure' of predator patches, and the existence of limit cycles in this case. In Section 4, a functional response of the prey is considered. Using the Hopf bifurcation theorem, limit cycles are shown to exist in this case, but are shown to be stable only for a range of parameter values by using the stability criterion of Hsü and Kazarinoff (1976) (see also Hassard and Wan, 1978). Some implications of the results are discussed in Section 5.

II. THE SIMPLEST MODEL

I will now derive the basic equations for the simplest model. The model will be stated in terms of the fraction of patches in various states, and will be deterministic. The following is a list of rules describing transitions between various patch states (cf. Caswell, 1978, Hastings, 1977, 1978):

(1) An *empty* patch can be invaded only by *prey*, creating a *prey* patch.

(2) The patch remains in this state until invaded by the predator, becoming then a *predator* patch.

(3) *Predator* patches are assumed to return to the *empty* state at a constant rate.

(4) For simplicity, it is assumed that prey disperse only from *prey* patches; hence they invade *empty* patches at a rate proportional to the fraction of patches that are *prey*.

(5) Predators invade *prey* patches at a rate proportional to the fraction of *predator* patches.

It is the effect of modifying rules (3) and (5) that will be considered in subsequent sections.

The form of the basic equations follows directly from the rules given above. Let x be the fraction of prey patches, y the fraction of predator patches, where

$$0 \leq x \leq 1$$
$$0 \leq y \leq 1 - x. \tag{1}$$

Let the decay rate of predator patches be k, and the invasion rates of the prey and predator be proportional to D_x and D_y, respectively. Then the system clearly abeys the following equations (Hastings, 1978):

$$\frac{dx}{dt} = D_x(1-x-y)x - D_y xy \quad \text{and} \quad \frac{dy}{dt} = D_y xy - ky. \tag{2}$$

As shown in Hastings (1978) the equilibrium

$$\hat{x} = 1, \quad \hat{y} = 0 \tag{3}$$

is globally stable if $D_y < k$. On the other hand, if $k < D_y$ there is a unique equilibrium with x and y positive, satisfying (1), which is globally stable. The proof of this rests on using the Poincare-Bendixson theorem, and using Bendixson's criterion (after a change to logarithmic variables) to eliminate the possibility of limit cycles.

Although it does demonstrate the powerful stabilizing effect of a patchy environment, the global stability of the model for all parameter values is unreasonable biologically. For this reason, in the next two sections, the effects of modifying rules 3 and 5 will be considered.

III. AN 'AGE DEPENDENT' MODEL

The assumption that the prey patches return to the empty state at a constant rate will be altered in this section. This assumption will be strictly true only if the probability that a *predator* patch returns to the empty state is independent of the time since it was created. Given that the predator-prey dynamics are unstable, this is unlikely to be true. Instead, I will consider a modification used in Hastings (1977, 1978). Here, the *predator* patches all last a fixed time and then return to the empty state - the probability of return to the empty state is totally determined by the 'age' of the patch. Without loss of generality, the "life-time" of the predator patches can be normalized to be 1, so the populations will obey the equations:

$$\begin{aligned} \frac{dx}{dt} &= D_x(1-x-y)x - D_y xy \\ y(t) &= D_y \int_{t-1}^{t} y(s)x(s)ds. \end{aligned} \tag{4}$$

When initial conditions are specified on an interval, a well-posed problem results (Hastings, 1978).

The local stability of the nontrivial equilibrium point of (4) (which is the same as the equilibrium point of (2) with $k = 1$), was determined in Hastings (1978), through a combination of analytic and numerical techniques. For large enough values of the invasion rates, the equilibrium point was unstable. Of more interest is the result that stability was lost when a pair of complex eigenvalues crossed the imaginary axis with non-zero speed (when, for example, D_y was increased). The equation for the eigenvalues of the linearization of (4) (see Hastings, 1978) is of a form that always has isolated roots, and zero is never an eigenvalue. The preceding discussion implies, using the Hopf bifurcation theorem (Marsden and McCracken, 1976), that periodic orbits exist in the vicinity of the equilibrium point near the parameter values where local stability is lost. Numerical integration of (4) indicates the existence of stable periodic orbits, for a range of parameter values. A more complete study of this bifurcation is possible using the techniques of Chow and Mallet-Paret (1977) and will be pursued elsewhere.

IV. FUNCTIONAL RESPONSE OF PREDATORS

The concept of a nonlinear response of predators to prey density has been incorporated in a number of biological models (see May, 1973 for a review). Mathematical analyses of limit cycle behavior in these models, which ignore spatial factors, have been performed by Kazarinoff and van den Driessche (1978) and Lin and Kahn (1976). In this section, I will focus on the effect of introducing a functional response into the model in Section 2.

In a 'patch' model, if the predators exhibit any searching behavior, the probability that a particular patch will be invaded will always be a decreasing function of patch density. For this reason, consider the class of models where the term $D_y x$ in equation (2) is replaced by Px^n, where n is some number between 0 and 1. (This is just one of many possible functional forms.) This yields the equations (setting $k = 1$):

$$\frac{dx}{dt} = D_x(1-x-y)x - Px^n y$$
$$\frac{dy}{dt} = Px^n y - y. \tag{5}$$

Although the qualitative behavior is similar for different values of n, I will concentrate on the case $n = \frac{1}{2}$, which simplifies greatly the calculations. In this case, the equations (5) have a nontrivial equilibrium point,

$$\hat{x} = P^{-2}, \quad \hat{y} = D_x(P^2-1)/(D_x P^2 + P^4) \tag{6}$$

if $P > 1$. The linearization of equations (5) about this equilibrium point is

$$\begin{pmatrix} D_x - 2D_x\hat{x} - D_x\hat{y} - \frac{1}{2}P\hat{x}^{-\frac{1}{2}}\hat{y} & -D_x\hat{x} - P\hat{x}^{\frac{1}{2}} \\ \frac{1}{2}P\hat{x}^{-\frac{1}{2}}\hat{y} & P\hat{x}^{\frac{1}{2}} - 1 \end{pmatrix} \tag{7}$$

It is easy to see that the equilibrium point will be locally stable if

$$D_x > (P^4 - 3P^2)/2 \tag{8}$$

and unstable otherwise.

The rest of this section is devoted to determining the behavior of the system when local stability is lost. The approach here follows closely that of Kazarinoff and van den Driessche (1978). Let D_x be the bifurcation parameter; the dynamics of the model will be investigated as this parameter varies for a fixed value of P. If $P > \sqrt{3}$, simple calculations show that stability is lost as D_x decreases below the critical value given in (8) and that at this point a pair of complex eigenvalues cross the imaginary axis with nonzero speed.

The statements of the previous paragraph imply, using the Hopf bifurcation theorem, that periodic orbits exist for D_x near the critical value. The stability of these orbits can then be calculated using the algorithm of Hsü and Kazarinoff (1976) (see also Hassard and Wan, 1978). The approach involves computing a Taylor series approximation, to third order, for the differential equations at the bifurcation point. Then, after a conversion to Poincare normal form, the stability of the bifurcating orbits can be determined using a formula due to the authors mentioned above (see those papers for more details) (see Appendix). In the present case, the bifurcating orbits turn out to be stable only if

$$3P^6 - 20P^4 + 40P^2 + 1 < 0 \tag{9}$$

This inequality is satisfied for $\sqrt{3} < P < P_{crit}$, where P_{crit} is approximately 2.73. Remembering that P must be greater than $\sqrt{3}$ for bifurcations to occur, the possible dynamic behavior can be summarized. For $P < \sqrt{3}$, no bifurcations are possible. For $\sqrt{3} < P < P_{crit}$, there is a critical value of D_x at which there is a bifurcation to a stable periodic orbit. For $P > P_{crit}$, the bifurcations are unstable and subcritical.

V. DISCUSSION

Here, the biological implications of the results of the previous sections will be considered. It is difficult to generalize from several examples to a whole class of models, and the question of the global dynamic behavior of the models considered in Sections 3 and 4 is still an open question. Nevertheless, some comments on the dynamical behavior of these predator prey models in patch environments are worthwhile.

This paper has demonstrated another mechanism for generating population fluctuations in predator prey systems in patchy environments, without the need to invoke the stochastic features used by Gurney and Nisbet (1978). The qualitative features of the role of a functional response in contributing to oscillatory behavior have been elucidated by Kazarinoff and van der Driessche (1978) in a similar model. However, in contrast to their model, where the bifurcation always leads to stable cycles, for sufficiently high levels of predation, the bifurcation does not lead to stable periodic orbits. In other words, if P is too large, the stabilizing effect of spatial heterogeneity is lost because the predator is too mobile. Then the unstable predator prey interaction dominates.

The qualitative features of the model presented here agree well with the experiments of Huffaker (1958) and Luckinbill (1974). Luckinbill, in particular, had to slow down the predator so as to achieve coexistence in his system and avoid extinctions.

ACKNOWLEDGEMENTS

I would like to thank Hal Caswell, Nicholas Kazarinoff and David Wollkind for helpful discussions. I would also like to thank Brian Hassard, Hall Caswell, and Nicholas Kazarinoff for access to unpublished material.

APPENDIX

Here the procedure for determining the nature of the bifurcation is described. Let u and v represent deviations away from an equilibrium point $\hat{x},\hat{y}$ for equation (5), with $n = \frac{1}{2}$. Then, u and v satisfy

$$\begin{pmatrix} \dot{u} \\ \dot{v} \end{pmatrix} = A \begin{pmatrix} u \\ v \end{pmatrix} + \begin{pmatrix} q_1(u,v) \\ q_2(u,v) \end{pmatrix} \tag{A1}$$

where, to third order,

$$q_1(u,v) = (-2D_x + \frac{1}{4}P\hat{x}^{-3/2}\hat{y})\frac{u^2}{2} - (\frac{3}{8}P\hat{x}^{-5/2}\hat{y})\frac{u^3}{6} - (D_x + \frac{1}{2}P\hat{x}^{-1/2})uv + (\frac{1}{4}P\hat{x}^{-3/2})\frac{u^2v}{2} \tag{A2}$$

and

$$q_2(u,v) = (\frac{1}{2}P\hat{x}^{-1/2})uv - (\frac{1}{4}P\hat{x}^{-3/2}\hat{y})\frac{u^2}{2} + (\frac{3}{8}P\hat{x}^{-5/2}\hat{y})\frac{u^3}{6} - (\frac{1}{4}P\hat{x}^{-3/2})\frac{u^2v}{2} \tag{A3}$$

and A is the matrix in (7).

Now, restrict attention to equilibrium points at which a bifurcation occurs, i.e., where

$$\hat{y} = \frac{2D_x}{P^4} \tag{A4}$$

and x is given by (6). Then (A2) and (A3) become

$$\begin{aligned} q_1(u,v) &= (-\frac{3}{4}D_x)\frac{u^2}{2} - (\frac{3}{4}P^2D_x)\frac{u^3}{6} - (D_x + \frac{1}{2}P^2)uv + (P^4/4)\frac{u^2v}{2} \\ q_2(u,v) &= (P^2/2)uv - (D_x/2)\frac{u^2}{2} + (\frac{3}{4}P^2D_x)\frac{u^3}{6} - (P^4/4)\frac{u^2v}{2} \end{aligned} \tag{A5}$$

and the matrix A becomes

$$\begin{pmatrix} 0 & -D_x/P^2 - 1 \\ D_x/P^2 & 0 \end{pmatrix} \tag{A6}$$

Let

$$\omega = [(D_x/P^2)(D_x/P^2 - 1)]^{\frac{1}{2}} \tag{A7}$$

Then the transformation $r = u$, $s = \omega v$ yields

$$\begin{pmatrix} \dot{r} \\ \dot{s} \end{pmatrix} = \begin{pmatrix} 0 & -\omega \\ \omega & 0 \end{pmatrix} \begin{pmatrix} r \\ s \end{pmatrix} + \begin{pmatrix} -\frac{3}{4}D_x r^2 - [(D_x + P^2)/(2\omega)]rs + [P^4/(8\omega)]r^2s - (P^2D_x/8)r^3 \\ (P^2/2)rs - (\omega D_x/4)r^2 - (\omega P^2 D_x/8)r^3 - (P^4/8)r^2s \end{pmatrix} \tag{A8}$$

The formula of Hsü and Kazarinoff (1976), or that of Hassard and Wan (1978) can now be directly applied to (A8), which yields equation (9).

REFERENCES

[1] Caswell, H. (1978). "Predator mediated coexistence: a non equilibrium model", *Amer. Nat. 112*, 127-154.

[2] Chow, S. N., and Mallet-Paret, J. (1977). "Integral averaging and bifurcation", *Jour. Diff. Eq. 26*, 112-159.

[3] Gurney, W. S. C., Nisbet, R. M. (1978). "Predator-prey fluctuations in patchy environments", *Jour. Anim. Ecol. 47*, 85-102.

[4] Hassard, B., and Wan, Y. H. (1978). "Bifurcation formulae derived from center manifold theory", *JMAA 63*, 297-312.

[5] Hastings, A. (1977). "Spatial heterogeneity and the stability of predator prey systems", *Theo. Pop. Bio. 12*, 37-48.

[6] Hastings, A. (1978). "Spatial heterogeneity and the stability of predator prey systems: predator-mediated coexistence", *Theo. Pop. Bio.*, (in press).

[7] Hilborn, R. (1975). "The effect of spatial heterogeneity on the persistence of predator-prey interactions", *Theo. Pop. Bio. 8*, 346-355.

[8] Hsü, I. D., and Kazarinoff, N. D. (1976). "An applicable Hopf bifurcation formula and instability of small periodic solutions of the Field-Noyes model", *JMAA 55*, 61-89.

[9] Hsü, I. D., and Kazarinoff, N. D. (1977). "Existence and stability of periodic solutions of a third-order non-linear autonomous system simulating immune response in animals", *Proc. Roy. Soc. Edin. 77A*, 163-175.

[10] Huffaker, C. B. (1958). "Experimental studies on predation: dispersion factors and predator-prey oscillations", *Hilgardia 27*, 343-383.

[11] Hutchinson, G. E. (1951). "Copedology for the ornithologist", *Ecology 32*, 571-577.

[12] Hutchinson, G. E. (1959). "Homage to Santa Rosalia, or why are there so many kinds of animals?", *Amer. Nat. 93*, 145-154.

[13] Kazarinoff, N. D., and van der Driesche, P. (1978). "A model predator-prey system with functional response", *Math. Biosci. 39*, 125-134.

[14] Levin, S. A. (1974). "Dispersion and population interactions", *Amer. Nat. 108*, 207-228.

[15] Levin, S. A. (1976a). "Spatial patterning and the structure of ecological communities", in Some Mathematical Questions in Biology, VII (S. A. Levin, Ed.), Lectures on Mathematics in the Life Sciences, Vol. 8, American Mathematical Society, Providence, R.I.

[16] Levin, S. A. (1976b). "Population dynamics in heterogeneous environments", *Ann. Rev. Ecol. Systematics 7*, 287-310.

[17] Lin, J., and Kahn, P. (1976). "Averaging methods in predator-prey systems and related biological models", *J. Theo. Bio. 57*, 73-102.

[18] Luckinbill, L. S. (1974). "The effects of space and enrichment on a predator-prey system", *Ecol. 55*, 1142-1147.

[19] Marsden, J. E., and McCracken, M. (1976). "The Hopf Bifurcation and Its Applications", Springer-Verlag, New York.

[20] Maynard Smith, J. (1974). "Models in Ecology", Cambridge University Press, New York.

[21] Skellam, J. G. (1951). "Random dispersal in theoretical populations", *Biometrika 38*, 196-218.

[22] Slatkin, M. (1974). "Competition and regional coexistence", *Ecology 55*, 128-134.

[23] Vandermeer, J. M. (1973). "On the regional stabilization of locally unstable predator-prey relationships", *J. Theo. Bio. 41*, 161-170.

[24] Ziegler, B. P. (1977). "Persistence and patchiness of predator-prey systems induced by discrete event population exchange mechanisms", *J. Theo. Bio. 67*, 687.

CAUCHY SYSTEM FOR THE NONLINEAR BOUNDARY VALUE PROBLEM OF A SHALLOW ARCH

R. E. Kalaba

Biomedical Engineering Department
University of Southern California
Los Angeles, California

E. A. Zagustin

Department of Civil Engineering
California State University
Long Beach, California

I. INTEGRO-DIFFERENTIAL EQUATION FOR A SHALLOW ARCH

The deflection $w(x)$ of a shallow arch under transversal load $q(x)$, shown in Figure 1, in dimensionless form is described by a fourth order integro-differential equation given [1,2] by

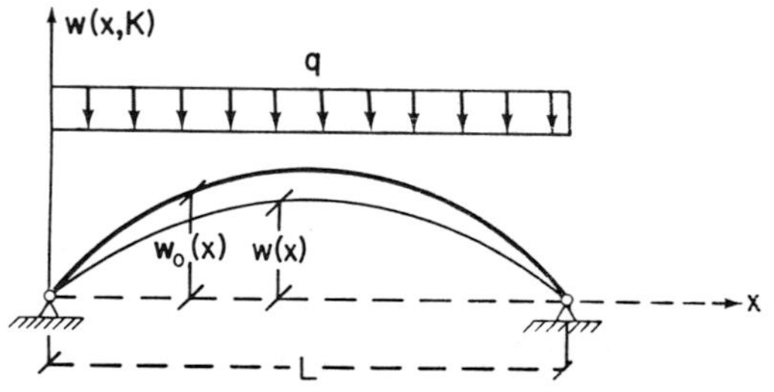

FIGURE 1. Shallow Nonlinear Arch.

ISBN 0-12-434180-2

$$\frac{d^4w}{dx^4} + k\,\frac{d^2w}{dx^2}\left[\int_0^1 (w_0'(y))^2 dy - \int_0^1 (w'(y))^2 dy\right] = F(x,L), \tag{1}$$

$$0 \leq x \leq 1,$$

where

$$F(x,L) = q(x) - \frac{d^4w_0}{dx^4}, \tag{2}$$

$$k = \frac{AL^2}{2I}, \quad \text{or} \quad k = 6\left(\frac{L}{h}\right)^2. \tag{3}$$

The function $w_0(x)$ is the initial shape of the shallow arch, $w(x)$ is the deflection of the arch under load $q(x)$, A is the cross-sectional area of rectangular section of the arch, L is the span of the arch, I is the moment of inertia of the cross-section, h is the thickness of the arch, and $q(x)$ is the transversal loading acting on the arch.

II. BOUNDARY CONDITIONS

In the case of a simply supported arch the boundary conditions at each end are zero deflection and zero bending moment or

$$\begin{aligned} w(0) &= w''(0) = 0, \\ w(1) &= w''(1) = 0. \end{aligned} \tag{4}$$

For the case of an arch clamped at both ends, the boundary conditions are: zero deflection and zero slope at each enc, i.e.,

$$\begin{aligned} w(0) &= w'(0) = 0 \\ w(1) &= w'(1) = 0. \end{aligned} \tag{5}$$

III. DERIVATION OF AN INITIAL VALUE PROBLEM

Let us assume that $w(x,k)$, where the constant k is the imbedding parameter. We differentiate equation (1) with respect to k which gives

$$w_k^{iv} + kw_k''\left[\int_0^1 (w_0'(y))^2 dy - \int_0^1 (w'(y))^2 dy\right] - 2kw''\int_0^1 w'(y)w_k(y)dy$$

$$= -w''\left[\int_0^1 (w_0'(y))^2 dy - \int_0^1 (w'(y))^2 dy\right], \qquad (6)$$

where

$$w'(x,k) = \frac{\partial w}{\partial x}, \quad w_k(x,k) = \frac{\partial w}{\partial k}.$$

Equation (6) is a linear integro-differential equation for the function w_k.

Consider now a differential equation for the auxiliary function $u(x,k)$ as follows

$$u^{iv}(x,k) + ku''\left[\int_0^1 (w_0'(y))^2 dy - \int_0^1 (w'(y))^2 dy\right]$$

$$- 2kw''\int_0^1 w'(y,k)u(y,k)dy = f(x,k) \qquad (7)$$

with the boundary conditions given by

$$u(0,k) = u''(0,k) = 0,$$
$$u(1,k) = u''(1,k) = 0. \qquad (8)$$

If $G(x,t,k)$ is the appropriate Green's function, then

$$u(x,k) = \int_0^1 G(x,t,k)f(t,k)dt. \qquad (9)$$

If

$$f(x,k) = -w''(x,k)\left[\int_0^1 (w_0'(y))^2 dy - \int_0^1 (w'(y))^2 dy\right], \tag{10}$$

then since w_k satisfies the differential equation (6) with boundary conditions (4) which is analogous to the differential equation (7) and (8) satisfied by $u(x,k)$, we have

$$w_k(x,k) = -\int_0^1 G(x,t,k)w''(t,k)\left[\int_0^1 (w_0'(y))^2 dy - \int_0^1 (w'(y))^2 dy\right] dt. \tag{11}$$

Let us differentiate equations (7) and (8) with respect to k which gives

$$u_k^{iv}(x,k) + ku_k''\left[\int_0^1 (w_0'(y))^2 dy - \int_0^1 (w'(y))^2 dy\right] - 2kw''\int_0^1 w'u_k dy$$

$$= \Psi(x,k), \tag{12}$$

where

$$\Psi(x,k) = f_k(x,k) + 2kw''\int_0^1 w_k'u' dy + 2kw_k''\int_0^1 w'u' dy, \tag{13}$$

and

$$u_k(0,k) = u_k''(0,k) = 0,$$

$$u_k(1,k) = u_k''(1,k) = 0. \tag{14}$$

Since $u(x,k)$ and $u_k(x,k)$ satisfy analogous differential equations we have

$$u_k(x,k) = \int_0^1 G(x,t,k)\Psi(t,k)dt. \tag{15}$$

Next, we differentiate equation (9) with respect to k which gives

$$u_k(x,k) = \int_0^1 G_k(x,t,k)f(t,k)dt + \int_0^1 G(x,t,k)f_k(t,k)dt. \tag{16}$$

Since the left hand sides of equations (15,16) are identical by equating the right hand sides of equations (15) and (16) and replacing $\Psi(t,k)$ by equation (13) we obtain an equation for $G_k(x,z,k)$.

$$G_k(x,z,k) = 2k\int_0^1 G(x,t,k)w''(t,k)\left[\int_0^1 w_k'(y,k)G'(y,z,k)dy\right]dt$$

$$+ 2\int_0^1 G(x,t,k)\left[kw_k''(t,k)+w''(t,k)\right]\left[\int_0^1 w_k'(y,k)G'(y,z,k)dy\right]dt$$

$$+ 2k\int_0^1 G(x,t,k)\left[\int_0^1 w'w_k'(y,k)dy-\int_0^1 (w_0')^2dy+\int_0^1 (w')^2dy\right]$$

$$\times G''(t,z,k)dt. \tag{17}$$

Since G' and G'' appear on the right hand side of equation (17) we have to get expressions for G_k', G_k'' by differentiating equation (17) with respect to x once and twice which gives

$$G_k'(x,z,k) = 2k\int_0^1 G'(x,t,k)w''(t,k)\left[\int_0^1 w_k'(y,k)G'(y,z,k)dy\right]dt$$

$$+ 2\int_0^1 G'(x,t,k)\left[kw_k''(t,k)+w''(t,k)\right]\int_0^1 w'(y,k)G'(y,z,k)dy\ dt$$

$$+ 2k\int_0^1 G'(x,t,k)\left[\int_0^1 w'w_k'(y,k)dy-\int_0^1 (w_0')^2dy+\int_0^1 (w')^2dy\right]$$

$$\times G''(t,z,k)dt, \tag{18}$$

$$G_k''(x,z,k) = 2k\int_0^1 G''(x,t,k)w''(t,k)\left[\int_0^1 w_k'(y,k)G'(y,z,k)dy\right]dt$$

$$+ 2\int_0^1 G''(x,t,k)\left[kw_k''(t,k)+w''(t,k)\right]\int_0^1 w'(y,k)G'(y,z,k)dy\ dt$$

$$+ 2k\int_0^1 G''(x,t,k)\left[\int_0^1 w'w_k(y,k)dy-\int_0^1 (w_0')^2 dy+\int_0^1 (w')^2 dy\right]$$

$$\times\ G''(t,z,k)dt. \tag{19}$$

To this system of equations (17,18,19) for G, G', G'' we adjoin equation for $w_k(x,k)$ given by (11). Since $w'(y,k)$, $w''(y,k)$ appear under the integral in equation (11) we have to differentiate both sides of equation (11) once and twice with respect to x which yields

$$w_k(x,k) = -\int_0^1 G(x,t,k)w''(t,k)\left[\int_0^1 (w_0'(y))^2 dy-\int_0^1 (w'(y)dy\right]dt, \tag{20}$$

$$w_k'(x,k) = -\int_0^1 G'(x,t,k)w''(t,k)\left[\int_0^1 (w_0'(y))^2 dy-\int_0^1 (w'(y))^2 dy\right]dt, \tag{21}$$

$$w_k''(x,k) = -\int_0^1 G''(x,t,k)w''(t,k)\left[\int_0^1 (w_0'(y))^2 dy-\int_0^1 (w'(y))^2 dy\right]dt. \tag{22}$$

When $k = 0$, we easily solve Equation (1) for $w(x,0)$, $w'(x,0)$ and $w''(x,0)$. Furthermore, the needed functions $G(x,z,0)$, $G'(x,z,0)$ and $G''(x,z,0)$ are easily obtained in analytical form. These functions, of course, depend upon the exact boundary conditions being used. Notice that the differential equations for w, w', w'', G, G', G'' do not involve the boundary conditions.

IV. NUMERICAL METHOD

The initial value problem for the functions $w(x,k)$, $w'(x,k)$, $w''(x,k)$ and $G(x,t,k)$, $G'(x,t,k)$, $G''(x,t,k)$ given by equations (17,18,19,20,21,22) with initial conditions at $k = 0$ yield a Cauchy system which is well suited for numerical integration. Basically, the integrals are approximated by means of a quadrature formula so that we finally have a system of ordinary differential equations subject to known initial conditions, the independent variable being k. Related results are given in references [4,5,6]. This problem has practical applications in the design of shallow arches as well as for the calculation of buckling loads of a shallow arch.

V. REFERENCES

[1] Fung, Y. C., and Kaplan, A. (1952). "Buckling of low arches or curved beams of small curvature", *NACA Tech. Note 2840.*

[2] Doyle, T. C., and Ericksen, J. L. (1956). "Nonlinear Elasticity", *Advances in Appl. Mech. 4,* 53-115.

[3] Tsien, H. S. (1947). "Lower buckling load in the nonlinear buckling theory for thin shells", *Quart. Appl. Math., 5,* 236-237.

[4] Kalaba, R., Spingarn, K., and Zagustin, E. (June 1976). "Imbedding Methods for Bifurcation Problems and Post-Buckling Behavior of Nonlinear Columns", Proceedings of Nonlinear Systems and Applications, University of Texas at Arlington.

[5] Casti, J., and Kalaba, R. (1973). "Imbedding Methods in Applied Mathematics", Addison-Wesley Co.

[6] Kalaba, R., and Zagustin, E. (Oct., 1976). "On the Conical Punch Pressing Into an Elastic Layer", *J. of Elasticity, Vol. 6, No. 4,* 441-449.

A SUMMARY OF RECENT EXPERIMENTS TO COMPUTE THE TOPOLOGICAL DEGREE

Baker Kearfott

Department of Mathematics and Statistics
University of Southwestern Louisiana
Lafayette, Louisiana

Classically, we obtain numerical solutions of systems of nonlinear algebraic equations by Newton's method or by related gradient methods. In the past decade, however, several new approaches have been explored. In particular, fixed points (or roots) of continuous functions can be approximated by constructive methods of combinatorial topology ([2], [3], [5], [7], [13], [14]). Resulting topological (or combinatorial) algorithms may be employed to locate reliably all roots of a map ([2], [3]).

To ascertain a priori the number of such roots within a given volume of $\mathbb{R}^n$, it is possible to compute the Brouwer degree of the map. Related to other combinatorial methods, new ideas for degree computation have recently been investigated ([8], [9], [11], [15], [16]). Below, we define the Brouwer degree, summarize our method of computation, discuss the algorithm's salient characteristics, and give c.p.u. times for test problems. References are provided where necessary.

ISBN 0-12-434180-2

I. THE CONCEPT AND THE COMPUTATION FORMULA

Consider a continuous map F from a compact n-polygon D into $\mathbb{R}^n$, such that $F(X) \neq 0$ if $X \in b(D)$, where $b(D)$ is the boundary of D. Suppose $\det(F'(X)) \neq 0$ when $F(X) = 0$. Then the degree of F at 0 relative to D, denoted $d(F,D,0)$, is equal to the number of zeros of F where $\det(F'(X)) > 0$, minus the number of zeros of F where $\det(F'(X)) < 0$.

We generalize $d(F,D,0)$ to non-differentiable continuous maps by defining the index of a zero X of F. Given a sufficiently small neighborhood M of X, $f(M)$ covers a neighborhood N of 0 in an m-to-one fashion. Considering the m branches of f/M, the index of X is the number of coverings of N with positive orientation minus the number of coverings of N with negative orientation. Then, $d(F,D,0)$ is the sum of the indices of the zeros of F in D ([4], ch. 1).

The above definitions for $d(F,D,0)$ coincide when F is differentiable with no multiple roots.

To compute $d(F,D,0)$ we consider $F/\|F\| \;|b(D)$. If $Y \in F(b(D))$, it can be shown that $d(F,D,0)$ is equal to the number of times $F/\|F\| \;|b(D)$ covers Y with a positive orientation in $\mathbb{R}^{n-1}$, minus the number of times $F/\|F\| \;|b(D)$ covers Y with a negative orientation ([4], ch. 1). Assume $b(D)$ is polygonal, and triangulate $b(D)$ into simplexes $\{S_i\}_{i=1}^k$ such that at least one component of F does not vanish on each S_i. Choose Y to be the intersection of the unit n-sphere with the positive first coordinate axis, and assume appropriate components of $F = (f_1, f_2, \ldots, f_n)$ do not vanish on the $(n-2)$-dimensional boundaries $b(S_i)$ ([8], ch. 3, etc.). We then have:

$$d(F,D,0) = \sum_{j=1}^{m} d(F\ ,S_{i_j},0).$$

Above, the sum is over all S_i on which $f_1 > 0$, and $F_1 = (f_2, \ldots, f_n)\colon\ S_{i_j} \to \mathbb{R}^{n-1}$ ([8], [11], [15]).

Formulas not involving recursion (in the computer programming sense) have also been presented ([8], [9], [15]). However, the recursion formula has been easiest to implement if we do not allow heuristic determination of the mesh on $b(D)$ ([11]).

II. THE ALGORITHM

We proceed as follows:

(1) Triangulate $b(D)$ to obtain a sufficiently small mesh, as in Section 1, then continue to step (2).

(2) Replace F by F_1 and D by S_{i_j}, and do step (3), for $j = 1,\ldots,m$.

(3) (a) If the S_{i_j} are one-dimensional, compute $d(F,S_{i_j},0)$ directly.

(b) If the dimension of the S_{i_j} is greater than 1, repeat steps (1) and (2) with S_{i_j} and F_1 in place of D and F.

In one dimension, $d(f,S,0) = d(f,<a,b>,0) = \frac{1}{2}\{\text{sgn}[f(b)] - \text{sgn}[f(a)]\}$. Also, observe that a stack of executions of steps (1), (2), and (3) is generated if $n > 2$.

It is convenient to construct the triangulation by "generalized bisection", which we define as follows: given the $(n-1)$-simplex $S = < X_1,X_2,\ldots,X_n >$, we define two new simplexes by replacing X_k by $(X_k + X_m)/2$ or X_m by $(X_k + X_m)/2$, where $< X_k,X_m >$ is the longest side of S ([8], [10], [15]). The simplexes in such triangulations correspond to nodes in binary trees, and the elements in the final triangulation can be considered with a minimum of computation and storage by a depth-first search of such trees ([11]). The depth of each path is set by examining the moduli of continuity of the components of F ([11]).

A more detailed exposition of the algorithm appears in [11].

III. PERFORMANCE OF THE ALGORITHM AND SCOPE OF APPLICATION

To date, few other methods for computing $d(F,D,0)$ have appeared. Erdelsky ([6]) described an efficient method, equivalent to ours except for the triangulation, for $n = 2$. O'Neil and Thomas ([12]) computed the degree for arbitrary n by quadrature involving the Kronecker integral ([1], p. 465). These computations involved probabilistic estimates for the accuracy of the result.

Our approach lends itself naturally to root-finding ([8], [9], [11]). Assume $D = < X_0, X_1, \ldots, X_n >$ is an n-simplex, and compute $(F,D,0)$, triangulating $b(D)$ by bisection. If $d(F,D,0) \neq 0$, bisect the n-simplex D, forming S_1 and S_2, and compute $d(F,S_1,0)$ and $d(F,S_2,0) = d(F,D,0) - d(F,S_1,0)$. This computation is expedited with information retained from computation of $d(F,S,0)$ ([11]). We repeat the process, bisecting the first S_i over which F has non-zero degree and storing the other S_i in a list if F also has non-zero degree on it ([11]). The procedure continues until a simplex with diameter less than a specified tolerance is found. We then repeat the bisection-degree computation process on the stored simplexes until the list is empty (ibid).

Our root-finding algorithm shares properties with other combinatorial fixed point algorithms. Function values only are required, and only rough accuracy is needed. Moreover, all roots, including ones difficult to obtain with gradient methods, may often be located. Our degree-computation method, however, gives lower bounds on the number of roots within the search region, while other methods may find approximate zeros which are not near true roots ([2], [3], etc.).

Degree computation-bisection has several disadvantages. The diameters of the resulting simplexes decrease linearly as bisection proceeds, and the rate of decrease increases with n ([10]). Furthermore, due to the recursive nature, execution time for functions of comparable smoothness increases exponentially with n. Lastly, we must assume that there is no root of F on the boundary of any n-simplex produced by bisection; also, there must be no roots of the truncated functions on the boundaries of any of the lower-dimensional simplexes (when $n > 2$). When such roots exist, they are found in the process of degree computation, preventing the algorithm from proceeding further.

In practice, it is possible to avoid roots on boundaries by changing the vertices of D slightly. Also, the algorithm is not necessarily too costly in small dimensions.

IV. NUMERICAL RESULTS

We present results for several test examples in 2 and 3 dimensions.

The experimental program involved root-finding by bisection. It contained a parameter controlling information storage between successive degree computations ([11]), but we present c.p.u. times for optimal values of that parameter. In all cases, the stopping diameter (tolerance) was .1, and all roots and corresponding indices within D were found.

The PL/I program was run interactively on a Multics 68/80 system. The results in 2 dimensions appear in Table 1. The trial function in 3 dimensions was: $f_1 = x_1^2 - x_2$, $f_2 = x_2^2 - x_3$, $f_3 = x_3^2 - x_1$, and $D = < (.9,.1,-.1), (0,1,0), (.1,0,1.1), (-.9,-.8,-.7) >$. The c.p.u. time for that example was 94.1 seconds.

TABLE I. Two Dimensional Examples
D = <(-4.1,-3.9),(4,-4),(-.15,4)>

function	*c.p.u. time*
z^2	3.5
z^2+1	3.7
z^3	13.6
z^3+1	11.9
z^4	37.3
z^4+1	37.2

REFERENCES

[1] Alexandroff, P., and Hopf, H. (1973, orig. 1935). "Topologie", Chelsea, New York.

[2] Allgower, E. L. (1974). "Numerische Approximation von Lösungen nichtlinearer Randuerts-aufgaben mit meheren Lösungen", *ZAMM, Tagung 54,* München.

[3] Allgower, E. L., and Keller, K. L. (1971). "A search routine for a Sperner simplex", *Computing 8,* 157-165.

[4] Cronin, J. (1964). "Fixed points and topological degree in nonlinear analysis", *A.M.S. Surveys No. 11.*

[5] Eaves, B. C., and Saigal, R. (1972). "Homotopies for the computation of fixed points", Mathematical Programming 13, Nos. 1 and 2.

[6] Erdelsky, P. J. (1973). "Computing the Brouwer degree in $\mathbb{R}^2$", *Math. Comp. 27, #121.*

[7] Jeppson, M. (1972). "A search for the fixed points of a continuous mapping", Mathematical Topics in Economic Theory and Computation, *SIAM,* Philadelphia, 122-125.

[8] Kearfott, R. B. (1977). "Computing the degree of maps and a generalized method of bisection", Ph.D. dissertation, University of Utah.

[9] Kearfott, R. B. "An efficient degree-computation method for a generalized method of bisection", submitted to *Numer. Math.*

[10] Kearfott, R. B. "A proof of convergence and an error bound for the method of bisection in $\mathbb{R}^n$", to appear in *Math. Comp.*

[11] Kearfott, R. B. "Root-finding experiments using direct computation of the topological degree", submitted to *Math. Comp.*

[12] O'Neil, T., and Thomas, J. (1975). "The calculation of the topological degree by quadrature", *SIAM J. Numer. Anal. 12*, 637-689.

[13] Saigal, R. (1976). "Fixed point computing methods", preprint, the Center for Mathematical Studies in Economics and Management Science, Northwestern University.

[14] Scarf, H. (1967). "The approximation of fixed points of a continuous mapping", *SIAM J. Appl. Math. 15, No. 5.*

[15] Stenger, F. (1976). "An algorithm for the topological degree of a mapping in $\mathbb{R}^n$", *Numer. Math. 25*, 23-28.

[16] Stynes, M. (1977). "An algorithm for the numerical computation of the degree of a mapping", Ph.D. dissertation, University of Oregon.

COMPUTATION OF EIGENVALUES/EIGENFUNCTIONS FOR TWO POINT BOUNDARY VALUE PROBLEMS[1]

M. E. Lord[2]
M. R. Scott
H. A. Watts

Applied Mathematics Division 2623
Sandia Laboratories
Albuquerque, New Mexico

I. INTRODUCTION

In solving linear two-point boundary value problems, the code *SUPORT* [1] uses the method of superposition together with orthonormalization of the base solutions to the homogeneous equation when linear dependence threatens. This report considers extensions of the procedure used by *SUPORT* to allow computation of solutions of eigenvalue problems.

The technique is iterative on the eigenvalue parameter and requires a nonlinear equation solver as a driver routine. Two such root finders were investigated - a quasi-Newton technique and a combination secant and interval halving method.

When solving eigenvalue problems by means of an initial value technique, the nonlinear function which is evaluated by the root finder is dependent on the boundary conditions at the final end

[1]*This work was supported by the U.S. Department of Energy.*
[2]*On leave from the University of Texas at Arlington.*

ISBN 0-12-434180-2

point. Several choices for this function were implemented and tested. The iteration schemes proceed by adjusting the eigenvalue parameter until a certain boundary condition matrix is singular. One choice defines the function as the determinant of the final boundary condition matrix. Another method involves computing the minimum singular value of the boundary condition matrix. Other choices involve satisfying certain boundary equations exactly and driving the remaining boundary equations to zero.

Two error tolerances arise naturally in our technique for solving the eigenvalue problem. One tolerance involves the convergence test in the nonlinear equation solver; that is, the error tolerance in the iteration on the eigenvalue parameter. The other error tolerance to be specified is that used in the initial value solver (e.g., Runge-Kutta or Adams type). Problems involved in selecting these error tolerances are discussed.

The effects of the orthonormalization process on the iterative scheme are mentioned and the advantages of preassigning orthonormalization points [2] are considered. Finally, several examples are presented which demonstrate the applicability of the code and allow comparison of the various optional features mentioned. We emphasize that, at the time of this writing, the code is to be regarded as a research tool in its early stages of development and this paper constitutes a preliminary report on our work.

II. STATEMENT OF PROBLEM

Consider the linear two-point boundary value problem involving an eigenvalue parameter

$$\frac{dy}{dx} = F(x,\lambda)y(x), \tag{2.1}$$

$$Ay(a) = 0, \tag{2.2}$$

$$By(b) = 0, \tag{2.3}$$

where solutions y are in R^n, F is an $n \times n$ real matrix function, A is an $(n-k) \times n$ real matrix of rank $n-k$ and B is a $k \times n$ real matrix of rank k.

For the homogeneous equation (2.1) the method of superposition [3] assumes a solution of the form

$$y(x) = c_1 u_1(x) + \ldots c_k u_k(x) = U(x)c, \tag{2.4}$$

where $U(x)$ is an $n \times k$ matrix whose columns $u_1(x),\ldots,u_k(x)$ are linearly independent solutions of

$$u'(x) = F(x,\lambda)u(x). \tag{2.5}$$

The matrix U will be referred to as the set of base solutions.

The method of superposition as presented here does not use n (the dimension of the solution space) vectors in the set of base solutions as in classical superposition. Instead, the method uses only k solutions for the following reasons. Since A is $(n-k) \times n$ and of rank $n-k$, the dimension of the null space of A is k. Therefore, k linearly independent starting vectors can be obtained which satisfy the boundary condition (2.2). (The *SUPORT* code actually produces an orthonormal basis for the null space of A.) These k starting vectors are then used to generate $U(x)$ by numerical integration of the differential equation (2.1). The function $y(x)$ in (2.4) is then a solution of the boundary value problem if the superposition coefficients $c_1,\ldots,c_k$ can be determined so that the boundary condition at $x = b$, equation (2.3), is satisfied.

We are interested only in nontrivial solutions, but this is equivalent to determining a nonzero vector c since the $u_i(x)$ are assumed to be independent. Thus we consider

$$By(b) = BU(b)c = 0. \tag{2.6}$$

The matrix $BU(b)$ is a $k \times k$ matrix and the homogeneous system (2.6) has a nontrivial solution for c if and only if $BU(b)$ is singular. Since the function $y(x)$ will be dependent on the eigenvalue parameter λ, the boundary condition matrix at the final end point will also depend on λ. We denote this dependence by writing (2.6) as

$$BU(b,\lambda)c = 0. \tag{2.7}$$

Thus, $\lambda = \lambda_0$ is an eigenvalue and $y(x) = U(x,\lambda_0)c$ is an associated eigenfunction if $BU(b,\lambda_0)$ is a singular matrix and c is a nontrivial solution of (2.7). The strategy then is to adjust the eigenvalue parameter $(\lambda \to \lambda_0)$ so that $BU(b,\lambda_0)$ is singular. When the eigenvalue has thus been computed, (2.7) can be solved for the superposition coefficient vector c and finally $y(x) = U(x,\lambda_0)c$ yields an eigenfunction. This does not determine the solution uniquely, as any appropriate normalization may be applied.

As discussed in [1,4] it is imperative that the numerical linear independence of the vectors in the base set be maintained during the integration procedure. This problem is overcome by the orthonormalization of the base vectors when they are near numerical linear dependence. The analysis of the orthonormalization procedure is not discussed here, but we reference [1]. Problems associated with orthonormalizations will be discussed in Section 6.

III. FINAL BOUNDARY CONDITION FUNCTION

As discussed in the previous section, it is necessary to adjust the eigenvalue parameter until the matrix $BU(b,\lambda)$ is singular. This defines the eigenvalue $\lambda = \lambda_0$ and then we can solve the system

$$BU(b,\lambda_0)c = 0 \tag{3.1}$$

for a nontrivial superposition coefficient vector. In this section we discuss several ways of defining an eigenvalue iteration function $f(\lambda)$ which, effectively, measures the singularity of the matrix $BU(b,\lambda)$. An eigenvalue is, therefore, defined to be a zero of the equation $f(\lambda) = 0$.

The most obvious and straightforward method is to evaluate the determinant of $BU(b,\lambda)$; that is, $f(\lambda) = \det[BU(b,\lambda)]$. For these experiments the $\det[BU(b,\lambda)]$ was evaluated by means of subroutine *RDET*, available from Sandia Laboratories Mathematical Subroutine Library. *RDET* uses Gaussian elimination.

Another method involved the singular value decomposition [5] of the matrix $BU(b,\lambda)$,

$$BU(b,\lambda) = PSQ^T.$$

Since $BU(b,\lambda)$ is $k \times k$, P and Q are $k \times k$ orthogonal matrices and S is a $k \times k$ matrix of the form

$$S = \begin{pmatrix} \mu_1 & & 0 \\ & \mu_2 \ddots & \\ 0 & & \mu_k \end{pmatrix},$$

where the μ_i are called the singular values and satisfy $\mu_1 \geq \mu_2 \geq \cdots \geq \mu_k \geq 0$. Theoretically, the matrix $BU(b,\lambda)$ being singular is equivalent to $\mu_k = 0$. Thus, $f(\lambda) = \mu_k(\lambda)$ represents our next boundary condition function. In the examples the singular value decomposition was accomplished by subroutine *SVDRS* as described in [6].

One problem associated with trying to drive the minimum singular value to zero is the fact that it is always nonnegative and, hence, the iteration function $f(\lambda)$ is nonnegative. An advantage of the singular value decomposition occurs for the case in which the eigenvalue has multiplicity greater than one. If λ_0 has multiplicity r, then $BU(b,\lambda_0)$ has rank $k-r$ and the singular values satisfy $\mu_1 \geq \cdots \geq \mu_{k-r} > 0$ and $\mu_{k-r+1} = \cdots = \mu_k = 0$. That is, the last r singular values will be zero and the first $k-r$ are non-zero. Therefore, the singular value decomposition could be used to monitor the multiplicity of the eigenvalue.

In order to define a third iteration function, let $H(\lambda) = BU(b,\lambda)$. Then the system $Hc = BU(b,\lambda)c = 0$ can be written as

$$\begin{aligned} h_{11}c_1 + \dots + h_{1k}c_k &= 0 \\ &\vdots \\ h_{k1}c_1 + \dots + h_{kk}c_k &= 0. \end{aligned} \tag{3.2}$$

Since the eigenvector is determined only to within a constant factor, it is clear from (2.4) that the superposition coefficient vector $c = (c_1,\dots,c_k)^T$ is determined to within a constant factor. For sake of argument, let us assume that $c_k \neq 0$ and that the first $k-1$ equations in (3.2) can be solved for $c_1,\dots,c_{k-1}$ in terms of c_k (which we shall take as $c_k = 1$). This determines the first $k-1$ boundary conditions exactly. When $\lambda = \lambda_0$ is an eigenvalue, the non-zero solution for c will also satisfy the last boundary equation. Thus, a third strategy then is to adjust the eigenvalue parameter so that in the above procedure the last boundary equation is satisfied. That is,

$$f(\lambda) = \sum_{i=1}^{k} h_{ki}(\lambda)c_i(\lambda)$$

where $c_k = 1$ and the remaining c_i are obtained by solving the associated $(k-1) \times (k-1)$ system.

There are some problems associated with this last approach. The normalization $c_k = 1$ is not valid if in the solution to the eigenvalue problem it is necessary to take $c_k = 0$. Even when $c_k \neq 0$, the $(k-1) \times (k-1)$ nonhomogeneous system

$$\begin{aligned} h_{11}c_1 + \dots + h_{1,k-1}c_{k-1} &= -h_{1k} \\ &\vdots \\ h_{k-1,1}c_1 + \dots + h_{k-1,k-1}c_{k-1} &= -h_{k-1,k} \end{aligned} \tag{3.3}$$

may not have a solution. This occurs if the coefficient matrix of the system in (3.3) is singular, which can certainly happen. This obviously occurs if the eigenvalue has multiplicity greater than one, but even for simple eigenvalues there is no guarantee that the above procedure is valid. Difficulties with this approach are illustrated in Section 7 when solving the Boltzmann equation.

Next, we consider a fourth procedure as a variation of the previous method which overcomes the above difficulties. In considering the system of equations (3.2) we would like to guarantee that the matrix of the subsystem in (3.3) is nonsingular. In order to accomplish this we first reorder the rows of H so that the first $k-1$ rows are linearly independent. This can be done, for example, by means of the modified Gram-Schmidt process using pivoting. Next, we apply modified Gram-Schmidt with pivoting to the columns of the new matrix. If H is of rank $k-1$ then the above procedure will move a $(k-1) \times (k-1)$ nonsingular submatrix in H to the upper left corner. We now proceed as in the previous method. Take $c_k = 1$ and solve the first $k-1$ equations for $c_1,\ldots,c_{k-1}$. Then the function value returned to the nonlinear equation solver will be the value of the last boundary equation.

However, some additional problems with this last procedure must be considered. If on different iterations of the eigenvalue parameter the row interchange results in a different boundary equation for the last row, this changes the iteration function. To see this more clearly, let us define

$$f_r(\lambda) = \sum_{i=1}^{k} h_{ri}(\lambda)c_i(\lambda)$$

as the rth row sum in the multiplication of H by c. Interchanging the rth row with the kth row would result in using $f_r(\lambda) = 0$ as the eigenvalue iteration equation. However, for given λ, it is clearly possible for $f_r(\lambda)$ to be scaled differently than $f_k(\lambda)$ and to even have opposite signs. One could attempt to remedy the problem by determining the row interchange strategy on the first iteration and leaving it fixed throughout subsequent iterations. This will likely be successful if the eigenvalue parameter does not vary too much in the iteration process. Otherwise, special attention (some form of restart) must be given to the root finder when the row interchange strategy dictates a new function for defining the root.

Now let us consider a problem associated with the column interchanges. We again suppose that the iteration function is basically defined by

$$f(\lambda) = \sum_{i=1}^{k} h_{ki}(\lambda)c_i(\lambda)$$

where we take $c_k = 1$. If column j is interchanged with column k, we obtain the function

$$g(\lambda) = \sum_{i=1}^{k} h_{ki}(\lambda)d_i(\lambda)$$

where $d_j = 1$. When solving linear systems, an interchange of columns in the matrix is reflected merely by a reordering of the solution components. Similarly, if $H(\lambda)$ is of rank $k-1$, then all nontrivial solutions of $Hc = 0$ are multiples of each other. In this case

$$\frac{1}{d_k} g(\lambda) = f(\lambda)$$

and, hence, if a column interchange takes place, a simple rescaling of the newly defined function is sufficient. If the divisor d_k actually turns out to be zero, the root finding procedure should be reinitialized to use the iteration function $g(\lambda)$. While this may be unlikely in practice, in principle this could occur anytime the iterations use values of λ which differ greatly from the eigenvalue λ_0. That is, the rank structure of the submatrices of $H(\lambda)$ could vary substantially.

In the above analysis, especially in connection with the remarks about the rank of $H(\lambda)$, we must remember that λ actually represents an approximation to the eigenvalue and, hence, the rank of H will nearly always be equal to k from a computational viewpoint. Also, we have tacitly assumed that rank $H(\lambda_0) = k - 1$. However, if λ_0 is an eigenvalue having multiplicity greater than one, or, if we have more than one eigenvalue parameter to obtain (as in the Orr-Sommerfeld problem which we examine in Section 7), the rank of H will be less than $k-1$. This problem requires an extension of the above ideas which we shall not elaborate on at this time. We just note that there are added complications in defining appropriate eigenvalue iteration functions. One final observation concerning the above methods is in order. If there is only one boundary equation to be satisfied at the final point, then $H(\lambda) = BU(b,\lambda)$ is just a scalar. The iteration function corresponding to the singular value approach is $f(\lambda) = |H(\lambda)|$, whereas for all the other techniques it is $f(\lambda) = H(\lambda)$.

IV. NONLINEAR EQUATION SOLVER

To compute solutions to the eigenvalue problems (2.1)-(2.3) we iterate on the eigenvalue parameter until one of the boundary condition functions discussed in the previous section is zero. A root finding method is necessary as a driving routine. Two such methods were implemented.

One technique uses a modification of subroutine *ZEROIN* available in the Sandia Laboratories Mathematical Subroutine Library. *ZEROIN* is a combination secant/interval halving method which is based on an algorithm due to Dekker [7]. Normally *ZEROIN* proceeds via the secant method to obtain a root. However, in certain instances the algorithm resorts to interval halving; as for example, when the iterate value computed falls outside the bracketing interval containing the root.

Initial input to *ZEROIN* was modified to include not only an upper and lower bound of the eigenvalue but a "sophisticated" guess of the eigenvalue. Many times a good initial guess to the eigenvalue is available either from analytical or experimental results or from previous computational results. However, no a priori information about the sign of the iteration function at the initial guess is generally known. For guaranteed success, *ZEROIN* requires an upper and lower bound of the eigenvalue between which the iteration function has a sign change. The "modified *ZEROIN*" checks initially for a sign change between the "sophisticated" initial guess and the bound nearest this guess. It proceeds with the usual algorithm if a sign change is detected. If not, it proceeds to the other subinterval. This change is designed to take advantage of good initial estimates of the eigenvalue and, on the average, should speed the iteration in the first few steps.

One obvious problem occurs in using "*ZEROIN*" when the function value returned is the minimum singular value of the matrix $BU(b,\lambda)$. The function value is always non-negative and does not exhibit a sign change regardless of the interval bracket on λ. In this case, "*ZEROIN*" will attempt to find the minimum of such a function. However there is no guarantee that it will be successful. This is demonstrated by the examples in Section 7.

Another limitation of "*ZEROIN*" is the fact that it computes only roots of a real valued function of a single real variable. Since it is important to solve complex eigenvalue problems (in real arithmetic this yields two eigenvalue parameters), a Newton type root finding method applicable for systems of equations was implemented. We actually use a quasi-Newton method which is a modification of subroutine *QN*, available in Sandia Laboratories Mathematical Subroutine Library. The modifications have been minor thus far, principally allowing for communication with the boundary problem solver.

It is difficult to attempt any direct comparison between the two rootfinders. The Newton scheme will often exhibit a superior convergence rate (once a local neighborhood of a root is entered) when compared to "*ZEROIN*." However, "*ZEROIN*" is guaranteed to converge to a root provided the endpoints of the initial bracket interval show a change in sign of the function. On the other hand, "*QN*" is less satisfactory with respect to global convergence behavior. This can be an important matter for some problems. The results produced by "*QN*" will often be more accurate than those of "*ZEROIN*" as it is typical for Newton schemes to overshoot on the requested accuracy of the roots, a fact attributed to the quadratic (or nearly so) convergence behavior. Also, it should be pointed out that the Newton scheme can easily converge to a different root (or perhaps not at all), depending on the shape of the function and location of the initial guess, while this effect can be more easily controlled when using "*ZEROIN*."

V. ERROR TOLERANCES

This section discusses the relationship between the error tolerance in the convergence of the nonlinear equation solver and the error tolerance in the ordinary differential equation integrator. Our comments will apply to all four of the eigenvalue iteration functions which were examined in Section 3.

It is generally believed that in the initial stages of the iteration process, when $|\lambda - \lambda_0|$ is relatively large, a rather crude integrator error tolerance can be used. Then as the iteration process begins to converge $(\lambda \to \lambda_0)$, smaller integration tolerances will likely be necessary. The typical behavior found in evaluating an eigenvalue iteration function (excluding the minimum singular value function) can be seen in Figure 5.1, which is meant to illustrate the following. For λ sufficiently far

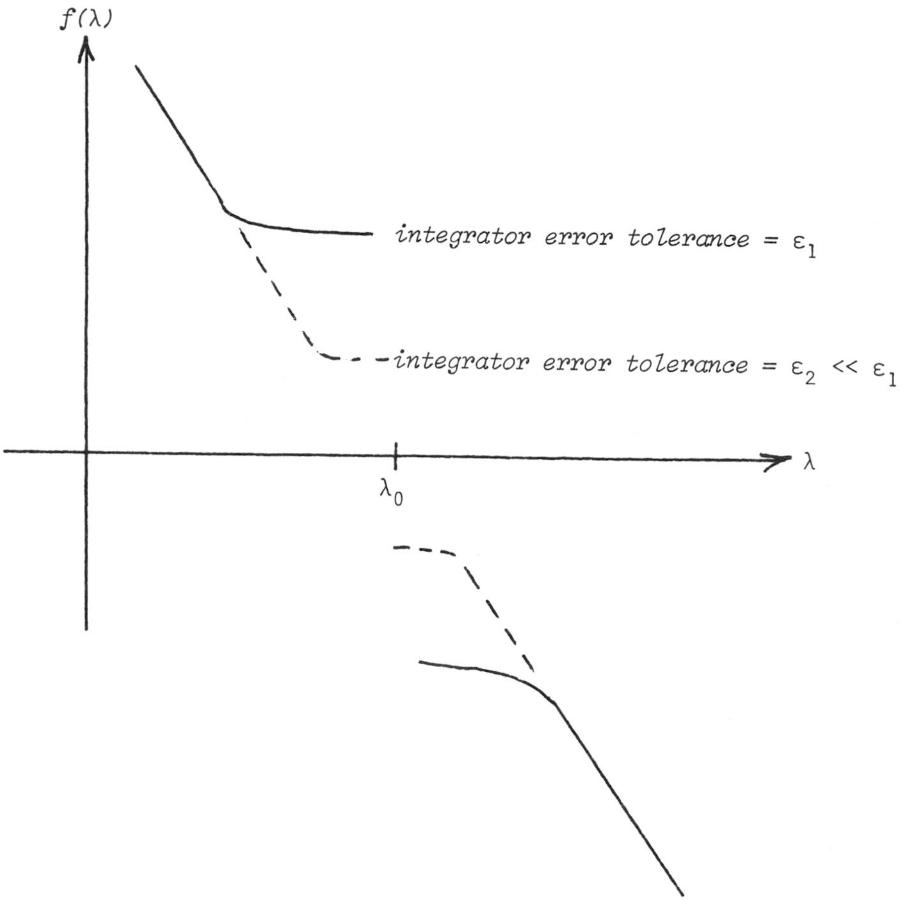

FIGURE 5.1

from the eigenvalue λ_0, the iteration function $f(\lambda)$ can be evaluated accurately enough for the search algorithm using any reasonable integration tolerance ε. Near λ_0 the function exhibits noticeable noise due mainly to errors in the integration process. As the integration errors are decreased, the amplitude of the noise level and the length of the noise interval about λ_0 decreases.

Naturally, the precise behavior of the computationally defined iteration function is somewhat problem dependent. We have not completed our investigations for an algorithm which appropriately defines the error tolerances. Here, we shall mention a couple of approaches which we have tried although it is clear that they are far from representing the optimal strategy. One scheme started the integration error tolerance off with $\varepsilon = 0.01$ (regardless of the user requested level of accuracy for the eigenvalue and eigenfunction). Then for each λ in the iteration two function values were computed, one corresponding to the current value of ε and another derived by using an integration tolerance of $\varepsilon/10$. If these two function values agreed reasonably well, the (last) function value was accepted. Otherwise, the integration tolerance was decreased by another factor of ten and so on until satisfactory agreement was achieved. Although this method appeared to be rather effective, it proved to be too inefficient, resulting in at least two sweeps of the integration process for each function evaluation requested by the nonlinear equation solver.

At the other extreme, we examined the following simple scheme. Corresponding to the user requested level of accuracy ε (which was used in the convergence criteria for the eigenvalue iteration process), the integration error tolerance was taken to be $\varepsilon/10$. In general, this appears to be quite satisfactory for crude tolerances ε but becomes less efficient than alternatives for more stringent tolerances. Most of the data reported in the examples was obtained with this scheme although we do illustrate the scheme described first.

VI. ORTHONORMALIZATION

The algorithm in *SUPORT* utilizes orthonormalization during the integration when linear dependence of the base solution set threatens. This overcomes the inherent numerical difficulties associated with superposition [1]. Since the code dynamically picks the points at which orthonormalization occurs, it is likely that on different iterations these points will vary somewhat. That is, the position of the orthonormalization points will in general be dependent on the eigenvalue parameter and integration error tolerance. The question then arises as to how much does this affect the iteration function. Although the examples given in the next section presented no apparent difficulties, problems involving large numbers of orthonormalizations were not extensively analyzed.

Errors introduced in the integration process generally dominate the discrepancies occurring in the iteration function. However, it is important to scale the functions in a consistent manner. When orthonormalizations occur on the interior of the integration interval, it seems necessary to orthonormalize at the final boundary point as well [4]. As this effectively introduces a form of scaling for the iteration function, orthonormalization was performed at the final point for all homogeneous (eigenvalue) problems.

Furthermore, it seems desirable to fix the orthonormalization point sequence from one iteration to the next so that additional noise is not introduced into the iteration funtion. This can be achieved through a recent addition to *SUPORT* [2], which allows the user to preassign the number and position of these points, thereby overriding the dynamic selection process. From prior experience with the automatic selection of orthonormalization points or with estimates involving the growth rates of solutions to the differential equation, the user may be able to preassign the orthonormalization points in a satisfactory manner. A better approach is possible in the context of the present problem. We suggest letting the code automatically determine the necessary orthonormalizations in the early iterations. Then after some stage we use the same orthonormalization points on subsequent iterations. This is feasible since as convergence begins to take place, the orthonormalization points will remain approximately the same from one iteration to the next.

VII. EXAMPLES

All of the results in this section were obtained on a CDC-6600 system. A variable step Runge-Kutta-Fehlberg code (*RKF*), one of the integrators available in *SUPORT*, was used. The relative and absolute error tolerances used in the rootfinders were taken to be 10^{-3} throughout. Except where otherwise stated, the integration tolerances used in the iteration sequence were taken to be 10^{-4}. The parameter $IFLAG = 54$ indicates "*ZEROIN*" collapsed the interval to the given tolerance and found no change in the sign of the function. No guarantee is given that the minimum of the function is found; in fact, see Example 1. The parameter $IFLAG = 53$ indicates a singularity in the function.

Example 1

The first example appears in [8] and arises from the structural analysis of the buckling of a vertical beam under a compression load. The differential equation with boundary conditions is given by

$$y^{(4)} = \lambda y,$$

$$y(0) = y^{(2)}(0) = 0,$$

$$y(\pi) = y^{(2)}(\pi) = 0.$$

This problem although relatively easy to solve, was studied because an exact solution is known: the first eigenvalue is $\lambda_0 = 1$ and the associated eigenfunction is $c \sin x$. The boundary condition matrix $BU(b,\lambda)$ can also be evaluated analytically and from this the four iteration functions mentioned in Section 3 can be studied. For the above reasons this example was quite useful in the early development and testing. Table 7.1 presents the results for this example. The column labeled "no. of iterations" counts the number of new eigenvalue approximations generated

TABLE 7.1: Example 1

Iteration Function	*Root Solver*	*Time (secs)*	*No. of Iterations*	*No. of Function Evaluations*	*Converged Values*
Determinant	*"QN"*	*0.24*	*4*	*7*	*1.0000*
	"ZEROIN"	*0.25*	*4*	*8*	*1.0001*
Minimum Singular Value	*"QN"*	*0.24*	*4*	*7*	*1.0000*
	"ZEROIN"	*0.41*	*8*	*11*	*1.05 (IFLAG=54)*
Last Boundary Equation	*"QN"*	*0.18*	*3*	*6*	*1.0000*
	"ZEROIN"	*0.26*	*4*	*8*	*1.0003*
Boundary Equation with Interchanges	*"QN"*	*0.22*	*4*	*7*	*1.0000*
	"ZEROIN"	*0.26*	*4*	*8*	*0.9996*

within the root finding scheme. The column labeled "no. of function evaluations" refers to the number of integration sweeps required in the total solution process. The column labeled "converged value" represents the eigenvalue approximation which the root finder converged to using the specified tolerances of 10^{-3}. The initial guess was provided as $\lambda = 1.1$ with the additional interval bracket information [*.6*, *1.5*] given to "*ZEROIN*." For this example, no interior orthonormalizations were required.

Example 2

The next example has been discussed in [9]. It is a buckling problem for a cylindrical structure, reinforced with longitudinal stringers, having an internal pressure and an axial load. The problem reduces to the following equations

$$w^{(4)} - 2(\eta + p)w^{(2)} + (\eta^2 - 2\eta\sigma)w - \eta f = 0,$$

$$f^{(4)} - 2\eta f^{(2)} + \eta^2 f + \eta w = 0,$$

$$w(0) = w^{(2)}(0) = f(0) = f^{(2)}(0) = 0,$$

$$w'(b) = w^{(3)}(b) = f'(b) = f^{(3)}(b) = 0,$$

TABLE 7.2: Example 2

Iteration Function	Root Solver	Time (secs)	No. of Iterations	No. of Function Evaluations	Converged Values
Determinant	"QN"	0.53	4	7	2.0654
	"ZEROIN"	0.63	4	8	2.0654
Minimum Singular Value	"QN"	0.85	7	10	2.0654
	"ZEROIN"	1.11	10	13	2.505 (IFLAG=54)
Last Boundary Equation	"QN"	2.48	9	13	DIVERGENT
	"ZEROIN"	1.02	10	13	2.505 (IFLAG=54)
Boundary Equation with Interchanges	"QN"	0.47	3	6	2.0654
	"ZEROIN"	0.55	3	7	2.0668

where $\eta = (\lambda/2\theta)^2$ and the eigenvalue parameter is σ. For a complete physical description of the parameters see [9]. The results in Table 7.2 were obtained with $\lambda = 1.$, $\theta = .7$, $p = .8$, and $b = .7\pi$. The initial guess for the eigenvalue was $\sigma = 1.$ along with the interval bracket $[0., 5.]$ for "*ZEROIN*." Again, no interior orthonormalizations were needed. Note the added difficulty, especially with "*QN*", when defining the iteration function from the last boundary equation.

Example 3

The next example is referred to as Boltzman's equation and is discussed in [1]. This equation models the particle transport in a slab. We state here the reduced or final form of the equation as the boundary value problem

$$\frac{\partial u}{\partial x}(x,s) + a(s)u(x,s) = \lambda k(s)\left\{ \int_0^1 u(x,t)dt + \int_{-1}^0 v(x,t)dt \right\},$$

$$-\frac{\partial v}{\partial x}(x,s) + a(s)v(x,s) = \lambda k(s)\left\{ \int_0^1 u(x,t)dt + \int_{-1}^0 v(x,t)dt \right\},$$

$$u(0,s) = 0,$$

$$v(L,s) = 0.$$

The functions $a(s)$ and $k(s)$ are known real piecewise continuous functions of s on $[-1,1]$. This problem can be considered an interval length problem, if λ is given and L is unknown. We solve it as an eigenvalue problem for which the interval length L is known and λ is to be determined. The computations presented here were performed with

$$a(s) = 1/|s|,$$

$$k(s) = 1/2\ |s|,$$

and $L = .62204$ which is the value used in [1] corresponding to $\lambda = 2$. Eight point Gaussian quadrature was used in evaluating

TABLE 7.3: Example

Iteration Function	*Root Solver*	*Time (secs)*	*No. of Iterations*	*No. of Function Evaluations*	*Converged Values*
Determinant	*"QN"*	*15.0*	*2*	*5*	*2.0000*
	"ZEROIN"	*21.4*	*3*	*7*	*2.0003*
Minimum	*"QN"*	*15.9*	*2*	*5*	*2.0000*
Singular	*"ZEROIN"*	*15.6*	*2*	*5*	*2.003*
Value					*(IFLAG=54)*
Last	*"QN"*	-	-	-	*NONCON-*
Boundary					*VERGENT*
Equation	*"ZEROIN"*	-	-	-	*IFLAG= 53*
Boundary					
Equation	*"QN"*	*18.7*	*3*	*6*	*2.0000*
with	*"ZEROIN"*	*22.6*	*3*	*7*	*1.9986*
Interchanges					
Determinant	*"QN"*	*21.*	*2*	*9*	*2.0000*
	"ZEROIN"	*31.*	*3*	*13*	*2.0000*
Boundary					
Equation	*"QN"*	*27.*	*3*	*11*	*2.0000*
with	*"ZEROIN"*	*31.*	*3*	*13*	*1.9986*
Interchanges					

the integral portions of the equations and this produced a differential equation system of order sixteen.

The initial guess for the eigenvalue was $\lambda = 1.9$ along with the bracket interval of $[1.5, 2.5]$ for "*ZEROIN*." An interesting result for this problem occurred when the last boundary equation function was used. In this case the iteration function possessed an infinite discontinuity near the eigenvalue and both "*QN*" and

"*ZEROIN*" failed to converge. This problem was caused by using a nearly singular submatrix as described in Section 3 but was averted when the row and column interchange strategy was employed. Table 7.3 first shows results when the integration tolerance is taken to be an order of magnitude smaller than the convergence tolerance for the root finder. We also include some comparisons when the integration tolerance is repeatedly reduced until the iteration function achieves at least one digit of relative accuracy. These results are given at the bottom of the table. In all cases, one interior orthonormalization point was required.

Example 4

The last example considered is a problem with a complex eigenvalue parameter and is referred to as the Orr-Sommerfeld equation. This problem has been studied extensively in the literature. We refer to the discussion in [1]. The Orr-Sommerfeld equation for plane Poiseuille flow is

$$y^{(4)} - 2k^2y^{(2)} + k^4y - ikR[(u(x)-\lambda)(y^{(2)} - k^2y) - u^{(2)}y] = 0,$$

where $y(x)$ is the amplitude of the stream function, R is the Reynolds number, k is the wave number of the disturbance, $i = \sqrt{-1}$, and λ is the eigenvalue parameter. The laminar velocity profile was taken to be $u(x) = 1 - x^2$. The boundary conditions are chosen to be

$$y'(0) = y^{(3)}(0) = 0,$$

$$y(1) = y'(1) = 0.$$

Since the solution is complex valued, the problem is transformed to an eighth order real system. The "*ZEROIN*" root finder is not applicable since we now have a pair of eigenvalue parameters and thus the quasi-Newton method "*QN*" is used. Only one technique for defining the iteration functions was attempted for this problem - namely, the last two boundary equations were split

TABLE 7.4: Example 4

R	Time (secs)	No. of Iterations	No. of Function Evaluations	Starting Guess	Converged Values
2500	3.6	2	6	$0.3 - 0.01\,i$	$0.30115-0.01418\,i$
10^6	151.1	9	13	$0.06-0.01\,i$	$0.06655-0.01402\,i$

off to define the iteration functions needed. Thus, in considering the 4×4 system $BU(b,\lambda)c = 0$, c_3 and c_4 were taken to be one and the first two equations were then solved for c_1 and c_2. Next, the resulting vector c is used to evaluate the last two boundary equations in $BU(b,\lambda)c = 0$.

The results for two Reynolds numbers, R, are presented in Table 7.4. For $R = 2500$, one interior orthonormalization was performed and for $R = 10^6$, 34 interior orthonormalizations were needed. For these computations the code was allowed to select the orthonormalization points automatically on each iteration.

REFERENCES

[1] Scott, M. R., and Watts, H. A. (1977). "Computational Solution of Linear Two-Point Boundary Value Problems via Orthonormalization", *SIAM J. Numer. Anal.*, *14*, 40-70. Also published as Sandia Laboratories Report SAND75-0198.

[2] Darlow, B. L., Scott, M. R., and Watts, H. A. (1977). "Modifications of SUPORT, A Linear Boundary Value Problem Solver Part I - Pre-Assigning Orthonormalization Points, Auxiliary Initial Value Problem, Disk or Tape Storage", Sandia Laboratories Report SAND77-1328.

[3] Godunov, S. (1961). "On the Numerical Solution of Boundary-Value Problems for Systems of Linear Ordinary Differential Equations", *Uspekhi Mat. Nauk.* *16*, 171-174.

[4] Conte, S. D. (1966). "The Numerical Solution of Linear Boundary Value Problems", *SIAM Review 8*, 309-321.

[5] Forsythe, G., and Moler, C. (1967). "Computer Solution of Linear Algebraic Systems", Prentice-Hall, Inc., Englewood Cliffs, NJ.

[6] Lawson, C., and Hanson, R. (1967). "Solving Least Squares Problems", Prentice-Hall, Inc., Englewood Cliffs, NJ.

[7] Dekker, T. J. (1969). "Finding a Zero by Means of Successive Linear Interpolation", Constructive Aspects of the Fundamental Theorem of Algebra, edited by B. Dejon and P. Henrici, Wiley-Interscience.

[8] Scott, M. R. (1973). "Invariant Imbedding and Its Applications to Ordinary Differential Equations", Addison-Wesley Publishing Co. Inc., Reading, MA.

[9] Stephens, W. B. (1971). "Imperfection Sensitivity of Axially Compressed Stringer Reinforced Cylindrical Panels Under Internal Pressure", *AIAA J. 9*, 1713-1719.

A CONTINUUM MODEL APPROPRIATE FOR NONLINEAR ANALYSIS OF THE SOLIDIFICATION OF A PURE METAL

David J. Wollkind
Ronald D. Notestine

Department of Pure and Applied Mathematics
Washington State University
Pullman, Washington

Robert N. Maurer

Department of Mathematics
Worcester Polytechnic Institute
Worcester, Massachusetts

I. INTRODUCTION

Most authors in treating controlled solidification situations involving alloys or pure metals [1,2] in the absence of convection have traditionally adopted a particular conservation of *heat* condition at the moving boundary separating the phases. Using a three-dimensional laboratory coordinate system this boundary condition at the solid-liquid interface,

$$\Sigma: \quad f(x,y,z,t) = z - \tilde{\zeta}(x,y,t) = 0,$$

relating the temperatures T_L and T_S of the liquid and solid phases measured in $°K$ can be written in the form [3,4]

$$\kappa_S \partial T_S/\partial n - \kappa_L \partial T_L/\partial n = \mathcal{L}\tilde{v}_n, \tag{1.1}$$

ISBN 0-12-434180-2

where κ_S and κ_L are the thermal diffusivities in the solid and liquid, respectively, $\mathcal{L} = L/\rho c_p$ is the latent heat of fusion per unit volume divided by the density and the specific heat at constant pressure, $\partial(\ )/\partial n = \nabla(\ ) \cdot \underline{n}$ in which $\underline{n} = \nabla f/|\nabla f|$ is the unit normal to the interface Σ pointing into the liquid, and [5] $\tilde{v}_n = (\partial\tilde{\zeta}/\partial t)/|\nabla f|$ is the relative normal speed of the interface Σ in the absence of convection. In writing (1.1) it is assumed that the product ρc_p is the same in both phases and temperature is continuous at the interface - i.e.,

$$T_L = T_S = \tilde{T} \quad \text{on} \quad \Sigma. \tag{1.2}$$

The left-hand side of (1.1) represents net heat flux through the interface while the right-hand side takes into account the heat released by solidification at that interface.

Recently Wollkind and Maurer [6] re-examined this boundary condition by carefully applying a general continuum mechanical balance equation for surfaces of discontinuity to *energy* conservation at that interface. Their analysis yielded the following conservation of *energy* condition:

$$\kappa_S \partial T_S/\partial n - \kappa_L \partial T_L/\partial n = \mathcal{L}\tilde{v}_n[1 + (\Delta\Gamma)(\tilde{T})\tilde{\eta}] \quad \text{on} \quad \Sigma, \tag{1.3a}$$

where $(\Delta\Gamma)(\tilde{T}) = [\gamma(\tilde{T}) - \gamma_0]/L$ in which $\gamma(\tilde{T})$ is interfacial *surface free energy* assumed to be a function of interface temperature $\tilde{T}$ and γ_0 is interfacial *surface energy*, both measured per unit area, while [7] $\tilde{\eta} = \nabla_2 \cdot [\nabla_2\tilde{\zeta}/|\nabla f|]$ is the curvature of Σ. Equation (1.3a) differs from (1.1) due to the presence of the extra curvature term which physically represents the difference between the *work* needed to create a new surface by stretching the interface Σ from a plane into a larger nonplanar area and the *surface energy* which can be stored or contained in this increased area. Since [8,9]

$$\gamma(\tilde{T}) = \gamma_0 - \tilde{T}\Delta S \quad \text{on} \quad \Sigma, \tag{1.3b}$$

where ΔS is the change of *surface entropy* associated with a unit area - increasing extention of the original planar interface,

$\Delta\Gamma$ is proportional to $\tilde{T}\Delta S$ and hence, consistent with Gibbs [9], that difference is representative of the *heat* absorbed at the interface during this process. Thus (1.3) is actually a new conservation of *heat* condition in which the extra term serves the same role with respect to latent heat that the curvature term does for melting temperature in the Gibbs-Thomson equation. For a pure metal the latter describes the alteration of the interface temperature from the normal melting temperature at a planar interface due to the curvature of the interface itself and is given by [2]

$$\tilde{T} = T_M[1 + \Gamma(\tilde{T})\tilde{\eta}] \quad \text{on} \quad \Sigma, \tag{1.4}$$

where $T_M \equiv$ normal melting temperature of the pure metal at a flat interface and $\Gamma(\tilde{T}) = \gamma(\tilde{T})/L$ is capillarity. In his derivation of this relation, Delves [2] makes use of the pressure condition

$$P_L - P_S = \gamma\tilde{\eta} \quad \text{on} \quad \Sigma, \tag{1.5}$$

which is a direct consequence of conservation of momentum at Σ in the absence of convection once one adopts the constitutive relations [3]

$$\underline{t} = -P\underline{n} \quad \text{and} \quad \underline{\ell} = \gamma\underline{N}, \tag{1.6a,b}$$

where $\underline{t} \equiv$ stress vector and $\underline{\ell} \equiv$ surface "tension" vector while $\underline{N}$ is the unit normal to C (the bounding curve of that section of Σ under consideration) tangent to Σ and pointing outward. Implicit in this formulation is the neglect of the effect of interface attachment kinetics which is very small for metals and alloys [10].

II. SURFACE FREE ENERGY AND THE NONLINEAR SOLIDIFICATION MODEL

Analogous to the concept of surface tension at a liquid-fluid interface, an equivalent effect in terms of surface free energy can be rigorously developed at an interface between a solid and a liquid [11]. That is why the $\underline{\ell}$ introduced at the end of the previous section was referred to as a surface "tension" vector. It seems appropriate at this time to discuss the constitutive relation of (1.6b) in conjunction with the interfacial surface free energy. In its most general form this relationship would be given by

$$\underline{\ell} = \underline{N} \cdot \underline{\underline{L}}\,, \tag{2.1}$$

where $\underline{\underline{L}}$ is the surface "tension" tensor. For an isotropic substance

$$\underline{\underline{L}} = \gamma \underline{\underline{I}}, \tag{2.2}$$

where $\underline{\underline{I}}$ is the identity tensor, and we recover the constitutive relation of (1.6b). Hence by taking $\underline{\ell} = \gamma \underline{N}$ we have assumed isotropy implicitly and in this instance one often says the surface free energy is isotropic [2]. Many authors, in addition, have taken this surface free energy to be constant [2]. That is, in terms of our notation, they consider

$$\gamma \equiv \gamma_M = \gamma_0 - T_M \Delta S, \tag{2.3}$$

while we allow for the possibility of the variation of γ with interfacial temperature. Since the surface energy γ_0 is independent of temperature [8], the temperature dependence of the surface free energy is thus represented by the second term in the expression on the right-hand side of (1.3b). Although the ΔS of that equation is always positive at liquid-fluid interfaces resulting in the decrease of the effect of surface tension with increasing temperature in such situations, this interfacial surface entropy change can be either positive or negative for the solid-liquid interface under examination. Further, since $\gamma(\tilde{T})$ can also be written as

$$\gamma(\tilde{T}) = \gamma(T_M) + \gamma'(T_M)(\tilde{T} - T_M), \tag{2.4}$$

upon comparision of (2.4) with (1.3b) and (2.3), we can make the following associations:

$$\Delta S = -\frac{d\gamma}{d\tilde{T}} = -\gamma'(T_M) \quad \text{and} \quad \gamma_M = \gamma(T_M). \tag{2.5}$$

Finally, it should be mentioned that as soon as γ is considered to be a function of temperature additional complications inevitably arise. For a fluid-liquid situation such dependence of surface tension effects upon temperature at the interface separating them, invariably gives rise to convection in at least one of the fluids. For a solid-liquid situation since the solid can support a shear force while the liquid remains stationary, such convection does not necessarily occur. From the tangential component of the conservation of momentum condition at the interface [12], we would then find in the absence of convection that

$$\underline{\tau} = -\nabla_{(\Sigma)}\gamma \quad \text{on} \quad \Sigma, \tag{2.6}$$

where $\underline{\tau}$ is the elastic surface stress vector and $\nabla_{(\Sigma)}\gamma$ is the surface gradient of γ. Thus for γ satisfying (1.3b),

$$\underline{\tau} = \Delta S \nabla_{(\Sigma)}\tilde{T} \quad \text{on} \quad \Sigma, \tag{2.7}$$

where $\nabla_{(\Sigma)}(\ ) = \underline{t}_1 \partial(\ )/\partial t_1 + \underline{t}_2 \partial(\ )/\partial t_2$ for the unit orthogonal tangent vectors $\underline{t}_1$ and $\underline{t}_2$ given in this instance by $\underline{t}_1 = (1, 0, \partial\tilde{\zeta}/\partial x)/[1 + \{\partial\tilde{\zeta}/\partial x\}^2]^{1/2}$ and

$$\underline{t}_2 = \frac{(-\{\partial\tilde{\zeta}/\partial x\}\{\partial\tilde{\zeta}/\partial y\},\ 1+\{\partial\tilde{\zeta}/\partial x\}^2,\ \partial\tilde{\zeta}/\partial y)}{[1+\{\partial\tilde{\zeta}/\partial x\}^2]^{1/2}[1+|\nabla_2\tilde{\zeta}|^2]^{1/2}}$$

In what follows we have adopted condition (2.7) while neglecting convection in the liquid phase and prescribed $\underline{\tau}$ so that it holds identically. Since such $\underline{\tau}$ is nonconstant it should be realized that such a procedure from the outset neglects the elastic properties of the solid as well. Also implicit in the constitutive relation of (2.2) is the assumption, which can be taken without loss of generality, that the tangential components of interface velocity are zero.

Returning to our formulation of the previous section we find that upon consolidation equations (1.3a,b) yield

$$\kappa_S \partial T_S/\partial n - \kappa_L \partial T_L/\partial n = \mathcal{L}\tilde{v}_n(1 - rT^*\tilde{\eta}) \quad \text{on} \quad \Sigma, \tag{2.8}$$

where $T^* = \tilde{T}/T_M$ and $r = \Delta S/\Delta S_f$ with $\Delta S_f = L/T_M \equiv$ entropy of fusion per unit volume. Then substituting the expression for γ given by (1.3b) into the Gibbs-Thomson equation of (1.4) and solving for T^* we obtain

$$T^* = 1 + \Gamma_M\tilde{\eta}/(1 + r\tilde{\eta}) \quad \text{on} \quad \Sigma, \tag{2.9}$$

where $\Gamma_M = \Gamma(T_M) = \gamma_M/L$. This is a modified Gibbs-Thomson equation differing from the traditional relation

$$T^* = \tilde{T}/T_M = 1 + \Gamma_M\tilde{\eta} \quad \text{on} \quad \Sigma, \tag{2.10}$$

used by those authors who assume $\gamma \equiv \gamma_M$. We note that since from (2.9)

$$T^* = 1 + \Gamma_M\tilde{\eta} + O(\tilde{\eta}^2) \quad \text{for} \quad |\tilde{\eta}| < 1/|r|, \tag{2.11}$$

then (2.8) implies that

$$\kappa_S \partial T_S/\partial n - \kappa_L \partial T_L/\partial n = \mathcal{L}\tilde{v}_n[1 - r\tilde{\eta} + O(\tilde{\eta}^2)] \quad \text{for} \quad |\tilde{\eta}| < 1/|r|. \tag{2.12}$$

For a linear analysis (2.11) would reduce to the ordinary Gibbs-Thomson equation of (2.10) while (2.12) reduces to

$$\kappa_S \partial T_S/\partial n - \kappa_L \partial T_L/\partial n = \mathcal{L}\tilde{v}_n(1 - r\tilde{\eta}) \quad \text{on} \quad \Sigma, \tag{2.13}$$

since terms of $O(\tilde{\eta}^2)$ are nonlinear. Hence in effect for such a linear analysis we could simply have taken $\gamma \equiv \gamma_M$ in equations (1.3) and (1.4) and thus deduced (2.13) and (2.10) directly which was the approach implicitly taken in [6].

Turning to the basic governing equations in the bulk phases from conservation of heat energy we have the usual diffusion equations

$$\partial T_L/\partial t = \nabla\cdot(\kappa_L \nabla T_L) \text{ for } f(x,y,z,t) > 0 \text{ (in the liquid)} \tag{2.14a}$$

and

$$\partial T_S/\partial t = \nabla\cdot(\kappa_S \nabla T_S) \text{ for } f(x,y,z,t) < 0 \text{ (in the solid).} \tag{2.14b}$$

In the prototype solidification problem presented in the next section we shall take $\kappa_L = \kappa_S \equiv \kappa$ and hence (2.14) reduces to

$$\partial T_{L,S}/\partial t = \kappa \nabla^2 T_{L,S} \, . \tag{2.15}$$

We also note that conservation of mass at the interface in the absence of convection would require the constant density of each phase to be equal and this would in turn necessitate that the constant specific heats be the same in both phases due to the assumption mentioned below (1.1).

III. THE GOVERNING NONDIMENSIONALIZED EQUATIONS FOR A PROTOTYPE PROBLEM

In order to exhibit the appropriate equations to be used for a nonlinear analysis, we consider the prototype solidification problem of the controlled two-dimensional growth of a pure solid metal into a thermally undercooled or supercooled liquid metallic bath of temperature $T_B < T_M$ under the assumptions introduced in the previous sections. The term *controlled* refers to the fact that the mean position of the interface between the liquid and solid phases is uniformly advanced into the pure melt at a constant specified rate, V. Using a moving nondimensional coordinate system (x, z) traveling with the mean position $(z = 0)$ of the interface; considering all independent and dependent variables in dimensionless form with κ/V, κ/V^2, $\mathcal{L}$, and δ as scale factors for distance, time, temperature, and deviation of the interface from its mean planar shape respectively; and employing the additional nomenclature given below:

$T(T') \equiv$ nondimensional temperature in the liquid (solid) metallic phase measured from a zero level corresponding to $T_B(\mathcal{L} + T_B)$, $\varepsilon = \delta V/\kappa$, $\Delta\theta = (T_M - T_B)/\mathcal{L}$, $\beta = \mathcal{L}/T_M$, $U = (\Gamma_M/\beta\kappa)\dot{V}$, $p = -\beta r/\Gamma_M$; the interface satisfies the equation $z = \varepsilon\zeta(x, t)$ and we have the following governing diffusion equations in two dimensions:

For $z > \varepsilon\zeta(x,t)$ (in the liquid): $\nabla^2 T + \partial T/\partial z - \partial T/\partial t = 0.$ (3.1)

For $z < \varepsilon\zeta(x,t)$ (in the solid): $\nabla^2 T' + \partial T'/\partial z - \partial T'/\partial t = 0.$ (3.2)

With boundary conditions:

For $z = \varepsilon\zeta(x,t)$ (at the interface): $T = T'+1 = \Delta\theta + U\eta/(1-pU\eta)$, (3.3a,b)

$$\partial T'/\partial\nu - \partial T/\partial\nu = [1+\varepsilon(\partial\zeta/\partial t)]\left\{1+pU[\beta(T-\Delta\theta)+1]\eta\right\}, \quad (3.3c)$$

where $\eta = \varepsilon(\partial^2\zeta/\partial x^2)/[1 + \varepsilon^2(\partial\zeta/\partial x)^2]^{3/2}$ and

$\partial(\ )/\partial\nu = \partial(\ )/\partial z - \varepsilon(\partial\zeta/\partial x)\partial(\ )/\partial x.$

As $|z| \to \infty$ (far from the interface): $T \to T_0(z)$ as $z \to \infty$,

$T' \to T_0'(z)$ as $z \to -\infty$,

where $T_0(z) = \Delta\theta e^{-z}$, $z > 0$; $T_0'(z) = (\Delta\theta - 1)e^{-z}$, $z < 0$.

In the above (3.3a) through (3.3c) are nondimensionalized versions of (1.2), (2.9), and (2.8) respectively, while $T_0(z)$ and $T_0'(z)$ represent the *planar interface solution* to the basic system (3.1)-(3.3) corresponding to the uniform growth of a planar interface $(\zeta = 0)$ of solid metal into the molten phase.

In an actual experiment the extent of the liquid and solid phases is naturally finite. A simplifying assumption in this model is that z extends to positive and negative infinity. We expect that far from the interface the influence of the shape of that interface on the temperature fields will become negligible; hence we adopt the far field conditions of (3.4). Observe that $T_0(z) \to 0$ (i.e., $T_L \to T_B$ in dimensional variables) while $|T_0'(z)| \to \infty$ as $|z| \to \infty$, the latter result being a consequence of the simplifying assumption that the phases are infinite in extent. In addition, x extends to positive and negative infinity and we also adopt the implicit requirement that the dependent variables remain bounded as $|x| \to \infty$.

Our basic equations contain three types of parameters which we shall designate by material, solidification, and perturbation respectively. Once a particular metal has been chosen the material parameters are established uniquely: for instance if tin is used $\mathcal{L} = 232\,^\circ K$ and $T_M = 504\,^\circ K$ hence $\beta = 29/63$. Γ_M, κ, ΔS_f, and ΔS are other such material parameters. T_B and V are the solidification parameters and for a particular metal $\Delta\theta$ and U represent nondimensionalized measures of the amount of supercooling and rate of solidification respectively. These are experimentally controllable quantities. The perturbation parameter is δ and for a given experimental situation involving a particular metal the nondimensional quantity ε is a measure of the maximum deviation of the interface from its planar position.

We have deferred until now a discussion of the initial conditions appropriate for our boundary value problem. Because of the infinite extent of the phases when performing a linear stability analysis of the planar interface solution to our basic equations by a normal mode technique, we do not obtain a complete set of eigenfunctions in the z direction from the point spectrum eigenvalues [6]. In order to accomplish this it would be necessary to use transform methods and obtain the continuous spectrum as well. There is no such problem in the x direction; hence it is possible to synthesize an arbitrary initial spatial shape for the interface. Since from the work of Sekerka [13] and Delves [14] it can be concluded that the most dangerous mode of linear theory lies in the point spectrum, most authors in investigating morphological stability of the interface have neglected intial conditions [1,2,4,6]. For a complete numerical solution of the problem, however, it would be necessary to impose initial conditions on the temperature fields in addition to those on the interface shape.

The mathematical model of (3.1)-(3.4) plus any such initial conditions can be classified as a *Stefan Problem* because it involves parabolic diffusion equations which must be satisfied in a region or regions whose boundaries are to be determined. There has been some conjecture about the desirability of using a coordinate system for such moving boundary problems which remains at rest with respect to the interface *itself* rather than just its *mean* position [15]. Introducing such a coordinate system (x,z) for our two-demensional problem, where

$$z_{\text{new}} = z_{\text{old}} - \varepsilon\zeta(x,t), \tag{3.5}$$

we would find the interface now satisfying the equation $z = 0$, while the governing diffusion equation for T would become

$$\nabla^2 T + \partial T/\partial z - \partial T/\partial t = \varepsilon[(\partial^2\zeta/\partial x^2 - \partial\zeta/\partial t)\partial/\partial z - 2(\partial\zeta/\partial x)\partial^2/\partial x\partial z]T - \varepsilon^2(\partial\zeta/\partial x)^2\partial^2 T/\partial z^2 \quad \text{for} \quad z > 0 \tag{3.6}$$

with an identical equation for T' valid when $z < 0$. The form of the boundary conditions would remain exactly as before except the interface ones would now be evaluated at $z = 0$ instead of at $z = \varepsilon\zeta$. Although this transformation has introduced nonlinear terms involving ζ into the originally linear diffusion equations it has fixed the position of the moving boundary at $z = 0$. The numerical solution of such nonlinear equations generally presents no additional serious difficulties not inherent in the original system and has the advantage that no special numerical techniques are necessary in the vicinity of the unknown boundaries as would be the case in the old space variable z [15]. For various analytical procedures involving nonlinear stability analyses, work currently in progress demonstrates that though care must be taken in interpreting the far field conditions as well as in determining the required adjoint linear eigenvalue problem because of the nonlinearities introduced in the diffusion equations, the cumbersome but usually necessary Taylor series expansions for the temperature functions at the boundary can now be avoided (see [4] for an explanation of the specialized techniques mentioned here).

In conclusion we would like to discuss briefly the dependence of our basic equations on the parameter p which is proportional to ΔS. The presence of such a parameter proportional to entropy in a solidification model may seem unusual at first but should not be surprising once one recalls that the Marangoni number characteristic of surface tension-driven convection has a similar dependence [16]. We observe that by setting $p = 0$ in our governing equations we would obtain the nondimensionalized versions of equations (1.1) and (2.10) appropriate for this situation, which are those generally used [1,2]. Hence by analyzing our model for the prototype problem with $p = 0$ and comparing this with a similar investigation for $p \neq 0$, we can examine the effect on a model which originally employs the traditional formulation of using our modified equations instead.

REFERENCES

[1] Sekerka, R. F. (1973). "Crystal Growth: An Introduction", Ed. P. Hartman, North-Holland, Amsterdam, 403.

[2] Delves, R. T. (1975). "Crystal Growth", Ed. B. R. Pamplin, Pergamon, Oxford, 40.

[3] Hurle, D. T. J., Jakeman, E., and Pike, E. R. (1968). *J. Crystal Growth 3/4*, 633.

[4] Wollkind, D. J., Segel, L. A. (1970). *Phil. Trans. Roy. Soc. London 268*, 351.

[5] Segel, L. A. (1977). "Mathematics Applied to Continuum Mechanics", MacMillan, New York.

[6] Wollkind, D. J., Maurer, R. N. (1977). *J. Crystal Growth 42*, 24.

[7] Langer, J. S. (1977). *Acta Met. 25*, 1121.

[8] Frenkel, J. (1955). "Kinetic Theory of Liquids", Dover, New York.

[9] Gibbs, J. W. (1928). "Collected Works, Vol. 1", Longmans, Green, and Co., New York.

[10] Tarshis, L. A. (1967). "Interface Morphology Consideration During Solidification", Thesis, Stanford University.

[11] Davies, J. T., and Rideal, E. K. (1961). "Interfacial Phenomena", Academic Press, New York.

[12] Batchelor, G. K. (1967). "An Introduction to Fluid Dynamics", Cambridge University Press, Cambridge.

[13] Sekerka, R. F. (1967). "Crystal Growth", Ed. H. S. Peiser, Pergamon, Oxford, 691.

[14] Delves, R. T. (1966). *Phys. Status Solidi 17*, 119.

[15] Ockendon, J. R., and Hodgkins, W. R., Eds. (1975). "Moving Boundary Problems in Heat Flow and Diffusion", Clarendon, Oxford.

[16] Scanlon, J. W., and Segel, L. A. (1967). *J. Fluid Mech. 30*, 149.

QUALITATIVE DYNAMICS FROM ASYMPTOTIC EXPANSIONS*

J. A. Murdock

Mathematics Department
Iowa State University
Ames, Iowa

In this report of ongoing joint work with R. Clark Robinson, we address the question: Given an asymptotic expansion of the general solution of a differential equation (calculated, say, to m terms in a small parameter), what can be said about the qualitative features of the exact solutions? It is general practice in applied mathematics to assume that the exact solution shows the same behavior as the approximation, at least when this is not "obviously unrealistic". We shall see, however, that it is necessary to be careful.

It is possible to pose our problem "locally" (example: given that the approximation has an asymptotically stable periodic solution, is the same true for the exact equation?) or "globally" (does the collection of all solutions in a certain region behave similarly for the exact and approximate systems?). Our local results are completed and will be presented first; then we discuss an approach to the global problem which we are developing.

In order to focus the ideas we shall consider a nonlinear oscillation problem

**This research was partially supported by the Science and Humanities Research Institute of Iowa State University.*

ISBN 0-12-434180-2

$$\dot{x} = F(x,t,\varepsilon) \tag{1}$$

where $x \in \mathbb{R}^n$, ε is a small parameter, and F is T-periodic in t. The *period map* $f_\varepsilon: \mathbb{R}^n \to \mathbb{R}^n$ is the diffeomorphism carrying $x(0)$ to $x(T)$ for every solution $x(t)$ of (1); note that f_0 is the identity. Generally only an asymptotic approximation to f_ε is available, by a method such as averaging or multiple scales, hence we assume

$$\begin{aligned} g_\varepsilon(x) &= x + \varepsilon g_1(x) + \ldots + \varepsilon^m g_m(x) \\ f_\varepsilon(x) &= g_\varepsilon(x) + \varepsilon^{m+1}\hat{f}_\varepsilon(x) \end{aligned} \tag{2}$$

where g_ε is known and $\hat{f}_\varepsilon$ is not. The following classical result, in which $m = 1$, is the basis for most regorous perturbation theory of nonlinear oscillators:

Theorem 1. Suppose (2) holds with $m = 1$, and suppose $g_1(x_0) = 0$ and $g_1'(x_0)$, the Jacobian matrix of g_1 at x_0, is nonsingular. Then there exists a unique fixed point $\bar{x}(\varepsilon)$ of f_ε satisfying $\bar{x}(\varepsilon) \to x_0$ as $\varepsilon \to 0$. If in addition $g_\varepsilon'(x_0)$ is hyperbolic (has no eigenvalues on the unit circle) for all small positive ε, then $f_\varepsilon'(\bar{x}(\varepsilon))$ is hyperbolic of the same type (i.e., has the same number of eigenvalues on either side of the unit circle as does $g_\varepsilon'(x_0)$) for sufficiently small positive ε. In particular if all eigenvalues of $g_\varepsilon'(x_0)$ lie inside the unit circle then the fixed point of f_ε is an attractor.

This theorem does not cover all cases which arise in practice. In my work on spin/orbit resonance in celestial mechanics ([1]) there is an example of $f_\varepsilon: \mathbb{R}^2 \to \mathbb{R}^2$ in which the first-order approximation $x + \varepsilon g_1(x)$ has a fixed point with $g_1'(x_0)$ nonsingular, but it is a center (both eigenvalues on the unit circle) surrounded by a nest of periodic solutions. This is a case in which it is "obviously unrealistic" to believe that f_ε must look the same, and in fact if we go to the second approximation $x + \varepsilon g_1(x) + \varepsilon^2 g_2(x)$ we find that the fixed point has become an attractor. It seems reasonable to guess that f_ε also has an attractor.

The following example shows that such a conclusion is not always warranted. Let f_ε be the linear map whose matrix is

$$\begin{bmatrix} 1 & 0 \\ 0 & 1 \end{bmatrix} + \varepsilon \begin{bmatrix} 0 & 0 \\ 1 & 0 \end{bmatrix} + \varepsilon^2 \begin{bmatrix} 1 & 0 \\ 0 & 1 \end{bmatrix} + \varepsilon^3 \begin{bmatrix} 0 & a^2 \\ 0 & 0 \end{bmatrix},$$

and let g_ε be the same map with the ε^3 term deleted. Then g_ε has a double eigenvalue $1 + \varepsilon^2$, so that the origin is a source, yet f_ε has eigenvalues $1 + (1 \pm a)\varepsilon^2$, which is a saddle if $a > 1$. The map g_ε is hyperbolic but is not "2-hyperbolic" in the sense of:

Definition. A continuous matrix function of ε, L_ε, with $L_0 = I$, is *k-hyperbolic* if for every matrix function N_ε continuous in an interval $0 \leq \varepsilon < \varepsilon_0$ there exists an interval $0 < \varepsilon < \varepsilon_1$ in which $L_\varepsilon + \varepsilon^{k+1} N_\varepsilon$ is hyperbolic.

The correct generalization of Theorem 1 for $k > 1$ may now be given.

Theorem 2. Let (2) hold and suppose $g_1(x_0) = 0$ and $g_1'(x_0)$ is nonsingular. Then (by Theorem 1) there exist unique fixed points $x^*(\varepsilon)$ of g_ε and $\bar{x}(\varepsilon)$ of f_ε tending to x_0 as $\varepsilon \to 0$. Suppose that $g_\varepsilon'(x^*(\varepsilon)) = L_\varepsilon + O(\varepsilon^{k+1})$ with $k \leq m$, and assume L_ε is k-hyperbolic. Then $\bar{x}(\varepsilon)$ is a hyperbolic fixed point of f_ε of the same type.

The hypotheses of Theorem 2 are somewhat difficult to verify, although in the spin/orbit resonance example in [1] this is essentially what was done. The following theorem gives an easy test for k-hyperbolicity.

Theorem 3. If $L_\varepsilon = I + \varepsilon L_1 + \ldots + \varepsilon^k L_k$ and L_1 has distinct eigenvalues, and if furthermore the eigenvalues $\lambda_i(\varepsilon)$ of L_ε are smooth (C^{k+1}) functions of ε each of which satisfies either $|\lambda_i(\varepsilon)| > 1 + C\varepsilon^k$ or $|\lambda_i(\varepsilon)| < 1 - C\varepsilon^k$ for some constant $C > 0$, then L_ε is k-hyperbolic.

This is a corollary of a more general result obtained by Clark Robinson. In the setting of Theorem 2, $L_1 = g_1'(x_0)$, the same matrix which we have assumed is nonsingular.

Turning to the global problem suppose that in a certain compact region $K \subset R^n$ whose boundary is a smooth manifold, we have located every fixed point and periodic point (fixed point of an iterate) of g_ε. Suppose further that g_ε is a Morse-Smale system in K, meaning that (a) each fixed or periodic point of g_ε is hyperbolic; (b) for each $x \in K$, $g_\varepsilon^i(s)$ either leaves K or tends to one of the fixed or periodic orbits as $i \to +\infty$ and as $i \to -\infty$; (c) stable and unstable manifolds of the fixed and periodic points intersect transversally (when they intersect at all). We ask for conditions under which this structure carries over to f_ε in some appropriate sense.

First of all, the local theory gives conditions under which each of the fixed and periodic points of g_ε carries over to f_ε. Next one seeks to show that f_ε does not possess any additional such points or, more generally, any nonwandering points (in the sense of [3]) not possessed by g_2. (Is f_ε "Ω-conjugate" to g_ε?). The next question is whether the stable and unstable manifolds for f_ε intersect in the same way as they do for g_ε, i.e. whether f_ε is "diagram-equivalent" to g_ε. Finally, one may ask if f_ε and g_ε are "topologically conjugate". Clark Robinson, motivated by this problem, has given ([2]) a definition of topological conjugacy suitable for manifolds - with - boundary such as K, and has proved a structural stability theorem which is the first step toward a solution of our problem.

To illustrate one of the approaches we are taking, suppose that $n = 2$ and that g_ε has a saddle whose unstable manifold (on one side) falls into a certain sink. Suppose that from local theory f_ε has a corresponding saddle and sink, and we want to show that they connect in the same way. There is a point near the saddle which is mapped, after many iterates of g_ε, to a point near the sink. We would like to assert the same for f_ε, and after all f_ε is arbitrarily near g_ε for ε small. However as $\varepsilon \to 0$ the number of iterates required approaches infinity, since g_ε approaches the identity. Hence it is necessary to have g_ε close to f_ε "on expanding time intervals", i.e. to have control over $\|g_\varepsilon^i(x) - f_\varepsilon^i(x)\|$ for $0 \leq i < c/\varepsilon^j$ for some suitable j. Fortunately such asymptotic estimates are available from the method of averaging, and we are hopeful of being able to give conditions for diagram equivalence or even conjugacy. In special cases a great deal of help can be obtained by using Lyapunov functions in combination with the methods we have discussed.

REFERENCES

[1] Murdock, J. A. "Some Mathematical Aspects of Spin/Orbit Resonance", to appear in Celestial Mechanics.

[2] Robinson, R. C. ( ). "Structural Stability on Manifolds with Boundary", preprint from Northwestern University.

[3] Smale, S. (1967). "Differentiable Dynamical Systems", *Bull. A.M.S.* *73*, 747-817.

A SECOND STAGE EDDY-VISCOSITY MODEL FOR TURBULENT FLUID FLOWS: OR, A UNIVERSAL STATISTICAL TOOL?

*Fred R. Payne**

Aerospace Engineering
The University of Texas at Arlington
Arlington, Texas

I. INTRODUCTION

In the past decade much empiric and modelling effort has addressed the "large eddy" component of turbulent flows; in direct contrast to more traditional, statistical models of the strongly non-linear processes of mixing, transport and dissipation in turbulence, the concept of (quasi-) coherent structures has become rather fadish. This recent development in the eternal process of attempting finite closure for the denumerable hierarchy of dynamical equations is partly due to recent "conditional sampling" techniques of flow field measurements. These techniques permit an isolation of narrow-band structures from the rest of the turbulence spectra which is continuous rather than discrete.

Some results are given of Lumley's Proper Orthogonalization Decomposition Theorem (PODT)[1] as developed by Payne[2,3] into a Structural Analysis System (SAS) and applied to the flat-plate

**Supported, in part, by NASA/Ames Grant NSG-2077, Dr. M. W. Rubesin, Technical Monitor).*

ISBN 0-12-434180-2

boundary-layer by Lemmerman[4]. Succeeding paper[6] summarizes new calculations[5] using Lemmerman's results to construct a 3-D velocity of the "large-eddy," i.e., dominant eigen-functions of the two-point velocity co-variance as measured by hot-wire anemometer.

Chuang[5] has calculated the Reynolds' stress contribution, $\underline{\underline{B}}$, of the biggest eddies and a "small-eddy viscosity," ν_{se} defined by

$$\overline{u_i(\underline{x})u_j(\underline{x}')} = B_{ij}(\underline{x},\underline{x}') + \nu_{se}\,\frac{\partial U_i}{\partial x_j}$$

which is the second-stage eddy-viscosity model alluded to in title and its LHS is the two-Point Reynolds' stress tensor, the activator and sustainer of turbulence via "feeding" upon mean flow kinetic energy and conversion of same into fluctuating kinetic energy. Speculation is made as to probable applicability of method to any set of statistical data, discrete or continuous.

This paper is a status report on a Structural Analysis System developed by the author[2,3] and students[4,5] over a period of years, and based upon Lumley's mathematical definition[1] of "Large Eddy" as interpreted in turbulent flow. The term "Large Eddy" has been used by turbulence workers for more than half a century but until Lumley's Proper Orthogonal Decomposition Theorem[1] (PODT) there existed no rational definition of these large scale structures which occur in a statistical ensemble. Hence, analysts[7-11] were forced to play a guessing game in their interpretation of experimental data and its revelation or concealment of large scale structures. The methodology described, in the author's opinion, is applicable to any set of statistical data, continuous or discrete, no matter the size of the data base. The method can be applied subject only to the usual restrictions of ingenuity of the applier and availability of good precise data with minimal inherent error.

A brief summary of the historical development of PODT-SAS (Proper Orthogonal Decomposition Theorem[1] and an associated Structural Analysis System[2-4]) is followed by the extension of the methodology into other fields, namely scalar-valued covariances and higher-order tensor covariances. A brief conjectural part simply lists probable fields of applicability of the SAS methodology. Mention is made of further extension of PODT-SAS by Chuang[5] who presents more details of his work in the immediately following joint paper of a similar title[6]. Paper concludes with an appeal for applied mathematicians to consider the PODT-SAS methodology as a possible candidate for any case in which the researcher seeks to isolate or identify a possible structure or structures in any statistical data set. One can consider PODT-SAS as a methodology filter or a synthesizer; in any case, the method provides an alternative approach to analysis of, specifically, turbulent velocity convariance data and is complimentary in the sense of providing a more rational approach to the problem of identifying narrow band quasi-coherent structures in turbulence.[7,8]

As indicated above such structures, particularly the large scale structures in turbulent shear flows, have been qualitatively identified, discussed and speculated upon by a number of workers from the time of Richardson[9].

Big whirls have little whirls,
That feed upon their velocity;
Little whirls have lesser whirls,
And so on to viscosity,
In the molecular sense. --Richardson, circa 1915.

II. PODT-SAS HISTORY

In the early 1960's John Lumley[1] addressed the problem of non-specificity and non-rationality of the "Large Eddies" in turbulence which has been variously interpreted as Fourier components, vortex rings, horseshoe vortices, and various other structures[10]. As indicated in Payne's dissertation[2], Lumley was led to look more deeply into the problem of large eddy identification and isolation from experimental data by the two diametrically opposed structures inferred from experimental data by Townsend in his monograph[10] and by Grant[11].

As a consequence, Grant's induced "large eddies" were a vortex pair counter-rotating with planes of circulation approximately parallel to the center-line of a two-dimensional wake combined with a "re-entrant jet", whereas Townsend's eddies[10] had their orientation approximately 60 degrees to those of Grant[11]. Lumley applied Loeve's[12] Harmonic Orthogonal Decomposition Theorem (HODP) to a vector-valued stochastic process, namely, turbulence.

Lumley's arguments are basically as follows: (See Box 1.) Given a random vector field $\underline{u}$ as a function of three-space and time, one wishes to extract from that field some information, namely, a structure of some sort. Hence, one is led, in a Hilbert space, to select a deterministic, normalized candidate, ϕ, and apply a criterion to this candidate to select an optimal ϕ. With this candidate and the random vector field, $\underline{u}$, Lumley formed the inner product, that is, projected the deterministic candidate upon the random field and summed up the contributions via an integral over the three-space and time variables upon which the random vector field $\underline{u}$ is functionally dependent. This leads us to equation (1) where I is defined to be the integral of the innerproduct of $\underline{u}$ and ϕ over the entire space. Since $\underline{u}$ is statistical, I can be positive or negative; hence, to apply some extremum principle one forms the mean square of I. Although the random vector $\underline{u}$ in the case of turbulence will be a real field, the generalization of $\underline{u}$ being complex is of no fundamental difficulty. ϕ may also be complex, complex in the sense of a phase relationship between ϕ and $\underline{u}$, the three components of the stochastic field. One defines λ as the mean square of

PODT-SAS: (Proper Orthogonal Decomposition Theorem - Structural Analysis System)

(A UNIVERSAL STATISTICAL TOOL?)

- PODT (Lumley, 1965)

1. Given: A random vector field, $u(\underline{x},t)$
2. Select: A deterministic, normalized candidate, ϕ, in Hilbert space, H: apply some criteria, e.g. inner product to test "parallelism":

$$I = \int_\Omega \underline{u} \cdot \phi \, d\underline{x} \, dt \qquad -(1)$$

3. Since $\underline{U}$ is statistical, form the mean square:

$$\lambda \equiv \overline{II^*} = \overline{\int_\Omega \underline{u} \cdot \underline{\phi} \int_\Omega (\underline{u} \cdot \underline{\phi})^*} \qquad -(2)$$

4. Extermize, $\frac{\delta\lambda}{\delta\phi} = 0$,

$$\therefore \int_\Omega R_{ik}\phi_k = \lambda\phi_i$$

where $R_{ik} = \overline{u_i u_k}$ is co-variance

SAS (Payne, 1966)

1. Ditto except can be tensor of arbitrary order, v
2. Ditto except candidate is χ, $(\chi,\chi) = 1$

$$I = (v,\chi) \qquad -(4)$$

3. Ditto:

$$\Lambda \equiv |\overline{(v,\chi)}| \qquad -(5)$$

4. Ditto: Except $\frac{\delta\Lambda}{\delta\chi} = 0$, $=>$

$$(R,x) = \Lambda\chi \qquad -(6)$$

when R = any order co-variance, e.g.

$$R = \overline{v \, v' \, v''} \qquad -(7)$$

I times it complex conjugate, I^* by equation (2) where the overbar denotes an ensemble or time average. Then the solution becomes apparent; one extremizes λ with respect to the candidate ϕ, and, by standard calculus of variations means, obtains a maximum or a minimum or an inflection point. By the nature of the dot product the minimum occurs when ϕ is orthogonal to $\underline{u}$ and λ itself would vanish. This trivial case is of no interest. Physically one expects a maximum rather than merely an inflection point. One can, by standard methods, show that the second variation of λ would in this case be positive. This approach then leads from equation (2) to equation (3) which is a classical eigen-value problem of the integral type where R_{ik} is the velocity covariance; that is, the random vector field at a point $(\underline{x},t)$ multiplied by its value at a different point $(\underline{x}',t')$ yields R_{ik} when averaged.

Equation (3) is Lumley's eigen-value problem. At the time of Lumley's paper[1] (1965), as presented in Moscow, he had solved the case for isotropic flow in which, due to the full homogeneity in all three space directions, the solutions are the circular functions, i.e., harmonic functions. This reduced PODT simply to HODT (Harmonic Orthogonal Decomposition Theorem).

The extension for the random vector field to be either a scalar or tensor of arbitrary order V is trival (See Box 1). The deterministic candidate, χ, could be other than a vector which is of no particular difficulty. I can be defined as before (eq. 4) as the inner product of V with χ, the deterministic candidate. Again, since I is statistical, one would form, say, Λ as in equation (5) as the mean square of I. Extermization via calculus of variations proceeds precisely as before except that one takes the first variation of Λ with respect to χ, equates to zero, and gets a generalized eigenvalue problem in the form of R, a covariance of any order inner producted with χ equal to $\Lambda\,\chi$, as shown in equation (6). Here R can be an arbitrary order covariance as in equation (7).

A. *Consequences of Statistical Homogeneity in Space (Stationarity in Time)*

Statistical homogeneity in one direction in space is mathematically equivalent to stationarity in time, a scalar variable. Assume that the random vector field $\underline{u}$ is statistically homogeneous in all space variables in some arbitrary but finite vector space of dimension N. Then

$$\underline{\underline{R}}(\underline{x},t;\underline{x}',t') \rightarrow \underline{\underline{R}}(\underline{r}, \ ); \quad \underline{r} = \underline{x}' - \underline{x}, \quad \tau = t' - t \qquad -(8)$$

which states that the covariance R at a space point $\underline{x}$ and time t with a different point $(\underline{x}',t')$ is a function of $(2N + 2)$ variables is reduced to a function of N space separation variables $\underline{r}$ and a single scalar variable time. Hence, the number of independent variables is reduced by half. Lumley also shows[1] that in this particular case (eq. 8) his PODT reduces to HODT (Harmonic Orthogonal Decomposition Theorem) where the solutions of Lumley's eigenvalue problem (eq. 3) or, equivalently, (7) are simply the circular functions as shown in equation (9):

$$\phi = \exp[i(\underline{k} \cdot \underline{r}) + \omega t] \qquad -(9)$$

where $\underline{k}$ is a wavenumber vector and ω is the frequence. Hence the harmonic functions are the eigen-functions of R if one has complete homogeneity in all space variables and stationary in the time variable. One then has the full power of usual probability theory, that is, that the random vector $\underline{u}$ is merely the Fourier transform of the deterministic candidate ϕ. So one can use the Weiner-Khintohine transform-pair theorem[12]. If on the other hand, partial homogeneity obtains, as in all applications to date of Lumley's methodology (namely, Payne[2] for the two dimensional wake, Lemmerman[4] for two dimensional flatplate boundary layer and Reed[13] in the quasi-two dimensional round jet) the R covariance becomes partially reduced in its complexity of dependence upon the space parameters $\underline{x}, \underline{x}'$. For example, if $\underline{x}, \underline{x}'$ are three dimensional vectors, an ordinary, physical three space, the full time averaged

R would be a function of 6 variables; partial homogeneity in two directions would reduce this to a function of a three component separation vector, $\underline{r}$ at a single space variable y which measures distance from the wall in the case of flat-plate boundary layer or distance from the centerline plane for a two dimensional wake or distance from the axis in the case of a round jet as is shown in equation (10):

$$\underline{\underline{R}}(\underline{x},\underline{x}') \rightarrow \underline{\underline{R}}(\underline{r};\underline{y}); \quad \underline{r} = \underline{x}' - \underline{x}$$

and

$$\phi(\underline{r}) \rightarrow \psi(y,\underline{k})\exp[i\underline{k} \cdot \underline{r}], \quad \underline{k} = (k_1,0,k_3) \qquad -(10)$$

where y is the distance from the wall, wake center plane or the jet axis, respectively. A further consequence of Lumley's theorem[1] is that in the nonhomogeneous direction namely, y, the spectrum of eigen-values is no longer continuous as is in the case of full homogeneity but is discrete and there exists a countable number of eigen-values as shown in equation (11).

$$\lambda = \lambda^{(n)}(k), \quad n = 1,2,\ldots \;. \qquad -(11)$$

λ becomes a $\lambda^{(n)}$ which is a function then of the wave number, in this particular case the wave number vector in the $(1,3)$ plane.

Lumley's PODT results are summarized in Box 2. The random vector field $\underline{u}$ may be expanded into a denumerable sum of eigen-functions as shown in equation (12) where the $\phi^{(n)}$ are deterministic eigen-functions arising from solutions of Lumley's eigen-value problem equation (3) or as simplied in equation (10). All the statistics in the random vector field $\underline{u}$ occur in the a_n random coefficients which are statistically orthogonal and uncorrelated as in equation (13). The sum of the eigen-values is finite although the eigen-values are discrete and denumerable, equation (14). All eigen-values are real and positive and can be ordered relative to zero, equation (14). The covariance itself

LUMLEY'S PODT, Summary:

$u \sim \sum_n a_n \phi^{(n)}$, a_n	*Random Coefficients*	--(12)
$\overline{a_n a_m} = \lambda^{(n)} \delta_{nm}$	*Statistically uncorrelated*	-(13)
$\sum \lambda^{(n)} < \infty$, $\lambda^{(1)} \geq \lambda^{(2)} \geq \ldots \lambda^{(n)} \ldots \geq 0$		-(14)
$R = \sum_1^N \lambda^{(n)} \phi^{(n)} \phi^{*(n)}$	*is optimal* $\forall\ 1 < N < \infty$	-(15)

may be expanded into a partial sum as shown in equation (15) where this decomposition is optimal in the sense that truncation of this series of any order N retains a maximum amount of information in the number of terms N. Schematically, one can summarize the methodology as shown in Box 3 which simply states that if one has available a set of theorectical or experimental stochastics such that one can define a covariance (according to Lumley of second order) in the velocity field $\underline{u}$ with three spatial components) then this information can be processed via a "black box" called PODT to extract eigen-values and eigen-vectors of the covariance R.

B. *Physical Interpretation of PODT-SAS*

1. PODT-SAS is a multi-dimensional filter which isolates any dominant structure(s) of the averaged stochastic process.

2. PODT is a generalized "Fourier" analysis which succeeds where the usual Fourier or Fourier-Stieltjes transform fails.

3. PODT-SAS extracts the largest amplitude eigen-values/vectors relative to background "noise".

4. For turbulence analysis:

(a) λ = mean square kinetic energy of "large eddy" components[1,2,4,14]

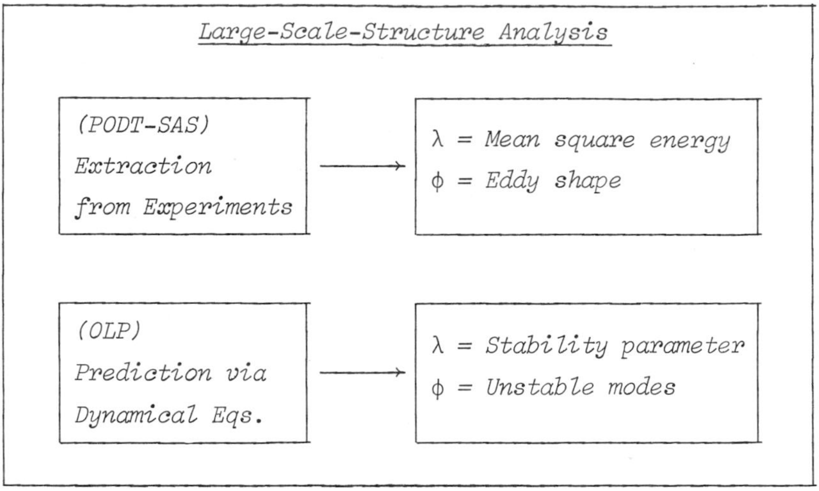

(b) ϕ = shape of normalized "eddy"[1,2]

(c) R can be expanded in a series of ϕ which is optimal for any truncation order[1].

(d) PODT defines in a rational way the experimental large-scale structure[1,2,4].

(e) There exists a predictive scheme (OLP)[15-18] for comparison[17] to PODT results[2,4,6] which should provide improved turbulence modeling[19,20]. OLP[16] is another variational method[15] which yields an eigen-value problem but of slightly different physical interpretation.

III. LIST OF CONJECTURED FIELDS OF APPLICABILITY OF PODT-SAS

A. Stock market price history (non-stationary, moving averages)

B. Demographics (trend analysis?)

C. Economics (macro vs. micro)

D. Bio-statistics

E. Ecological systems

F. Planetary systems

G. Ω - ? (A big system!)

Note that one needs *only* a definable co-variance in order to apply PODT-SAS [eq. (3) or eq. (10)].

IV. A SECOND-STAGE EDDY-VISCOSITY MODEL FOR TURBULENTS: (Payne 1968-77, Chuang[5] 1978, Chuang and Payne[6])

First: Extract λ, ϕ via PODT-SAS[2,4]

Second: Construct the "Big Eddy" co-variance, $\underline{\underline{B}}$:[3,5]

$$\underline{\underline{B}}^{(N)} = \sum^{N} \lambda^{(n)} \phi^{(n)} \phi^{*(n)}$$

Third: Form the difference of measured $\underline{\underline{R}}$ and $\underline{\underline{B}}^{(N)}$[3,5] to define ν_{se}:

$$\underline{\underline{R}} - \underline{\underline{B}}^{(N)} \equiv \nu_{se} \underline{\underline{D}}$$

where ν_{se} is the second-stage eddy-viscosity model for the "rest of the turbulence", i.e., "small eddies" and $D = \nabla \underline{U}$, known, $\underline{U}$ mean velocity

Fourth: ν_{se} is now calculable (see joint paper by Chuang and Payne[6] immediately following this one).

Fifth: The pay-off - put λ, ϕ, and ν_{se} into Navier-Stokes and *calculate* turbulence (in progress at the University of Texas at Arlington).

V. CLOSURE

A. *PODT-SAS*

1. Will extract the structure in any statistical data set.

2. Is *independent* of any dynamical model of the stochastic process under study - it is a *structural* analysis system.

B. *OLP*

1. Is predictive from "first principles" (for assumed dynamical model, i.e., Newtonian fluid behaviro at an instant).

2. Is *not* restricted to linearity assumptions.

3. Does *not* assume a form of disturbance.

4. Is a *global* criterion for stability (of Newtonian flow).

Conjecture: The usual (Heisenberg-Tollmein-Schlichting-Lin[21,22], et al) linearized stability analysis of (assumed) parallel shear flows leads to the linear Orr-Sommerfeld equation[16,21,22] and a *local* criterion (P.D.E.) and an *upper bound* on stability. However, OLP's *global* criterion leads to a *lower bound.*

C. *Extensions*

1. *PODT-SAS*: There are *no* inherent limits.

2. *OLP*: Since "dynamic" but global, the integral (or ODE) approach is far easier to implement but is model dependent.

Note: Both methods are (integral) eigen-value problems?

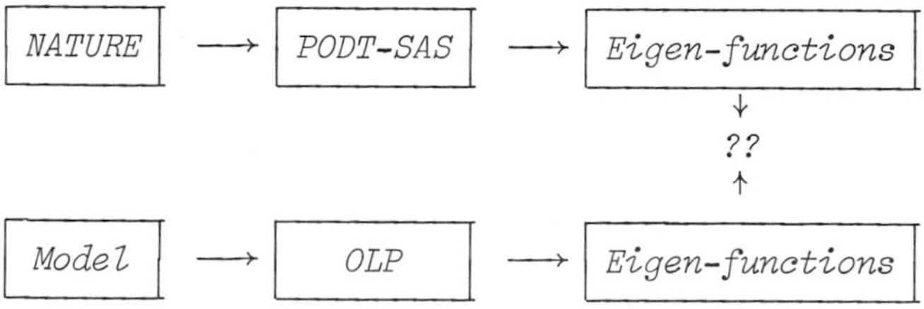

REFERENCES

[1] Lumley, J. L. (1966). *Dok. Akad. Nauk SSSR,* Moscow, published in 1967.

[2] Payne, F. R. (1966). Ph.D. Dissertation, Penn State University.

[3] Payne, F. R. (1977). Siam Fall/1977 Meeting, Albuquerque.

[4] Lemmerman, L. A. (1976). Ph.D. Dissertation, The University of Texas at Arlington and AIAA Paper No. 77-717, 10th Fluid and Plasma Conference, Albuquerque.

[5] Chuang, S. L. (1978). MSAE Thesis, The University of Texas at Arlington.

[6] Chuang, S. L., and Payne, F. R. (1978). "A Second Stage Eddy-Viscosity Calculation for the Flat Plate Turbulent Boundary-Layer", *Proc. Applied Nonlinear Analysis Conf., RCAS/UTA,* April, 1978 (in this volume as immediately succeeding paper).

[7] Willmarth, W. W. (1975). "Structure of Turbulence in Boundary Layers", Advances in Applied Mechanics, *15,* 159-254.

[8] Blackwelder, R. F., and Kaplan, R. E. (1972). *Int. Union Theor. Applied, Mech., 12th.*

[9] Monin, A. S., and Yagloma, A. M. (1975). "Statistical Fluid Mechanics, Vol. 1 and 2", Massachusetts Institute of Technology.

[10] Townsend, A. A. (1956). "The Structure of Turbulent Shear Flows", Cambridge Press and 2nd Edition, 1976.

[11] Grant, H. L. (1958). *J. Fluid Mech. 4,* 149.

[12] Loeve, M. (1955). "Probability Theory", 2nd Ed., Van Nostrand Co., Princeton.

[13] Reed, X. B., et al (1977). *Proc. 1st Int. Symp. on Turb. Shear Flows, Vol. I,* 2.33.

[14] Lemmerman, L. A., and Payne, F. R. (1977). *AIAA Paper No. 77-717,* Albuquerque, June.

[15] Lumley, J. L. (1965). Internal Memo, ORL/Penn State, August.

[16] Payne, F. R. (1977). "The OLP Method of Non-Linear Stability Analysis of Turbulence in Newtonian Fluids", Proc. Nonlinear Equations in Abstract Spaces, The University of Texas at Arlington, June, Academic Press in 1978.

[17] Payne, F. R. (1968). "Predicted Large-Eddy Structure of a Turbulent Wake", contractor report to *USN/ONR* (Fluid Mechanics Branch Contract *NONR 656 (33)*.)

[18] Hong, S. K. MSAE Thesis, The University of Texas at Arlington (in progress).

[19] Payne, F. R. (1977). Fall/1977 SIAM Meeting, Albuquerque.

[20] Payne, F. R. (1977). "Future Computer Requirement for Computation Aerodynamics", *NASA CP-2032*, 260-266.

[21] Betchov, R., and Criminale, W. O. (1967). "Stability of Parallel Flows", Academic Press, N.Y.

[22] Lin, C. C. (1955). "The Theory of Hydrodynamic Stability", Cambridge Univ. Press, London.

FIXED POINT ITERATIONS USING INFINITE MATRICES

K. L. Singh

Department of Mathematics
Texas A&M University
College Station, Texas

Let X be a normed linear space and C be a nonempty, closed, bounded and convex subset of X. Let $T: C \to C$ be a mapping with at least one fixed point. Let A be an infinite matrix. Given the iteration scheme

$$\bar{x}_0 = x_0 \quad \text{in } C, \tag{1}$$

$$\overline{x_{n+1}} = Tx_n, \quad n = 0,1,2,\ldots, \tag{2}$$

$$x_n = \sum_{k=0}^{n} a_{nk}\bar{x}_k, \quad n = 1,2,3,\ldots, \tag{3}$$

it is natural to ask what restriction on the matrix A are necessary and/or sufficient to guarantee that the above iteration scheme converges to a fixed point of T.

Recently several mathematicians, using iteration schemes (1) - (3) have obtained results for certain class of infinite matrices. In this paper we establish the generalizations of several of these results.

An infinite matrix A is called *regular* if it is limit preserving over c, the space of convergent sequences; i.e., if x in c, $x_n \to y$ then $(Ax)_n = \sum_{k=1}^{\infty} a_{nk}x_k \to y$. A matrix A is

ISBN 0-12-434180-2

called *triangular* if $a_{nk} = 0$ for $k > n$. A is called a *triangle* if $a_{nk} = 0$ for $k > n$, $a_{nn} \neq 0$ for all n. We shall be concerned to regular triangular matrices A satisfying

$$0 \leq a_{nk} \leq 1, \quad k = 0,1,2,\ldots, \tag{4}$$

$$\sum_{k=0}^{n} a_{nk} = 1, \quad n = 0,1,2,\ldots . \tag{5}$$

Conditions (4) and (5) are obviously necessary to ensure that x_n and $\bar{x}_n$ in (2) and (3) remain in C. The scheme (1)-(3) is known as *Mann Process*.

Definition 1. A *weighted mean matrix* is a regular triangular matrix $A = (a_{nk})$ defined by $a_{nk} = p_k/P_n$, where the sequence $\{p_n\}$ satisfies $p_0 > 0$, $p_n \geq 0$ for $n \geq 0$, $P_n = \sum_{k=0}^{n} p_k$ and $P_n \to \infty$ as $n \to \infty$.

Following J. Reinermann [13] we define summability matrix A by

$$a_{nk} = \begin{cases} d_k \prod_{J=k+1}^{n} (1-d_j) & k < n \\ d_n & k = n \\ 0 & k > n, \end{cases} \tag{I}$$

where the real sequences $\{d_n\}$ satisfies (i) $d_0 = 1$, (ii) $0 < d_n \leq 1$ for $n \geq 1$ and (iii) $\sum_k d_k$ diverges.

Remark 1. For A defined by (I) we will also use condition (iv) $\sum_k d_k(1-d_k)$ diverges. In fact condition (iv) on $\{d_n\}$ implies condition (iii). It can be easily seen that A is regular and satisfies conditions (4) and (5). The matrix of (I) with $\{d_n\}$ satisfying (i)-(iii) is a regular weighted mean matrix ([11], pp. 163-164).

Let X be a Banach space and C be a nonempty, convex

subset of X. Given an initial value x in C, we consider the iteration scheme $x_{n+1} = \sum_{k=0}^{n} a_{nk} Tx_k$, which can be written as

$$x_{n+1} = (1-d_n)x_n + d_n Tx_n. \qquad \text{(II)}$$

Let us remark that even though the matrices involved are the same, the iteration schemes (1)-(3) and (II) are different. Scheme (1)-(3) takes the form $x = Az$, where $z = \{x_0, Tx_0, Tx_1, \ldots\}$; where as (II) becomes $x = Aw$, where $w = \{Tx_0, Tx_1, Tx_2, \ldots\}$. In other words the first scheme uses a translate of w.

Definition 1. Let X be a normed linear space and C be a non-empty subset of X. Let α, β, γ be nonnegative real numbers satisfying $\alpha < 1$, $\beta, \gamma < 1/2$. We shall say that $T: C \to C$ satisfies *condition (z)* if, for each pair of points x, y in C, at least one of the following conditions is satisfied:

(a) $\|Tx - Ty\| \leq \alpha\|x - y\|$,

(b) $\|Tx - Ty\| \leq \beta[\|x - Tx\| + \|y - Ty\|]$,

(c) $\|Tx - Ty\| \leq \gamma[\|x - Ty\| + \|y - Tx\|]$.

Definition 2. A mapping $T: C \to C$ is said to satisfy *condition (L)* if for all x, y in C and $0 < k < 1$ we have

$$\|Tx-Ty\| \leq k \max\{\|x-y\|, [\|x-Tx\|+\|y-Ty\|]/2, \|x-Ty\|, \|y-Tx\|\}.$$

Defintion 3. A mapping $T: C \to C$ is called a *generalized contraction* if $\|Tx-Ty\| \leq k \max\{\|x-y\|, \|x-Tx\|, \|y-Ty\|, \|x-Ty\|, \|y-Tx\|\}$ for all x, y in C and $0 < k < 1$.

Definition 4. A mapping $T: C \to C$ is said to satisfy *condition (K)* if $\|Tx-Ty\| < \max\{\|x-y\|, k\|x-Tx\|, \|y-Ty\|, \|y-Tx\|, \|x-Ty\|\}$ for all $x, y \in X$, $x \neq y$ and $0 < k < 1$.

Definition 5. A mapping $T: C \to C$ is said to be *generalized contractive* if for each $x, y \in X$, $x \neq y$,

$$\|Tx-Ty\| < \max\{\|x-y\|, \|x-Tx\|, \|y-Ty\|, \|x-Ty\|, \|y-Tx\|\}.$$

Definition 6. Let $\alpha_i(t): (0,\infty) \to [0,1]$ be a monotone decreasing function satisfying $\sum_{i=1}^{5} \alpha_i(t) \leq 1$ for all $t > 0$. Let T be an operator mapping the Banach space X into itself. We say that T satisfies *condition (β)* if for each x, y in X, $x \neq y$, $t = \|x-y\|$ we have

$$\|Tx-Ty\| \leq \alpha_1(t)\|x-y\| + \alpha_2(t)\|x-Tx\| + \alpha_3(t)\|y-Ty\| + \alpha_4(t)\|x-Ty\| + \alpha_5(t)\|y-Tx\|.$$

Because of symmetry of x and y above implies

$$\|Tx-Ty\| \leq a(t)\|x-y\| + b(t)[\|x-Tx\|+\|y-Ty\|] + c(t)[\|x-Ty\|+\|y-Tx\|],$$

where $a = \alpha_1$, $2b = \alpha_2 + \alpha_3$ and $2c = \alpha_4 + \alpha_5$.

Remark 2. It is clear that any mapping satisfying *condition (z)* or *condition (β)* also satisfies *condition (L)*. The following example shows that a mapping satisfying *condition (L)* need not satisfy either *condition (z)* or *condition (β)*.

Example 1. Let

$$M_1 = \{m/n: \ m = 0,1,3,9,\ldots; \ n = 1,4,\ldots,3k+1,\ldots\},$$

$$M_2 = \{m/n: \ m = 1,3,9,\ldots; \ n = 2,5,\ldots,3k+2,\ldots\},$$

and let $M = M_1 \cup M_2$ with the usual metric. Define the mapping $T: M \to M$ by

$$T(x) = \begin{cases} 4x/5 & \text{for } x \text{ in } M_1 \\ x/3 & \text{for } x \text{ in } M_2. \end{cases}$$

The mapping T satisfies *condition (L)*. To see that T does not satisfy *condition (z)* and *(β)*, take $x = 1$ and $y = 1/2$.

Remark 3. It is clear that any generalized contraction mapping also satisfies *condition (K)* and hence is generalized contractive.

The following example shows that a mapping satisfying *condition (K)* and a generalized contractive mapping need not be a generalized contraction.

Example 2. Let $T(x) = 0,\ 0 \le x \le 1/2,\ T(x) = 1/2,\ 1/2 < x \le 1.$ Then T satisfies *condition (K)* and hence is genalized contractive. However T is not a generalized contraction.

Definition 7. Let X be a normed linear space and C be a nonempty subset of X. Suppose for each x, y in X, there exist nonnegative numbers $q(x,y)$, $s(x,y)$, $r(x,y)$ and $t(x,y)$ such that $\sup_{x,y\in X}\{q(x,y)+r(x,y)+s(x,y)+2t(x,y)\} = \lambda < 1$. Let $T\colon C \to C$ be a mapping. We say that T satisfies *condition* (α) if

$$\|Tx-Ty\| \le q(x,y)\|x-y\| + r(x,y)\|x-Tx\| + s(x,y)\|y-Ty\| + t(x,y)[\|x-Ty\| + \|y-Tx\|]$$

for all x, y in C.

Remark 4. Clearly any mapping satisfying *condition* (α) also is a generalized contractive mapping, but the converse is not true, as follows from example 1.

Lemma 1 [5]. Let X be a uniformly convex Banach space. Suppose x, y in X and $\|x\| \le 1$, $\|y\| \le 1$. Then for $0 < \lambda < 1$ we have $\|\lambda x+(1-\lambda)y\| \le 1 - 2\lambda(1-\lambda)\delta(\varepsilon)$.

Theorem 1. Let X be a uniformly convex Banach space, C be a nonempty, closed and convex subset of X. Let $T\colon C \to C$ be a mapping satisfying *condition (L)* with at least one fixed point. Let A be defined by (I), with $\{d_n\}$ satisfying (i), (ii), and (iv). Then $\{x_n\}$ of (II) converges to a fixed point of T.

Proof. Let p be the fixed point of T in C. For any x in C

$$\|x_{n+1}-p\| = \|(1-d_n)(x_n-p)+d_n(Tx_n-p)\| \le (1-d_n)\|x_n-p\| + d_n\|Tx_n-p\|. \quad (1)$$

Using the definition of T we have

$$\|Tx_n-p\| = \|Tx_n-Tp\| \leq k \max\{\|x_n-p\|, [\|x_n-Tx_n\|+\|p-Tp\|]/2, \|p-Tx_n\|, \|x_n-Tp\|\} \tag{2}$$

Now $\|Tx_n-p\| \leq k\|Tx_n-p\|$ for $0 < k < 1$ implies $\|Tx_n-p\| = 0$. Also $\|Tx_n-p\| \leq \frac{1}{2} \|x_n-Tx_n\| \leq \frac{1}{2} [\|x_n-p\|+\|p-Tx_n\|]$ implies that $\|Tx_n-p\| \leq \|x_n-p\|$. Thus we conclude from (2) that

$$\|Tx_n-p\| \leq k\|x_n-p\| \leq \|x_n-p\|. \tag{3}$$

Using (3) we can write (1) as

$$\|x_{n+1}-p\| \leq \|x_n-p\|. \tag{4}$$

Hence $\{\|x_n-p\|\}$ is nonincreasing for all n. Also $\|x_n-Tx_n\| \leq \|x_n-p\| + \|p-Tx_n\| \leq 2\|x_n-p\|$. We may assume that there is a number $\alpha > 0$ such that $\|x_n-p\| \geq \alpha$. Suppose $\{\|x_n-Tx_n\|\}$ does not converge to zero. Then we have the following two possibilities. Either there exists a $\varepsilon > 0$ such that $\|x_n-Tx_n\| \geq \varepsilon$ for all n, or $\underset{n}{\underline{\lim}} \|x_n-Tx_n\| = 0$. In the first case using Lemma 1 and (4) we have

$$\|x_{n+1}-p\| \leq \|x_n-p\| - \|x_n-p\|d_n(1-d_n)b, \text{ where} \tag{5}$$

$$b = 2\delta(\varepsilon/\|x_0-p\|).$$

Also

$$\|x_n-p\| \leq \|x_{n-1}-p\| - \|x_{n-1}-p\|d_{n-1}(1-d_{n-1})b. \tag{6}$$

Substituting the values from (6) into (5) we have

$$\|x_{n+1}-p\| \leq \|x_{n-1}-p\| - \|x_{n-1}-p\|d_{n-1}(1-d_{n-1})b - \|x_n-p\|d_n(1-d_n)b. \tag{7}$$

Now $\|x_n-p\| \leq \|x_{n-1}-p\|$ implies $-\|x_n-p\| \geq -\|x_{n-1}-p\|$. Thus we can write (7) as

$$\|x_{n+1}-p\| \le \|x_{n-1}-p\| - \|x_n-p\|d_{n-1}(1-d_{n-1})b \quad (8)$$
$$- \|x_n-p\|d_n(1-d_n)b$$
$$= \|x_{n-1}-p\| - \|x_n-p\|[d_n(1-d_n)b+d_{n-1}(1-d_{n-1})b].$$

By induction we have

$$\alpha \le \|x_{n+1}-p\| \le \|x_1-p\| - \|x_{n-p}\|b\sum_{k=1}^{n} d_k(1-d_k).$$

Therefore;

$$\alpha + \|x_n-p\|b\sum_{k=1}^{n} d_k(1-d_k) \le \|x_1-p\|.$$

But by assumption, $\|x_n-p\| \ge \alpha$, hence

$$\alpha + \alpha b\sum_{k=1}^{n} d_k(1-d_k) \le \|x_1-p\|, \quad \text{or}$$

$$\alpha[1 + b\sum_{k=1}^{n} d_k(1-d_k)] \le \|x_1-p\|,$$

a contradiction, since the series on the left diverges.

In the second case there exists a subsequence such that $\lim_k \|x_{n_k}-Tx_{n_k}\| = 0$. Now

$$\|Tx_{n_k}-Tx_{n_l}\| \le k \max\{\|x_{n_k}-x_{n_l}\|, [\|x_{n_k}-Tx_{n_k}\|+\|x_{n_l}-Tx_{n_l}\|]/2,$$
$$\|x_{n_k}-Tx_{n_l}\|, \|x_{n_l}-Tx_{n_k}\|\}. \quad (9)$$

Now

$$\|x_{n_k}-x_{n_l}\| \le \|x_{n_k}-Tx_{n_k}\| + \|Tx_{n_k}-Tx_{n_l}\| + \|Tx_{n_l}-x_{n_l}\|,$$

$$\|x_{n_k}-Tx_{n_l}\| \le \|x_{n_k}-Tx_{n_k}\| + \|Tx_{n_k}-Tx_{n_l}\|,$$

$$\|x_{n_l}-Tx_{n_k}\| \leq \|x_{n_l}-Tx_{n_l}\| + \|Tx_{n_l}-Tx_{n_k}\|.$$

Thus we can write (9) as

$$\|Tx_{n_k}-Tx_{n_l}\| \leq \frac{k}{1-k}\,[\|x_{n_k}-Tx_{n_k}\|+\|x_{n_l}-Tx_{n_l}\|]. \tag{10}$$

Therefore $\{Tx_{n_k}\}$ is a Cauchy sequence, hence convergent. Call the limit u. Then $\lim_k x_{n_k} = \lim_k Tx_{n_k} = u$. For each k,

$$\|u-Tu\| \leq \|u-x_{n_k}\| + \|x_{n_k}-Tx_{n_k}\| + \|Tx_{n_k}-Tu\|. \tag{11}$$

A calculation similar to above yields

$$\|Tx_{n_k}-Tu\| \leq \frac{k}{1-k}\,[\|x_{n_k}-u\|+\|x_{n_k}-Tx_{n_k}\|]. \tag{12}$$

Using (12) we can write (11) as

$$\|u-Tu\| \leq \frac{1}{1-k}\,[\|u-x_{n_k}\|+\|x_{n_k}-Tx_{n_k}\|]. \tag{13}$$

Hence $u = Tu$. Since p is the unique fixed point of T, $p = u$. Thus two conditions $\lim_k x_{n_k} = u = p$ and $\{\|x_n-p\|\}$ is decreasing in n yield $\lim_n x_n = p$.

Corollary 1 (Rhoades [11], Theorem 4). Let X be a uniformly convex Banach space, E a closed convex subset of X, $T\colon E \to E$, T in z. Let A be defined by (I) with $\{d_n\}$ satisfying (i), (ii) and (iv). Then $\{x_n\}$ of (II) converges to the fixed point of T.

Corollary 2 (Rhoades [12], Theorem 2). Let K be a nonempty, bounded, closed and convex subset of a uniformly convex Banach space X, $T\colon X \to X$ and satisfying (β) on K with $b(\infty) \neq 0$. Pick $x_0 \in X$ and define x_n, $n > 0$ by $x_{n+1} = (1-\alpha_n)x_n + \alpha_n Tx_n$, where $\{\alpha_n\}$ satisfies $\alpha_0 = 1$, $0 \leq \alpha_n \leq 1$ for $n > 0$ and $\sum \alpha_n(1-\alpha_n)$ diverges. Then $\{x_n\}$ converges strongly to the fixed point of T in K.

Theorem 2. Let X be a Banach space. Let $T: X \to X$ be a generalized contractive mapping. Let A be defined by (I) with $\{d_n\}$ satisfying (i) and (ii) and bounded away from zero. Then if $\{x_n\}$ defined by (II) converges to a point p, p is the unique fixed point of T.

Proof. For each n, $x_{n+1} - x_n = (1-d_n)x_n + d_n Tx_n - x_n = d_n(Tx_n - x_n)$. Since by hypothesis $\lim_n x_n = p$, it follows that $\lim_n \|x_{n+1}-x_n\| \leq \lim_n \|x_{n+1}-p\| + \lim_n \|x_n - p\| = 0$. Since $\{d_n\}$ is bounded away from zero, $\lim_n \|Tx_n - x_n\| = 0$. Since T is generalized contractive it follows that

$$\|Tx_n-Tp\| < \max\{\|x_n-p\|, \|x_n-Tx_n\|, \|p-Tp\|, \|p-Tx_n\|, \|x_n-Tp\|\}. \tag{14}$$

Now

$$\|p-Tp\| \leq \|p-x_n\| + \|x_n-Tx_n\| + \|Tx_n-Tp\|,$$

$$\|p-Tx_n\| \leq \|p-x_n\| + \|x_n-Tx_n\|,$$

$$\|x_n-Tp\| \leq \|x_n-Tx_n\| + \|Tx_n-Tp\|.$$

Thus we can write (14) as

$$\|Tx_n-Tp\| < \max\{\|x_n-p\|+\|x_n-Tx_n\|+\|Tx_n-Tp\|\}. \tag{15}$$

Taking limit in (15) as $n \to \infty$ we obtain, $Tx_n = Tp$. Now

$$\|p-Tp\| \leq \|p-x_n\| + \|x_n-Tx_n\| + \|Tx_n-Tp\|. \tag{16}$$

Tanking limit in (16) as $n \to \infty$ we obtain $p = Tp$. The uniqueness of p follows from the definition of T.

Corollary 3 (Rhoades [11], Theorem 5). Let X be a Banach space, $T: X \to X$, T satisfyinc *condition* (α). Let A be defined by (I) with $\{d_n\}$ satisfying (i), (ii) and bounded away from zero. Then, if $\{x_n\}$ defined by (II) converges to a point p, p is the unique fixed point of T.

Corollary 4 (Achari [1], Theorem 4). Let X be a Banach space, $T: X \to X$ be a generalized contraction mapping. Let A be defined by (I) with $\{d_n\}$ satisfying (i), (ii) and bounded away from zero. Then, if $\{x_n\}$ defined by (II) converges to a point p, p is the unique fixed point of T.

Theorem 3. Let H be a Hilbert space and C be a closed convex subset of H. Let $T: C \to C$ be a mapping satisfying *condition (K)* with nonempty fixed points set. Let A be defined by (1) with $\{d_n\}$ satisfying (i)-(iii) and $\overline{\lim}\, d_n < 1 - k^2$. Then the iteration scheme (II) converges to the fixed point of T.

Proof. Ishikawa [7] has shown that for any x, y, z in a Hilbert space and any real number k, $\|kx+(1-k)y-z\|^2 = k\|x-z\|^2 + (1-k)\|y-z\|^2 - k(1-k)\|x-y\|^2$. Thus for each $y \in F(T)$ and each integer n, we have

$$\|x_{n+1}-y\|^2 = \|(1-d_n)x_n+d_nTx_n-y\|^2 \tag{17}$$

$$= (1-d_n)\|x_n-y\|^2+d_n\|Tx_n-y\|^2-d_n(1-d_n)\|x_n-Tx_n\|^2.$$

Using definition of T we have

$$\|Tx_n-y\| \leq \max\{\|x_n-y\|,\ k\|T_n-Tx_n\|,\ \|y-Ty\|,\ \|y-Tx_n\|,\ \|x_n-y\|\}$$

$$= \max\{\|x_n-y\|,\ k\|x_n-Tx_n\|\}.$$

For each n such that the maximum is $\|x_n-y\|$, we have, using (17)

$$\|x_{n+1}-y\|^2 \leq \|x_n-y\|^2 - d_n(1-d_n)\|x_n-Tx_n\|^2.$$

•For each n such that the maximum is $\|x_n-Tx_n\|$, we have using (17)

$$\|x_{n+1}-y\|^2 \leq (1-d_n)\|x_n-y\|^2 - d_n(1-d_n-k^2)\|x_n-Tx_n\|^2.$$

In either case, we have

$$\|x_{n+} -y\|^2 \leq \|x_n-y\|^2 - d_n(1-d_n-k^2)\|x_n-Tx_n\|^2.$$

The above inequality implies that $\{\|x_n-y\|\}$ is decreasing for all sufficiently large n. Also, since $\{d_n\}$ satisfies (iii) and $\overline{\lim_n}\, d_n < 1 - k^2$, there exists a subsequence $\{x_{n_k}\}$ such that $\lim_k\|x_{n_k}-Tx_{n_k}\| = 0$. We claim that $\{Tx_{n_k}\}$ is a Cauchy sequence. Indeed,

$$\|Tx_{n_k}-Tx_{n_l}\| < \max\{\|x_{n_k}-x_{n_l}\|,\ k\|x_{n_k}-Tx_{n_k}\|, \tag{18}$$

$$\|x_{n_l}-Tx_{n_l}\|,\ \|x_{n_l}-Tx_{n_k}\|,\ \|x_{n_k}-Tx_{n_l}\|\}.$$

$$\leq \max\{\|x_{n_k}-Tx_{n_k}\|+\|Tx_{n_k}-Tx_{n_l}\|+\|Tx_{n_l}-x_{n_l}\|\}.$$

Taking limit as k and $l \to \infty$ we have

$$\|Tx_{n_k}-Tx_{n_l}\| = 0$$

Thus $\{Tx_{n_k}\}$ is a Cauchy sequence, hence convergent. Call the limit p. Then $\lim_k Tx_{n_k} = x_{n_k} = p$. We claim that $\lim_k\|Tp-Tx_{n_k}\| = 0$. In fact, using definition of T we have

$$\|Tp-Tx_{n_k}\| < \max\{\|p-x_{n_k}\|,\ \|p-Tp\|,\ \|x_{n_k}-Tx_{n_k}\|, \tag{19}$$

$$\|x_{n_k}-Tp\|,\ \|p-Tx_{n_k}\|\}$$

$$\leq \max\{\|p-x_{n_k}\|+\|x_{n_k}-Tx_{n_k}\|+\|Tp-Tx_{n_k}\|\}.$$

Taking limit in (19) as $k \to \infty$ we have $\|Tp-Tx_{n_k}\| = 0$. Finally

$$\|p-Tp\| \leq \|p-x_{n_k}\| + \|x_{n_k}-Tx_{n_k}\| + \|Tx_{n_k}-Tp\|.$$

Taking limit as $k \to \infty$, we have $\|p-Tp\| = 0$, i.e. $p = Tp$.

Corollary 5 (Rhoades [11], Theorem 7). Let H be a Hilbert space, E be a closed convex subset of H. Let $T: E \to E$ be a mapping satisfying condition (β). Let A be defined by (I) with $\{d_n\}$ satisfying (i)-(ii) and $\overline{\lim_n}\, d_n < 1 - k^2$. Then the sequence $\{x_n\}$ defined by (II) converges to the fixed point of T.

Corollary 6 (Achari [1], Theorem 5). Let H be a Hilbert space, C be a closed convex subset of H. Let $T: C \to C$ be a generalized contraction with nonempty fixed points set. Let A be defined by (I) with $\{d_n\}$ satisfying (i)-(iii) and $\overline{\lim_n}\, d_n < 1 - k^2$. Then the sequence $\{x_n\}$ defined by (II) converges to the fixed point of T.

Finally we prove a theorem for ths solution of operator equations in a Banach space involving generalized contraction mappings and obtain few interesting results as corollaries.

Theorem 4. Let $\{f_n\}$ be a sequence of elements in a Banach space X. Let g_n be the unique solution of the equation $h - T(h) = f_n$, where $T: X \to X$ is a generalized contraction mapping. If $\|f_n\| \to 0$ as $n \to \infty$, then the sequence $\{g_n\}$ converges to the solution of the equation $h = T(h)$.

Proof. Using the definition of generalized contraction ampping, we will show that $\{g_n\}$ is a Cauchy sequence.

$$\|g_n-g_m\| \leq \|g_n-Tg_n\| + \|Tg_n-Tg_m\| + \|Tg_m-g_m\|. \tag{20}$$

Since T is a generalized contraction

$$\|Tg_n-Tg_m\| \leq k \max\{\|g_n-g_m\|, \|g_n-Tg_n\|, \|g_m-Tg_m\|, \|g_m-Tg_n\|, \|g_n-Tg_m\|\} \tag{21}$$

$$\leq \frac{k}{1-k} [\|g_n-Tg_n\| + \|g_m-Tg_m\|].$$

Substitution from (21) into (20) yields

$$\|g_n-g_m\| \leq \frac{1}{1-k} [\|g_n-Tg_n\|+\|g_m-Tg_m\|].$$

Thus

$$\|g_n-g_m\| \leq \frac{1}{1-k} [\|f_n\|+\|f_m\|].$$

It follows therefore, that $\{g_n\}$ is a Cauchy sequence in X. Hence it converges, say to g in X. Also

$$\|g-Tg\| \leq \|g-g_n\| + \|g_n-Tg_n\| + \|Tg_n-Tg\|. \tag{22}$$

Since T is a generalized contraction we have

$$\begin{aligned}\|Tg_n-Tg\| &\leq k \max\{\|g_n-g\|, \|g_n-Tg_n\|, \|g-Tg\|, \\ &\qquad \|g-Tg_n\|, \|g_n-Tg\|\} \\ &\leq \frac{k}{1-k} [\|g-g_n\|+\|g_n-Tg_n\|].\end{aligned}$$

Using this value in (22), we obtain

$$\|g-Tg\| \leq \frac{1}{1-k}[\|g-g_n\|+\|g_n-Tg_n\|]$$

for arbitrary large n. Hence taking the limit as $n \to \infty$ we get $\|g-Tg\| = 0$, or $g = Tg$.

Corollary 7 (Rhoades [12], Theorem 4). Let $\{g_n\}$ be a sequence of elements in a Banach space X. Let u_n be the unique solution of the equation $u - Tu = g_n$ for each n, where $T: X \to X$ satisfies *condition* (β) with $b(\infty) \neq 0$. If $\lim_n \|g_n\| = 0$, then $\{u_n\}$ converges to the solution of $u = Tu$.

Corollary 8 (Kannan [8], Theorem 6). Let $\{f_n\}$ be a sequence of elements in a Banach space X. Let V_n be the unique solution of the equation $u - \phi(u) = f_n$, where $\phi: X \to X$ satisfies (b) of *z* with $\beta = 1/2$. If $\|f_n\| \to 0$ as $n \to \infty$, then the sequence $\{V_n\}$ converges to the solution of the equation $u = \phi(u)$.

Corollary 9 (Singh [16], Theorem 3.2). Let $\{f_n\}$ be a sequence of elements in a Banach X. Let g_n be the unique solution of the equation $h - T(h) = f_n$, where $T: X \to X$ satisfies $\|Tx-Ty\| \leq \frac{1}{3} [\|x-Tx\|+\|y-Ty\|+\|x-y\|]$ x, y in X. If $\|f_n\| \to 0$ as $n \to \infty$, then the sequence $\{g_n\}$ converges to the solution of the equation $h = T(h)$.

REFERENCES

[1] Achari, J. (1977). "Some results on Ciric's quasi-contraction mappings", *Publ. Inst. Math. (Beograd), 21(35)*, 9-14.

[2] Ciric, Lj. B. (1974). "A generalization of Banach's contraction principle", *Proc. Amer. Math. Soc., 45*, 7-10.

[3] DeFigueiredo, D. G. (1967). "Topics in nonlinear functional analysis", Lecture series No. 48, University of Maryland.

[4] Doston, W. G., Jr. (1973). "On the Mann iteration process", *Trans. Amer. Math. Soc., 149*, 65-73.

[5] Groetsch, G. W. (1972). "A note on segmenting Mann iterates", *Jour. Math. Anal. and Appl., 40*, 369-372.

[6] Hicks, Troy L., and Kubicek, John D. (1977). "On the Mann iteration in a Hilbert space", *Jour. Math. Anal. and Appl.*, 498-504.

[7] Ishikawa, S. (1974). "Fixed points by a new iteration", *Proc. Amer. Math. Soc., 44*, 147-150.

[8] Kannan, R. (1971). "Some results on fixed points III", *Fund. Math. Vol. LXX*, 169-177.

[9] Mann, W. R. (1953). "Mean value methods in iterations", *Proc. Amer. Math. Soc. 4*, 506-510.

[10] Opial, Z. (1967). "Nonexpansive and monotone mappings in Banach spaces", Lecture series No. 1, Brown University.

[11] Rhoades, B. E. (1974). "Fixed point iteration using infinite matrices", *Trans. Amer. Math. Soc., 196*, 161-175.

[12] Rhoades, B. E. (1975). "Some fixed point theorems in a Banach space", *Comment. Math. Univ. St. Pauli, Vol. XXIV, No. 1,* 13-16.

[13] Reinermann, J. (1969). "Uber Toeplitze iterations verfahrenund einige ihre anwendungen in der konstruktiven fixpunk theorie", *Studia Math., 32,* 209-227.

[14] Singh, K. L. (1977). "Fixed and common fixed points for generalized contractions", *Bull. Del'Acaddemie Polonaise Des Sciences, Vol. XXV, No. 8,* 767-773.

[15] Singh, K. L. "Sequence of iterates for generalized contractions", *Fund. Math.* (to appear).

[16] Singh, K. L. (1976). "Fixed point theorems for quasi-nonexpansive mappings", *Rend. Acad. Naz. Lincei, LXI, No. 8,* 354-363.

[17] Singh, K. L. (1977). "Generalized contractions and sequence of iterates", Proceedings Nonlinear Equations in Banach Spaces, University of Texas at Arlington, June 8-10, (to appear).

A NUMERICAL METHOD FOR SOLVING THE HAMILTON-JACOBI INITIAL VALUE PROBLEM

Michael Tamburro

Department of Mathematics
Georgia Institute of Technology
Atlanta, Georgia

ABSTRACT

A numerical method for solving the initial value problem for the Hamilton-Jacobi equation, i.e.

$$v_t(t,x) + f(x, v_x(t,x)) = 0 \tag{HJ}$$

$$v(0,x) = u(x)$$

is given. The method, for $f \in C^2$ and convex in v_x, is based on a constructive version of the author's evolution operator solution

$$s(t)u = \lim_{m\to\infty}(J_{t/m})^m u \tag{SG}$$

where $J_\lambda = (I + \lambda A)^{-1}$, $\lambda > 0$; A being an accretive extension of an operator A_0: $u \mapsto f(\cdot, u_x)$ defined on a dense subset of BU, the bounded uniformly continuous functions on $\mathbb{R}^n$.

The method consists of two parts. First approximating J_λ by a discrete operator. $Q_\lambda(\varepsilon,h)$, where $Q_\lambda(\varepsilon,h)u$ is a discrete (mesh-size h) version of the (unique) solution $v \in BU$ of

ISBN 0-12-434180-2

$$v - \varepsilon\Delta v + \lambda f(\cdot, v_x) = u, \quad \varepsilon > 0 \qquad (\text{BE})_\varepsilon$$

and second, an extrapolation, $\widetilde{\lim_{m\to\infty}}$, of a family $\{S_{t,m}u \mid m \in M_t$, a finite set$\}$ where

$$S_{t,m}u \equiv (Q_{t/m}(\varepsilon(m), h(m)))^m u.$$

Using the convergence properties of (SG) and $(\text{BE})_\varepsilon$ (as $\varepsilon \downarrow 0$) together with error estimates for the $Q_\lambda(\varepsilon, h)$ algorithm, an extrapolation set $\{(m, \varepsilon(m), h(m) \mid m \in M_t\}$ is chosen to give efficient estimates of

$$e(t) = \|S(t)u - \widetilde{\lim_{m\to\infty}} S_{t,m}u\|_\infty.$$

DIFFERENTIAL GEOMETRIC METHODS IN NONLINEAR PROGRAMMING

Kunio Tanabe

The Institute of Statistical Mathematics
Minamizatu, Minatoku
Tokyo, Japan
and
Applied Mathematics Department
Brookhaven National Laboratory
Upton, New York

I. INTRODUCTION

This paper is concerned with robust methods for nonlinear constrained optimization, which can produce convergence from a very poor initial estimate of the optimal solution. A differential geometric method is developed to obtain robust algorithms without resorting to the penalty-type approach. In particular, a generic class of "feasibility-improving gradient acute projection methods" and their Levemberg-Marquardt-type modification is developed for solving the general nonlinear constrained minimization problems. Each method in this class is an amalgamation of a generalized gradient projection method and a generalized Newton-Raphson method which respectively take care of reducing the value of the objective function and satisfying constraint equations at the same time. The class of the methods contains various new methods such as the second-order feasibility-improving GRG method

ISBN 0-12-434180-2

as well as many of the existing methods. The concept of various generalized inverses and related projectors facilitates geometric interpretation of the methods in the class. Analysis is given to the continuous analogues of the methods to obtain robust algorithms, which also gives an insight into the global behavior of the related algorithms. Quasi-Newton algorithm which estimates projected Hessian matrix and requires only approximation of a nonnegative definite matrix of size $n-m$, is developed to enhance the local convergence, where n and m are numbers of variables and constraint equations respectively.

The following is a summary of a part of the paper [10] which extends works [5-9].

II. FEASIBILITY-IMPROVING GRADIENT ACUTE PROJECTION METHODS

We consider the constrained minimization problem,

$$\text{minimize } f(x);\ \text{subject to } g(x) = (g_1(x), \ldots, g_m(x))^t = 0 \quad (2.1)$$

where $f\colon R^n \to R$ and $g\colon R^n \to R^m$ $(m \leqq n)$ are twice continuously differentiable. The set of feasible points will be denoted by V_g, i.e.,

$$V_g = \{x \in R^n\colon\ g(x) = 0\}. \quad (2.2)$$

Define the Lagrangian

$$L(x,\lambda) = f(x) + \lambda^t g(x). \quad (2.3)$$

Under mild conditions there exists a multiplier λ^* corresponding to the optimal solution x^* such that (x^*,λ^*) is a solution of

$$\nabla_x L(x,\lambda) = \nabla_x f(x) + J_g^t(x)\lambda = 0 \quad \text{and} \quad (2.4)$$

$$g(x) = 0 \quad (2.5)$$

where $J_g(x)$ is the m by n Jacobian matrix of $g(x)$ with respect to x.

Any numerical method for solving the problem (2.1) eventually solves this first order necessary condition,

$$F(x,\lambda) = (\nabla_x L(x,\lambda), g(x))^t = 0 \tag{2.6}$$

The Newton-Raphson method for solving Equation (2.6) is described by

$$x^+ = x + \Delta x \quad \text{and} \quad \lambda^+ = \lambda + \Delta\lambda, \tag{2.7}$$

where $(\Delta x, \Delta\lambda)$ is the solution of

$$\begin{bmatrix} H(x,\lambda) & J_g^t(x) \\ J_g(x) & 0 \end{bmatrix} \begin{bmatrix} \Delta x \\ \Delta\lambda \end{bmatrix} = - \begin{bmatrix} \nabla_x f(x) + J_g^t(x)\lambda \\ g(x) \end{bmatrix} \tag{2.8}$$

where $H(x,\lambda)$ is the n by n Hessian matrix of $L(x,\lambda)$ with respect to x. The convergence of the method is guaranteed only for a very restricted set of starting points. A primary concern of this paper is to enlarge the region of convergence. Somewhat similar treatment of the problem was given in Tapia [3-4] and Powell [2]. We emphasize, however, that the penalty-type function can (and should) be avoided to obtain robust methods by taking a (differential) geometric approach.

Replacing $H(x,\lambda)$ in Equation (2.8) by an n by n symmetric matrix $N(x,\lambda)$ we obtain a modified Newton-Raphson method for solving Equation (2.6), which is described by Equations (2.7), where $(\Delta x, \Delta\lambda)$ is the solution of

$$\begin{bmatrix} N(x,\lambda) & J_g^t(x) \\ J_g(x) & 0 \end{bmatrix} \begin{bmatrix} \Delta x \\ \Delta\lambda \end{bmatrix} = - \begin{bmatrix} \nabla_x f(x) + J_g^t(x)\lambda \\ g(x) \end{bmatrix} \tag{2.9}$$

This method is equivalently described by

$$x^+ = x + \Delta x \quad \text{and} \tag{2.10}$$

$$\begin{bmatrix} N(x,\lambda) & J_g^t(x) \\ J_g(x) & 0 \end{bmatrix} \begin{bmatrix} \Delta x \\ \lambda^+ \end{bmatrix} = - \begin{bmatrix} \nabla_x f(x) \\ g(x) \end{bmatrix} \tag{2.11}$$

Note that the right-hand side of the equation does not depend on λ. The form of Equation (2.11) suggests that the two variables x and λ be treated differently. In this paper we are concerned with the behavior of the methods in x-space rather than (x,λ)-space.

Different methods are obtained by different choices of the matrix $N(x,\lambda)$. In many of the existing methods $N(x,\lambda)$ are chosen to be a positive definite matrix. However, it can be a non-definite singular matrix, as is shown later.

Theorem 2.1. Let $P(x)$ be a projector onto the null space Ker $J_g(x)$ of $J_g(x)$. If

$$\text{Rank } J_g(x) = m \quad \text{and} \tag{2.12}$$

$$\text{Rank } P^t(x)N(x,\lambda)P(x) = n - m \tag{2.13}$$

then Equation (2.11) has a unique solution described by

$$\begin{aligned} \Delta x = \Psi(x) &= -(I - J_g^-(x)J_g(x))N\#(x)(I - J_g^-(x)J_g(x))^t\nabla f(x) \\ &\quad - J_g^-(x)g(x), \end{aligned} \tag{2.14}$$

$$\lambda^+ = \Lambda(x) = -(J_g^-(x))^t\nabla f(x) + R(x)g(x), \tag{2.15}$$

where $J_g^-(x)$, $N\#(x)$ and $R(x)$ are given by

$$J_g^- = (Q - P(P^tNP)^+P^tNQ)J^= \tag{2.15}$$

$$N\# = P(P^tNP)^+P^t \tag{2.16}$$

$$R = (J^=)^t(Q^tNQ - (P^tNQ)^t(P^tNP)^+(P^tNQ))J^=. \tag{2.17}$$

Here the argument x is dropped and A^+ is the Moore-Penrose inverse of A, $J^=$ is an arbitrary generalized inverse of J_g

such that $J_g J_g^{=} J_g = J_g$, and $P = I - J^{-}J_g$ and $Q = J^{=}J_g$ are projectors such that

$$\text{Im } P = \text{Ker } J_g \quad \text{and} \quad \text{Ker } Q = \text{Ker } J_g. \tag{2.18}$$

Note that $J_g^{-}(x)$ is a reflexive generalized inverse of $J_g(x)$ and that $I - J_g^{-}(x)J_g(x)$ is a projector onto the null space $\text{Ker } J_g(x)$ of $J_g(x)$. The formulas (2.14-15) are of little value in practical computation, their main use being to give a geometric insight into the method (2.10-11).

The search direction Δx is the sum of the two terms,

$$\begin{aligned}\Psi_{\text{gap}}(x) &= -(I - J_g^{-}(x)J_g(x))N\#(x)(I - J_g^{-}(x)J_g(x))^t\nabla f(x) \qquad (2.19)\\ &= -(I - J_g^{-}(x)J_g(x))N\#(x)\nabla f(x)\\ &= N\#(x)(I - J_g^{-}(x)J_g(x))^t\nabla f(x)\\ &= N\#(x)\nabla f(x), \quad \text{and}\end{aligned}$$

$$\Psi_{\text{nr}}(x) = -J_g^{-}(x)g(x). \tag{2.20}$$

Proposition 2.2. The matrix $N\#(x)$ is nonnegative definite if and only if the condition

(a) $P^t(x)N(x,\lambda)P(x)$ is a nonnegative definite matrix,

is satisfied, in which case the direction $\Psi_{\text{gap}}(x)$ forms a descent direction of $f(x)$. Hence, $\Psi_{\text{gap}}(x)$ will be called a "gradient acute projection". If feasibility is maintained by some other procedure, Δx forms a gradient acute projection onto the tangent space of V_g at the point x.

$\Psi_{\text{nr}}(x)$ is a Newton-Raphson direction for solving an under-determined system of nonlinear equations (2.5), which was studied in [6-7].

Proposition 2.3. Δx forms a descent direction of each of the function $g_i^2(x)$ $(i = 1,\ldots,m)$.

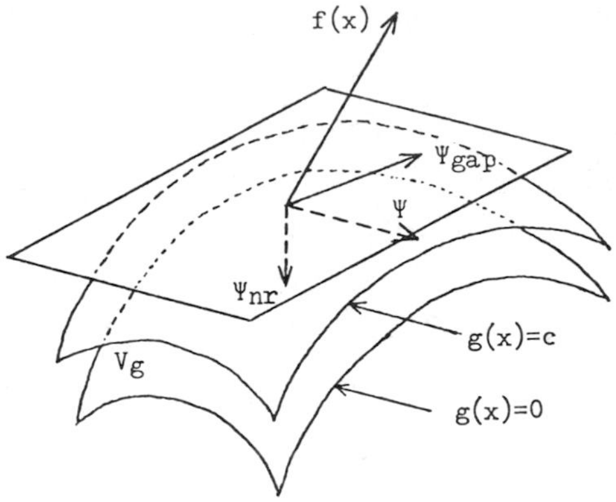

FIGURE 2.1.

A geometric configuration of Δx is shown schematically in Figure 2.1. $\Psi_{gap}(x)$ and $\Psi_{nr}(x)$ respectively take care of reducing the value of the objective function and satisfying the constraint equations. We will call $\Psi(x)$ a "feasibility-improving gradient acute projection" if the condition (a) is satisfied besides (2.12-13). It is interesting to note that Δx is derived from the formula (2.11) which solves such a local condition as Equation (2.6) and still it gives a meaningful direction even for a point which is remote from the optimal point. We will study the methods which use this particular class of search directions in the following way;

$$x^+ = x + \rho\Delta x, \quad \Delta x = \Psi(x) \quad \text{and} \quad \rho > 0. \tag{2.21}$$

We will call a method of this form a "feasibility-improving gradient acute projection method". This class of methods contains the feasibility-improving gradient projection method [6] and the feasibility-improving GRG method [9].

A. *Feasibility-Improving Gradient Projection Method*

$$N(x,\lambda) = I \tag{2.22}$$

$$\Delta x = -(I - J_g^+(x)J_g(x))\nabla f(x) - J_g^+(x)g(x) \tag{2.23}$$

$$\lambda^{+} = -(J_g^{+}(x))^t \nabla f(x) + (J_g(x) J_g^t(x))^{-1} g(x) \tag{2.24}$$

The condition (a) is always satisfied in this case.

B. Feasibility-Improving GRG Method

$$N(x,\lambda) = \begin{bmatrix} 0 & 0 \\ 0 & I_{n-m} \end{bmatrix} \tag{2.25}$$

$$\Delta x = -(I - J_g^{-}(x) J_g(x))(I - J_g^{-}(x) J_g(x))^t \nabla f(x) - J_g^{-}(x) g(x) \tag{2.26}$$

$$\lambda^{+} = -(J_g^{-}(x))^t \nabla f(x) \tag{2.27}$$

where $J_g(x) = ((J_g(x))_1 : (J_g(x))_2)$, $(J_g(x))_1$ is a nonsingular sub-matrix of $J_g(x)$, and $J_g^{-}(x)$ is a generalized inverse of $J_g(x)$ defined by

$$J_g^{-}(x) = \begin{bmatrix} (J_g(x))_1^{-1} \\ 0 \end{bmatrix} \tag{2.28}$$

The condition (a) is always satisfied in this case.

Usually $N(x,\lambda)$ is chosen to be a good approximation to the Hessian matrix $H(x,\lambda)$ to obtain a locally rapid convergence around the optimal solution. It is, however, important to note that fast convergence is achieved only if $P^t(x)N(x,\lambda)P(x)$ is close enough to the projected Hessian matrix $P^t(x)H(x,\lambda)P(x)$, where $P(x)$ is an arbitrary projector onto Ker $J_g(x)$. Note also that the conditions (2.12-13) and (a) are satisfied around the optimal solution which satisfies the second-order sufficient conditions. The following method was given in [9].

C. Second-Order Feasibility-Improving GRG Method

$$N(x,\lambda) = (I - J_g^{-}(x) J_g(x))^t H(x,\lambda)(I - J_g^{-}(x) J_g(x)) \tag{2.29}$$

$$\Delta x = -P_2 N_c^{-1} P_2^t \nabla f(x) - J_g^{-}(x) g(x) \tag{2.30}$$

$$\lambda^{+} = -(J_g^{-}(x))^{t}\nabla f(x), \tag{2.31}$$

where $J_g^{-}(x)$ is the generalized inverse defined in (2.28), and the matrix $N(x,\lambda)$ is implicitly represented by an $(n-m)$ by $(n-m)$ matrix N_c which is defined by

$$N_c = P_2^{t} H(x,\lambda) P_2 \tag{2.32}$$

$$P_2 = \begin{bmatrix} -(J_g(x))_1^{-1}(J_g(x))_2 \\ I_{n-m} \end{bmatrix} \tag{2.33}$$

Next we discuss how to choose the step length ρ in (2.21).

Theorem 2.4. Let $\Phi(x,\lambda) = \nabla_x L(x,\lambda) = \nabla f(x) + J_g^{t}(x)\lambda$. We have

$$(\Delta x)^{t}\Phi(x,\lambda^{+}) \leqq 0, \tag{2.34}$$

where Δx and λ^{+} are given either in (2.23-24) or in (2.26-27). This implies that in general the search direction Δx is a descent direction of the Lagrangian $L(x,\lambda^{+})$ for λ^{+}. The inequality (2.34) holds for (2.30-31) in a neighborhood of the optimal point which satisfies the second-order sufficient condition.

This theorem provides us with a criterion to determine ρ.

D. *Line Search Method*

Choose the step length by a line search for minimizing $L(x+\rho\Delta x, \lambda^{+})$ subject to $0 \leqq \rho < u$, where u is prescribed upper limit for ρ.

Finally we introduce a quasi-Newton method which approximates the projected Hessian matrix. Before describing a specific algorithm, we give the general Quasi-Newton method.

E. General Quasi-Newton Method

Put $\rho = 1$ and update $N = N(x,\lambda)$ at each iteration by a secant updating algorithm which preserve the nonnegative definiteness of N, in the following way;

$$N^+ = \text{Secant Update } (s,y,N), \tag{2.35}$$

$$s = P(x)\Delta x = \Psi_{\text{gap}}(x) \tag{2.36}$$

$$y = P^t(x)(\Phi(x',\lambda) - \Phi(x,\lambda)) \tag{2.37}$$

$$x' = x + s. \tag{2.38}$$

The following method is closely related to the method (2.29-33).

F. Quasi-Newton Feasibility Improving GRG Method

The formulas for Δx and λ^+ is the same as (2.30-31). Instead of (2.32) we use the following updating formula for the $(n-m)$ by $(n-m)$ matrix N_c;

$$N_c^+ = \text{Secant Update } (s,y,N_c) \tag{2.39}$$

$$s = -N_c^{-1}P_2^t\nabla f(x) \in R^{n-m} \tag{2.40}$$

$$y = P_2^t(\Phi(x',\lambda) - \Phi(x,\lambda)) \in R^{n-m} \tag{2.41}$$

$$x' = x + P_2 s, \tag{2.42}$$

where P_2 is defined in (2.33).

Many other algorithms are derived from the general method by using the QR decomposition, the singular value decomposition and the LU decomposition of the Jacobian matrix $J_g(x)$ of $g(x)$. See [10-11] for further detail.

III. CONTINUOUS ANALOGUE OF FIGAP METHOD

In this section we analyze the global behavior of a continuous analogue of the feasibility-improving gradient acute projection method by taking a differential geometric approach. We assume that the condition (2.13) is satisfied on the feasible set V_g. Hence, V_g is an $(n-m)$-dimensional differentiable manifold in R^n and $f(x)$ is a differentiable function on V_g. The function f will be denoted by f_v when it is considered as a function on V_g. Let $(df_v)_x$ be the differential of f_v at a point $x \in V_g$. A point $x \in V_g$ is called a critical point of f_v, if $(df_v)_x = 0$. The set of all the critical points of f_v will be denoted by C_f which consists of minimal points, saddle points and maximal points of f_v.

Theorem 3.1. Under the assumptions (2.13-14), the following three conditions are equivalent, where $\Psi(x)$ and $F(x,\lambda)$ are defined in (2.14) and (2.6) respectively.

(b) $(df_v)_x = 0$,

(c) $\Psi(x) = 0$,

(d) $F(x,\lambda) = 0$ for some λ.

A critical point $x \in C_f$ is called regular if the Hessian matrix of f_v with respect to a local coordinate system of V_g is nonsingular at the point x.

Lemma 3.2. A critical point $x \in C_f$ is regular if and only if the $(n-m)$ by $(n-m)$ matrix

$$K^t H(x,\Lambda(x))K \tag{3.1}$$

is nonsingular for some n by $(n-m)$ matrix such that

$$J_g(x)K = 0, \tag{3.2}$$

where $\Lambda(x)$ and $H(x,\lambda)$ are defined in (2.15) and (2.8) respectively.

Theorem 3.3. A regular critical point $x \in C_f$ is a minimal point of the function f_v if and only if the matrix (3.1) is positive definite.

A continuous analogue of the FIGAP method is described by the autonomous system

$$dx/dt = \Psi(x) = -N^{\#}(x)\nabla f(x) - J_g^{-}(x)g(x) \tag{3.3}$$

where $\Psi(x)$ is given in (2.14). We assume that $N(x,\lambda)$ depends only on the variable x, hence it will be denoted by $N(x)$ and that $N(x)$ is a continuous mapping of x, which satisfies the conditions (2.13) and (a) everywhere. Note that the system (3.3) is defined on the set of points which satisfy the condition (2.12).

Under the assumptions (2.12-13), the set of equilibrium points of the system (3.3) coincides with C_f. In general it contains C_f.

Lemma 3.4. If the initial point x^0 satisfies the condition (2.12), there exists a solution $x(t,x^0)$ of (3.3) for some interval $0 \leqq t < M$ with $x(0,x^0) = x^0$. Further the solution can be prolonged until its trajectory approaches the set S of singular points,

$$S = \{x \in R^n: \text{ rank } J_g(x) < m\}; \tag{3.4}$$

otherwise the solution exists for $0 \leqq t < \infty$, or diverges.

Lemma 3.5. A solution $x(t,x^0)$, $0 \leqq t < M$ of (3.3) satisfies the "first integral",

$$g(x(t,x^0)) = \exp(-t)g(x^0) \quad \text{for } 0 \leqq t < M. \tag{3.5}$$

Theorem 3.6. If a solution $x(t,x^0)$, $0 \leqq t < M$ of (3.3) is bounded and if its trajectory does not approach the singular set S, then $M = \infty$ and the positive limit set Γ of the solution is a compact connected set contained in C_f.

Theorem 3.7. The system (3.3) is asymptotically stable at a regular minimal point $x \in C_f$ of the function f_v.

Applying a numerical integration process to (3.3) we obtain an algorithm for solving the problem (2.1). Since V_g is a local attractor and the optimal point is generally an asymptotically stable point of (3.3) we can use larger step-size for the numerical integration than usual without worrying about the violation of constraints. In fact, the Newton-Raphson correction term $\Psi_{nr}(x)$ in (3.3) can obviate the difficulties with the complicated and time-consuming treatment with violated constraints in a practical implementation of the usual method, and make it easy to control step-sizes automatically. See [6] for further detail.

Difficulties with the FIGAP methods are encountered when their iterations approach the singular set S. However, applying the Levenberg-Marquardt-type modification, we can overcome the difficulty. For example, we can use

$$\Psi_{lm}(x) = -J_g^t(x)(J_g(x)J_g^t(x) + \delta I)^{-1}g(x), \quad (\delta > 0) \tag{3.6}$$

in place of $\Psi_{nr}(x) = -J_g^+(x)g(x)$ in (2.23).

REFERENCES

[1] Fletcher, R. (1971). "Minimizing general functions subject to linear constraints", Numerical Methods for Nonlinear Optimization, 279-296.

[2] Powell, M. J. D. (1978). "Algorithms for nonlinear constraints that use Lagrangian functions", Mathematical Programming *14*, 224-248.

[3] Tapia, R. A. (1977). "Diagonalized multiplier methods and quasi-Newton methods for constrained optimization", Journal of Optimization Theory and Applications *22*, 135-194.

[4] Tapia, R. A. (1977). "Quasi-Newton methods for equality constrained optimization: Equivalence of existing methods and a new implementation", Manuscript presented at Nonlinear Programming Symposium 3, Madison, Wisconsin.

[5] Tanabe, K. (1974). "An algorithm for the constrained maximization in nonlinear programming", Journal of Operations Research Society of Japan *17*, 184-201.

[6] Tanabe, K. (1977). "A geometric method in nonlinear programming", Stanford University Computer Science Report, STAN-CS-643, 1977, presented at the Gatlinburg VII Conference, Asilomar, 1977.

[7] Tanabe, K. (1977). "Continuous Newton-Raphson method for solving an underdetermined system of nonlinear equations", Brookhaven National Laboratory Report 23573 AMD784, to appear in Nonlinear Analysis.

[8] Tanabe, K. (1978). "Global analysis of continuous analogues of the Levenberg-Marquardt and Newton-Raphson methods for solving nonlinear equations", Brookhaven National Laboratory Report 24022 AMD 789, 1978, presented at the *SIAM* 1978 Spring Meeting, Madison.

[9] Tanabe, K. (1978). "Differential geometric approach to extended GRG methods with enforced feasibility in nonlinear programming: Global analysis", Brookhaven National Laboratory Report 24497 AMD 797, 1978, to appear in Recent Applications of Generalized Inverses, Ed. M. Z. Nashed, Pitman.

[10] Tanabe, K. (1978). "A unified theory of feasibility-improving gradient acute projection methods for nonlinear constrained optimization", to appear in Brookhaven National Laboratory Report.

[11] Tanabe, K. "Differential geometric methods for solving nonlinear constrained optimization problems and a related system of nonlinear equations: Global analysis and implementation", to appear in the Proceedings of the International Congress on Numerical Methods for Engineering, Paris.

LIMITING EQUATIONS AND TOTAL STABILITY

F. Visentin

Istituto Matematico
"R. Caccioppoli"
dell'Università di Napoli
Naples, Italy

I.

Some authors [1,2,4,5,6,8,11,12] use the techniques of topological dynamics to study the behavior of solutions of a differential equation

$$\dot{x} = f(t,x), \quad f(t,0) \equiv 0, \tag{1.1}$$

where $f\colon \mathbb{R}^+ \times W \to \mathbb{R}^n$ is continuous and such that uniqueness of noncontinuable solutions $x_f(t,t_0,x_0)$ of (1.1) is assured for every $(t_0,x_0) \in \mathbb{R}^+ \times W$; and W is an open set of $\mathbb{R}^n$. In this context one introduce the concept of limiting equations of (1.1). Let $F = \{f_\tau\colon \ \tau \in R^+\}$ be the set of the translate of the function f defined by $f_\tau(t,x) = f(t+\tau,x)$, $\forall\ (t,x) \in \mathbb{R}^+ \times W$. Consider the sets

$$\Omega_f^* = \{f^* \in C(\mathbb{R}^+ \times W, \mathbb{R}^n)\colon \ \exists\ \{\tau_n\} \subset \mathbb{R}^+,\ \ \tau_n \to +\infty,\ \ f_{\tau_n} \to f^* \text{ in the compact open topology}\},$$

$$\Lambda_f^* = \{f^* \in C(\mathbb{R}^+ \times W, \mathbb{R}^n)\colon \ \exists\ \{\tau_n\} \subset \mathbb{R}^+,\ \ \tau_n \to +\infty,\ \ f_{\tau_n} \to f^* \text{ in the Bohr topology}\}.$$

ISBN 0-12-434180-2

We have $\Lambda_f^* \subset \Omega_f^*$. Limiting equation of (1.1), [12], is every equation

$$\dot{x} = f^*(t,x), \tag{1.2}$$

with $f^* \in \Omega_f^*$.

A fundamental problem is the relation between stability properties of (1.1) and stability properties of limiting equations. Uniform stability properties of (1.1) generally imply analogous stability properties of limiting equations [12]. The converse is more complicated to solve. This converse problem, for the uniform asymptotic stability, was discussed in different ways in the papers [2,4]. Under reasonable conditions it was proved that the uniform asymptotic stability of the origin for equations (1.2) implies the same property for the original one. Instead the uniform stability of the origin for the limiting equations doesn't imply the same property for the equation (1.1), also in very simple cases [12]. Then it is natural to search for some intermediate property between the uniform stability and the uniform asymptotic stability which is transferable from the limiting equations to equation (1.1). This problem also involves the topology in the functional space $C(\mathbb{R}^+ \times W, \mathbb{R}^n)$. Here we are concerned with the property of total stability in Dubosin's sense and another concept of total stability which we shall call S-total stability. Both these properties are between uniform stability and uniform asymptotic stability.

II.

The following theorem holds

Theorem 2.1. If f is Lipschitzian in x, uniformly in t, and $\Lambda_f^* \neq \emptyset$, and there exists $f^* \in \Lambda_f^*$ such that the null solution of (1.2) is uniformly totally stable in Dubosin's sense, i.e.:

$$\forall\ \varepsilon > 0\quad \exists\ \delta_1, \delta_2 > 0:\quad \forall\ t_0 \in \mathbb{R}^+,\quad \forall\ x_0 \in S_{\delta_1},$$
$$\forall\ g \in C(\mathbb{R}^+ \times W, \mathbb{R}^n):\quad \|g(t,x) - f^*(t,x)\| < \delta_2 \tag{2.1}$$
$$\forall\ (t,x) \in \mathbb{R}^+ \times S_\varepsilon \Rightarrow x_g(t,t_0,x_0) \in S_\varepsilon\quad \forall\ t \geq t_0,$$

then also the null solution of (1.1) is uniformly totally stable in Dubosin's sense.

Line of the Proof. If $f_{\tau_n} \to f^*$ in the Bohr topology, then

$$\forall\ \varepsilon > 0\quad \forall\ \eta > 0\quad \exists\ v > 0:\quad \forall\ n \geq v\quad \|f_{\tau_n}(t,x) - f^*(t,x)\| < \eta$$
$$\forall\ (t,x) \in \mathbb{R}^+ \times \overline{S}_\varepsilon.$$

By using this notion we prove that the null solution of every equation

$$\dot{x} = f_{\tau_n}(t,x),\quad \tau_n \geq \tau_v,$$

is uniformly totally stable. Therefore the null solution of (1.1) satisfies the condition (2.1) for every $t_0 \geq \tau_v$, uniformly in t_0. Clearly this implies that the null solution of (1.1) is stable, uniformly in $t_0 \in \mathbb{R}^+$. Then, choosing properly $\delta_1' < \delta_1$ and $\delta_2' < \delta_2/2$ (δ_1 and δ_2 relative to condition (2.1)), we can prove that

$$x_0 \in S_{\delta_1'},\quad t_0 \in [0,\tau_v],\quad g \in C(\mathbb{R}^+ \times W, \mathbb{R}^n):$$
$$\|g(t,x) - f(t,x)\| < \delta_2'$$
$$\forall\ (t,x) \in \mathbb{R}^+ \times S_\varepsilon \Rightarrow x_g(t,t_0,x_0) \in S_{\delta_1}\quad \forall\ t \in [t_0,\tau_v].$$

This result admits a converse.

Theorem 2.2. If f is Lipschitzian in x, uniformly in t, and $\Lambda_f^* \neq \emptyset$, and the null solution of (1.1) is uniformly totally stable in Dubosin's sense, then the null solution of every equation (1.2) (with $f^* \in \Lambda_f^*$) is uniformly totally stable in Dubosin's sense, and δ_1, δ_2 can be chosen as independent of $f^* \in \Lambda_f^*$.

Line of the Proof. We prove that the null solution of every equation

$$\dot{x} = f_\tau(t,x), \quad \tau > 0,$$

is uniformly totally stable. Then by using the notion of convergence in the Bohr topology, we show that the null solution of every equation (1.2) with $f^* \in \Lambda_f^*$ is also uniformly totally stable.

III.

A similar result can be given in terms of S-uniform total stability in the compact open topology. We give the following definition.

Definition 3.1. The null solution of the equation (1.1) is said to be S-uniformly totally stable if:

$$\forall\, \bar{t} > 0,\ \exists\, \varepsilon > 0,\ \exists\, \delta_1,\delta_2 > 0:\ \forall\, t_0 \in \mathbb{R}^+,\ \forall\, x_0 \in S_{\delta_1},$$

$$\forall\, g \in C(\mathbb{R}^+ \times W, \mathbb{R}^n) \text{ with } \|x_f(t,t_0,z) - x_g(t,t_0,z)\| < \delta_2$$

$$\forall\, (t,z) \in [t_0,t_0+\bar{t}] \times S_\varepsilon \Rightarrow x_g(t,t_0,x_0) \in S_\varepsilon \quad \forall\, t \geq t_0.$$

This concept of S-total stability is stronger than Dubosin's definition and is also stronger than a Seibert definition [10]. The concept of S-total stability was introduced in [7] in the setup of autonomous dynamical systems in connection with bifurcation problems and, successively, generalized to general dynamical systems in [3]. We have the following result.

Theorem 3.2. If F is relatively compact in the compact open topology, and f is Lipschitzian in x, uniformly in t, and the null solution of every equation (1.2) is S-uniformly totally stable, uniformly also with respect to $f^* \in \Omega_f^*$; then the null solution of (1.1) is "eventually" S-uniformly totally stable.

The proof is analogous to Theorem 2.1, by using in this case the convergence in the compact open topology.

Remark 3.3. If the null solution of (1.1) is eventually S-uniformly totally stable, then it is also uniformly totally stable in Dubosin's sense.

Theorem 3.2 also has a converse.

Theorem 3.4. If f is Lipschitzian in x, uniformly in t, and the null solution of (1.1) is eventually S-uniformly totally stable, then the null solution of every equation (1.2) is S-uniformly totally stable, and δ_1, δ_2 can be chosen as independent of $f^* \in \Omega_f^*$.

REFERENCES

[1] Artstein, Z. (1977). "The limiting equations of nonautonomous ordinary differential equations", *J. Diff. Eqs.*, *25 (2)*.

[2] Artstein, Z. (1978). "Uniform asymptotic stability via the limiting equations", *J. Diff. Eqs.*, *27 (2)*.

[3] Bondi, P., and Moauro, V. (1976). "Total stability for general dynamical systems", *Ric. di Mat.*, *Napoli*, *XXV*.

[4] Bondi, P., Moauro, V., and Visentin, F. (1977). "Limiting equations in the stability problem", *Nonlin. An.: Th.*, *Meth.*, *Appl.*, *1 (2)*.

[5] Bondi, P., Moauro, V., and Visentin, F. (1977). "Addendum to the paper 'Limiting equations in the stability problem'", *Nonlin. An.: Th.*, *Meth.*, *Appl.*, *1 (6)*.

[6] Kato, J., and Yoshizawa, T. (1976). "Stability under the perturbation by a class of functions", Dynamical Systems, International Symposium (2).

[7] Marchetti, F., Negrini, P., Salvadori, L., and Scalia, M. (1976). "Liapunov direct method in approaching bifurcation problems", *Ann. Mat. Pura e Appl., 108 (4).*

[8] Markus, L. (1956). "Asymptotically autonomous differential systems", *Contr. to Nonlin. Oscill., 3,* Princeton Univ. Press, N. J.

[9] Miller, R. K. (1965). "Almost periodic differential equations as dynamical systems with applications to the existence of a. p. solutions", *J. Diff. Eqs., 1.*

[10] Seibert, P. (1963). "Stability under perturbations in generalized dynamical systems", Int. Symp. Nonlin. Diff. Eqs. and Nonlin. Mech., Academic Press.

[11] Sell, G. R. (1967). "Nonautonomous differential equations and topological dynamics I. The basic theory", *Trans. Am. Math. Soc., 127.*

[12] Sell, G. R. (1967). "Nonautonomous differential equations and topological dynamics II. Limiting equations", *Trans. Am. Math. Soc., 127.*